Mechanical Engineering Series

Series Editor

Francis A. Kulacki
Department of Mechanical Engineering
University of Minnesota
Minneapolis, MN, USA

The Mechanical Engineering Series presents advanced level treatment of topics on the cutting edge of mechanical engineering. Designed for use by students, researchers and practicing engineers, the series presents modern developments in mechanical engineering and its innovative applications in applied mechanics, bio-engineering, dynamic systems and control, energy, energy conversion and energy systems, fluid mechanics and fluid machinery, heat and mass transfer, manufacturing science and technology, mechanical design, mechanics of materials, micro- and nano-science technology, thermal physics, tribology, and vibration and acoustics. The series features graduate-level texts, professional books, and research monographs in key engineering science concentrations.

More information about this series at http://www.springer.com/series/1161

Rakesh Kumar Maurya

Reciprocating Engine Combustion Diagnostics

In-Cylinder Pressure Measurement and Analysis

 Springer

Rakesh Kumar Maurya
Department of Mechanical Engineering
Indian Institute of Technology Ropar
Rupnagar, Punjab, India

ISSN 0941-5122   ISSN 2192-063X   (electronic)
Mechanical Engineering Series
ISBN 978-3-030-11956-0   ISBN 978-3-030-11954-6   (eBook)
https://doi.org/10.1007/978-3-030-11954-6

Library of Congress Control Number: 2018968537

This Springer imprint is published by the registered company Springer Nature Switzerland AG
The registered company address is: Gewerbestrasse 11, 6330 Cham, Switzerland

Preface

The global economy and modern societies are heavily dependent on the availability of reliable transportation systems. Reciprocating internal combustion engines are the primary source of power for the automotive vehicles, ships, construction equipment, and agricultural machines, among others. Transportation sector is one of the major sources of air pollution throughout the world, which is increasing with the increase in number of vehicles due to urbanization and expansion of the economy globally. To fulfill the increasingly stringent emission regulations and fuel economy requirements, cleaner and efficient combustion engines are required. For the development of fuel-efficient and clean (low exhaust emissions) engines, a sound understanding of the combustion process is essential. Understanding the combustion processes would not be possible without diagnostics. Combustion diagnostics is always used in engine development when unexploited potential compared with thermodynamically possible targets is ascertained during the measurement of fuel consumption, power output, and exhaust emissions. Given the high targets set for modern reciprocating engines, thermodynamic combustion analysis using the cylinder pressure measurement is a fixed element in the engine development process.

The engine performance development process has two aspects: hardware development and engine calibration. The hardware development needs robust and accurate combustion measurements including work and heat release analysis that provides the basis to both explain and evaluate the results. From the engine calibration perspective, accurate combustion measurements are required for optimization of spark timing or injection timings, fuel control, combustion stability, knock and combustion noise, etc. Professionals responsible for the development tasks feel advantageous to have one sensor accomplishing all of these requirements. The in-cylinder pressure signal provides a valuable and rich source of information for engine monitoring, diagnostics, and control because it is the direct feedback from the combustion status in the engine combustion chamber.

Since in-cylinder pressure measurement systems have become almost standard equipment in engine test environments, researchers/engineers need to have in-depth understanding of different aspects of in-cylinder pressure measurement and analysis.

A comprehensive source is required which provides the details of all the information that can be extracted by processing of measured in-cylinder pressure signal along with the estimation algorithm. The present book fulfills this gap and provides information on measuring, analyzing, and qualifying in-cylinder pressure data, as well as details on hardware requirements and system components. Additionally, the book will provide the models/algorithms for processing the in-cylinder pressure data for the combustion quality analysis and estimation of various engine and combustion parameters. Historically, the measurement and analysis of in-cylinder pressure have been a key tool for offline combustion diagnosis in reciprocating engines, but recently, online applications for real-time combustion management and condition monitoring have become popular. Thus, the real-time estimation algorithm of various combustion parameters is also discussed in the present book. Information provided in the book can be effectively used for further development, optimization, and calibration of modern reciprocating engines.

Typically, the workflow of a measurement chain consists of sensors (signal generation), data acquisition, signal processing, and data analysis. The present book provides a detailed discussion on all these aspects of combustion pressure measurement and enables the reader to understand the different transducers for combustion analysis, data acquisition, and data analysis to determine the different combustion parameters. The characteristics details of cylinder pressure sensors along with peripheral sensors used for combustion diagnostics are provided in Chaps. 2 and 3. The acquisition of the experimental cylinder pressure signal and their processing (phasing with crank angle position, absolute pressure referencing, smoothing/filtering, and cycle averaging) is presented in Chaps. 4 and 5, respectively. The detailed data analysis of the measured pressure signal is presented in Chaps. 6–10. In Chaps. 6–9, the detailed discussion on performance analysis (torque, power, and work), heat release analysis (mass fraction burned, burn rate, combustion phasing, and combustion duration), combustion stability analysis (cycle-to-cycle variations) with statistical and nonlinear dynamics methods, and knocking and combustion noise analysis along with determination of different metrics are provided. The last Chap. 10 presents the discussion on estimation methods of different engine operating parameters (TDC, compression ratio, air-fuel ratio, wall temperature, trapped mass, and residual gas fraction) using the measured in-cylinder pressure data. For those interested in reading the book, Sects. 1.3 and 1.4 provide an overview of the subject.

In summary, the book provides a comprehensive discussion on the in-cylinder pressure measurement and its processing for analyzing the combustion quality in reciprocating engines. This book is written lucidly and in an easily understandable manner with more than 350 illustrations and 215 discussion questions/problems. Researchers and scholars can get recent updates and future directions for in-cylinder pressure signal processing for combustion analysis of conventional as well as advanced engine combustion modes. Instructors teaching courses on engine instrumentation and combustion diagnostics or advance automotive/engine courses can find it as a good reference or textbook. The book can be used as a resource for researchers and graduate students, professors dealing with engine combustion, engine development engineers/scientists, and anyone wishing to understand and update the knowledge on the state-of-the-art engine combustion diagnostics using in-cylinder pressure measurement and analysis.

This book is the outcome of the research and teaching material that has been collected for several years of my teaching and research on the subject. Presentation of the book is influenced by technical literature published by SAE International, Elsevier, SAGE, ASME, and Taylor & Francis Group. I wish to acknowledge with thanks for the permissions given by Elsevier, SAGE, Taylor & Francis Group, AVL, and KISTLER to reprint some of the figures from their publications.

I wish to express my thanks to my students Mr. Mohit Raj Saxena and Mr. Pankaj Kumar Yadav for helping in the preparation of some of the illustrations for the book. I thank my wife, Suneeta Maurya, and my son, Shashwat Maurya, for their patience, kind cooperation, and moral support during the writing of this book. Lastly, I would like to express my thanks to all those who helped me directly or indirectly for the successful completion of this book.

I warmly welcome you to read this book. I am confident that you could find the important information pertaining to the area of engine combustion diagnostics employing in-cylinder pressure measurement. I will be thankful for any constructive criticism for improvement in the future edition of the book.

Rupnagar, Punjab, India   Rakesh Kumar Maurya
December 2018

Contents

1 Introduction 1
  1.1   Introduction to Reciprocating Engines 2
        1.1.1   Reciprocating Engine Fundamentals 5
        1.1.2   Spark Ignition Engine 9
        1.1.3   Compression Ignition Engine 15
        1.1.4   Low-Temperature Combustion Engine 20
  1.2   Engine Testing and Combustion Diagnostics 25
  1.3   In-Cylinder Pressure-Based Combustion Diagnostics 28
  1.4   Organization of the Book 29
  Discussion/Investigation Questions 32
  References 34

2 In-Cylinder Pressure Measurement in Reciprocating Engines 37
  2.1   In-Cylinder Pressure Measurement Setup 38
  2.2   Piezoelectric Pressure Transducer 41
        2.2.1   Functional Principle 43
        2.2.2   Transducer Materials and Construction 46
        2.2.3   Transducer Design 51
        2.2.4   Transducer Properties and Specifications 59
        2.2.5   Transducer Adaptors and Mounting Position 79
        2.2.6   Transducer Selection 91
  2.3   Alternatives to Piezoelectric Pressure Transducer 95
        2.3.1   Ion Current Sensor 95
        2.3.2   Optical Sensors 104
        2.3.3   Strain Gauges and Indirect Methods 105
  2.4   Crank Angle Encoder 107
        2.4.1   Working Principle and Output Signal 108
        2.4.2   Resolution Requirement 112
  Discussion/Investigation Questions 114
  References 118

3 Additional Sensors for Combustion Analysis 123
   3.1   Exhaust and Intake Pressure Sensors 124
   3.2   Fuel Line Pressure Sensor 131
   3.3   Needle Lift Sensor 135
   3.4   Mass Flow Sensors 140
   3.5   Temperature Sensor 145
   3.6   Valve Lift Sensor 145
   3.7   Ignition Current Sensor 147
   3.8   Sensors for Oxygen Detection and Air-Fuel Ratio Control 148
   Discussion/Investigation Questions 150
   References 151

4 Computer-Aided Data Acquisition 153
   4.1   Introduction 154
   4.2   Data Acquisition Principle 155
   4.3   Transducer and Signal Conditioning 158
          4.3.1   Sensors and Transducers 158
          4.3.2   Signal Conditionings 160
   4.4   Sampling and Digitization of Experimental Signal 162
   4.5   Data Display and Storage 167
   Discussion/Investigation Questions 169
   References 170

5 Digital Signal Processing of Experimental Pressure Signal 171
   5.1   Introduction 172
   5.2   Crank Angle Phasing of Pressure Signal 173
          5.2.1   Phasing Methods Using Additional Hardware 174
          5.2.2   Phasing Methods Using Measured Pressure Data 179
   5.3   Absolute Pressure Referencing (Pegging) 181
          5.3.1   Inlet and Outlet Manifold Pressure Referencing 185
          5.3.2   Two- and Three-Point Referencing 188
          5.3.3   Referencing Using Least-Square Methods 191
          5.3.4   Referencing Using Polytropic Coefficient
                   Estimation 193
   5.4   Smoothing/Filtering of Experimental Data 194
          5.4.1   Moving Average Filters 199
          5.4.2   Low-Pass FIR Filter 202
          5.4.3   Low-Pass IIR (Butterworth) Filters 204
          5.4.4   Thermodynamic Method-Based Filter 208
          5.4.5   Wavelet Filtering 210
   5.5   Cycle Averaging of Measured Data 210
          5.5.1   Method Based on Standard Deviation Variations 211
          5.5.2   Method Based on Statistical Levene's Test 216
   Discussion/Investigation Questions 217
   References 220

6   Engine Performance Analysis 223
   6.1   Indicating Diagram Analysis 225
   6.2   Engine Geometry and Kinematics 232
   6.3   Indicated Torque Calculation and Analysis 240
   6.4   Work and Mean Effective Pressure Calculation
       and Analysis 244
   6.5   Engine Efficiency Analysis 254
   6.6   Gas Exchange Analysis 260
   6.7   Torque, Power and Engine Map 262
   Discussion/Investigation Questions 269
   Problems 271
   References 278

7   Combustion Characteristic Analysis 281
   7.1   Introduction 283
   7.2   Burned Mass Fraction Estimation and Analysis 285
       7.2.1   Marvin's Graphical Method 285
       7.2.2   Rassweiler and Withrow Method 286
       7.2.3   Pressure Ratio and Pressure Departure
             Ratio Method 289
       7.2.4   Vibe's Function-Based Method 292
       7.2.5   Apparent Heat Release-Based Method 293
   7.3   Combustion Gas Temperature Estimation 294
   7.4   Wall Heat Transfer Estimation 300
   7.5   Heat Release Rate Analysis 311
       7.5.1   Rassweiler and Withrow Model 311
       7.5.2   Single-Zone Thermodynamic Model 313
       7.5.3   Real-Time Heat Release Estimation 318
       7.5.4   Characteristics of Heat Release Rate
             in Different Engines 319
   7.6   Tuning of Heat Release 324
       7.6.1   Motoring Pressure-Based Method 324
       7.6.2   Self-Tuning Method 327
   7.7   Estimation of Various Combustion Parameters 329
       7.7.1   Start of Combustion 329
       7.7.2   Ignition Delay 338
       7.7.3   Combustion Phasing 340
       7.7.4   End of Combustion and Combustion Duration 343
   7.8   Thermal Stratification Analysis 348
   Discussion/Investigation Questions 352
   References 355

8   Combustion Stability Analysis 361
   8.1   Combustion Stability in Reciprocating Engines 363
       8.1.1   Manifestation of Combustion Variability 367
       8.1.2   Sources of Cyclic Combustion Variability 369

8.2   Characterization of Cyclic Combustion Variations   380
       8.2.1   Indicators of Cyclic Combustion Variations   381
       8.2.2   Recognition of Partial Burn and Misfire Cycles   383
8.3   Combustion Stability Analysis by Statistical Methods   390
       8.3.1   Time Series Analysis   392
       8.3.2   Frequency Distribution and Histograms   395
       8.3.3   Normal Distribution Analysis   400
       8.3.4   Coefficient of Variability and Standard Deviation   402
       8.3.5   Lowest Normalized Value   408
       8.3.6   Autocorrelation and Cross-Correlation   409
       8.3.7   Principal Component Analysis   412
8.4   Combustion Stability Analysis Using Wavelets   413
8.5   Nonlinear and Chaotic Analysis of Combustion Stability   424
       8.5.1   Phase Space Reconstruction   425
       8.5.2   Poincaré Section   428
       8.5.3   Return Maps   430
       8.5.4   Symbol Sequence Statistics   435
       8.5.5   Recurrence Plot and Its Quantification   441
       8.5.6   0–1 Test   446
       8.5.7   Multifractal Analysis   449
8.6   Steps to Improve Combustion Stability   451
Discussion/Investigation Questions   451
References   453

9   Knocking and Combustion Noise Analysis   461
    9.1   Introduction   464
    9.2   Knock Fundamentals   468
           9.2.1   Knock Onset   468
           9.2.2   Modes of Knock   474
           9.2.3   Super-Knock and Preignition   482
           9.2.4   Characteristic Knock Frequencies   489
    9.3   Cylinder Pressure-Based Knock Analysis   495
           9.3.1   Signal Processing   498
           9.3.2   Knock Indices   499
    9.4   Methods of Knock Detection and Characterization   504
    9.5   Knock Detection by Alternative Sensors   513
           9.5.1   Optical Methods   515
           9.5.2   Ion Current Sensor   516
           9.5.3   Engine Block Vibration   517
           9.5.4   Microphone   518
    9.6   Knock Mitigation Methods   518
    9.7   Combustion Noise Determination   521
           9.7.1   Combustion Noise Calculation and Metrics   522
           9.7.2   Combustion Noise Characteristics   528
    Discussion/Investigation Questions   532
    References   537

10   Estimation of Engine Parameters from Measured Cylinder
    Pressure   543
    10.1   TDC Determination   545
            10.1.1   Polytropic Exponent Method   546
            10.1.2   Symmetry-Based Method   548
            10.1.3   Loss Function-Based Method   550
            10.1.4   Temperature-Entropy Diagram Method   552
            10.1.5   Inflection Point Analysis-Based Method   553
            10.1.6   TDC Calibration Using IMEP   554
            10.1.7   Model-Based TDC Determination Method   555
    10.2   Compression Ratio Determination   557
            10.2.1   Temperature-Entropy-Based Method   557
            10.2.2   Heat Loss Estimation Method   559
            10.2.3   Polytropic Model Method   560
            10.2.4   Determination of Effective Compression Ratio   561
    10.3   Blowby Estimation   565
    10.4   Wall Temperature Estimation   566
            10.4.1   Heat Transfer Inversion Method   567
            10.4.2   Observer Model-Based Method   570
    10.5   Trapped Mass Estimation   571
            10.5.1   Fitting Cylinder Pressure Curve During
                     Compression   571
            10.5.2   The ΔP Method   574
            10.5.3   Mass Fraction Burned Method   575
            10.5.4   Physical Model-Based Method   578
            10.5.5   Resonance Frequency Analysis-Based Method   579
    10.6   Residual Gas Fraction Estimation   582
            10.6.1   Pressure Resonance Analysis Method   584
            10.6.2   Iterative and Adaptive Methods   585
    10.7   Air-Fuel Ratio Estimation   589
            10.7.1   Statistical Moment Method   589
            10.7.2 G-Ratio Method   593
            10.7.3   Net Heat Release-Based Method   594
            10.7.4   Other Estimation Methods   595
    Discussion/Investigation Questions   596
    References   598

Bibliography   603

Index   605

About the Author

Rakesh Kumar Maurya has been a faculty member in the Department of Mechanical Engineering, Indian Institute of Technology Ropar, since August 2013. Before joining IIT Ropar, he was working as senior research associate (Pool Scientist-CSIR) at IIT Kanpur. He received his bachelor's, master's, and Ph.D. degrees in Mechanical Engineering from Indian Institute of Technology Kanpur, India. He received the Early Career Research Award from Science and Engineering Research Board (SERB), Government of India, New Delhi. He is also a recipient of the Young Scientist Award (2016) from the International Society for Energy, Environment and Sustainability. He is the author of the book titled *Characteristics and Control of Low Temperature Combustion Engines: Employing Gasoline, Ethanol and Methanol* published by Springer International Publishing AG in 2018. He has also edited 1 book published by Springer and published 15 book chapters and more than 40 international peer-reviewed journal papers. He teaches and conducts research in the area of internal combustion engines. His areas of interest are low-temperature engine combustion, alternative fuels, engine combustion diagnostics, engine instrumentation, combustion and emission control, particulate matter characterization, engine management systems, engineering ethics, and philosophy of science.

Chapter 1
Introduction

Abbreviations and Symbols

BDC	Bottom dead center
BEV	Battery electric vehicle
CA	Crank angle
CAD	Crank angle degree
CI	Compression ignition
CNG	Compression natural gas
CO	Carbon monoxide
CRDI	Common rail direct injection
DDFS	Dual direct injection fuel stratification
DI	Direct injection
DISI	Direct injection spark ignition
DME	Dimethyl ether
DPF	Diesel particulate filter
ECE	External combustion engine
ECU	Electronic control unit
EGR	Exhaust gas recirculation
EMS	Engine management system
FMEP	Friction mean effective pressure
GCI	Gasoline compression ignition
GDCI	Gasoline direct injection compression ignition
GDI	Gasoline direct injection
HC	Unburned hydrocarbon
HCCI	Homogeneous charge compression ignition
HEV	Hybrid electric vehicle
HTC	High-temperature combustion
IC	Internal combustion
ICE	Internal combustion engine

© Springer Nature Switzerland AG 2019
R. K. Maurya, *Reciprocating Engine Combustion Diagnostics*, Mechanical
Engineering Series, https://doi.org/10.1007/978-3-030-11954-6_1

IMEP   Indicated mean effective pressure
IT   Ignition timing
LPG   Liquefied petroleum gas
LTC   Low-temperature combustion
MBT   Maximum brake torque
N Engine speed
NO_x Nitrogen oxides (NO and NO_2)
OBD   Onboard diagnostics
OEM   Original equipment manufacturer
PCCI   Premixed charge compression ignition
PFI   Port fuel injection
PM   Particulate matter
PMEP   Pumping mean effective pressure
PPC   Partially premixed combustion
RCCI   Reactivity controlled compression ignition
SACI   Spark-assisted compression ignition
SCCI   Stratified charge compression ignition
SCR   Selective catalytic reduction
SI   Spark ignition
SN   Swirl number
SOC   Start of combustion
TBI   Throttle body injection
TDC   Top dead center
VCR   Variable compression ratio
V_d Displacement volume
VGT   Variable geometry turbocharger
VVT   Variable valve timing
η Efficiency
ρ Density
φ Equivalence ratio

1.1   Introduction to Reciprocating Engines

Modern society is to a large extent built on the transportation of both people and goods. Transport is almost entirely (>99.9%) powered by internal combustion engines (ICEs), which typically burn fossil petroleum-based liquid fuels. Reciprocating internal combustion engines mainly perform land and marine transportation, and air transport is mainly powered by jet engines [1]. Several factors are responsible for wide-range utilization of liquid fuels for transportation including (1) high energy density, (2) easy transportation and storage, and (3) large global infrastructure developed over time. Reciprocating ICEs are well accepted and the most significant source of energy since the last century due to their superior performance, robustness, controllability, durability, and absence of other viable alternatives.

   The electric vehicle and fuel cell-operated vehicles are considered as an alternative to internal combustion engines for automotive applications. Presently, there is much interest in electric vehicles, and several governments are supporting this initiative. However, battery electric vehicles (BEVs) seem to have zero local pollution, but they can actually have higher total greenhouse gas emission (CO_2) than similar reciprocating engine-operated vehicle [1–3]. Thus, promoting BEVs can be counterproductive until the electricity production is sufficiently decarbonized [2]. The hybrid electric vehicle (HEV) is a better option in terms of CO_2 emission reduction rather instead of the BEVs [1, 3]. Even if electricity generation is decarbonized, the BEVs have a very significant impact on human toxicity, freshwater eco-toxicity, and freshwater eutrophication, mainly caused by the production of metals required for batteries [1, 4].

   Worldwide several initiatives are being developed and proposed for implementation including alternative fuels and alternative combustion mode in reciprocating engines, and BEVs with green power alternatives. However, the importance of a particular alternative is different at different times and in a different country or region. Kalghatgi [1] proposed that the evolution of transportation energy is dependent on the complex interplay between several drivers for the change, which is illustrated in Fig. 1.1. Various factors affecting the energy policy include the energy security and local pollution concerns, climate change concerns, support to farmers and increase in rural employment, and aspiration for leadership in newer technologies. Transport policy should be based on a balanced approach using all available technologies, considering local and global environmental and greenhouse gas impacts, security of supply, and social, economic, political, and ethical impacts [1]. Reliable projections indicate that even by 2040 about 90% of transportation energy will come from combustion engines operated on petroleum fuels. The reciprocating combustion engines will continue to power transport (especially

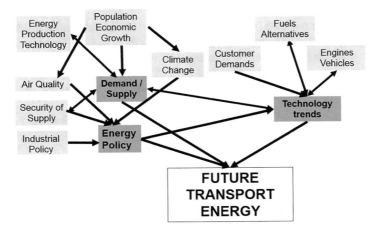

Fig. 1.1 The evolution of future transport energy system [1]

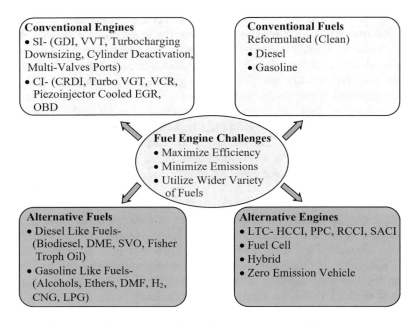

Fig. 1.2 Fuel-engine challenges and possible technological solutions [5]

commercial sector) by a large extent for several decades, and it will also continue to improve [1].

With the increasing concern on achieving a substantial improvement in automotive fuel economy, automotive engineers are striving to develop engines that provide both significant reduction in the brake-specific fuel consumption (BSFC) and compliance with future stringent emission requirements. To meet these requirements, four possible approaches (improving conventional engines and fuels or using alternative fuels and engines) and their combination are proposed. The major fuel-engine challenges are summarized in Fig. 1.2 along with possible technological solutions. The alternative fuels can be used in conventional engines with the latest technologies as well as in advanced low-temperature combustion concepts [5]. There are plenty of options for improving the reciprocating ICEs (Fig. 1.2). Furthermore, to meet the emissions legislation, aftertreatment technologies can be utilized along with improved engines.

At the automotive vehicle level, several efforts are made to operate them as efficiently and cleanly as possible. The major trends are downsizing, hybridization, driver support system, and newer infrastructure development [6]. To meet the emission legislation requirement, downsizing is one of the options. Downsizing includes two approaches: (1) developing smaller and lightweight vehicles leading to lower fuel consumption and (2) developing smaller engines with lower fuel consumption (often using turbocharger). Hybridization of the vehicle is another technique used to improve fuel consumption. Hybrid vehicles use the battery as well as smaller size ICE. Several factors contribute to efficiency improvements in a hybrid vehicle that include avoiding transients, running the engine at optimal conditions, and regenerative braking. Hybrid vehicles are mainly relevant to spark ignition engines in stop/start city driving conditions [1]. Fuel consumption and quantity of emissions are highly governed by the driving of the vehicles. There is a strong interest in developing a system which assists the driver in optimal driving which leads to fuel saving. A driver support system proposes the speed and gear selection to the driver and can also evaluate and educate the driver [6]. The technological developments such as GPS and map database can be beneficial for optimal driving in all circumstances, which include informing traffic conditions, roadside information, weather, etc. These developments can potentially lead to fuel saving and improving the environmental conditions by reducing the exhaust emissions.

1.1.1 Reciprocating Engine Fundamentals

An engine is a device to convert the fuel energy into mechanical energy (useful work). In engines, chemical energy bounded in fuel is converted into heat energy by combustion, and the heat energy is converted to useful work by the driving mechanism. The driving mechanism can be reciprocating or rotary. There are two basic types of engines: (1) internal combustion engines (ICEs) and (2) external combustion engines (ECEs). In ICEs, the combustion products act as a working fluid, while in ECEs, the combustion products transfer heat (using heat transfer device) to another fluid that act as working fluid. Presently, most of the engines used for transportation are internal combustion type. Reciprocating ICE offers several advantages over the ECE (steam turbines) such as (1) mechanical simplicity (due to absence of heat exchangers (boiler and condenser) in the path of working fluid); (2) works as lower average temperature than the maximum temperature of working fluid (high temperature occurs only of fraction of cycle), leading to possibility of employing higher working fluid temperature for higher efficiency; (3) lower weight-to-power ratio of the engine; and (4) possibility to develop ICE of very small power output [7].

There are four general types of internal combustion engines, namely, (1) four-stroke cycles, (2) two-stroke cycles, (3) rotary engines, and (4) continuous combustion gas turbine engines [8]. Reciprocating piston engines operate on four- or two-stroke combustion cycles. The basic components of the reciprocating piston engine are shown in Fig. 1.3. Reciprocating engine is characterized by a slider-crank mechanism which converts the reciprocating motion of the piston into rotating motion of crankshaft [9].

The piston moves back and forth in a cyclic manner in the engine cylinder and transmits power to the crankshaft through connecting rod. The piston speed becomes zero at topmost position, before its direction changes, and this crankshaft position is known as top dead center (TDC). Similarly, the bottommost position of the piston is known as bottom dead center (BDC). Further details about each of the engine components can be found in any textbook [7, 9, 10].

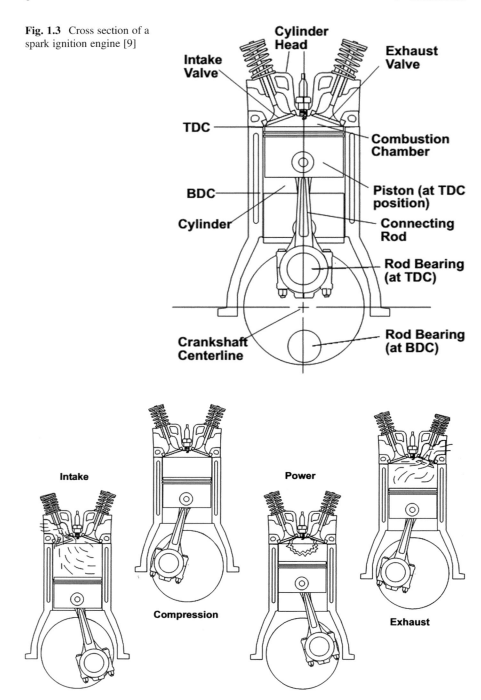

Fig. 1.3 Cross section of a spark ignition engine [9]

Fig. 1.4 A four-stroke reciprocating internal combustion engine cycle [9]

Presently, most automotive vehicles used for road transport operates on four-stroke cycles. Figure 1.4 illustrates a four-stroke reciprocating internal combustion engine cycle. A stroke consists of piston motion from TDC to BDC position, and during this period, crankshaft travels 180° (half round). Thus, four-stroke cycle completes in two rounds (720°) of crankshaft motion. Four-stroke cycle consists of intake, compression, expansion (power), and exhaust stroke. During the intake stroke, the piston moves from TDC to BDC position and inducts air and fuel (in SI engine) or only air (CI engines) in the cylinder. The intake valve opens slightly before TDC position and closes after BDC position to improve the volumetric efficiency or increase the mass of inducted air [10]. In compression stroke, the piston moves from BDC to TDC positions and increasing the pressure and temperature of the gases in the cylinder. Toward the end of compression stroke, combustion starts in the cylinder either by autoignition (CI engine) or by spark (SI engine). High-pressure and high-temperature gases expand during the power stroke, and useful work is extracted at the crankshaft. In the exhaust stroke, burned gases are pushed out of the cylinder past the exhaust valve as the piston moves from BDC to TDC position. The exhaust valves typically open slightly before the start of exhaust stroke leading to exhaust blowdown (rapid decrease of pressure in the cylinder). This process reduces the work done by the piston during the exhaust stroke. This cycle repeats again, and the piston starts moving again from TDC to BDC position in the next cycle.

The reciprocating combustion engines can be classified in several ways including basic design (reciprocating or rotary), types of ignition and combustion (SI and CI; homogeneous and heterogeneous), working cycle (four- and two-stroke), air intake (boosting) process (naturally aspirated, turbo-charged, super-charged), valve locations (overhead, valve in block, valve in head), position and number of cylinders (single-cylinder, in-line, V engine, opposed cylinder engine, opposed piston engine, radial engine), cooling (air cooled, liquid cooled, water cooled), fuel used (diesel, gasoline, CNG, DME, ethanol, methanol), and method of fuel input (DI, PFI, TBI) [7, 10].

For designing the new engine, four basic questions need to be answered to define its characteristics. The questions are as follows: (1) what is the required displacement volume of engine to produce desired power?, (2) how many cylinders will the displacement be distributed?, (3) what should be the bore and stroke of each cylinder?, and (4) what will be the configuration of the engine?.

The displacement volume required for the engine is calculated by determining the air required corresponding to the maximum power produced. For estimation of air requirement, the full-load curve of the engine along with expected specific fuel consumption needs to be determined. In spark ignition engine, the work output is typically limited by displacement over its entire speed range while in diesel engine displacement limited only at maximum torque condition [9]. Based on the expected maximum power and specific fuel consumption, the mass flow rate of fuel can be computed. Air-fuel ratio of engine operation depends on the type of engine (SI or CI), and the mass flow rate of air is calculated by considering the air-fuel ratio. The displacement volume required for the engine can be calculated using Eq. (1.1) at engine speed (N) corresponding to maximum power output.

$$V_{\mathrm{d}} = \frac{\dot{m}_{\text{air}}}{\rho_{\text{air}} \cdot \left(\frac{N}{X}\right) \cdot \eta_{\text{vol}}} \tag{1.1}$$

where ρ is the air density at the inlet, η_{vol} is the expected volumetric efficiency, and the value of X is 2 for four-stroke engine and 1 for a two-stroke engine. Detailed discussion on displacement volume determination can be found in reference [9].

Typically, when estimated displacement volume exceeds the 500 cm^3 for high-speed engines, the multicylinder engines are preferred. In multicylinder engines, the number of cylinders and their configuration is two major parameters from a design perspective. Several factors affect the selection of the number of cylinders including cost and complexity, engine balancing and vibration, surface-to-volume ratio of the combustion chamber, engine speed requirements (mean piston speed), and engine breathing characteristics (pumping losses) [9]. The configuration of the engine is governed by balancing of forces and packaging consideration of engine. Typically, the in-line configuration is used up to four cylinders, and V configuration is used for greater than four cylinders.

After deciding the displacement volume per cylinder, the bore-to-stroke ratio is an important parameter that also defines the shape of the combustion chamber. The major factors affected by bore-to-stroke ratio are heat transfer losses (surface-to-volume ratio near TDC position), piston speed, and valve flow area (decides the size of valves). These factors need to be optimized for determination of bore-to-stroke ratio [9]. Figure 1.5 illustrates the trade-offs associated with the three parameters for the determination of optimal bore-to-stroke ratio (please note the scales of the figure). Engine bore must be made sufficiently large to maintain the mean piston speed lower than the design target and to minimize pressure drop across the valves.

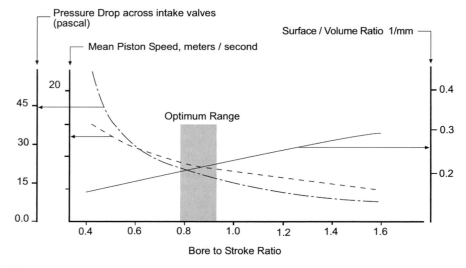

Fig. 1.5 Estimation of optimum bore-to-stroke ratio for a particular engine [9]

The resulting bore-to-stroke ratio is dependent on the design criteria for the specific engine. The bore-to-stroke ratio is typically just below unity to provide the best balance of performance in automotive applications where fuel efficiency and emission requirements dominate. In high-performance applications such as racing engine (racing rules often limit displacement), the bore-to-stroke ratio tends to be larger to maximize breathing and operate at acceptable piston speeds during engine operation at high speeds.

1.1.2 Spark Ignition Engine

The spark ignition (SI) engines are used in applications where higher engine speed and the lightweight engine are required. The SI engines operate at high engine speeds, and thus, the reciprocating components are made lightweight to reduce the inertia forces at high speeds. The SI engines can be designed for two-stroke as well as four-stroke cycles. Due to the urban air pollution concerns, four-stroke engines are largely used for small and large power requirements.

In conventional SI engines, the premixed homogeneous mixture of fuel and air is ignited by a positive source of ignition. The positive source of ignition is often an electric discharge spark plug although alternatives such as laser-induced ignition, corona ignition, etc. are also existing [10]. For the preparation of premixed fuel-air mixture, fuel is introduced in the intake manifold with a carburetor or fuel injection system. Ideally, the stoichiometric air-fuel ratio is maintained in conventional SI engines at all the engine load conditions. Thus, engine load is varied by changing the amount of air using the throttle in the inlet manifold. The spark initiates a flame kernel, which grows with time, and a turbulent flame propagates in the combustion chamber and consumes the entire charge. The flame reaches the cylinder walls by consuming all the charge and extinguishes. Typically, this process is called normal spark ignition combustion. Figure 1.6 illustrates the ignition and flame propagation in SI engine using the conventional spark (left) and corona ignition (right). The first image corresponds to the point of ignition, and subsequent images are at a constant distance of 2.5° between frames. Corona ignition creates a significantly larger high-intensity plasma ignition source, which spreads throughout the combustion chamber early when compared to conventional spark ignition systems.

In conventional spark ignition case (Fig. 1.6), spark flashover at ignition timing (−14.3° CA) and bright light from the spark is reflected. After 2.5° CA of spark, an initial hemispheric flame kernel can be detected that further moves slightly off-center in the chamber due to turbulence. At −6.8° CA, i.e., after 7.5° of ignition, the large-scale wrinkling of the flame front surface increases the active flame front surface and thus enlarges and accelerates its propagation, increasing light intensity due to a magnified flame volume and higher pressure/temperature. However, in the corona ignition system, after 2.5° of ignition, five individual kernels developed around streamers, and the relatively larger area is enflamed as flame propagates [11].

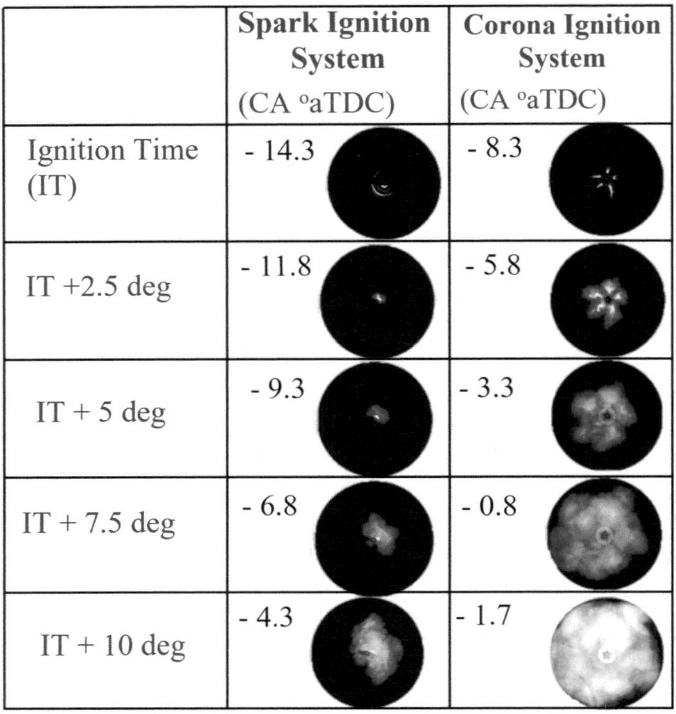

	Spark Ignition System (CA °aTDC)	Corona Ignition System (CA °aTDC)
Ignition Time (IT)	- 14.3	- 8.3
IT +2.5 deg	- 11.8	- 5.8
IT + 5 deg	- 9.3	- 3.3
IT + 7.5 deg	- 6.8	- 0.8
IT + 10 deg	- 4.3	- 1.7

Fig. 1.6 Flame propagation in SI Engine [11]

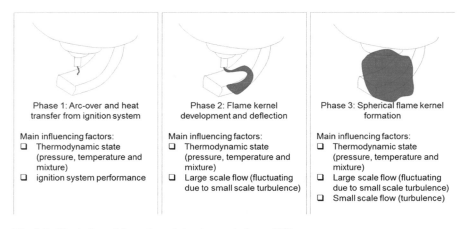

Phase 1: Arc-over and heat transfer from ignition system

Main influencing factors:
☐ Thermodynamic state (pressure, temperature and mixture)
☐ ignition system performance

Phase 2: Flame kernel development and deflection

Main influencing factors:
☐ Thermodynamic state (pressure, temperature and mixture)
☐ Large scale flow (fluctuating due to small scale turbulence)

Phase 3: Spherical flame kernel formation

Main influencing factors:
☐ Thermodynamic state (pressure, temperature and mixture)
☐ Large scale flow (fluctuating due to small scale turbulence)
☐ Small scale flow (turbulence)

Fig. 1.7 Illustration of flame kernel development phases [12]

The charge motion around the spark plug and charge composition at the time of spark discharge is decisive for the flame development and subsequent flame propagation [5]. The process from ignition to early combustion can be divided into different phases as illustrated in Fig. 1.7. The arc-over phase is primarily affected

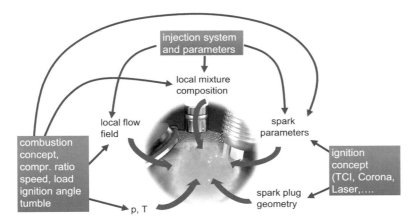

Fig. 1.8 Factors affecting the flame kernel formation in SI engine [13]

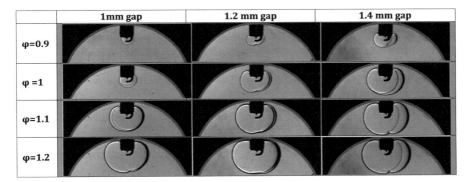

Fig. 1.9 Comparison of typical flame growth for different spark plug gaps and charge composition [14]

by the performance of the ignition system and the thermodynamic state of the charge. In this phase, chemical reactions result into an early flame kernel development. Approaching charge flow will deflect such a flame structure, but the flame kernel is too small to be influenced by small-scale turbulence. However, turbulence will affect the deflection indirectly by having an influence on large-scale flow. This deflection affects the volume activated by energy supplied from the ignition system. The early flame kernel develops to a shape which could be approximated as a sphere. The transition from a laminar to a turbulent flame occurs during this phase [12]. The flame kernel formation is affected by several factors such as local flow field and mixture composition, spark parameters, spark plug geometry, etc. which are illustrated in Fig. 1.8.

Figure 1.9 shows the effect of spark plug gap and charge composition on flame growth at 6 ms after the start of flame kernel initiation. The figures show that the difference in flame kernel growth is high between the three spark plug gaps for lean

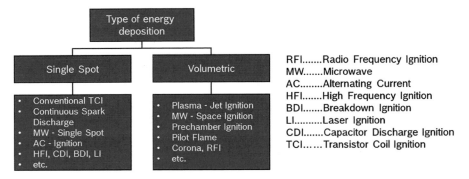

Fig. 1.10 Ignition systems in SI engines [13]

and stoichiometric conditions and flame kernel growth is not significantly affected by the spark plug gap at rich condition ($\varphi = 1.2$). Figure 1.9 also depicts that spark plug gap of 1.2 and 1.4 mm has a significantly larger flame kernel in comparison to gap 1 mm. At the beginning of the flame kernel initiation and up to 1 ms from the start of spark, the difference in the flame kernel size between different spark plug gaps is relatively small especially for larger gaps such as 1.2 and 1.4 mm. It was demonstrated that at the beginning of the flame kernel initiation and up to 1 ms from the spark timing, the difference in the flame kernel is small, but as the time after ignition progresses, the larger spark plug gaps produce a significantly larger flame kernel areas in comparison to the spark plug gap of 1 mm [14]. Thus, spark parameter and charge conditions affect the initial flame development, which leads to variations in flame propagation and heat release in SI engine.

Figure 1.10 summarizes the different types of ignition systems for SI engines based on how the flame kernel volume is activated during the ignition process. The "single spot ignition" means the deposition of ignition energy within a fixed location in the cylinder, and the energy distribution is not actively controlled by the ignition system itself [13]. However, a spark volume is characterized by the spark channel and even enlarged by charge motion (e.g., deflecting the spark). The deflection of spark leads to activation of flame volume, which results in potential to ignite the dilute mixtures in gasoline engines. Volumetric ignition concepts such as corona or plasma-jet ignition systems actively provide a flame kernel volume even in low charge motion conditions. The ignition systems can also be classified based on how the primary energy for operating the circuit is made available, i.e., battery ignition systems and magneto ignition systems [7].

The overall combustion process in SI engines can be considered to consist of three phases: (1) flame development phase, (2) flame propagation phase, and (3) flame termination phase [10, 15]. On the firing of spark, plasma discharge ignites the air-fuel mixture in a small volume between and around the spark plug gap. Sustained combustion reactions result in the development of a turbulent flame which propagates outward from the spark plug. The combustion reactions depend on both temperature and pressure, the nature of the fuel and proportion of the exhaust residual gas. At first, the flame propagates very slow, and very small pressure rise in the cylinder pressure is noticed until the flame kernel grows into a fully developed flame. A fully developed flame becomes turbulent and like a spherical wave propagates at high speed across the combustion chamber. A variety of fluid motions such as swirl, tumble and squish, and turbulence is generated in the cylinder by proper design of intake ports and the combustion chamber shape to accelerate the combustion process (increase flame speed). The turbulent flame speed depends on the turbulence intensity and is several times higher than the laminar flame speed of a flame propagating through a fuel-air mixture of similar composition and thermodynamic state. The combustion rate depends on various factors such as temperature and pressure of charge, mixture strength, residual gas fraction or charge dilution by external EGR, turbulence, and fuel chemistry. With the progress of flame propagation, cylinder pressure rises, and the unburned mixture in the front of the flame gets compressed. The pressure and temperature of unburned charge rise till the peak pressure is achieved in the cylinder. The unburned charge can autoignite if it remains at sufficiently high temperatures for a significant time period. The autoignition of charge leads to knocking, which is a different (abnormal) process than the combustion by normal flame propagation. The detailed discussion on heat release rate and knocking is provided in Chaps. 7 and 9, respectively.

Well-mixed (normally called homogeneous) stoichiometric operation is the leading combustion mode for gasoline SI engines used for automotive applications. However, several factors limit the engine efficiency of stoichiometric operation (without exhaust gas recirculation, EGR) particularly at low and intermediate loads [16]. First, throttling used for load control increases the pumping loss at part-load conditions. Second, high combustion temperatures result in both high heat transfer losses and unfavorable thermodynamic properties of the combustion products (lower γ). The lower ratio of specific heats (γ) leads to reduction of the work extraction efficiency. Third, the stoichiometric combustion is not able to fully complete near TDC due to dissociation of CO_2 in the hot O_2-depleted gases. A limiting 10–90% burn duration of 30 CAD was found at the location of peak efficiency irrespective of operating conditions, which means that the 10–90% burn duration determines efficiency and that combustion deteriorates significantly beyond this 30 CAD offsetting any efficiency gains from thermodynamics effects, reduced pumping work, or reduced heat transfer losses [17].

To overcome some of the demerits of well-mixed stoichiometric SI engines, gasoline direct injection (GDI)-based engines are developed [18]. The direct injection spark ignition (DISI) engines can operate unthrottled as well as on the leaner mixture. In this engine, the fuel spray plume is injected directly into the cylinder, generating a fuel-air mixture with an ignitable composition at the spark gap at the time of ignition. The GDI engines have improved fuel economy over PFI engines due to a substantial reduction in pumping loss, reduction in heat loss, higher compression ratio, increased volumetric efficiency, and less acceleration-enrichment requirement [18].

The fuel-lean operation can improve engine efficiency. However, the challenge is to maintain stable and efficient combustion despite a reduction of flame speeds in

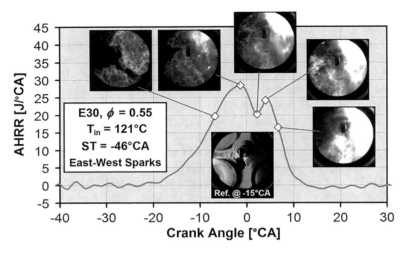

Fig. 1.11 Mixed-mode operation in DISI at lean E30 operation with two gable-mounted spark plugs without swirl [19]

fuel-lean mixtures. The reduction of flame speeds becomes a particular problem from the perspective of ignition, and effective flame spreads throughout the charge. During lean engine operation, high combustion efficiency, shorter burn duration, and faster inflammation provide higher thermal efficiency [16]. It was also found that short delay from spark to main combustion (i.e., fast inflammation) reduces cyclic variations of the deflagration-based combustion for both lean and dilute operation. To obtain short combustion duration, several possibilities are explored including multi-spark plug and advanced volumetric ignition systems such as corona, increasing the turbulence, intake heating, etc.

To maintain a sufficiently short burn duration, mixed-mode combustion is proposed for $\phi < 0.6$ in DISI engine [19]. In the mixed-mode combustion, a combination of deflagration and end-gas autoignition occurs, and the end-gas reactants reach to the point of autoignition by the compression of pressure rise due to the deflagration-based combustion [16, 19]. Figure 1.11 illustrates the mixed-mode combustion process using the flame imaging of transition from turbulent deflagration to end-gas autoignition. The mode can ensure sufficiently short burn duration for ultra-lean SI operation such that relative efficiency gain of roughly 20% can be realized [19]. The mixed-mode combustion is conceptually the same as spark-assisted compression ignition (SACI). However, in this mode, a larger fraction of the combustion is based on flame propagation, and the level of internal residuals (or external EGR) is much lower. This combustion mode introduces a noise concern, and it should be restricted to very lean ($\phi \leq 0.55$) or highly dilute ($\phi_m \leq 0.65$) operation while ensuring very repeatable deflagration-based combustion to avoid occasional knocking cycles [16].

1.1.3 Compression Ignition Engine

A four-stroke compression ignition engine also experiences the four strokes of intake, compression, expansion, and exhaust similar to four-stroke gasoline engines. However, due to differences in fuel characteristics, the formation and ignition of fuel-air mixture differ from the gasoline (spark-ignition) engines. Diesel fuel has a higher viscosity, and lower autoignition temperature than gasoline, and it is also not susceptible to vaporizing.

In a diesel engine, only air is inducted during the intake stroke and compressed in the compression stroke that increases the pressure and temperature of the air. Toward the end of compression stroke, diesel is injected into the cylinder through a fuel injector in the hot and dense air. Fuel is injected at 400–2000 bar pressure through three to eight hole injector nozzle depending on the engine design and size. In a diesel engine, the load is controlled by varying the amount of fuel, and inlet air quantity remains the same at particular engine speed in a naturally aspirated engine. Injected fuel in the cylinder atomizes into fine droplets, and air entrainment takes place in the fuel spray. The fuel droplet evaporates and mixes with air. The combustion initiates by autoignition of premixed charge. The time period between the start of injection and start of combustion (SOC) is defined as ignition delay. After SOC, the flame spreads rapidly in the combustible mixture prepared during ignition delay period. This phase of diesel combustion is typically known as premixed combustion phase. At higher engine loads, fuel injection continues even after the start of combustion. The combustion of fuel injected after SOC depends on how fast it gets evaporated and mixed with air. During this period, turbulent diffusion process governs the fuel-air mixing and combustion rate. Thus, this phase of combustion is known as mixing controlled or diffusion combustion phase [10].

The fuel distribution in the cylinder is not uniform, and local fuel-fuel ratio also varies from zero (only pure air present) to the infinity (only fuel present). Images in Fig. 1.12 illustrate the heterogeneous nature of diesel combustion and diffusion flame development and its progress. At high combustion temperatures (2000–2500 °C), carbon particles in the diffusion flame have sufficient luminosity and appear as a yellow region, and radiation from particle varies as flames cool down [15]. The combustion has just initiated in the first image (Fig. 1.12), and all the fuel sprays have developed diffusion flame in the second image. Subsequent two combustion images have the highest intensity, depending on the heat release rate. The heat release rate is also affected by the different swirl number (SN) as illustrated in Fig. 1.12 [20]. The premixed combustion phase is not observed in modern diesel engines with high fuel injection pressures (Fig. 1.12), which typically occurs in the low-pressure mechanical fuel injection (see details in Chap. 6). Based on engine load, only a very small portion of premixed heat release appears in modern diesel engines due to the accelerated mixing process and reduced ignition delay period because of high fuel injection pressures [5].

Figure 1.13 schematically illustrates a modern diesel engine showing different systems. Diesel engines used microprocessor control with direct fuel injection and wastegate turbochargers around the year 1989. Later, EGR with both high pressure

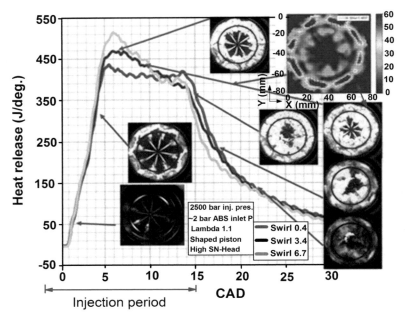

Fig. 1.12 Heat release rate and flame luminescence images of the combustion chamber in a modern diesel engine at a load of 20 bar IMEP, 2500 bar fuel injection pressure and 1000 rpm [20]

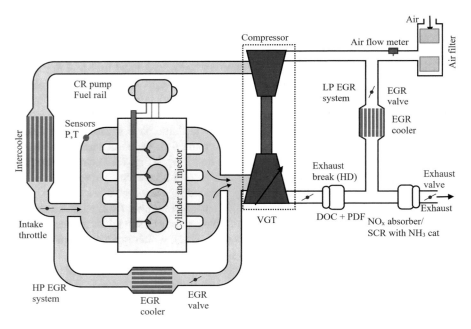

Fig. 1.13 Schematic diagram of a modern diesel engine showing different systems (adapted from [6])

(HP) and low pressure (LP), oxidation catalyst, and turbochargers with variable geometry (VGT) are developed. Presently diesel engines are equipped with common rail diesel injection (CRDI) system with very high fuel injection pressure, piezo injectors, multiple fuel injection capability, high EGR rates, twin turbochargers or VGT chargers, and recent aftertreatment technologies (DPF, SCR, DeNO$_x$-catalyst) as illustrated in Fig. 1.13 [5].

The combustion is a very complex process in the diesel engines, where fuel injection, atomization, vaporization, mixing, as well as combustion can simultaneously occur in the cylinder. Vaporization and mixing are the slowest processes and thus control the combustion rate. The speed of diesel combustion is limited by the mixing between the injected fuel and the air in the cylinder, and thus, it also limits the maximum engine operating speed. Diesel engine characteristics be can be roughly classified into six groups: (1) fuel injection characteristics, (2) fuel spray characteristics, (3) combustion characteristics, (4) engine performance characteristics, (5) ecology characteristics, and (6) economy characteristics [21]. Figure 1.14 shows the diesel engine characteristics and their interrelationship. All these characteristics are dependent on the most basic parameters such as fuel type or injection system type and on several process characteristics such as the injection process, fuel spray development, atomization, mixture fuel/air formation, ignition and combustion, etc. [21]. Figure 1.14 depicts that fuel injections' characteristics significantly affect the performance, combustion, and emissions' characteristics of the diesel engine.

The most important injection characteristics are fuel injection pressure, injection duration, injection timing, and injection rate history [21]. All these parameters affect the diesel engine characteristics (Fig. 1.14). In conventional diesel engines, fuel is injected through a mechanical fuel injection system, which is illustrated in Fig. 1.15. In this system, the pump pressurizes the fuel which reaches to the fuel injector via a

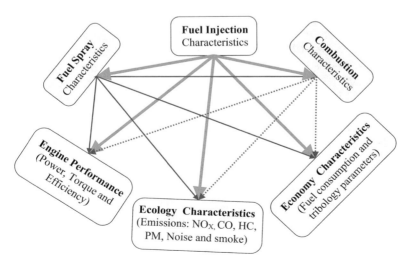

Fig. 1.14 Diesel engine characteristics and their interrelationships [21]

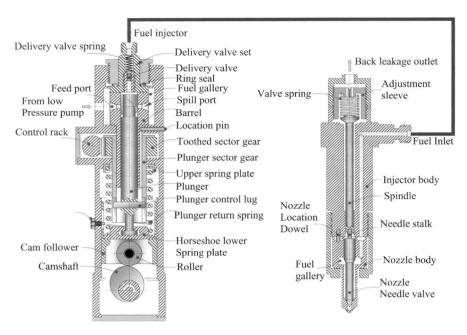

Fig. 1.15   Schematic diagram of mechanical fuel injection system in diesel engine [21]

high-pressure line. The fuel injector is a spring-loaded system; when the line pressure overcomes the spring force, the fuel injection occurs in the cylinder. In this case, the pressure is not constant throughout the engine cycles, and high pressure is generated when fuel injection is required. The timing of fuel injection is governed by the cam, which drives the plunger of the fuel pump. The quantity of fuel injected is governed by plunger lug, which rotates the plunger and the orientation of helix changes. In the mechanical fuel injection system, the fuel injection pressure is relatively lower (~400 bar or less). In a modern diesel engine, electronic fuel injection systems are used, which is often referred as common rail direct injection (CRDI) systems. The CRDI system mainly fulfills the three requirements of modern diesel engine, i.e., (1) high-pressure capability and injection pressure control, (2) flexible timing control, and (3) injection rate control. The schematic of a typical CRDI system is presented in Fig. 1.16.

In CRDI systems, the generation of high fuel pressure and fuel injection events is separated. The fuel pressure is independent of the engine speed and load unlike the mechanical in-line jerk and distributor pumps. The high-pressure fuel is fed to a rail (accumulator), which delivers the fuel to individual injectors (Fig. 1.16). The CRDI systems consist of four main components, i.e., (1) high-pressure pump, (2) high-pressure distribution rail and pipes, (3) electronic fuel injectors, and (4) electronic control unit (ECU). The common rail is connected to the fuel injectors through short pipes. The main advantage of the CRDI system over the conventional in-line jerk pumps is that the fuel injection pressure is constant and independent of the engine

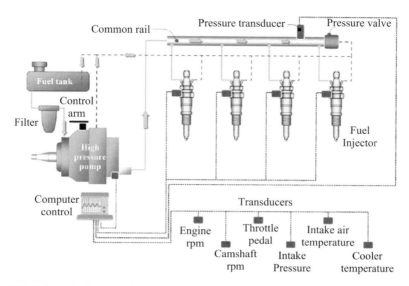

Fig. 1.16 Schematic diagram of CRDI system in modern diesel engines [21]

load and speed. Due to this independence, the maximum flow rate (and maximum torque required to drive the pump) does not have to coincide with the injection event (which is the case with the distributor pump), and thus, CRDI system requires a lower fuel pump maximum torque.

Due to electronic solenoid-based control of fuel injectors, the multiple injection events in a cycle are possible. The pilot injection is typically used to control the diesel engine noise. Figure 1.17 illustrates the components of electronic CRDI fuel injector and their associated signals. The injection quantity and injection timings are controlled by the opening and closing of the solenoid valve (Fig. 1.17). In the closed condition of the solenoid valve, the fuel pressures in the working chamber and in the needle chamber are equal to the rail pressure, and the needle is in the "closed" position. When the solenoid valve opens, the pressure in the working chamber falls, and the common rail pressure in the needle chamber lifts the needle, which initiates the fuel injection in the cylinder. The working chamber is pressurized again when solenoid valve closes [21].

The characteristics of the fuel injection system affect the fuel spray characteristics (droplet size distribution, spray angle, and spray tip penetration), which affect the combustion and emissions' characteristics of the diesel engine. The diesel engines are widely known for their higher fuel economy. The main advantages of using diesel engine are higher fuel conversion efficiency, high torque, low HC and CO emissions, durability and reliability, and typically low fuel and maintenance cost [22]. However, there are several demerits in diesel engines, which include cold start difficulty, noisy-sharp pressure rise, inherently slower combustion (low engine speed), lower power-to-weight ratio, expensive components, NO_x and particulate matter emissions, and low air utilization.

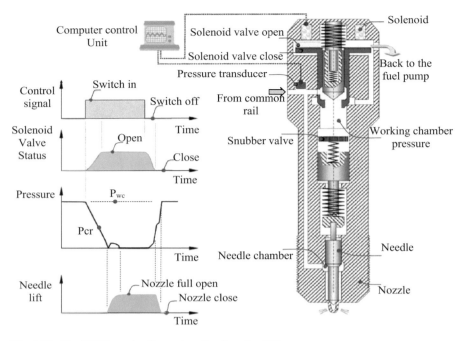

Fig. 1.17 The CRDI nozzle of a modern diesel engine [21]

1.1.4 Low-Temperature Combustion Engine

Spark ignition (SI) and compression ignition (CI) engines powered by gasoline and diesel, respectively, are the most widely used reciprocating engines. Diesel CI engines are preferred choice for medium- and heavy-duty applications due to their higher fuel conversion efficiency as it is operated at higher compression ratio, on the lean fuel-air mixture and unthrottled. However, the CI engines emit higher nitrogen oxides (NO_x) and particulate matter (PM), and there exists a trade-off in the emission of two species. The aftertreatment of NO_x and PM is difficult in diesel engines due to leaner engine operation. To meet the future emission legislation, a combination of in-cylinder strategies as well as exhaust aftertreatment devices is proposed for emission reduction. Employing currently available aftertreatment devices has limitations of higher cost, fuel economy penalties, durability issues, and larger space requirements [23]. Moreover, to compete in the market, higher efficiency at minimal cost is required. Thus, there is a need for drastic improvement in the in-cylinder combustion strategies to achieve higher fuel conversion efficiency along with a decrease in engine emissions which can reduce dependence on the exhaust aftertreatment technologies.

The NO_x formation in the engine cylinder is mainly dependent on the combustion temperature and air-fuel ratio (oxygen availability) [15]. The NO_x formation increases exponentially after the combustion temperature reaches around 2000 K.

In conventional engines (SI and CI), the maximum combustion temperature can occur in the range 2500–3000 K, which leads to the higher NO_x formation in the cylinder. Due to the higher combustion temperature occurrence, the conventional SI and CI combustions are referred as high-temperature combustion (HTC). Soot formation tendency increases as the local fuel-air mixture becomes richer. Ideally, to avoid the soot and NO_x formation in the combustion chamber, the engine should be operated on the premixed fuel-air mixture and at lower combustion temperature. The homogeneous charge compression ignition (HCCI) strategy employed in reciprocating engines is precisely working on the same concept by burning well-mixed fuel-air mixture and leaner combustion (lower temperature) [5]. Due to the limitations of the power density of HCCI engines, several partially stratified charge concepts are developed, and the common name for all the premixed combustion technologies is low-temperature combustion (LTC). All the premixed LTC modes have the common characteristic of relatively higher fuel conversion efficiency and simultaneous drastic reduction in soot and NO_x emissions. Figure 1.18 illustrates the LTC technology on local equivalence ratio-temperature (ϕ-T) map.

Figure 1.18 depicts the dependency of the formation of NO_x, soot, HC, and CO on the local combustion temperature and local equivalence ratio in the combustion chamber. In the φ-T map, the shape and size of the soot formation area depend on the fuel characteristics [25]. However, the region of NO_x formation is dependent on φ-T and independent of fuel. The conventional SI engine typically operates at stoichiometric mixture and operates at significantly higher temperatures, which lead to mainly higher NO_x formation (Fig. 1.18). However, diesel combustion is heterogeneous, and there is a large variation in the local equivalence ratio. Thus, there will be

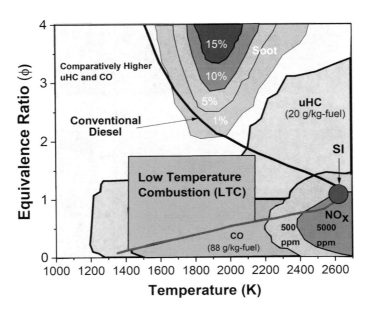

Fig. 1.18 Operating region of low-temperature combustion mode on the ϕ-T map (adapted from [5, 24–27])

higher soot formation along with the NO_x emission in conventional diesel engines (Fig. 1.18). When combustion temperature is decreased to reduce the NO_x formation, the soot formation kinetics is also slowed down [28, 29]. However, soot oxidation rate decreases more than the soot formation rate at relatively lower combustion temperatures [30], and therefore, soot increases in the engine exhaust. At lower combustion temperature, the HC and CO emissions increase, which is another major challenge. The HC and CO formation is also dependent on the combustion temperature and local equivalence ratio (Fig. 1.18). Near-complete oxidation of HC and CO occurs at higher combustion temperatures and leaner mixtures. In the φ-T map, there is a region which has relatively lower HC and CO emissions, and NO_x and soot emission can be avoided. This region is defined as low-temperature combustion (LTC) region (Fig. 1.18), and engine operation in this region leads to ultra-low NO_x and soot emissions [5]. The basic combustion mode in this region is HCCI combustion.

Figure 1.19 illustrates the working process of the HCCI engine and its comparison with conventional SI and CI engines. In HCCI combustion, the well-mixed

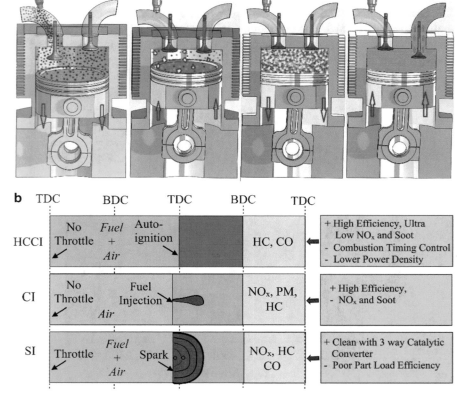

Fig. 1.19 (**a**) HCCI combustion process illustration, and (**b**) comparison of HCCI combustion process with conventional SI and CI combustion in the four-stroke cycle [5]

charge is prepared typically by PFI system. The fuel is inducted in the inlet manifold, and it gets mixed with air during intake and compression stroke. Toward the end of the compression stroke, when charge temperature reaches its autoignition temperature, the combustion starts in the entire combustion chamber (Fig. 1.19a).

In HCCI combustion, the autoignition of entire well-mixed charge occurs, and thus, the ignition timing is controlled by pressure and temperature in the combustion chamber. Therefore, there is no direct control of combustion timing in HCCI engine as spark timing in SI engine and fuel injection timing in the CI engines. There is no throttling and flame propagation in the HCCI combustion, which is there in conventional SI engines. The advantages and challenges of HCCI combustion engine are shown in Fig. 1.20. The solutions proposed to overcome the challenges are also presented in Fig. 1.20. The main advantages of the HCCI engine are the higher thermal efficiency and ultra-low NO_x and PM emissions. However, the major challenges are limited operating range and control of combustion phasing. Since combustion starts with autoignition of fuel, thus, there will be a cold start problem with high-octane fuels. There are several solutions proposed to overcome the challenges in HCCI combustion engines (Fig. 1.20). The detailed combustion process, autoignition chemistry, effect of different engine parameters on HCCI

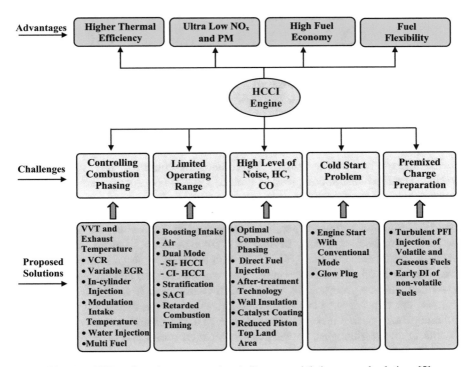

Fig. 1.20 The HCCI engine advantages, major challenges, and their proposed solutions [5]

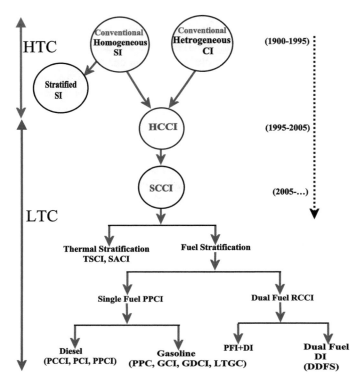

Fig. 1.21 Evolution of different combustion concepts for reciprocating internal combustion engines [5]

combustion, and control strategies of HCCI engines can be found in author's book [5].

To improve the power density and combustion control of the HCCI engine, various stratified charge compression ignition (SCCI) concepts are investigated by various researchers with different levels of stratification in fuel as well as temperature. Figure 1.21 illustrates the evolution of different combustion concepts in LTC and HTC regime. In high-temperature combustion (HTC) regime, the conventional SI and CI combustion occurs. Several LTC strategies have been investigated with various acronyms and names (Fig. 1.21), with different levels of inhomogeneity of the charge, but all of them are compression ignition. All the LTC strategies can be classified into two groups as HCCI and SCCI combustion. In the SCCI regime, the combustion depends on two possible stratifications of the charge by temperature inhomogeneity or fuel inhomogeneity (Fig. 1.21). The natural thermal stratification occurs in all the strategies. In thermal stratification group, the thermally stratified compression ignition (TSCI) and spark-assisted compression ignition (SACI) concepts are present. Fuel stratification can be created in the combustion chamber by single or dual fuel with different injection strategies. Using different direct injection strategies of single, fuel composition stratification is created in the combustion chamber; while using dual fuel, reactivity stratification is also created along with the composition stratification. In the fuel stratification regime, gasoline partially premixed combustion (PPC) and dual-fuel reactivity controlled compression ignition (RCCI) are the two major combustion strategies that are extensively investigated because of their potential of better combustion phasing control along with the advantages of HCCI engines. Combustion concepts in the PPC range are also termed as gasoline compression ignition (GCI) or gasoline direct injection compression ignition (GDCI) with a minor difference in injection strategies. In the dual-fuel stratification strategy, two fuel injections strategies are possible: (1) PFI injection of low reactivity fuel and direct injection (DI) of high reactivity fuel and (2) direct injection of both the fuels, which is termed as dual direct injection fuel stratification (DDFS). The detailed discussion of all these LTC concepts can be found in reference [5].

1.2 Engine Testing and Combustion Diagnostics

In general, engine testing is conducted to improve the design and configuration, to integrate the new technology, and to find out performance characteristics prior to the production and employing into a vehicle. Earlier, the engine tests were conducted to determine the power and fuel consumptions along with an additional test to find out the effectiveness of cooling, vibration, and noise, lubrication, controllability, etc. Present stringent emission legislation to limit the engine emissions leads to the more and more sophisticated engine testing. Test beds for internal combustion engines are generally differentiated by the following areas and types of application (objective of use): (1) research (single-cylinder engine test bed and flow test bed), (2) development (performance test bed, function test bed, endurance test bed, calibration test bed, emission certification test bed), and (3) production (end-of-line break-in test bed, quality assurance test bed) [31]. Single-cylinder engine test beds are mainly used for the purpose of research, which typically helps to assess the combustion process optically as well as thermodynamically. The flow test beds are utilized to investigate the charge motion, which has a significant role in the engine combustion process. The performance test bed is used for the determination of engine power output in the whole operating range in the predefined test conditions. The performance, combustion, and emission characteristics are analyzed using this type of test beds. Function test beds serve to optimize, verify and secure engine related overall system features. Dedicated calibration test beds are typically used in series development to determine a specific and optimal engine behavior on engine control units. Final homologation tests are conducted on the exhaust emission certification engine test bed. Endurance test beds are used to investigate and ensure the durability and long term stability of the engine and its components.

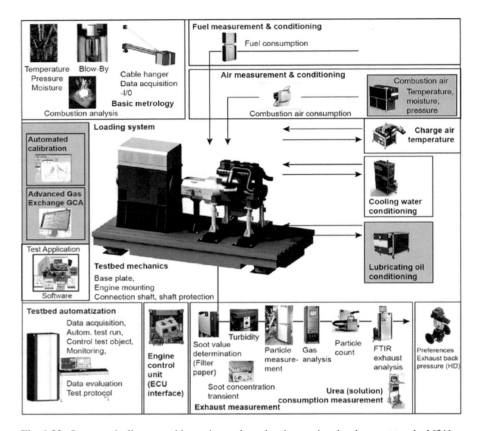

Fig. 1.22 Scope typically covered by an internal combustion engine development test bed [31]

To fulfill the broad range of objective, typical test beds for the development of reciprocating combustion engines contain several components. Figure 1.22 shows the typical test bed for internal combustion engines depicting the main components. The main components of a typical test bed are dynamometer (for torque measurement and control), engine media conditioning, fuel, air consumption measurement systems, instrumentation for combustion diagnostics, temperature and pressure measurement at the engine periphery, emission measurement devices, communication interface to the ECU, test bed automation (control/simulation), and calibration tools [31]. Test beds are classified as steady-state test beds and non-steady-state test beds based on the used test bed technologies, particularly engine dynamometer and test bed automation system.

Reciprocating automotive engine tests for product development and calibration can be classified as (1) steady-state test, (2) transient test, and (3) cold start tests. Another way of classifying the engine tests for research is based on whether test is conducted in-cylinder or out of the engine cylinder. Typical in-cylinder tests include in-cylinder pressure measurement, ion-current measurement, optical diagnostics, fast gas sampling, and flame ionization detector. Typical test conducted out to the engine cylinder includes the normal performance measurement (fuel consumption, engine efficiency, etc.), test for transient performance improvement, valvetrain testing, turbocharger performance test, friction (Morse test), and test related to aftertreatment technologies.

Understanding combustion processes would not be possible without diagnostics. Direct investigation of combustion systems reveals important information about the processes or properties including fuel spray, evaporation, mixing, ignition, flame speed, reactivity, and formation of various emission species. Combustion diagnostics provide the access to parameters such as species concentration, pressure, temperature, and flow velocity as well as to their spatial distribution and development with respect to time [32]. Combustion diagnostics can be used for serving various objectives including gaining fundamental insight of the combustion process, investigating and validating theoretical models, and real-time process optimization and control.

Combustion diagnostics is always used in engine development when unexploited potential compared with thermodynamically possible targets is ascertained during the measurement of consumption, output, and emissions [33]. Given the high target set for modern reciprocating engines, thermodynamic combustion analysis using the measurement of in-cylinder pressure is always a fixed element in the development sequence even along with the optical methods. For the fundamental studies, in-cylinder visualization of flow and combustion is typically performed using different optical methods. Optical methods can be used for various development tasks in spark ignition and compression ignition engines. In SI engines, optical methods can be used for determination of knock location, mixture formation study in the intake manifold as well as a direct injection in the cylinder, flame kernel formation and flame propagation, irregular combustion study, and noncontact temperature measurements. In diesel engines, optical methods can be used for flame image analysis, soot formation, and spray characteristics. Combustion chamber endoscopy and optical engines (optical window in piston and liner) are typically used for various optical investigations. Discussion on optical methods is out of the scope of the present book.

Figure 1.23 illustrates the measurement chain for generating various signals for combustion diagnostics other than the optical methods. Ion-current and cylinder pressure signals are the most direct methods to access the combustion process for the diagnostics. During the combustion process, ions are produce by oxidation reactions, and pressure increases due to the temperature increase by exothermic reactions. The major limitations of ion-current-based diagnostics are (1) measurement in a small portion of the combustion chamber and (2) measurement depends on the location of the sensor (mostly spark plug). The cylinder pressure is the most widely used method for combustion diagnostics in reciprocating engines. However, due to the higher cost of cylinder pressure-based measurement, other signals such as torque or instantaneous angular speed are also used for determination of combustion parameters. Instantaneous angular speed can be measured using crank angle encoder. However, some of the combustion information is attenuated in the torque and angular speed-

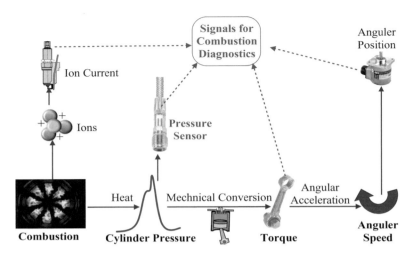

Fig. 1.23 Various signals for combustion diagnostics

based measurements. Thus, cylinder pressure is the best available signal containing most of the information of the combustion process, and presently, it can be measured easily using high-speed data acquisition and modern piezoelectric pressure sensors. Cylinder pressure-based combustion diagnostics is the main focus of the present book, and it is briefly introduced in the next section.

1.3   In-Cylinder Pressure-Based Combustion Diagnostics

In-cylinder pressure measurement and analysis is a fundamental practice used globally for research and development of reciprocating engines. Since the invention of internal combustion (IC) engines, the cylinder pressure measurement has been the principal diagnostic tool for experimenters. In-cylinder pressure measurement of the reciprocating engine has great potential for calibration, monitoring, diagnosis, validation of numerical modeling, and closed-loop control purposes. Calibration engineers use combustion analysis to improve emissions, fuel consumption, and performance calibrations of an engine. The application of methodologies based on the in-cylinder pressure measurement finds widespread applications as it provides direct combustion information with a high dynamical potentiality, which is fundamentally required for the diagnosis and control of the engine combustion process. Additionally, the in-cylinder pressure-based engine diagnostics may also lead to the reduction in a number of existing onboard sensors, which lower the equipment costs and the engine wiring complexity [34].

With the stringent legislation requirements on engine emissions and performance, the pressure-based combustion control systems have been regarded as the potential gain for the future automotive engines, which relies mostly on cylinder pressure based extracted information such as IMEP, heat release rate, combustion duration, compression condition, etc. In order to ensure the precision of the extracted parameters, high accuracy of results is required. The accuracy of the results depends on the accuracy of each component of the measuring chain, as well as the correct processing of the pressure signal.

Crank angle-based in-cylinder pressure measurement and analysis can be used to determine the variables such as piston and crankshaft loads, torque produced from combustion (equals the IMEP), gas exchange torque (PMEP), spark timing relative to MBT, time required for the combustion flame to develop and propagate, presence and magnitude of knock, cycle-to-cycle and cylinder-to-cylinder variability, etc. [35]. Additionally, the cylinder pressure-based combustion analysis can be used to achieve several development objectives such as evaluation of inlet and/or exhaust port and manifold geometries, quantifying compression ratio trade-offs, optimization of the combustion chamber geometry, selection of valve timing overlap and duration, optimization of fuel injector timing and on-time (opening duration), power/pressure rise/NO_x, automated mapping (MBT, knock, preignition control), inquiry of transient response, measurement of mechanical friction, and calibration optimization.

The various performance, combustion, and engine operating parameters that can be derived using cylinder pressure measurement and analysis are summarized in Table 1.1. Table 1.1 also briefly provides the signal processing methods along with the engine information generated. Detailed discussion on the derived parameters and signal processing methods is provided in different chapters of the present book.

1.4 Organization of the Book

Measuring and analyzing the in-cylinder pressure is an essential part of the combustion system development and calibration of an internal combustion engine. Cylinder pressure-based heat release analysis has been widely used in the field of engine research as an essential tool for understanding combustion behavior, as well as providing key data for engine and combustion modeling. Considering the advantages of in-cylinder pressure-based combustion diagnostics and a large amount of valuable information derived, this book presents a detailed discussion on different aspects of in-cylinder pressure measurement and its analysis.

The schematic of book organization is presented in Fig. 1.24. Typically, the workflow of a measurement chain consists of sensors (signal generation), data acquisition, signal processing, and data analysis. The present book provides a detailed discussion on all these aspects of measurement. The details of cylinder pressure sensors along with peripheral sensors for combustion diagnostics are provided in Chaps. 2 and 3. The acquisition of the cylinder pressure signal and their processing (phasing with crank angle position, absolute pressure referencing,

Table 1.1 Summary of performance and combustion parameters that can be estimated by processing of the measured in-cylinder pressure data [36]

Data derived from in-cylinder pressure signal	Engine information	Signal processing methods/equations
Cycle-averaged in-cylinder pressure	Firing cycle crank angle (CA)-based combustion event and quality of combustion	Averaging of cylinder pressure of different combustion cycle on crank angle (CA) basis
Motoring cylinder pressure, peak compression pressure, and its CA position	Used for tuning of heat release parameter, TDC correction, engine cylinder condition, and blowby	Measuring signal in non-firing engine cycle and averaging to a number of cycles. Calculation of TDC position using thermodynamic method using the measured pressure signal
Peak pressure	Mechanical load on the cylinder	Maximum value computation of in-cylinder pressure signal
Rate of pressure rise	Knock limit of the engine	Derivative of measured pressure signal
Crank angle at peak pressure, rate of heat release curve, and energy conversion points	Overall efficiency, combustion efficiency, qualitative exhaust values, quality of ignition system	Using first law of thermodynamics, rate of heat release (ROHR) calculation using the equation $$\frac{dQ}{d\theta} = \frac{\gamma}{\gamma-1}P\frac{dV}{d\theta} + \frac{1}{\gamma-1}V\frac{dP}{d\theta} + \frac{\partial Q_w}{d\theta}$$ Energy conversion points are calculated by integrating ROHR curve
IMEP, FMEP, PMEP	Cylinder work output, combustion stability (cyclical fluctuations), friction losses, gas exchange losses	Computation of work using the cylinder pressure signal and volume curve generated from engine geometry $$W_{net} = \frac{2\pi}{360}\int_{-360}^{360}\left(P(\theta)\frac{dV}{d\theta}\right)d\theta$$ Calculation of work in high-pressure component and low-pressure component to calculate IMEP, PMEP, and FMEP
PV and log (PV) diagram	Work output and pumping losses, determination of TDC position, the polytropic coefficient of the mixture during compression and expansion	Computation of area under the curve, calculation of logarithmic value of pressure and volume curve, the slope of the curve in compression and expansion stroke to find the polytropic coefficient
Gas temperature and wall heat transfer	Qualitative exhaust values, heat transfer	Using empirical equations developed using in-cylinder pressure and mean gas temperature

High-frequency component of vibration	Knocking, combustion noise, ringing intensity	Filtering using a bandpass filter, wavelet analysis, power spectrum
Ignition delay	Formation of air/fuel mixture	Calculated from ignition (SI) or injection point (CI) and start of combustion calculated from heat release curve
Mass flow rate	Air mass flow estimation, residual exhaust gas in cylinder, backflow	Application of the Δp-method for estimating the air mass, frequency analysis of the pressure trace
Compression ratio estimation	Actual compression ratio	Computation using polytropic compression model and optimization algorithm, temperature-entropy-based method
Air-fuel ratio	Cylinder mixture strength	Cylinder pressure moment-based approach, net heat release-based estimation, g-ratio method
Torque	Engine torque	Indicated torque is estimated from the peak pressure and its location, and the load torque is observed by the estimated indicated torque
COV$_{\text{IMEP}}$, CA position for different heat release and mass burn fraction	Cycle-to-cycle dispersion, misfire	Statistical analysis, symbol sequence analysis, return maps
Emission	Engine emissions, Quality of combustion	Neural network-based algorithm, regression-based approach
Control parameters calculations	EGR control, noise control, emission control, online combustion failure detection, start of injection control	Online signal processing (filtering) and computation of control parameter

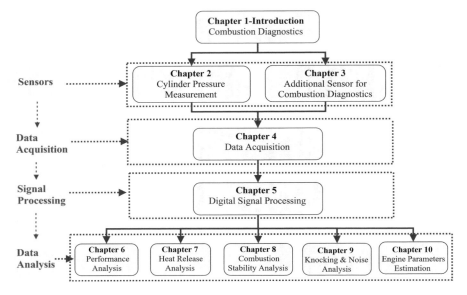

Fig. 1.24 Schematic of the book organization

smoothing/filtering, and cycle averaging) is presented in Chaps. 4 and 5, respectively. The detailed data analysis of the measured in-cylinder pressure signal is presented in Chaps. 6–10. Detailed discussion is provided on performance analysis (torque, power, and work), heat release analysis (mass fraction burned, combustion phasing, and combustion duration), combustion stability analysis (cycle-to-cycle variations) with statistical and nonlinear dynamics methods, and knocking and combustion noise analysis. The last chapter presents the discussion on estimation methods of different engine operating parameters (TDC, compression ratio, air-fuel ratio, wall temperature, trapped mass, and residual gas fraction) using the measured cylinder pressure data. Information provided in the proposed book can be effectively used for further development, optimization, and calibration of modern reciprocating engines.

Discussion/Investigation Questions

1. Describe the difference between an internal combustion engine (ICE) and an external combustion engine (ECE). Write the examples of ICE and ECE. Discuss the advantages and disadvantages of the ICE over the ECE.
2. Discuss the major environmental concerns arises due to extensive use of reciprocating ICEs powered by fossil fuels.
3. Discuss the major trends in vehicle developments to improve the engine efficiency and comply with the emissions legislation.

4. Describe the merits and demerits of ICE based vehicles over the battery electric vehicles (BEVs). Discuss the factors contributing to higher fuel economy in hybrid electric vehicles (HEV) over ICE vehicles. Justify the statement, "hybrid vehicles are mainly relevant to SI engines."

5. Discuss the reasons why the work output is typically limited by displacement over its entire speed range in SI engines while in diesel engine displacement limited only at maximum torque condition.

6. Describe why displacement volume of SI engine is lower than the diesel engine for similar power ratings.

7. Discuss the factors which govern the preference of multicylinder engine when total displacement volume is approximately above 500 cm^3 in high-speed engines.

8. Discuss the advantages and disadvantages of conventional diesel engines over the gasoline engines.

9. Describe the reasons why gasoline SI engines typically operate at stoichiometric air-fuel ratio condition.

10. Discuss the advantages of gasoline direct injection (GDI) engines over the port fuel injection (PFI) spark ignition engines.

11. Discuss the reasons why engine speed of a diesel engine is relatively lower than gasoline engines. Modern diesel engine has a relatively higher engine speed than conventional diesel engines. Investigate the development of technologies responsible for the increase in engine speed.

12. Discuss the factors governing the higher fuel conversion efficiency of diesel engines.

13. Discuss the advantages of common rail direct injection system over the mechanical fuel injection (in-line jerk type) system.

14. Write the advantages of HCCI engine over SI and CI engines. Discuss the factors contributing to higher indicated efficiency even though higher HC emissions and lower combustion efficiency found in HCCI engines.

15. Considering the ϕ-T map shown in Fig. 1.18, determine the typical local equivalence ratio and combustion temperature for LTC engines. Discuss the reasons, why NO_x and soot formation is very low in this temperature and equivalence ratio range.

16. Describe how the fuel stratification and thermal stratification help in increasing the power density of HCCI engines. Discuss the difference between fuel composition stratification and reactivity stratification.

17. Explain the importance of experiments in general? Describe the type of engine experiments performed for research and development purposed. Discuss the general instrumentation of engine for conducting various tests related to performance, combustion, and emissions?

18. Discuss the need of combustion diagnostics particularly in reciprocating combustion engines. Explain the different methods of combustion diagnostics in internal combustion engines.

19. Describe the different possible signals (other than optical signals) that can be used for combustion diagnostics in reciprocating engines. Discuss the

advantages and limitation of each signal with respect to combustion information revealed. Explain the factors contributing in the favor of cylinder pressure signal for the most widespread applications in reciprocating engine combustion diagnostics.

20. Describe the in-cylinder pressure measurement-based combustion diagnostic method. Write the typical information that can be produced by processing of measured in-cylinder pressure signal. Discuss the significance of the cylinder pressure-based diagnostics in engine research and development process.

References

1. Kalghatgi, G. (2018). Is it really the end of internal combustion engines and petroleum in transport? *Applied Energy, 225*, 965–974.
2. Hofmann, J., Guan, D., Chalvatzis, K., & Huo, H. (2016). Assessment of electrical vehicles as a successful driver for reducing CO2 emissions in China. *Applied Energy, 184*, 995–1003.
3. Doucette, R. T., & McCulloch, M. D. (2011). Modeling the CO2 emissions from battery electric vehicles given the power generation mixes of different countries. *Energy Policy, 39*(2), 803–811.
4. Hawkins, T. R., Singh, B., Majeau-Bettez, G., & Strømman, A. H. (2013). Comparative environmental life cycle assessment of conventional and electric vehicles. *Journal of Industrial Ecology, 17*(1), 53–64.
5. Maurya, R. K. (2018). *Characteristics and control of low temperature combustion engines: Employing gasoline, ethanol and methanol.* Cham: Springer.
6. Eriksson, L., & Nielsen, L. (2014). *Modeling and control of engines and drivelines.* Wiley.
7. Ganesan, V. (2012). *Internal combustion engines.* New Delhi: McGraw Hill Education (India) Pvt Ltd.
8. Wang, Y. (2007). *Introduction to engine valvetrains.* Warrendale, PA: SAE International.
9. Hoag, K., & Dondlinger, B. (2016). *Vehicular engine design* (2nd ed.). Vienna: Springer.
10. Pundir, B. P. (2010). *IC engines: Combustion and emissions.* New Delhi: Narosa Publishing House.
11. Marko, F., König, G., Schöffler, T., Bohne, S., & Dinkelacker, F. (2016, November). Comparative optical and thermodynamic investigations of high frequency corona-and spark-ignition on a CV natural gas research engine operated with charge dilution by exhaust gas recirculation. In *International Conference on Ignition Systems for Gasoline Engines* (pp. 293–314). Cham: Springer.
12. Morcinkowski, B., Hoppe, P., Hoppe, F., Mally, M., Adomeit, P., Uhlmann, T., ... Baumgarten, H. (2016, November). Simulating extreme lean gasoline combustion–flow effects on ignition. In *International Conference on Ignition Systems for Gasoline Engines* (pp. 87–105). Cham: Springer.
13. Brandt, M., Hettinger, A., Schneider, A., Senftleben, H., & Skowronek, T. (2016, November). Extension of operating window for modern combustion systems by high performance ignition. In *International Conference on Ignition Systems for Gasoline Engines* (pp. 26–51). Cham: Springer.
14. Badawy, T., Bao, X., & Xu, H. (2017). Impact of spark plug gap on flame kernel propagation and engine performance. *Applied Energy, 191*, 311–327.
15. Heywood, J. B. (1988). *Internal combustion engine fundamentals.* New York: McGraw-Hill.
16. Sjöberg, M., & Zeng, W. (2016). Combined effects of fuel and dilution type on efficiency gains of lean well-mixed DISI engine operation with enhanced ignition and intake heating for enabling mixed-mode combustion. *SAE International Journal of Engines, 9*(2), 750–767.

17. Ayala, F. A., Gerty, M. D., & Heywood, J. B. (2006). *Effects of combustion phasing, relative air-fuel ratio, compression ratio, and load on SI engine efficiency* (No. 2006-01-0229). SAE Technical Paper.
18. Zhao, F., Harrington, D. L., & Lai, M. C. D. (2002). *Automotive gasoline direct-injection engines* (p. 372). Warrendale, PA: Society of Automotive Engineers.
19. Sjöberg, M., & He, X. (2018). Combined effects of intake flow and spark-plug location on flame development, combustion stability and end-gas autoignition for lean spark-ignition engine operation using E30 fuel. *International Journal of Engine Research, 19*(1), 86–95.
20. Dembinski, H. W. (2014). The effects of injection pressure and swirl on in-cylinder flow pattern and combustion in a compression–ignition engine. *International Journal of Engine Research, 15*(4), 444–459.
21. Kegl, B., Kegl, M., & Pehan, S. (2013). *Green diesel engines. Biodiesel usage in diesel engines.* London: Springer.
22. Majewski, W. A., & Khair, M. K. (2006). *Diesel emissions and their control.* Warrendale, PA: SAE International.
23. Johnson, T. V. (2002). *Diesel emission control: 2001 in review* (No. 2002-01-0285). SAE Technical Paper.
24. Dec, J. E. (2009). Advanced compression-ignition engines—Understanding the in-cylinder processes. *Proceedings of the Combustion Institute, 32*(2), 2727–2742.
25. Kitamura, T., Ito, T., Senda, J., & Fujimoto, H. (2002). Mechanism of smokeless diesel combustion with oxygenated fuels based on the dependence of the equivalence ration and temperature on soot particle formation. *International Journal of Engine Research, 3*(4), 223–248.
26. Dempsey, A. B., Curran, S. J., & Wagner, R. M. (2016). A perspective on the range of gasoline compression ignition combustion strategies for high engine efficiency and low NOx and soot emissions: Effects of in-cylinder fuel stratification. *International Journal of Engine Research, 17*(8), 897–917.
27. Kim, D., Ekoto, I., Colban, W. F., & Miles, P. C. (2009). In-cylinder CO and UHC imaging in a light-duty diesel engine during PPCI low-temperature combustion. *SAE International Journal of Fuels and Lubricants, 1*(1), 933–956.
28. Dobbins, R. A. (2002). Soot inception temperature and the carbonization rate of precursor particles. *Combustion and Flame, 130*(3), 204–214.
29. Musculus, M. P., Miles, P. C., & Pickett, L. M. (2013). Conceptual models for partially premixed low-temperature diesel combustion. *Progress in Energy and Combustion Science, 39*(2–3), 246–283.
30. Huestis, E., Erickson, P. A., & Musculus, M. P. (2007). In-cylinder and exhaust soot in low-temperature combustion using a wide-range of EGR in a heavy-duty diesel engine. *SAE Transactions*, Paper no - 2007-01-4017, 860–870.
31. Paulweber, M., & Lebert, K. (2016). *Powertrain instrumentation and test systems. Development–hybridization–electrification.* Cham: Springer International Publishing.
32. Kohse-Höinghaus, K., Reimann, M., & Guzy, J. (2018). Clean combustion: Chemistry and diagnostics for a systems approach in transportation and energy conversion. *Progress in Energy and Combustion Science, 65*, 1–5.
33. Schafer, F., & Van Basshuysen, R. (Eds.). (2016). *Internal combustion engine handbook: Basics, components, systems, and perspectives* (2nd ed.). Warrendale, PA: SAE International.
34. Schiefer, D., Maennel, R., & Nardoni, W. (2003). *Advantages of diesel engine control using in-cylinder pressure information for closed loop control* (No. 2003-01-0364). SAE Technical Paper.
35. Atkins, R. D. (2009). *An introduction to engine testing and development.* Warrendale, PA: Society of Automotive Engineers.
36. Maurya, R. K., Pal, D. D., & Agarwal, A. K. (2013). Digital signal processing of cylinder pressure data for combustion diagnostics of HCCI engine. *Mechanical Systems and Signal Processing, 36*(1), 95–109.

Chapter 2
In-Cylinder Pressure Measurement in Reciprocating Engines

Abbreviations and Symbols

ATDC	After top dead center
AEAP	Average exhaust absolute pressure
ANN	Artificial neural networks
ARMA	Autoregressive moving average
ASE	Average signal envelope
BDC	Bottom dead center
BTDC	Before top dead center
CA	Crank angle
CA_{50}	Combustion phasing 50% heat release
CI	Compression ignition
COV	Coefficient of variation
ECU	Electronic control unit
EGR	Exhaust gas recirculation
EMS	Engine management systems
EVC	Exhaust valve closing
EVO	Exhaust valve opening
FSO	Full-scale output
$GaPO_4$	Gallium phosphate
GDI	Gasoline direct injection
HCCI	Homogeneous charge compression ignition
IEPE	Integrated electronics piezoelectric
IMEP	Indicated mean effective pressure
IVC	Inlet valve closing
IVO	Inlet valve opening
$LiNbO_3$	Lithium niobate
LP	Low pass
LSE	Lower signal envelope

© Springer Nature Switzerland AG 2019
R. K. Maurya, *Reciprocating Engine Combustion Diagnostics*, Mechanical
Engineering Series, https://doi.org/10.1007/978-3-030-11954-6_2

LTD	Long-term drift
NO	Nitric oxide
$PbTiO_3$	Lead titanate
PC	Personal computer
PE	Piezoelectric
P_{max}	Peak cylinder pressure
RPM	Revolution per minute
RTV	Room temperature vulcanizing
SI	Spark ignition
SiO_2	Silicon dioxide
TDC	Top dead center
USE	Upper signal envelope
WOT	Wide open throttle
L	Length of the air duct
Q	Calculation of charge output
R	Gain adjustment resistance
Δt	Time interval
a_b	Mounting base or reference acceleration
a_o	Output acceleration
D_i	Vector of electric flow density
$d_{i\mu}$	Tensor of piezoelectric coefficients
f_n	Undamped natural frequency
G_A	Charge amplifier gain
G_s	Piezoelectric pressure transducer sensitivity
L_T	Length of tube
T_μ	Tensor of mechanical stress
V_A	Amplifier output voltage
V_c	Clearance volume
V_{cv}	Volume of the cavity in front of the pressure sensor
V_{dead}	Dead volume of measuring bore
V_{disp}	Displacement volume
V_P	Passage volume
φ	Equivalence ratio

2.1 In-Cylinder Pressure Measurement Setup

Combustion pressure measurement was a topic of interest for researchers since the advent of reciprocating engines [1]. The mechanical work produced by reciprocating engines results from the action of gas pressure on the piston. The cylinder (combustion chamber) pressure is directly related to engine power output and the fuel conversion efficiency. The data of cylinder pressure development versus crank angle is used to calculate heat release rates and to analyze the progress of combustion process in the cycle. In engine development process, combustion diagnostics is always used when the unexploited potential in comparison to thermodynamically possible targets is determined during the measurement of fuel consumption, output, and emissions. Thermodynamic combustion analysis by in-cylinder pressure measurement is a fixed element in the development sequence of modern engines due to high-performance targets [2]. Currently, most of the combustion parameter required for the industrial application can be derived from cylinder pressure measurement. The measurement of cylinder pressure in reciprocating engines poses one of the most challenging tasks for instrument manufacturers. Pressure transducers not only need to be compact and stable with the very fast response as well as good dynamic range, but these characteristics must be realized and withstand explosive and intense transient thermal conditions of the combustion chamber. High accuracy of cylinder pressure measurement is required for combustion work and thermodynamic calculations such as determination of IMEP, efficiency, and friction losses. Typically, high repeatability of cylinder pressure measurement is required for engine calibration and component testing. Various types of pressure transducers were used including variable resistance, variable inductance, balanced disk type, and piezoelectric, with different levels of robustness and accuracy [3].

Early in-cylinder pressure measurements of reciprocating engines were conducted using several configurations of mechanical indicators [4, 5]. The term "indicating" is used to designate the measurement and depiction of the cylinder pressure plot with crank angle position or time [2]. With the development of high operating speed engines, the frequency response of mechanical indicators was found to be insufficient for in-cylinder pressure measurement. Thus, mechanical indicators become outdated in the mid-1960s [5]. To meet the demand of in-cylinder pressure measurement in high speed engines, electronic transducers were used to convert the deflection of a low inertia diaphragm into an electrical signal. Early versions of electrical pressure transducers were built using extensometers and piezoelectric crystals as sensing element, and these devices had sufficient frequency response to the combustion process in the engine cylinder. Toward the end of the 1960s, the complex analog systems having the potential of completely electronic pressure signal processing became available, which were used for indicated work calculation and study of knock and misfire. In the mid-1970s, analog-to-digital converters were included, and the signal from transducer amplifier was digitized and stored in the computer, which can be further post-processed [5]. Hence, flexibility in data analysis and higher storage capacity were achieved while maintaining the adequate level of accuracy.

Figure 2.1 schematically illustrates the typical in-cylinder pressure measurement setup using piezoelectric transducer along with supplementary measurements for combustion diagnosis in a diesel engine. Typically, piezoelectric transducers are used for in-cylinder pressure measurements in modern engine tests due to their small size, light weight, the potential of high-frequency response, and low sensitivity to environmental conditions. Cylinder pressure measurements are augmented with various measured variables which define the working fluid state and component functions. Such "indicated data," which are typically recorded on cycle-to-cycle

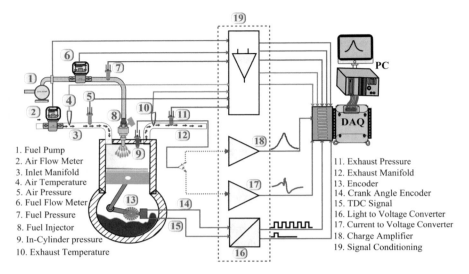

1. Fuel Pump
2. Air Flow Meter
3. Inlet Manifold
4. Air Temperature
5. Air Pressure
6. Fuel Flow Meter
7. Fuel Pressure
8. Fuel Injector
9. In-Cylinder pressure
10. Exhaust Temperature

11. Exhaust Pressure
12. Exhaust Manifold
13. Encoder
14. Crank Angle Encoder
15. TDC Signal
16. Light to Voltage Converter
17. Current to Voltage Converter
18. Charge Amplifier
19. Signal Conditioning

Fig. 2.1 Schematic diagram of typical in-cylinder pressure measurement setup using piezoelectric transducer (adapted from [5, 6])

basis or crank angle basis (depending on the test), form the basis for thermodynamic evaluation of combustion and the optimization of the adjustment parameters of the engine [2]. Thus, data logging of supplementary characteristic variables, such as measurement of air and fuel flow rates, measurement of the injection characteristics (injection pressure, needle lift), manifold pressures, ignition current, and thermal variables (intake and exhaust temperatures), proceeds as a natural progression from cylinder pressure indicating (Fig. 2.1).

High-pressure measurement in the engine cylinder is mostly used for combustion analysis. In addition to cylinder pressure measurement, low-pressure indicating on the intake and exhaust manifold constitutes the precondition for the gas exchange analysis and the estimation of intake gas mass used for combustion. Piezoresistive and piezoelectric pressure sensors are typically used for low-pressure indicating in reciprocating engines. The measured pressure traces in the intake and exhaust manifold form the basis for analyzing the gas exchange, which involves intake and exhaust duct design, controls design (cam profiles and control times), gas exchange work estimation, and mass flow analysis (charge, residual gas, reverse flows) [1]. Cylinder pressure data are indexed with the crank angle position using angle encoder (discussed in Sect. 2.4), with reference to the compression top dead center (TDC) position (Fig. 2.1). As illustrated in Fig. 2.1, the piezoelectric pressure signal generated can be conditioned for two measurement procedures: (1) in-cylinder pressure is acquired by using a charge amplifier for signal conditioning, and (2) in the second method, a current-to-voltage converter is employed for the measurement of derivative of in-cylinder pressure signal [5, 6].

The main structure of a cylinder pressure measurement system includes a piezo-electric pressure sensor, measuring amplifier, measurement wiring, data acquisition system, real-time characteristics calculator, system operation, and post-processing of data [2]. The piezoelectric sensor is typically installed either directly by a special boring in the engine cylinder or by special adapters in existing borings, such as those for the spark plugs (SI engines) or glow plug (CI engines). The functional principle of the piezoelectric pressure sensor is that a particular crystal creates a charge under mechanical strain (deformation) and thus, it acts as an active measuring element, where charge generated is proportional to the applied pressure [1]. More detailed discussion on piezoelectric pressure transducer is provided in Sect. 2.2. The low level of charge generated by piezoelectric pressure transducer is amplified and integrated into a voltage signal, which is typically processed for combustion analysis. The magnitude of amplified voltage range needs to ensure the transmission through long cable distance to data acquisition system while maintaining high signal-to-noise ratio. In addition to low-noise amplification, the long-term stability and short-circuit strength are very important for amplifier units [2]. The charge and voltage signals are transferred through measurement wirings. The length of measurement wiring between the pressure transducer and the charge amplifier is always kept as short as possible, to achieve high signal quality. Low noise levels, very high insulating values (10^{14} Ω), robustness, and simple handling are required for both charge amplifier and the measurement wiring [1]. Data acquisition system records the measured data based on crank angle position depending on the resolution of crank angle encoder. Detailed discussion on data acquisition system is presented in Chap. 4. Real-time characteristics calculator generates control signals on the basis of measured pressure progressions compared with reference values predefined as model characteristics for the particular operating conditions. The control signals affect the actuators (injector, ignition, timing settings, valve lift, etc.) through the engine control unit (ECU). Real-time analyses are subjected to continuous modification depending on needs and appropriateness and as a function of computing potentials. Real-time calculation of important combustion parameters is described in Chap. 7. System operation is typically achieved by a special PC software which performs the following functions: (1) parameterizes the entire measuring system and the measurement itself, (2) obtains characteristics data and calculations, (3) defines algorithms for the determination of characteristics data or the calculation of results from the measured data, and (4) displays the measured and calculated data values. Post-processing is applied for the presentation and processing of the measured data. More complex calculations, comparisons of results, and documentation procedures are conducted during post-processing using corresponding graphical and computing aids [2].

2.2 Piezoelectric Pressure Transducer

In general, combustion sensors (such as piezoelectric pressure sensors, ion current sensors, and optical sensors) are used for engine research and development. In modern engines, there is a need for installation of combustion sensors on production vehicles for closed-loop control of engine combustion. Combustion sensors are used for determination of various combustion parameters as well as different engine input and output parameters. Several design issues with combustion sensors need to be considered for engine applications [7]. Sensor cost, real-time signal processing, and working environment of the sensor are the major factors governing the design of combustion sensor. Signal processing is typically not a problem in an engine test cell as additional hardware performs signal processing. However, in a production vehicle, real-time signal processing is limited by the computational capability of the electronic control unit (ECU) of the engine. The engine compartment of the automotive vehicle is not sensor friendly, and the combustion chamber also has high temperature and pressure. In the laboratory conditions, the life expectancy or recalibration interval of a hundred hours may be acceptable, but on a production vehicle, the operation of combustion sensors is expected for hundreds of millions of cycles with minimal or no servicing requirements. These factors limit the widespread use of combustion sensor on a production vehicle. Ruggedness has an immediate bearing on the packaging. Choice and flexibility of sensor installation are heavily restricted due to space constraints and servicing/replacement requirements. Sensor installation must penetrate the combustion chamber, which is typically surrounded by oil and cooling water circuits. In modern multivalve engines, space is almost always very limited. Sensor installations considered as the best which require no additional access to the engine combustion chamber, and if it also needs to be installed on production vehicle, then awkward or costly modification should be avoided [7]. Ideal combustion sensor should be nonintrusive and non-perturbing. Any protrusion in the combustion chamber can locally distort the flow field of the charge, which affects the combustion characteristics. Thus, flush-mounted or recessed sensor installation is preferred. Electronics and electronic packaging are also important in packaging. In case of the production vehicle, the burden on ECU can be reduced when some aspects of signal conditionings are integrated with the sensor. Electronic package must guard against earth loops and electromagnetic interferences by appropriate shielding. Sensor fouling is another important design aspect to be considered for combustion sensors. Carbon deposition due to soot formation in the combustion chamber can contribute to fouling. A more complete description of design issues can be found in reference [7].

In-cylinder pressure measurement (high-pressure indicating) has been developed into highly sophisticated analytic methods for reciprocating engine combustion optimization and diagnosis. Modern sensors and computer-based high-speed data acquisition systems make indicating technology as an industrial measurement technology and also satisfy the accuracy demands for calculating extensive information from the measured pressure signal. Indicating is widely used because of the following reasons [1]: (1) it is a developmental tool for engine combustion optimization with high quality and speed; (2) it provides huge amount of information about the in-cylinder phenomena and engine operations, which is typically not available with other measurement techniques; and (3) it is a safe, reliable, and repeatable measurement technique and thus used as standard technology on development test beds [8]. Piezoelectric pressure transducers are typically used for combustion pressure measurement by directly installing the sensor into the engine head or installed into the spark or glow plug using adapters.

The main advantages of piezoelectric sensors are extremely wide measuring range (span-to-threshold ratio up to over 108), extremely high rigidity (measuring deflections are typically in the μm range), high natural frequency (up to over 500 kHz), high reproducibility, very high stability, wide operating temperature range, high linearity, and insensitivity to electric and magnetic fields and to radiation [9]. The main demerit of the piezoelectric sensor is that it is inherently not able to measure static signal over a long period of time due to self-discharge as there is no material with infinitely high insulation resistance and no semiconductor is completely free of leakage currents. Ideally, materials suitable for transduction elements in pressure sensors should have the following properties [9]: (1) high piezoelectric sensitivity, (2) high rigidity, (3) high electric insulation resistance, (4) high mechanical strength, (5) linear relationship between mechanical stress and electric polarization, (6) minimal hygroscopicity, (7) high stability of all properties, (8) absence of hysteresis, (9) low-temperature dependence of all properties within a wide temperature range, (10) low anisotropy of mechanical properties, (11) good machinability, and (12) low production cost. Operating principle, design, construction, and the mounting position of a piezoelectric pressure transducer used for engine indicating are discussed in the next subsections.

2.2.1 Functional Principle

Piezoelectric pressure transducers are the most commonly used sensors for combustion measurement, which rely on the piezoelectric effect that refers to the property of particular crystals to exhibit electrical charge under mechanical deformation. Figure 2.2 schematically illustrates the piezoelectric effect using quartz crystal.

Fig. 2.2 Illustration of direct piezoelectric effect (Courtesy of Kistler)

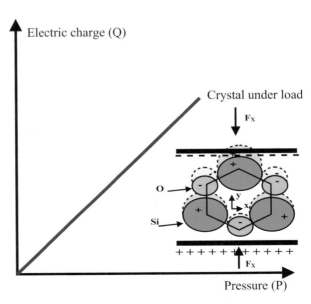

The piezoelectric materials generate positive or negative electrical charges on mechanical loading to their outer surfaces. An electric dipole is formed due to the charge generated by the displacement of positive and negative crystal lattice elements. The electrical charge produced is proportional to the force (pressure) applied to the piezoelectric crystal.

In order to produce a measurable electrical output from a cylinder pressure input, the pressure must first be converted into a proportional mechanical strain, which is transmitted to an electrical transduction element that creates the required electrical signal. Therefore, piezoelectric pressure transducers consist of two key components, i.e., one mechanical (diaphragm of the sensor) and one electrical (piezoelectric crystal). Figure 2.3 illustrates the functional principle of the piezoelectric pressure transducer. The diaphragm of pressure transducer experiences the change in the cylinder pressure (dP/dt) which is transmitted to a piezoelectric crystal through intermediate elements [5]. The rate of pressure change leads to the deformation in the piezoelectric crystal at a strain rate of $d\varepsilon/dt$. Deformation in piezoelectric crystal polarizes charge "q" in the transducer electrode, which generates an electric current "i" that establishes the output signal of pressure transducer as represented by Eq. (2.1) [6].

$$i = -\frac{dq}{dt} = -G_s\frac{dp}{dt} \tag{2.1}$$

where G_s is the sensitivity of piezoelectric pressure transducer.

Figure 2.4 presents a simplified schematic of signal conditioning of piezoelectric pressure transducer that is used to get in-cylinder pressure data in combustion engines. Typically, two methods can be used to obtain the pressure data, i.e., through

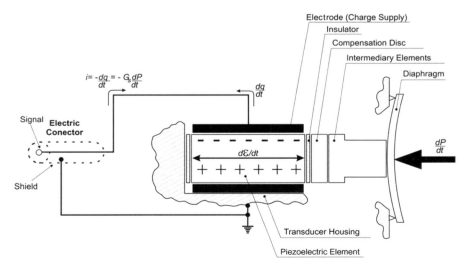

Fig. 2.3 The schematic diagram of piezoelectric pressure transducer [6]

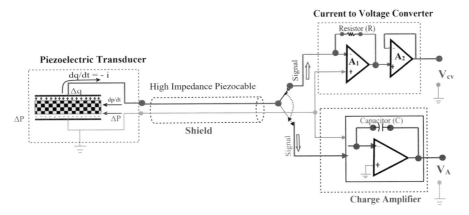

Fig. 2.4 Signal conditioning of piezoelectric pressure transducer by charge amplifier or current-to-voltage converter (adapted from [5, 6])

charge amplifier or current-to-voltage converter (Fig. 2.4). Charge amplifier is most commonly used for getting the pressure data. The charge amplifier-based method is preferred possibly due to the legacy of the mechanical indicators [5]. During early engine development, the mechanical indicators were used for pressure measurement as well as for determination of indicated work from measured pressure data or for evaluating the thermodynamic state of the charge. The charge polarized by piezo-electric pressure transducer is carried to the inlet of the charge amplifier by a shielded cable with high insulation resistance. The charge received from the piezoelectric pressure transducer is converted into a voltage signal in the charge amplifier, which mainly consists of an amplifier and a degenerative feedback capacitor (C) to inte-grate the charge (Fig. 2.4). The integrator circuit of charge amplifier produces an output voltage proportional to the time integration of the electric current produced by the transducer during a time interval Δt. The time interval is considered between the desired instant of measurement and the instant from which it was started (or reset). The change in measured pressure level during the interval Δt is given by Eq. (2.2) [5].

$$P - P_{\text{ref}} = \frac{V_A G_A}{G_s} \tag{2.2}$$

where G_A is the charge amplifier gain, V_A is amplifier output voltage, and P and P_{ref} are the pressure acting on the diaphragm during time interval Δt.

The charge amplifier has very large internal charge amplification (amplification factor up to about 100,000). The charge received from the piezoelectric pressure transducer is drawn off by the feedback capacitor, and it is not used for charging (i.e., to increase the voltage at the input capacitances). Feedback capacitor has been negatively pre-loaded to accept the charge from the sensor input [1]. The piezoelec-tric pressure transducer generates very small charge (just tens of picocoulombs per bar). Therefore, the charge amplifier output is highly sensitive to electronic circuit nonidealities; particularly leakage currents occur by insulation resistance of the measurement system, which slowly and continuously yields lower-voltage output leading to lower-pressure measurement values. In order to reduce this inaccuracy, high input impedance in the charge amplifier is used. The operation of signal conditioning system in the low-humidity environment along with clean electrical contacts also helps in reducing the measurement error. Long-term drift error (load change drift) along with instrumentation nonidealities also leads to instability in the pressure data baseline (up to several bars), which is inherent demerit of charge amplifier system. Therefore, it is mandatory to periodically reset the charge amplifier to avoid saturation during pressure measurement [6].

Another approach to obtaining the cylinder pressure data is through current-to-voltage converter (Fig. 2.4). This circuit fulfills the gain and the frequency response requirements for cylinder pressure measurements. This device has low input impedance, as provided by the ratio of the gain adjustment resistance (R) to the open-loop gain of the operational amplifier A_1, which removes the inaccuracies generated from the inherent capacitance of the sensor. A voltage follower amplifier (A_2) is used to isolate the converter with respect to the impedance of the instrument. The change in pressure rate is estimated by Eq. (2.3).

$$\frac{dP}{dt} = \frac{V_{cv}}{G_s R} \tag{2.3}$$

where V_{cv} is the voltage output of converter and R is the gain-adjusting resistance of the current-to-voltage converter.

The in-cylinder pressure can be obtained by integrating the measured pressure derivative data. The cylinder pressure measurement with this method eliminates the need for special care for insulation resistance and leakage currents because in this case current generated from the transducer flows to the ground unrestrictedly. Additionally, the need of periodically resetting the charge amplifier during measurements is also eliminated [5, 6]. One of the main objectives of cylinder pressure measurement is the determination of heat release for combustion diagnostics [10]. Since pressure derivative is used for computation of heat release, the inaccuracies in the pressure data get amplified and reflected in heat release data. Direct measurement of the pressure derivative data through current-to-voltage converter reduces the noise of pressure derivative data (~70 times) [6]. A similar strategy based on direct pressure derivative measurement for combustion detection is also demonstrated in reference [11].

2.2.2 Transducer Materials and Construction

Materials used for transduction elements in piezoelectric pressure sensors are expected to have good measuring behavior (high output signal, good linearity, high natural frequency), good resistance (high mechanical strength and high temperature resistance), stability of the measuring properties (against temperature and load variations), and low price (including material cost and easy machining) [8]. Quartz is the most frequently used piezoelectric material in pressure sensors for combustion measurements; however, researchers continue to investigate and develop alternative piezoelectric materials because of their manufacturing advantages over quartz [7], but the alternative materials can compromise the quality of the measurement data for engine combustion measurement application. Alternative piezoelectric materials are polycrystals of lead niobate ($LiNbO_3$) and lead titanate ($PbTiO_3$) and single crystals of gallium phosphate ($GaPO_4$), lithium niobate ($LiNbO_3$), and silicon dioxide (SiO_2) [12]. Thermal tolerance (sensor operating temperature) is one of the major limitations for selection of piezoelectric transduction materials as temperature affects the piezoelectric effect as well as durability of the sensor. Piezoelectric effect vanishes, and the sensor does not respond above a threshold temperature (Curie temperature) [7, 13]. During piezoelectric material selection, thermal tolerance of material is traded with pyroelectricity in which charge is produced due to thermal variations instead of mechanical strain. Pyroelectricity is also a linear function of the first differential of temperature similar to piezoelectricity which is a linear function of the first differential of force [7]. The pyroelectric effect needs to be small (preferably no) in the signal used for obtaining pressure sensor. Temperature effects are significant even with quartz crystals used in pressure sensors. Figure 2.5 depicts the temperature effect on the piezoelectric constant for quartz and gallium orthophosphate material, which are presently used in combustion pressure sensors. The figure illustrates that the maximum operating temperature limit of quartz crystal is around 250 °C. Thus, this simple crystal can be used in combustion pressure measured with the suitable cooling system in a cooled pressure

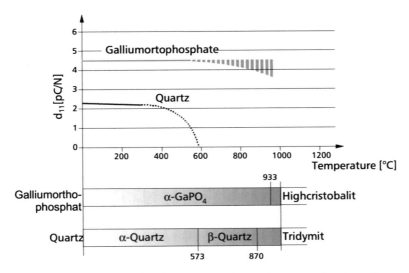

Fig. 2.5 Variations of the piezoelectric constant with operating temperature for gallium orthophosphate and quartz (Courtesy of AVL)

transducer. In typical reciprocating engine environment, the temperature of more than 400 °C can likely occur at transducer locations [1]. Engineered gallium orthophosphate has a typically higher piezoelectric sensitivity (nearly twice of quartz), and it is independent of temperature until much higher than 500 °C (Fig. 2.5). Therefore, gallium orthophosphate is suitable for construction of uncooled miniature pressure transducers for in-cylinder pressure measurement in reciprocating engines.

Piezoelectric pressure transducer is typically protected by enclosing it in sensor housing (due to limited temperature tolerance range), and the transducer indirectly receives the strain from the diaphragm. Thus, high-temperature resistance, small size, and invulnerability to electrical interference are essential requirements for transducers for combustion application. Pyroelectric effects can also be reduced by constructing electrodes that are parallel to polarization axis [14]. The optimization of the orientation of crystal cut can extend the temperature range of piezoelectric crystal (quartz). Three main types of operations can be distinguished as transversal, longitudinal, and shear depending on the way of piezoelectric material cut. Figure 2.6 illustrates the different types of cuts in the piezoelectric material used in sensor technology. Piezoelectric elements with longitudinal or transverse cut are typically used for combustion-measuring applications [15].

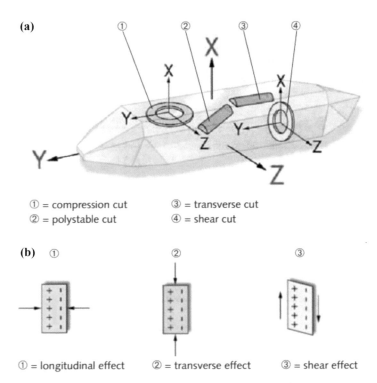

Fig. 2.6 Illustration of different piezoelectric crystal cuts (Courtesy of Kistler)

The piezoelectric effect is differentiated based on the direction of piezoelectric polarization with respect to the direction of mechanically introduced strain. The direct piezoelectric effect can be phenomenologically described using Eq. (2.4) [1].

$$D_i = d_{i\mu} \cdot T_\mu \qquad (2.4)$$

where D_i ($i = 1$ to 3) is a vector of electric flow density, $d_{i\mu}$ is the tensor of piezoelectric coefficients, and T_μ ($\mu = 1$ to 6) is the tensor of mechanical stress. The Eq. (2.4) is used for calculation of charge output (Q) using Eq. (2.5).

$$Q = A \cdot D_i \cdot n_i \qquad (2.5)$$

where A is the surface area and n_i ($i = 1$ to 3) are components of the normal vector of the face.

In longitudinal piezoelectric effect, the amount of charges generated is strictly proportional to the acting force (mechanical strain), and output is independent of shape and size of the piezoelectric element. However, in transverse piezoelectric effect, the quantity of charge output depends on the geometrical dimensions of the piezoelectric element, and the polarized charge is perpendicular to the line of applied force. Therefore, the crystal elements used in the pressure transducer can be cut to optimize charge output as a function of the force and stresses in the element. A "polystable" cut (Fig. 2.6) is employed by Kistler company for particular sensors, which is optimized to decrease the effect of operating temperature on the piezoelectric sensitivity of the quartz element. This cut enhances the maximum operating range of around 350 °C (piezoelectric behavior remains constant within range), which makes the possibility of using this crystal in uncooled sensors as well [15].

The current trend toward miniaturization and higher operating temperatures leads to the development of new types of crystals having greater sensitivity and higher temperature. The PiezoStar® crystal elements (KI85 and KI91) developed by Kistler are optimized for use in such demanding applications [16]. Figure 2.7 depicts the sensitivity and sensitivity shift as a function of temperature for different piezoelectric crystals. PiezoStar crystal has a very high piezoelectric sensitivity (~3–5 times of quartz), high stability of the properties, no twin formation, no phase transition up to the melting point, and no pyroelectric effect and can be used up to 600 °C temperature. The stated disadvantages of PiezoStar crystal are lower mechanical strength and higher cost than quartz [16]. The major drawbacks of quartz crystal are low sensitivity, twin formation, and phase transition at 573 °C. Peculiarities with different piezoelectric crystals along with their production process and operating temperatures are presented in Table 2.1.

The amount of electrical charge produced by a single crystal element under mechanical strain depends on the piezoelectric material. To fulfill the requirement of high-sensitivity sensor with lower-sensitivity piezoelectric material (such as quartz), several crystal disks are stacked and connected electrically in parallel, which is illustrated in Fig. 2.8.

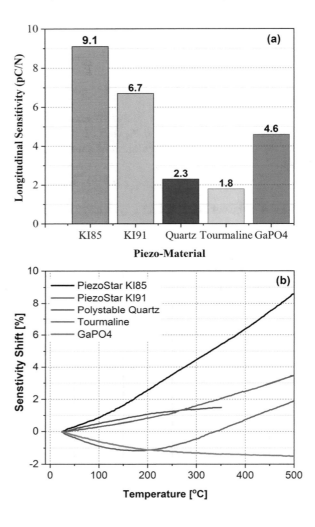

Piezoresistive materials can also be used as pressure-sensing devices [17]. In
engine research application, piezoresistive pressure sensors are used for
low-pressure indicating and only few applications for high-pressure combustion
measurements. The advantages of piezoresistive sensors over piezoelectric sensors
are the ease of signal processing (including temperature compensation) and better
noise rejection due to low output impedance [17]. In the piezoresistive sensors, the
electrical resistivity of sensing material (typically semiconductor) changes when
mechanical strain (force) is applied. The electrical resistance of material varies due
to change in conductivity of mater as well as geometry change.

Table 2.1   Comparison of different high-temperature piezoelectric materials (Courtesy of Kistler)

Crystal	Peculiarity	Pyroelectric effect	T_{max} [a]	Production		
				Process	Volume	Cost
Quartz	High mechanical	No	573 °C	Hydrothermal	Large	Low
KI85	strength	No	T_m	Czochralski	Average	High
KI91	High sensitivity Low-temperature coefficient	No	T_m	Czochralski	Average	High
GaPO4	Low-temperature coefficient	No	970 °C	Hydrothermal	Small	High
Tourmaline	High-tempera- ture stability	Yes	>900 °C	Natural	Small	High
Piezoceramic (PZT)	High sensitivity	Yes	250 °C	Sintering	Large	Low

Tm no transition below the melting point
[a]Phase transition temperature

Fig. 2.8   Arrangement of crystal elements to increase the sensitivity of transducer (Courtesy of Kistler)

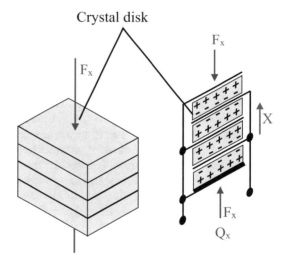

Crystal disk

2.2.3   Transducer Design

Piezoelectric pressure sensor for engine combustion application is available in a large number of designs (as illustrated in Fig. 2.9) depending on the complexity of the application. The variety of demands such as temperature stress, installation space, vibrations, mechanical deformation of the location of sensor installation, etc. leads to the development of different designs of pressure sensors. Mostly pressure sensors have a flush-welded diaphragm (eliminating any dead volume) which allows flush mounting of the transducer on the engine cylinder head. Typically, the measuring element is housed inside the transducer body so that installation, sealing, and heat transfer are away from the sensing element. Some of the sensors are

Fig. 2.9 Different types of
piezoelectric pressure
sensors based on their
construction (courtesy of
Kistler)

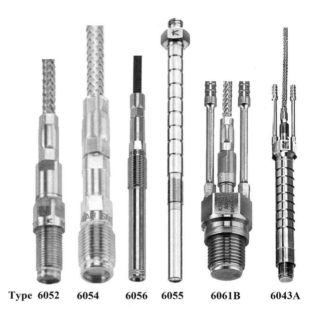

Type 6052   6054   6056   6055   6061B   6043A

constructed and designated as probe type (6043, 6055, and 6056 in Fig. 2.9), which
are suitable for cylinder pressure measurement in engines with very little space such
as multivalve engines, motorcycles, etc. A basic difference in design of the sensor is
based on with or without mounting threads on the sensor. Pressure sensors with
mounting threads can be directly screwed into the combustion chamber (engine
head) having mounting bore with suitable thread. Pressure sensors without mounting
thread are mounted into bores providing only a sealing shoulder, and sensors are
held by adapters, and mounting nuts or nipples. Both of these types of sensors have
corresponding merit and demerit (discussed in Sect. 2.2.4). A precise mounting bore
on cylinder head is essential along with tight tolerances for surface finish of the
sealing surface, and orthogonality between thread axis and sealing face. The sensor
can be strained during mounting with compromised tolerances, which leads to
variations in sensor characteristics such as sensitivity and linearity. Sensors without
mounting threads are mounted in an adapter that is installed in a corresponding bore
in the engine head. Sensor mounting with adapter sleeve needs more space, but this
method ensures the sensor specifications by well-defined mounting geometry.
Adapters can handle less stringent tolerances in the mounting bore due to their
massiveness and ruggedness [9].

   Thermal tolerance is critical constraint in the selection of the piezoelectric
material. Depending on the piezoelectric material, both cooled and uncooled sensors
are developed over a period of time for combustion analysis in reciprocating engines.
Figure 2.10 shows the typical cross section of cooled and uncooled piezoelectric
pressure transducers. Cooled type of piezoelectric pressure transducers is most
widely used for accurate combustion pressure measurement and analysis

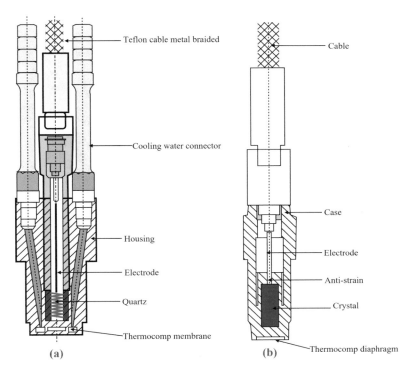

Fig. 2.10   Cross section of (**a**) cooled and (**b**) uncooled piezoelectric pressure sensor (Adapted from and courtesy of Kistler)

[15]. Cooled type of pressure transducer is relatively larger and available in mounting thread size from M8 to M14. Cooled sensors can accommodate relatively larger sensing element because of the relatively large size of the sensor, but the advantage of high sensitivity depends on crystal cut. During the operation of cooled pressure sensor, the sensing element and the membrane are directly cooled by surrounding water jacket, and thus, sensing element is only slightly warmer than the coolant (~10–20 °C) [1].

Water-cooled piezoelectric sensors are used for applications where sufficient mounting space is available and extreme precision is the top priority [18]. In the cooled pressure sensors, engine load change drift is small because the almost constant temperature is maintained using the cooling system by supplying optimal quantity of water at the right pressure. Thus, absolute reproducibility in the measurement can be ensured. The pressure transducer must be cooled using deionized water or a cooling agent-water mixture. Local supply water can lead to deposits that can affect the sensitivity or in worst case block the water lines. Therefore, a closed cooling circuit is typically used and recommended. Pressure pulsations in the water flow should be avoided because it can be picked by sensing elements and superimposed with measured cylinder pressure sensor (cross-talk phenomena) [15].

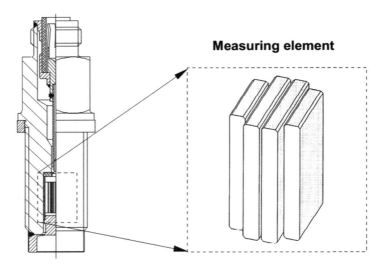

Fig. 2.11 Cutaway view of the uncooled pressure sensor with transverse piezoelectric effect (courtesy of AVL)

Modern reciprocating engines are very compact in terms of construction due to four-valve technology and direct fuel injection, which leads to very less availability of space for installation of pressure sensors for combustion analysis. Thus, miniature piezoelectric pressure sensors are used for such limited space applications. Figure 2.11 shows the cross section of such miniature piezoelectric pressure sensor using transverse crystal cut. This type of sensors is typically uncooled due to their smaller size. The uncooling requirement demands piezomaterials with high-temperature resistance and stability in properties over wide temperature range [1]. Current uncooled pressure sensors have reached to a comparable measurable quality similar to cooled piezoelectric pressure sensors.

ThermoComp® sensor from Kistler has a double diaphragm to minimize the thermal shock. The cyclical combustion process and heat flow between cylinder charge and cylinder head leads to the different temperatures in the surrounding of the pressure transducer. Thus, the temperature of pressure transducer depends on the engine operating conditions, the mounting position, and the type of sensor used. Figure 2.12 schematically shows the heat flow from the combustion chamber through pressure to the cooled cylinder head. For higher measuring accuracy and a long sensor service life, Kistler uses front sealing which keeps the measuring element at a lower temperature by heat dissipation in the front of the pressure sensor (Fig. 2.12).

AVL uses Double Shell™ design of piezoelectric pressure sensors (illustrated in Fig. 2.13) for combustion analysis which ensures premium signal quality by decoupling the piezoelectric crystals mechanically from the sensor housing. Piezoelectric crystals are susceptible to any kind of mechanical strain due to their high sensitivity. Double shell design of the pressure sensor makes sensing element

Fig. 2.12 Schematic of heat dissipation from the combustion chamber through pressure sensor to the cooled cylinder head (courtesy of Kistler)

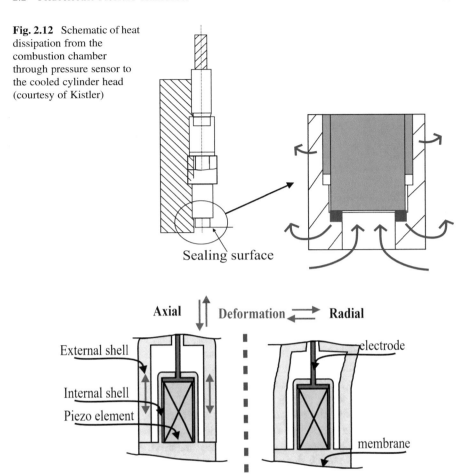

Sealing surface

Axial   Deformation   **Radial**

External shell

Internal shell

Piezo element

electrode

membrane

Fig. 2.13 Illustration of Double Shell™ technology in piezoelectric pressure sensors for combustion analysis (courtesy AVL)

isolated from any axial or radial deformation of sensor housing that can occur from different sources such as mechanical stress by sensor mounting, etc. Thus, double shell design helps in ensuring absolute pressure measurement precision.

The output signal of the piezoelectric sensor is transported out to an external connector, which is transmitted to the charge amplifier via a measuring cable. This signal can be ground isolated or ground referenced depending on the sensor construction. Figure 2.14 shows the schematic of piezoelectric pressure sensor without ground isolation and with ground insulation. Most of the sensors are not ground isolated [18]. One terminal of sensing element is connected to the ground of the engine (test stand ground), and the amplifier is connected to the equipment ground, which is also connected to the power system ground. Thus, ground-isolated pressure

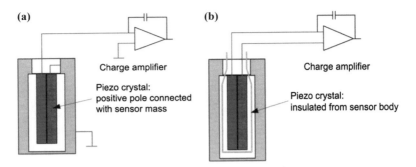

Fig. 2.14 Piezoelectric pressure sensor (**a**) without ground isolation and (**b**) with ground insulation (courtesy of Kistler)

transducers should be used with suitable charge amplifiers. Ground-insulated sensor design avoids problems with ground loops and increases immunity to electrical noise in the test environment.

Piezoelectric pressure transducers are sensitive to acceleration, particularly in the direction of transducer axis. The diaphragm mass and transmitting plates in front of the piezoelectric element act as a seismic mass (similar to acceleration sensor) which leads to acceleration sensitivity in the transducer. Hence, during vibration (acceleration), the output signal of the transducer is superimposed with the actual pressure signal. Typically, the acceleration error is in the order of a few mbar/g [9]. The acceleration error is increased in cooled piezoelectric sensors due to the additional mass of the cooling water present in cooling ducts in front of the sensing element. The acceleration error can be ignored in most applications, but it can be significant when the sensor is subjected to strong vibrations while measuring small pressure. This leads to the development of accelerated sensitivity compensated pressure sensors. Figure 2.15 shows the section through sensor and operation principle for active and passive acceleration sensitivity compensation designs. The Kistler company has developed these two concepts. Two piezoelectric elements of different piezoelectric sensitivities are used for active acceleration compensation techniques. An additional measuring element is used in conjunction with a seismic mass along with the sensing element. The additional element is connected with opposite polarity, and the signal produced by tuned vibration frequency gets canceled. In passive acceleration sensitivity compensation, the additional measuring element is not required, and the sensing element is supported by a sleeve which acts as a dynamic spring-mass system. In this system, the sensing element is tuned to the same natural frequency as the transducer diaphragm. Thus, the piezoelectric sensing element is effectively "sandwiched" between the two spring-mass systems, and both the diaphragm and the mounting support sleeve system respond by oscillating with the same amplitude and direction during acceleration of transducer. In this method, the sensing element does not receive any additional force due to vibration. The passive acceleration sensitivity compensation technique is more suitable for uncooled sensors as it does not require any additional compensating measurement elements.

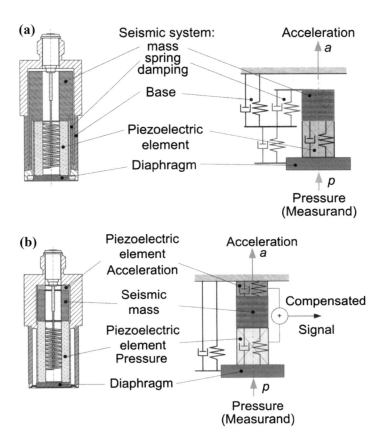

Fig. 2.15 Section through sensor and operation principle for (**a**) active acceleration sensitivity compensation and (**b**) passive acceleration sensitivity compensation (courtesy of Kistler)

The piezoelectric sensor can be divided into two categories based on the output signal. Pressure sensors are available as high impedance or charge output (PE) and low impedance or voltage output (IEPE). In voltage output sensors or integrated electronics piezoelectric (IEPE) sensor, the electronic circuit for converting the charge into a voltage is integrated with sensor body. Kistler company has registered trademark Piezotron® for IEPE sensors. Figure 2.16 illustrates the basic design of charge mode and voltage mode pressure sensor system. The charge output or PE sensors produce a high-impedance charge that needs to be converted into a usable low-impedance voltage signal (that can be recorded by data acquisition system) by the external charge amplifier. For transmitting the charge to amplifier, special low-noise high-impedance cable is required. The standard two-wire coaxial cable can have triboelectric noise generation (charge generation) due to friction between the conductors of the cable [19]. Since the charge produced by the sensor is very small, it is very difficult to differentiate it with charge produced by cable.

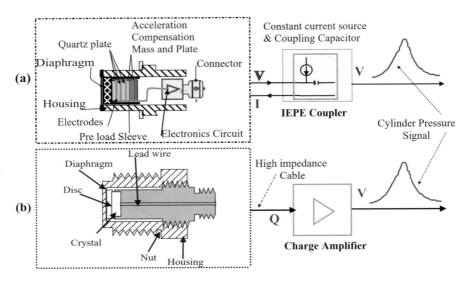

Fig. 2.16 Schematic illustration of pressure sensor system with (**a**) voltage output (IEPE) and (**b**) charge output (PE)

Additionally, environmental contaminants on the connector, such as moisture, dirt, oil, or grease, contribute to reduced insulation that can lead to inconsistency in results as well as signal. Therefore, high insulation resistance of the cables and connections is essential for PE sensors. The PE sensors need not be powered because the charge is generated by piezoelectric element when mechanical pressure is acting on it. The dynamic pressure traces and pressure pulsations can be measured with both PE and IEPE pressure sensors. Typically, high-impedance systems are more versatile than low impedance. The time constant, gain, normalization, and reset parameters are all controlled via an external charge amplifier. Adjustable measuring range can be achieved with one pressure sensor as the range is adjustable in the charge amplifier. Additionally, the time constants are generally longer with high-impedance systems, allowing easy short-term static calibration. Since electronic circuit is not attached to PE sensors, measurement of pressure in extremely low- or very-high-temperature conditions is possible.

Voltage output types (IEPE) of transducers also use the same piezoelectric sensing element and also have an integrated miniaturized electronic circuit for the charge-to-voltage converter (Fig. 2.16a). The IEPE sensors should be connected to a current (IEPE) coupler, which powers the sensor electronics and decouples the voltage signal from the power supply signal. Typically, low-impedance or low-voltage output systems are tailored to a particular application because the transducer has an internally fixed range and time constant.

2.2.4 Transducer Properties and Specifications

The piezoelectric pressure sensor for reciprocating engine combustion measurement has to operate in an extremely harsh environment. Figure 2.17 illustrates the typical working conditions of the piezoelectric in-cylinder pressure sensor, which affects the characteristics of the output signal. The pressure sensor is exposed to very high dynamic gas temperature (up to >2700 °C) in each combustion cycle. The cyclic heat flux exposed can be more than 1000 W/cm^2 during abnormal combustion (e.g., knocking) conditions. However, in normal operating conditions, the mean heat flux is about 50 W/cm^2 [1]. The temperature of the sensor needs to be maintained below the maximum working temperature of the transducer. Pressure transducer also typically faces acceleration of 1000 g due structure-borne vibrations produced by reciprocating engine parts, and the vibrations/acceleration can increase during abnormal combustion conditions (such as knocking/ringing) of engines. Large stress of up to 200 N/mm^2 in the cylinder head material at the transducer mounting

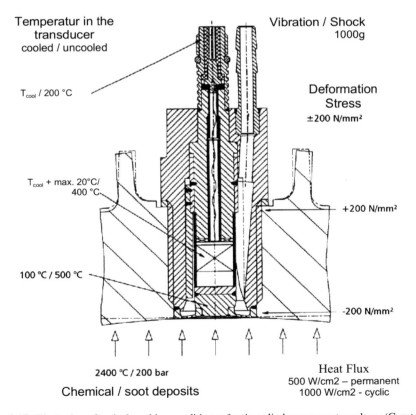

Fig. 2.17 Illustration of typical working conditions of an in-cylinder pressure transducer (Courtesy of AVL)

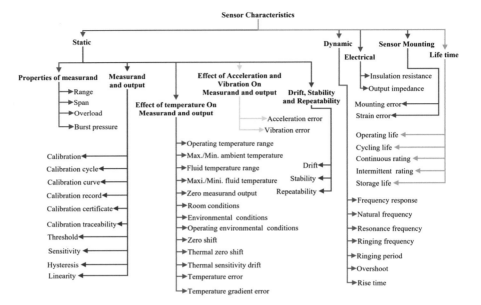

Fig. 2.18 Typical sensor characteristics need to be considered for selecting a piezoelectric sensor (Adapted from [9])

interface can be created by the gas pressure and the thermal stresses due to expansion and contraction. These imposed stresses on the transducer body can have a significant effect on the measured pressure signal. The signal quality of pressure sensor can also be affected by deposits on the transducer face. Carbon deposits can be a problem typically in high soot formation operating conditions depending on the fuel and combustion quality.

Various characteristics of sensors need to be considered for the selection and installation in particular working environment. The properties of sensor govern its effectiveness in generating the desired output signal with respect to working conditions of sensors. Figure 2.18 presents a vast list of sensor characteristics that need to be considered before the selection of transducer including static characteristics, dynamic characteristics, electrical characteristics, and sensor mounting and lifetime-related characteristics. Static properties of sensors are not a function of time, and it can be related to properties of the measurand, measurand and output, the effect of temperature on measurand and output, the effect of acceleration and vibration on measurand and output, drift, stability, and repeatability. Dynamic characteristics of a sensor are related to its response to variations of the measurand with time, which include frequency response, natural frequency, resonance frequency, ringing frequency and period, overshoot, and rise time. All the sensor properties are defined and discussed in the book [9]. However, a brief description of important properties of a piezoelectric pressure sensor for combustion measurement application is provided in this section.

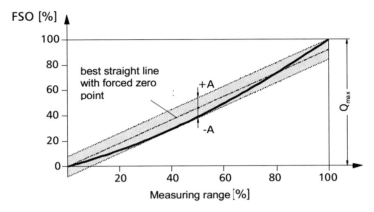

Fig. 2.19 Illustration of sensitivity and linearity of a sensor (Courtesy of AVL)

2.2.4.1 Properties Related to Transmission Performance

The important parameters related to transmission performance of piezoelectric pressure sensor are sensitivity, linearity deviation, natural frequency, insulation resistance, and natural capacitance [8]. The sensitivity of the sensor is defined as the ratio of the change in the output signal (i.e., charge) to the associated change in the measurand quantity (i.e., cylinder pressure). Usually, the manufacturer provides the nominal (or average) sensitivity value of pressure transducer in terms of pC/bar. Typically all the sensors have some amount of sensitivity deviation over the operating range. Figure 2.19 illustrates the sensitivity and linearity of a sensor. Sensitivity and linearity are quantities, which are determined using calibration.

Calibration of the sensor is a test conducted where known values of measurand are acting upon sensor and the associated output signal is recorded in specified operating conditions. In a calibration cycle, measurand values are applied starting from the lowest point of sensor range and increased up to the highest point of the range and back again. The recorded output signal of calibration cycle is processed for estimation of sensitivity, linearity, and hysteresis [9]. The smallest change in measurand which can be measured with the sensor is defined as threshold. The slope of the tangent of the calibration curve is termed as sensitivity. If calibration curve is not a straight line, then the sensitivity of sensor depends on measurand values. The linearity defines the deviation of a calibration curve from a particular straight line. The closeness of calibration curve to the "best straight line with forced zero point" is defined as linearity (Fig. 2.19). The linearity of a sensor is presented as percent of full-scale output (%FSO), which is shown in Eq. (2.6) [8].

$$\text{Linearity}(\%\text{FSO}) = \pm \frac{A}{Q_{\max}} \cdot 100 \qquad (2.6)$$

where A is the distance of straight lines enclosing the characteristic from the best line with forced zero point and Q_{max} is the maximum value of output signal (FSO). Piezoelectric sensors (particularly quartz) have a very good linearity and also an exceptionally high ratio of span-to-threshold.

Dynamic characteristics of a sensor are related with its response to variations in the measurand as a function of time. Piezoelectric pressure sensors can be considered as under-damped, spring-mass systems with a signal degree of freedom. This system is modeled by the classical second-order differential equation, whose solution is presented by Eq. (2.7).

$$\frac{a_o}{a_b} \cong \frac{1}{\sqrt{\left[1 - \left(\frac{f}{f_n}\right)^2\right]^2 + \left(\frac{1}{Q^2}\right)\left(\frac{f}{f_n}\right)^2}} \tag{2.7}$$

where f is the frequency at any point of the curve, f_n is the undamped natural frequency, a_o is the output acceleration, a_b is the mounting base or reference acceleration ($f/f_n = 1$), and Q is the factor of amplitude increase at resonance. Quartz sensors have a Q of approximately 10–40.

Figure 2.20 depicts the typical frequency response of piezoelectric pressure sensor (quartz). In a typical frequency response curve (Fig. 2.20), about 5% amplitude rise can be expected at one-fifth of the resonant frequency. Low-pass (LP) filtering can attenuate this effect. The flat response of transducers is seen as a useable range. The defining characteristic of this group is the high resonant frequency leading to a flat response throughout the operating range.

The natural frequency is described as the frequency of free oscillations of the sensing element of a fully assembled transducer, and the resonant frequency is the measurand frequency at which the transducer responds with maximum output signal amplitude. The natural frequency of the pressure transducer must be appropriately

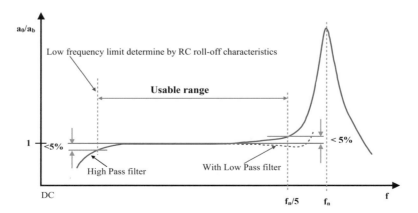

Fig. 2.20 Typical frequency response of piezoelectric pressure sensor (Courtesy of Kistler)

high such that oscillations from the measurand do not excite the transducer into resonance. The excitation of transducers resonance can superimpose false high-frequency components on the measured signal. As a thumb rule, the highest-frequency component expected in the measured pressure signal should not exceed 20% of the resonant frequency of the pressure sensor [15].

Piezoelectric sensors can be described as an electrical device having an active capacitor with a very high insulation resistance. Typically, one electrode of this capacitor is connected to the sensor housing (except ground-isolated transducer). The insulation resistance is the resistance measured between the output signal connector of the sensor and the sensor housing. The insulation resistance should be as high as possible ($>10^{13}$ Ω at room temperature) to keep the influence of the electrical drift as low as possible. Piezoelectric pressure sensors have a natural capacitance, primarily due to the electrodes of the measuring element, and the capacitances of the connector and the line to the measuring element. This natural capacitance can be ignored when a charge amplifier is used [15].

2.2.4.2 Measurement Range and Operating Life

The measurement range of piezoelectric pressure transducer is described as a range of pressure values where it fulfills the defined specifications. This is basically the operational pressure range of the sensor. The algebraic difference between the limits of the range is defined as the span of the sensor. The overload range becomes a concern when the pressure measurement exceeds the normal upper limit. The overload range is defined as the maximum magnitude of a measurand that can be applied to a sensor without causing a change in performance beyond a specified tolerance. Most transducers can withstand some degree of overload without causing any irreversible damage. However, the accuracy of the sensor is not ensured in the overload range. Burst pressure is defined as the pressure which may be applied to sensing element or sensor without rupture. Operating temperature range of pressure transducer is also defined as temperature range in which the defined specifications are fulfilled.

Depending on the quality of installation and operating conditions, the pressure sensor has a finite lifetime. The operating life of pressure sensor is defined as the number of load cycles (or engine combustion cycles) over which the sensor retains its technical performance properties [15]. A generally quoted value is around 10^7 cycles as the projected operating life of a piezoelectric pressure sensor in a typical operating conditions of reciprocating engines. An important means of optimizing the life of the sensor is the use of dummy units when pressure measurement is not required (e.g., warm-up or conditioning of engine).

2.2.4.3 Thermal and Acceleration Influences

Thermal characteristics are important in evaluating a sensor's suitability for accurate measurements of in-cylinder pressure during combustion. Pressure sensors are sensitive to temperature, and any deviations from the calibrated temperature of the sensor may lead to measurement error. The temperature sensitivity of pressure sensor can result into a signal drift. The change in sensitivity of piezoelectric pressure sensor is typically described by the temperature coefficient of the sensitivity, which indicates the actual change in sensitivity as a percentage of the nominal sensitivity per °C within a specific temperature range [8]. The change in sensitivity of pressure sensor is negligible for small change in operating temperature or with cooled pressure sensors.

The pressure transducer is exposed to the non-steady-state heating by the combustion gases on cyclic basis, which results in thermal drift. The amplitude and time characteristic of the temperature-related drift are functions of the type of pressure transducer and heat flow at installation position. Temperature drift in pressure signal is described as the "pressure indicating" that is caused solely by the temperature changes at the pressure sensor and mounting position. Two types of thermal drift phenomena are observed in reciprocating engines, namely, cyclic temperature drift (short-term drift, thermal shock) and the load change drift (long-term drift).

In short-term drift, the measurement error in pressure signal occurs due to cyclic combustion heating of the pressure transducer within a cycle. This cyclic heating problem is more severe at low engine speeds due to the larger duration of combustion time, and relatively more time is available for heat transfer. Figure 2.21 illustrates the cyclic temperature drift in an uncooled pressure sensor. The maximum error in the pressure data within a combustion cycle is related to a point at the start of the heating phase. The figure depicts the temperature distribution and deformation of the transducer at three different points in the cycle (25 °CA before ignition TDC as well as 25 °CA and 180 °CA after intake TDC). Significant deformations in the vicinity of the pressure transducer diaphragm can be clearly seen at the 25 °CA after intake TDC position. The loads on the sensing element due to the deformation lead to cyclic temperature drift (Fig. 2.21).

Thermal shock is one of the major problems limiting the accuracy of piezoelectric transducers for cylinder pressure measurements in reciprocating engines. Thermal shock is generated due to the temperature variation during the engine combustion cycle. The response of a piezoelectric pressure transducer is affected by thermal load variations in two ways: (1) through the corresponding deformation of the transducer/diaphragm and (2) through its effect on the sensitivity of the transducer [20]. A temperature gradient is set up in the transducer material and in the metal surrounding it, when heat is exchanged between the in-cylinder gases and a pressure transducer. The corresponding thermal expansion will deform part of the transducer. Under normal operation, the diaphragm of the piezoelectric transducer deflects toward the piezoelectric crystal of the transducer when the cylinder gas pressure is applied. When the intermittent flame (high thermal load) is exposed to its diaphragm of the transducer, mechanical deformation of the diaphragm occurs due to an abrupt change in temperature on the surface.

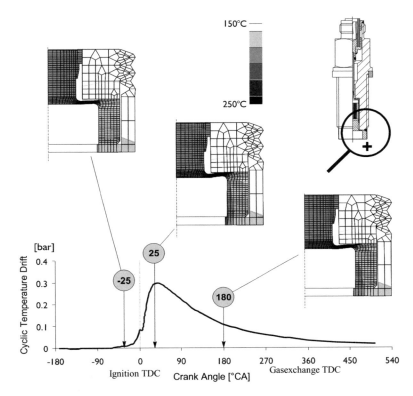

Fig. 2.21 Typical temperature distribution and deformation of a pressure transducer (Courtesy AVL)

Since both sides of the diaphragm are fixed completely, the diaphragm elongates and deflects in the direction of sensing element. To the extent that the piezoelectric crystal is touched by this deformation, this will result in a response of the transducer, even when the pressure remains constant. Therefore, the error in pressure data is strongly influenced by the amount of deflection of the sensor diaphragm due to thermal shock. The temperature variation corresponding to a cyclic thermal loading will penetrate the transducer only to a limited depth, and this depth increases with decreasing cycle frequency. In reciprocating engines, the lowest-frequency components of the thermal load on the transducer are correlated to the changes in engine working condition, and they will penetrate most of the transducer. The highest-frequency components are correlated to the intermittent combustion process, which will affect only the transducer diaphragm (and not the pressure-sensing piezoelectric element because it is located several mm away from the front diaphragm). A cyclic expansion and contraction of the diaphragm occur due to the intermittent combustion process. Transducer sensitivity is affected because thermal load variation influences the stiffness of the pressure-sensing diaphragm. With increasing temperature, the diaphragm weakens and sensitivity of the sensor increases. Of course, the

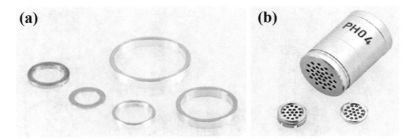

Fig. 2.22 Gaskets and flame arrestors for pressure transducer used in combustion measurements (Courtesy of AVL)

sensitivity of transducer also changes with the operating temperature variation of the piezoelectric crystal [20].

A number of thermal shock reduction methods have been proposed to reduce thermal shock error using mechanical and numerical techniques [21]. The mechanical methods of thermal shock reduction include the water cooling [22], use of heat shields [23], RTV (room temperature vulcanizing) silicones [24], front coating with silicon, and recess mounting [23]. Coating the transducer diaphragm with silicon and water cooling could damp the rapid changes in temperature at the diaphragm. When recess-mounted, the transducer communicates with the cylinder through a single passage or series of passages which quench the flame before its arrival on the transducer surface [23]. Typically, the smaller sensors use the interconnecting passage and face sealing to reduce the thermal load. The larger sensors use heat shield to quench the combustion gases [20, 23]. Figure 2.22 shows the typical gaskets and flame arrestors for the piezoelectric pressure sensors used for cylinder pressure measurement. Gaskets are used as sealing between the sensor and the cylinder head or adaptor. The gasket reduces the additional temperature stress during operation because of the optimized gasket material. During installation of the pressure sensor, the appropriate gasket is used for all sensors with shoulder sealing. The flame arrestors are used as thermo-protection for highly accurate cylinder pressure measurements. During extremely high-temperature operation of the sensor, the use of flame arrestors can result in a significant reduction of the cyclic drift as well as protection of the sensor. However, flame arrestors are not recommended for use in high soot-generating engines due to small holes at the front of flame arrestors.

Thermal shock occurs due to excessive temperatures at low speeds, whereas excessive temperatures at high speeds influence the sensitivity of the transducer, thereby its accuracy [25]. To investigate the thermal effect on the pressure measurement, the temperature of the diaphragm is measured (using fast response thermocouple), and the correlation between surface temperature and thermal shock is derived (Fig. 2.23). Figure 2.23a presents the diaphragm temperatures based upon heat shield design at different engine speeds. The study noted that the diaphragm temperature of all the sensors at idle is approximately the same as the engine coolant temperature, indicating negligible influence from combustion [25]. However, the

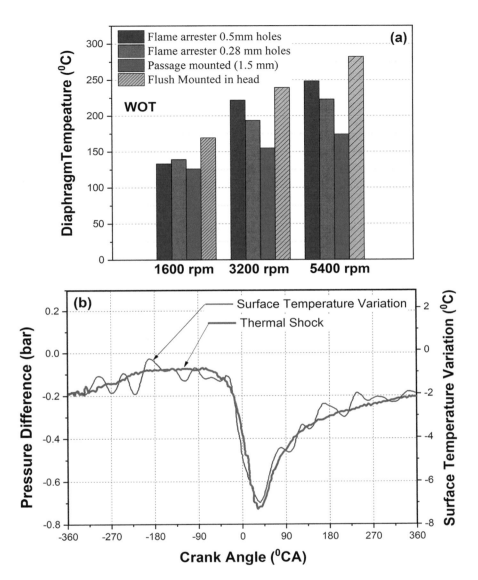

Fig. 2.23 (**a**) Transducer diaphragm temperatures based upon heat shield design (Adapted from [25]) and (**b**) correlation between surface temperature and thermal shock (Adapted from [21])

temperature of a flush-mounted sensor reaches up to 282 °C at higher engine speeds and loads, and this temperature range can influence the sensitivity of the pressure sensor. Figure 2.23a also depicts that the sensors protected by the heat shields have a lower diaphragm temperature. The reduction in passage diameter leads to the reduction in diaphragm temperature. Figure 2.23b shows the correlation between surface temperature and thermal shock in the diesel engine. The sign of the temperature fluctuation has been changed to allow assessment of the correlation with thermal shock (Fig. 2.23b). The existence of a strong correlation between the pressure error measurements and temperature variations is confirmed by Fig. 2.23b [21]. The location of pressure sensor installation also affects the diaphragm temperature, and thus, thermal shock effect can be reduced by selecting the favorable installation position.

In the published literature, the effect of thermal load on pressure transducer is described in terms of short-term thermal drift (or thermo-shock) and medium (or long)-term thermal drift. Short-term thermal drift is defined as the measurement error within a single cycle due to transducer deformation and sensitivity change caused by the change in thermal load within that cycle. Only the transducer deformation effect can be considered a drift in the strict sense as the effect of a cyclic change in sensitivity is minimal in modern transducers. Medium-term thermal drift is defined as the measurement error related to a change in engine working condition. Finally, long-term drift is caused by the change in sensitivity of the transducer over longer periods (it can be several hours for the good transducers) [20].

There are several methods to evaluate the sensor accuracy during the thermal shock, whether by another sensor or to itself. The methods for evaluating thermal shock include (1) comparing the output signal to a reference sensor, which has a much higher resistance to thermal shocks [15, 26]; (2) using a dedicated heating test rig to measure the response of the transducer to an external heat load [1, 15]; (3) a comparison of the measured signal with a calculated or simulated pressure [21, 27]; (4) a quasi-steady-state test, applying the average cylinder pressure during exhaust stroke as a function of the combustion phase [23, 28]; (5) comparison of pressure signal envelopes [28]; (6) cyclic pressure deviation at specific points in the engine cycle [23, 25]; (7) the relationship of IMEP to the location for 50% mass fraction burned [23, 25]; and (8) the difference between the pressure at 540° ATDC and at −180° BTDC (intake BDC position), called drift, and a value close to zero would be ideal [25, 29].

To evaluate the short-term drift of a sensor by comparison method, a reference sensor known for higher accuracy (typically water cooled) is required. In this method, the absolute value of measured cylinder pressure throughout the entire engine cycle is examined with respect to reference sensor. Figure 2.24 depicts the short-term drift of sensor by presenting the difference in pressure measurements due to the thermal shock effect and reference sensor. Typically, the average of few hundred cycles is computed, and the reference signal is subtracted to highlight the difference. The figure shows that thermal shock occurs during combustion when the flame contacts the diaphragm and persists during the expansion stroke. Effect of thermal shock on reference sensor is minimal and neglected. The difference between the outputs of the two sensors provides an indication of the magnitude of thermal shock. A study defined the thermal shock as the maximum pressure difference after TDC between the test and the reference sensor [26]. It is important to note that the duration of the thermal shock affects IMEP, not just the maximum thermal shock magnitude. This is because the IMEP calculation is an integral of the pressure with

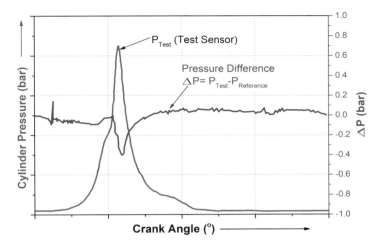

Fig. 2.24 Difference between the reference sensor (water-cooled) and test sensor outputs (Courtesy of Kistler)

respect to the volume change throughout the engine cycle, not just at a single point. The magnitude of thermal shock depends on the (1) duration of high-temperature shock (e.g., flame impingement time), which affected by engine speed, and (2) the peak pressure (or temperature) during engine cycle depending on engine operating conditions. Typically, greatest thermal shock errors occur at low engine speed, high load, advanced ignition timings, slightly rich mixtures, and low EGR operating conditions [26]. As noted, the severity of the thermal shock is controlled by the time history of the instantaneous heat flux into the sensor. The time history of the heat flux varies from cycle to cycle due to normal variations in heat release rates causing the timing and severity of the thermal shock to vary from cycle to cycle as well.

To evaluate the thermal shock on the pressure sensor, the measured signal can be compared with a calculated or simulated pressure data at the same operating conditions. Figure 2.25a presents a comparison between the simulated and experimental pressure data for the blowdown process at two different engine loads. The figure shows that the cylinder pressure level remains relatively constant during the blowdown process irrespective of the engine load in case of modeling data. However, in the case of the experimental data, the cylinder pressure at the higher engine load condition is lower than that at the lower engine load condition. This trend is possible due to the thermal shock which is much higher at high-load condition. Additionally, experimental pressure data should be close to the measured exhaust port pressure during the blowdown process. Once the exhaust valve opens during the exhaust stroke, the pressure in the cylinder should be identical to the pressure in the exhaust port. Any difference between the two can indicate an error with the cylinder pressure sensor. The measured cylinder pressure appears to be lower than the atmospheric pressure for the exhaust process, and it is also lower than the measure of exhaust port pressure (Fig. 2.25a). This can be attributed to thermal shock [21]. Figure 2.25b

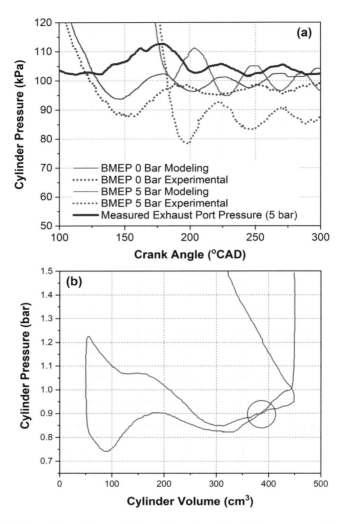

Fig. 2.25 (**a**) Comparison of the modeling and experimental cylinder pressure at the blowdown process at different engine loads (Adapted from [21]), (**b**) effect of thermal shock on *P-V* diagram (Adapted from [30])

illustrates the effect of thermal shock on intake and exhaust stroke on pressure-volume (*P-V*) diagram. The pressure curve of the intake stroke and that of immediately after the blowdown intersect each other (Fig. 2.25b) due to the thermal shock applied to the sensor diaphragm during combustion. The effect of thermal shock appeared after blowdown instead of during combustion period due to the heat capacity of the diaphragm, which causes the time lag of the temperature difference between inside the cylinder and the surface of the diaphragm [30].

The metrics based on analysis of the low pressure or pumping loop can be used to evaluate the effect of thermal shock. The thermal shock causes the transducer output to shift (either high or low) from the actual cylinder pressure depending upon the specific sensor design characteristics. This shift occurs with a relatively fast time constant such that, for typical engine (modern passenger car and light truck engine) speeds, the transducer will generally recover back to its normal state by the end of the intake stroke following the thermal shock event [28]. The average exhaust absolute pressure (AEAP) during exhaust stroke is one of the metrics used to evaluate the thermal shock. The average cylinder pressure during the exhaust stroke is calculated during a portion of the exhaust stroke from about $60°$ aBDC to about $60°$ bTDC. The averaging window needs to be selected such that interference from differences in the exhaust blowdown process can be avoided, which may lead to normal cyclic combustion variation. Figure 2.26 demonstrates the effect of thermal shock on AEAP (the average cylinder pressure during the exhaust stroke) as a function of the average CA_{50} (combustion phasing) position. Transducer A has been installed with the proper mounting techniques and in a location within the cylinder so as to minimize the occurrence and severity of thermal shock. The AEAP values during exhaust stroke from transducer "A" do not show any apparent sensitivity to the combustion phasing as expected. Additionally, during the exhaust stroke, the AEAP values from transducer "A" are just slightly higher than the ambient atmospheric pressure of 100 kPa (Fig. 2.26), which is also expected at the relatively low-speed, light load condition (experimental condition). The other transducers experience some level of thermal shock and show pressures either substantially higher than the expected pressure or pressures lower than ambient which is physically not possible. They also show varying amount of the sensitivity of the average cylinder pressure during the exhaust stroke to the combustion phasing [28].

Figure 2.26 also depicts a low value of standard deviation of the average cylinder pressure during the exhaust stroke that is not influenced by the combustion phasing for the transducer A, which is intended to have the least thermal shock. The remaining transducers show a higher standard deviation of the average cylinder pressure during the exhaust stroke and show sensitivity to the combustion phasing. The more advanced combustion phasing leads to the more severe thermal shock and higher standard deviation of the average cylinder pressure during the exhaust stroke (Fig. 2.26).

Thermal shock on transducer during the combustion pressure measurement can be detected by signal envelope method. In this method, three signal envelopes are defined: the average signal envelope (ASE), upper signal envelope (USE), and lower signal envelope (LSE). The ASE is determined by the average pressure recorded at each crank angle increment over the entire set of cycles in the data sample. The USE is calculated by taking the highest pressure recorded at each crank angle increment on any one of the sampled cycles. Similarly, the LSE is calculated by taking the lowest pressure recorded at each crank angle increment on any one of the sampled cycles [28]. It is important to note that USE and LSE traces are composite cycles, which contain the data from many individual cycles at different crank angle positions. Figure 2.27 illustrates the typical USE, ASE, and LSE curves for a transducer

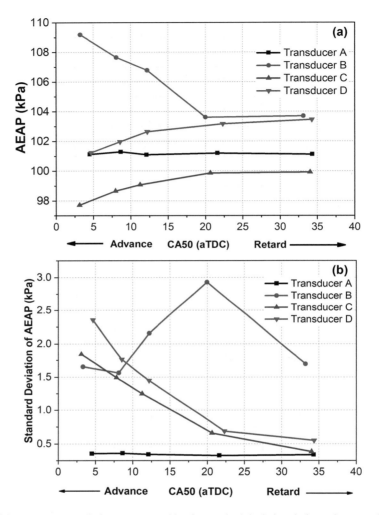

Fig. 2.26 (**a**) Average cylinder pressure and its (**b**) standard deviation during exhaust stroke as a function of combustion phasing (Adapted from [28])

with good (low) thermal shock characteristics and poor (high) thermal shock characteristics. The transducer with low/good thermal shock characteristics can be used as reference pressure sensor. The average spread between the USE and LSE curves at each crankshaft position is small (<3 kPa), which is mainly governed by the digitizer resolution and background noise on the data. Additionally, the comparison of the signal envelopes can also highlight the acceleration sensitivity of the transducer as indicated by the diverging signals during the valve closing events (Fig. 2.27a) [28]. The pressure transducer with poor thermal shock characteristics has the larger average spread between the USE and LSE curves at each crankshaft position, which indicates the presence of thermal shock.

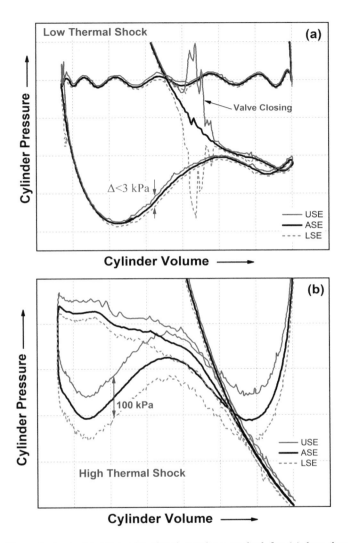

Fig. 2.27 Thermal shock detection with signal envelope method for (**a**) low thermal shock transducer (Adapted from [28]) and (**b**) high thermal shock transducer (Adapted from [27])

Another technique named intra-cycle variability analysis of pressure measurements is used for thermal shock detection, which is also a pressure deviation-based method. Metric in this technique compares the stability of the sensor to itself at specific locations in the engine cycle and shows the relationship of those points during particular sections of the cycle. This helps to show if the sensor is unstable to itself and if it may have recovered before the next portion of the engine cycle. In this method, the cycle-to-cycle variation of the cylinder pressure is compared at specified crankshaft positions (points shown in Fig. 2.28a) along the whole cycle. The cycle-

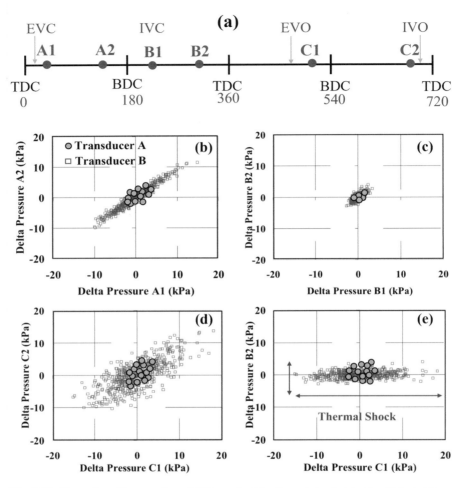

Fig. 2.28 Illustration of intra-cycle variability method for thermal shock detection (Adapted from [25])

to-cycle variability is normal due to the random variations in the combustion process. This variability leads to changes in the pressure transducer thermal load, and when a thermal shock occurs, the sensitivity of the transducer will be changed, and the scattering of pressure measurements will be enlarged. Thus, the pressure variance scattering is used to indicate the presence of a thermal shock [27]. - Figure 2.28 illustrates the intra-cyclic variations in pressure at different defined crank angle positions for two typical pressure transducers. Intra-cycle variability originates during the combustion event and drifts into the exhaust, intake, and compression segments of the pressure curve, while the pressure sensor simultaneously recovers from the thermal shock. With zero intra-cycle variability, all pressure differences would lie at the graph center (origin of the graph). Constant variability during any segment would cause all the pressure differences to lie along the graph diagonal, while a gradual change of variability during the segment would rotate the line around the graph origin away from the diagonal. The spread of the points along the graph diagonal is a measure of the intra-cycle variability range during the segment, while the spread of the points away from the diagonal quantifies measurement repeatability [31]. A slope of 1 means the sensor's accuracy does not change during the section; instead, the variability is the combustion process. During the intake stroke (Fig. 2.28b), the variability range is larger for transducer B, which suggests the transducer recovers from thermal shock impact as intake proceeds. The variability of the transducers is small in the compression stroke. The maximum amount of intra-cycle variability occurs during the exhaust stroke (Fig. 2.28d). This is expected because this segment of the pressure curve immediately follows the combustion event where the variability is originated. The large change in variability between C1 and B2 crank positions denotes the thermal shock (Fig. 2.28e). Data points are spread along the horizontal axis rather than along the diagonal for transducer B, clearly showing the effect of thermal shock.

Heat shield can avoid the extreme diaphragm temperatures during thermal shock on pressure transducers depending on the design of heat shield. Figure 2.29 depicts the effect of heat shield design on the reduction in AEAP pressure difference and drift as a function of the passage size. The figure shows that as the diameter of the passages decreases, the difference between the two measurements decreases, thereby supporting the flame-quenching process and reduced thermal shock. The drift (pressure difference at intake BDC position of consecutive engine cycles) for the different heat shields is all within the range of 1.6–1.7 kPa indicating sensor distortion is complete by the end of one engine cycle and the start of the next [25]. From an intra-cycle variability perspective, all heat shields reacted about equally.

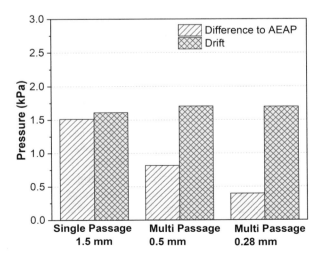

Fig. 2.29 Drift and AEAP difference in relation to passage sizes (Adapted from [25])

Thermal shock is generally reduced by mechanical methods, such as coating the transducer diaphragm, heat shields, recess mounting of the sensor, water cooling, etc. An empirical equation-based compensation method has also been developed, which provides greatly increased accuracy in the measured IMEP over a wide range of engine speed [21, 30]. From theoretical analysis using actual pressure data, a thermal shock error compensation equation was proposed [30]. The amount of deflection of the diaphragm due to the thermal shock could be expressed as a function of the temperature difference between the temperature of the diaphragm and the minimum temperature of each cycle. It is assumed that the diaphragm deflection is proportional to the additional pressure resulting from thermal shock. The error compensation equation obtained is presented in Eq. (2.8).

$$\Delta p = -\sqrt{A\Delta T^2 + B\Delta T} \tag{2.8}$$

The constants A and B are experimentally determined.

An IMEP thermal shock error correction equation which has been derived in the study [21] is presented by Eq. (2.9).

$$\text{IMEP}_{\text{corr}} = \text{IMEP}_{\text{meas}} + (F \times P_{\text{max}}) + \text{Offset};$$

$$F = 0.0000834\left(\frac{\text{rpm}}{1000}\right) - 0.00051\left(\frac{\text{rpm}}{1000}\right) + 0.00502 \tag{2.9}$$

$$\text{Offset} = 0.01534\left(\frac{\text{rpm}}{1000}\right)$$

The numerical approach for correction of IMEP thermal shock errors has a number of advantages: (1) increased measurement accuracy without expensive replacement of existing pressure sensor and extra machining of engine heads; (2) problems created by mechanical, thermal shock reduction techniques can be avoided; (3) the requirement of water-cooled sensor is reduced as it fits with difficulty in modern four-valve heads; and (4) sensors with good characteristics other than thermal shock resistance can also be used [26].

Typically a reciprocating combustion engine operates in transient conditions of speed and load. The variation in engine speed and load conditions affects the temperature as well as heat load experienced by the pressure sensor. The varying thermal conditions of pressure sensor have consequences on the output signal over a number of cycles, after a change in engine operating conditions. The load drift manifests as a slow variation in the pressure signal after a load change due to the altered thermal stresses in the sensor body. This shift in pressure level will only stop when the mean temperature in the pressure sensor no longer changes. Figure 2.30 illustrates the typical load drift on measured pressure signal to a step change in engine load condition. The figure depicts the two characteristics values: (1) maximum zero-line gradient (dp/dt) and (2) permanent zero-line deviation (shown in the bottom half of Fig. 2.30). The maximum zero-line gradient shows the change in pressure level per time unit caused by the heat flow, which will affect the combustion

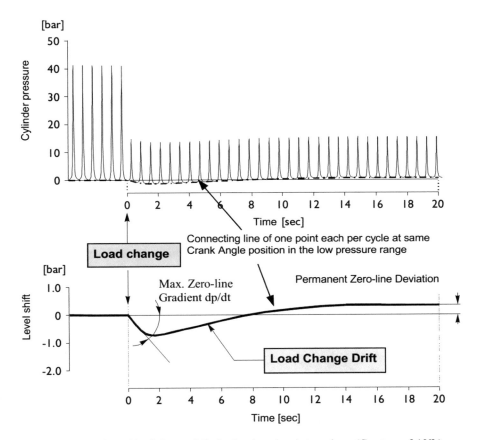

Fig. 2.30 Illustration of load change drifts in the piezoelectric transducer (Courtesy of AVL)

parameter calculation using measured pressure signal. The permanent zero-line deviation is shown as zero point after 20 s of load change (Fig. 2.30). This deviation does not affect measured value because the zero-line determination is always necessary for measurements using piezoelectric pressure sensors [8].

At constant engine working conditions, medium-term drift should not occur. In this case, a measure for long-term drift (LTD) can be obtained from the slope of the signal. To determine the long-term drift, the pressure trace (at a constant crank angle) is fit to a straight line (using the least squares method) as presented in Eq. (2.10) [20].

$$\hat{p}(\alpha)_i = a \cdot i + b$$
$$\text{LTD} = a \tag{2.10}$$

where "i" refers to the cycle number. It can be noted that the LTD errors are compensated if pegging of individual cycles is used.

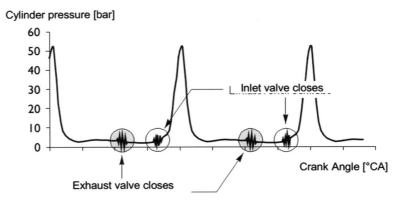

Fig. 2.31 Effect of structure-borne vibration on the measured pressure signal (Courtesy of AVL)

Apart from thermal shock, there are other effects which also lead to drift (error) in the pressure signal also on a short-term basis. The pressure changes during the engine working cycle result in cylinder head deformation which leads to deformation in pressure sensor together with the engine cylinder head [20]. Another major cause is the engine vibration. Acceleration (vibration) causes the quartz element and some of the other components to act as seismic masses leading to measurement error and can also result in a cyclic reduction of transducer cooling for water-cooled transducers.

The sensitivity of the pressure sensor to structure-borne vibrations of the engine can cause false frequency components to be imposed on the measured cylinder pressure signal. Thus, acceleration compensation is provided in pressure sensor by manufactures (illustrated in Fig. 2.15). The extent of the influence of acceleration also depends to a large extent on the installation site, on the direction of the accelerations that occur in relation to the pressure transducer axis, and on the engine speed [8]. Figure 2.31 illustrates the effect of structure-borne vibration/acceleration on the measured pressure signal. The high-frequency oscillations superimposed on the pressure signal are caused in this specific measurement arrangement by the impact of the intake and outlet valves on the valve seat and transmitted by structure-borne noise. Pressure measurements on revving racing engines often reflect a strong influence of structure-borne noise over the entire cycle [8].

2.2.4.4   Chemical Influences and Deposits

The combustion process in reciprocating engines produces hundreds of chemical species (combustion products) depending on the fuel used. Some of the species produced during combustion can lead to corrosive damage to the piezoelectric pressure sensor. Modern sensors use special coatings and corrosion-resistant materials, which makes corrosion effects insignificant [8]. The corrosion effect can be significant in very unusual conditions such as utilization of fuel with very high sulfur content or other newly developed alternative fuel (with corrosive combustion product).

Typically, deposits build up on the cylinder walls of both petrol and diesel engines. Direct injection diesel engines are well known for the sooting tendency. However, modern gasoline direct injection engines also have the soot formation issues. Soot formed during combustion tends to form deposits on cooler surfaces, and diaphragm of the pressure sensor is a favorable place for building up deposits particularly in cooled pressure sensors. The deposits can have an undesirable effect on measurement signal depending on the type of sensor and engine operating conditions. In extreme conditions, the altered pressure signal due to deposits can lead to more than 10% error in indicated mean effective pressure (IMEP) determination. The IMEP stability is a characteristic value, which provides the information about the sensitivity of pressure sensor to soot deposits. The IMEP stability is defined as the percentage change in IMEP over a defined runtime in relation to values determined with a reference pressure transducer [8].

2.2.5 Transducer Adaptors and Mounting Position

The mounting position of the pressure sensor and the mounting method to access the combustion chamber have a significant effect on the accuracy of measured combustion chamber pressure and the lifetime of the sensor itself. Pressure transducer must be small enough to be accommodated within the cylinder head without affecting the shape of the combustion chamber, and it must be fitted to produce minimal intrusion or disturbance. During installation process, a problem of proper mounting of pressure sensor usually arises, and the optimum is to flush mount the transducer such that it can be directly in contact with the gas. To access the combustion chamber for pressure measurement, there are basically two methods, namely, intrusive mounting (with intervention in test engine) and nonintrusive mounting (without intervention in test engine). Both of these methods have their merits and demerits. Typically, intrusive methods have relatively higher accuracy, and the measuring position can be selected by user. However, higher cost and substantial time and effort is required for intrusive installation of pressure sensor [8]. In nonintrusive mounting of the pressure transducer, installation time and cost is lower as it has to be installed in place of a standard engine component. The intrusive and nonintrusive methods of pressure sensor installation are discussed in the following subsections.

2.2.5.1 Intrusive Mounting

In an intrusive installation of the pressure transducer, intervention via precision modification of the cylinder head is performed in such a way that the transducer and measuring face are suitably positioned for exposure to the cylinder gas pressure.

Fig. 2.32 Illustration of
pressure transducer
mounting with (**a**) direct
installation and (**b**) adapter
installation (Courtesy of
Kistler)

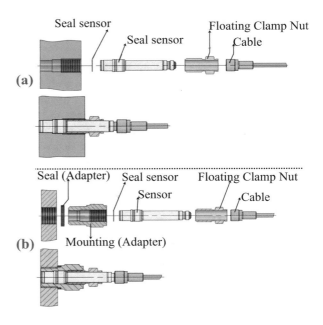

This type of pressure transducer has several variations due to a large number of
possibilities with respect to installation position [15]. The intrusive mounting of
pressure transducer can be executed with or without an adaptor. Figure 2.32 illustrates the pressure transducer mounting with (a) direct installation and (b) adapter
installation. Direct installation method is preferred for small spaces. However,
complex drilling with special tools is required for direct installation of the pressure
sensor. An adaptor sleeve is required when access to the combustion chamber
requires traversing oil or water passages. Installation with adapter is preferred
(requires adequate mounting space), and simple tapped hole in the engine head
can accept the adapter. Additionally, the adaptor sleeve effectively separates the
transducer body from the surrounding cylinder head material (mechanical
decoupling), thus isolating the sensor from deformation stresses that could cause a
shift in sensitivity during engine operation.

Measuring point accessibility and the measuring task are significant factors
determining the position in which sensor is mounted in the engine cylinder head.
The position of pressure transducer installation depends on a number of interrelated
factors which must be considered judiciously. Pressure sensor installation should not
affect the spaces of various engine parts in the combustion chamber such as valves,
injectors, and spark plugs. The structural integrity of the engine parts needs to be
retained, and the wall thicknesses of engine head casting must be ensured such that it
does not lead to failure. The transfer passages for oil or water in the cylinder head
should not be blocked by pressure sensor installation. Otherwise thermal conditions
of engine head and cylinder can vary drastically. The temperature of the sensor
(particularly its diaphragm) depends on the mounting position. Additionally,

measurement errors can also be introduced due to environmental factors, such as heat flow, temperature, and accelerations. Therefore, all these boundary conditions must be considered before choosing the installation site for the in-cylinder pressure sensor of engines.

Mounting positions near the exhaust valves results in increased sensor temperatures due to higher gas temperature and increased flow velocity during the exhaust stroke. Increased sensor temperature can affect the measurement accuracy and sensor service life, and thus, positions near the inlet valve area are preferred [18]. Gas dynamics during intake and exhaust stroke can also affect the measurement and can introduce error in the measured signal. The high flow rates near valves can lead to local pressure differences, which may not be representative of cylinder pressure [15]. The measurement errors due to the design of the measuring position can include dead volume when the transducer is installed in an inclined position or recessed, pipe oscillations for recessed installation, interference to the gas flow, and fuel deposits [8].

The optimal choice of pressure sensor installation is mounting flush with the combustion chamber, perpendicular to the surface, and if possible without the sort of flow pocket produced by inclined mounting. In this mounting, the transducer is recessed just enough to prevent deposit buildup on the piston from damaging the transducer, but not so much as to create a measuring pipe. The recessed installation can produce high-frequency oscillations at resonance or increase cylinder volume, which reduces the engine compression ratio. Figure 2.33 illustrated the flush and recessed mounting installation of the sensor for cylinder pressure measurement. Ideally, for best accuracy, a combustion sensor should be flush-mounted in the combustion chamber (to reduce heat flow load and temperature), but often this is not possible, and sensors are installed recessed. In the recessed installation, the cylinder pressure travels through an indicator passage to reach the sensor, which can lead to signal acoustic oscillations, and signal distortion can occur. Figure 2.34 shows the effect of the length of indicating channel on the measured pressure signal. Five pressure curves (shifted in level) from single cycle measurements are shown for each indicating channel length. The oscillations in pressure signal decrease with reduced length of the channel. The frequency of this interference depends not only on the length of the indicating channel but also on the gas state, which makes the use

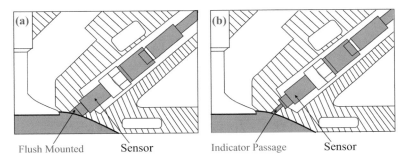

Fig. 2.33 Pressure sensor installation with (**a**) flush mounting and (**b**) recessed mounting

Pressure [bar]

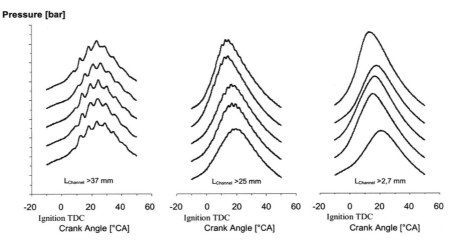

Fig. 2.34 Influence of the length of the recessed channel on the measured pressure signal (Courtesy of AVL)

of frequency filters for eliminating pipe oscillations difficult. Additionally, it is not easy to differentiate the pipe oscillations from actual combustion chamber oscillations [8].

Figure 2.35 presents the two different types (Fig. 2.35a, b) of pressure transducer installation passages. In the first design (Fig. 2.35a), only indicating channel with length (L) is present. However, in the second design (Fig. 2.35b), there is an additional volume (V_{cv}) in front of the pressure transducer with an indicating channel. For the first type of air duct (Fig. 2.35a), the lowest natural frequency of oscillations can be computed using Eq. (2.11) [32].

$$f_n = \frac{a}{4L} \tag{2.11}$$

where f_n is natural frequency, "a" is local acoustic velocity, and "L" is the length of the air duct.

For the second type of air duct with additional volume (Fig. 2.35b), the natural frequency of oscillations can be computed using Eq. (2.12) [32].

$$f_n = \frac{a}{2\pi L} \left(\sqrt{\frac{A \cdot L}{V_{cv}}} \right) \tag{2.12}$$

where A is passage cross-sectional area and V_{cv} is the volume of the cavity in front of the pressure sensor.

Figure 2.36 depicts typically achievable natural frequencies as a function of diameter and length of the indicator channel. It is shown that the natural frequency of the passage acoustic oscillation should be above 3 kHz by assuming that the engine has a minimum knock frequency of about 2 kHz. A passage with a diameter

Fig. 2.35 Illustration of installation passage types for piezoelectric pressure sensor (Courtesy of Kistler)

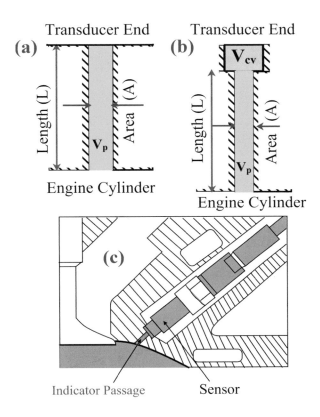

Fig. 2.36 Typical pipe oscillation frequency as a function of indicating channel length and diameter (Adapted from and courtesy of Kistler)

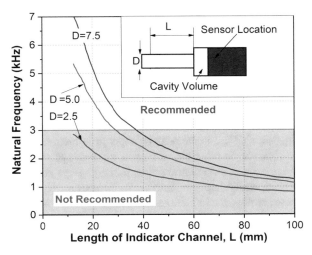

of 2.5 mm (or 5 mm with maximum length 30 mm) will not allow the minimum
frequency to be measured (Fig. 2.36). The shorter bores are recommended for
monitoring knocking conditions.

In the recessed mounting installation, the thermal shock can be significantly
reduced by isolating the transducer tip from the burning and burned gases.
Figure 2.37a presents the typical burned and unburned gas temperatures as a
function of crank position at wide open throttle (WOT) conditions in spark ignition
engine. The figure shows that the burned gases are at a much higher temperature
(over 2000 K) than the unburned gases. It is well known that the peak temperatures
of the first gases to burn are up to 300 °C hotter than the gases that burn in the middle
and toward the end of the combustion event. This results from the fact that as the
early gases combust, they expand considerably which effectively removes energy
from these burned gases. As additional heat is released from combustion of the
remaining charge, energy is added to these initial combustion products, raising their
temperature significantly higher than they would have otherwise obtained. The net
effect is that the gases that are burned before peak cylinder pressure being reached
will achieve a higher peak temperature than the mixture that burns as cylinder
pressure is falling. Thus, it is beneficial to mount the transducer away from the
ignition source in the end-gas region [33].

Considering the temperature difference of burned and unburned gases, the ability
of the connecting passage to quench a propagating flame is not enough to prevent
very hot gases from reaching the pressure sensor. The burned gases can be mechanically transported through the passage onto the diaphragm of the sensor because of
cylinder pressure rise after passing the flame front from sensor location. To prevent
the impingement of hot gases on the transducer face, the connecting passage must
have a volume sufficiently large relative to the sensor cavity volume to contain all
burned (and burning) gases that are compressed into it after the flame traverses the
passage entrance. This critical volume ratio is approximately defined by
Eq. (2.13) [33].

$$\left(\frac{V_\mathrm{p}}{V_\mathrm{cv}}\right)_{\text{critical}} = \left(\frac{P_\mathrm{max}}{P_\mathrm{FA}}\right)^{1/1.32} \tag{2.13}$$

where V_p is the passage volume, V_cv is the transducer cavity volume (as shown in
Fig. 2.35b), P_max is the peak cylinder pressure, and P_FA is the in-cylinder pressure at
the time the flame front arrives at the passage entrance. The Eq. (2.13) is derived
assuming that the volume of gas in the passage at the time of flame arrival must first
be completely displaced into the transducer cavity before burned gases can reach the
sensor cavity. It can be noted that the Eq. (2.13) is valid only if the cylinder pressure
is rising when the flame front arrives at the passage entrance, and it is also assumed
that design of passage quenches the flame.

Figure 2.37b presents a typical example of critical volume ratio as a function of
transducer radial location and combustion phasing. The critical volume ratio was
calculated for a full range of transducer radial locations for a combustion system with

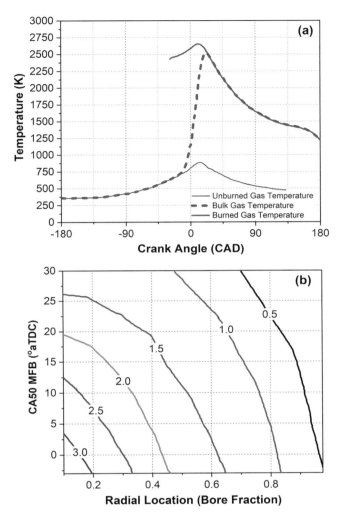

Fig. 2.37 (**a**) Typical burned and unburned gas temperatures as a function of crank position and (**b**) critical volume ratio as a function of transducer radial location and combustion phasing (Adapted from [33])

a centrally located ignition source. The figure depicts that the smaller radial location bore fraction (i.e., sensor is closer to the ignition source) requires the larger critical volume ratio. As the sensor location gets near to the bore wall/end-gas region, the critical volume ratio approaches zero. The critical volume ratio also depends on combustion phasing. A higher critical volume ratio is required for advancing combustion phasing due to higher peak pressure and a larger pressure ratio for a particular radial position of the sensor. A more complete discussion can be found in the original study [33].

Fig. 2.38 Change in
compression ratio as a
function of the dead volume
of measuring bore (Adapted
from and courtesy of
Kistler)

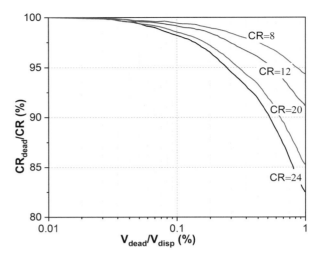

The recessed or inclined mounting of pressure sensor increases the compressed
volume of the engine. The dead volume (V_{dead}) of the measuring bore reduces the
engine compression ratio (CR). The reduced compression ratio (CR_{dead}) can be
calculated by Eq. (2.14) [18].

$$CR_{\text{dead}} = \frac{V_c + V_{\text{dead}} + V_{\text{disp}}}{V_c + V_{\text{dead}}} \tag{2.14}$$

where V_c is clearance volume, V_{disp} is displacement volume, and V_{dead} is the dead
volume of measuring bore.

Figure 2.38 depicts the change in engine compression ratio as a function of the
dead volume of measuring bore. The change in compression ratio is higher for
smaller displacement volume and higher compression ratio engines. For a typical
measuring bore (Fig. 2.35c) having a dead volume of 60 mm^3, the change in
compression ratio is 0.064 for a half liter engine with an original compression
ratio of 24. This change can be significant if the cylindrical bore of 8 mm is formed
(Fig. 2.35c) and not recessed as original measuring bore. The measuring bore
volume is quadrupled and change in compression ratio 0.289. This change in
compression ratio is substantial, which need to be taken care off.

There are special criteria for mounting position of pressure sensors in spark
ignition and compression ignition engines. In spark ignition engines, the measurement of engine knock modes depends on the position of the pressure sensor.
Centrally mounted pressure sensors typically provide the most representative data
for knock analysis. Pressure sensor installed near cylinder walls records different
knock amplitudes depending on their position relative to exothermic center during
knocking. The high-frequency pressure waves reflected from cylinder wall can cause
a surge wave, and high amplitude can be recorded by the sensor installed near
the cylinder wall [18]. However, depending on the knock detection algorithm, the

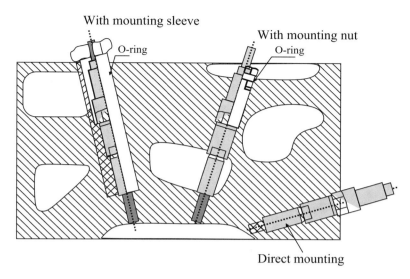

Fig. 2.39 Intrusive pressure sensor mounting examples (Courtesy of Kistler)

central position may not be preferred. Central mounting position may not reliably detect the first fundamental knock oscillation mode depending on combustion chamber shape.

In case of direct injection diesel engines, the sensor can be mounted either in the vicinity of squish clearance or over the piston bowl. Sensors over the bowl indicate the pressure without any delay. In this case, the pressure sensor will also expose to higher thermal load, which can be reduced by recessed mounting. Sensors mounted in the squish area can produce distorted pressure readings due to gas dynamics generated by the motion of the piston around TDC [15, 18]. However, thermal loading is reduced, and for monitoring applications, this position is ideal.

Considering all the factors discussed in this section, different intrusive mounting styles can be selected depending on the mounting location and space available. Figure 2.39 illustrates the different mounting example of an uncooled pressure sensor installed in the engine head. The installation of the pressure sensor can be done by using adapter mounting or direct mounting as illustrated in Fig. 2.39.

2.2.5.2   Nonintrusive Mounting

Cylinder pressure measurement in the modern engine can be possible without intervention (nonintrusive mounting) in the engine head. The nonintrusive mounting of the pressure sensor is possible with existing bores (for a spark plug or glow plug) in the engine head. With the development of piezoelectric crystal technology, it is possible to design miniature uncooled pressure sensor that can be accommodated in adapters which can fit in the space of standard engine components such as glow plug or spark plug. This method of installation of pressure sensor reduces the adaptation

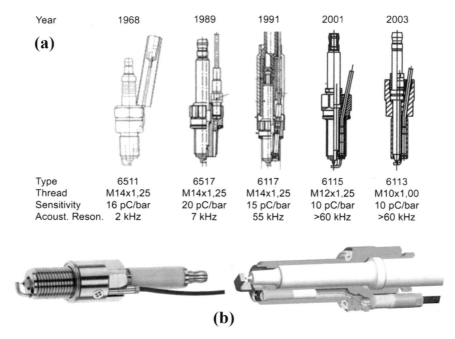

Year	1968	1989	1991	2001	2003
Type	6511	6517	6117	6115	6113
Thread	M14x1,25	M14x1,25	M14x1,25	M12x1,25	M10x1,00
Sensitivity	16 pC/bar	20 pC/bar	15 pC/bar	10 pC/bar	10 pC/bar
Acoust. Reson.	2 kHz	7 kHz	55 kHz	>60 kHz	>60 kHz

Fig. 2.40 (**a**) Stages of development of measuring spark plug and (**b**) structure of the measuring spark plug (Courtesy of Kistler)

time and cost. However, there is no choice of the installation position of a pressure sensor in the combustion chamber with nonintrusive mounting. The present measuring spark plug is able to match the combustion characteristics identical to the reference sensor (cooled sensor) [34]. The study suggested the following criteria to be considered for measuring spark plug: (1) measuring spark plugs can be installed very easily (i.e., no need of additional measuring bore); (2) turbocharged engines require high ignition voltage up to 40 kV, and thus, the gap of electrodes is very important to avoid misfire; and (3) engines with direct injection are sensitive to the spark position.

Figure 2.40 shows the stages of development of measuring spark plug and a typical structure of modern measuring spark plug. Historically, the measuring spark plugs were conversions of standard spark plugs, and pressure sensors are included using an adapted long and thin passages. This type of conversion required significantly more space in the plug shaft and had the disadvantage of very clear pipe oscillations occurring due to the longer distance between the sensor diaphragm and the combustion chamber. In this arrangement of the pressure sensor, the cyclic thermal drift is very low due to lower thermal load. Additionally, the spark plug remained largely unchanged, and the spark function was thus identical to that of the original spark plug. To prevent the pipe oscillations, there is need to bring the transducer diaphragm close to the combustion chamber. Development of compact sensors allows flush mounting of the pressure sensor in measuring spark plug (Fig. 2.40b). This arrangement avoids the pipe oscillations with less deterioration in the spark function due to the changed position of the spark distance as well as the maximum possible ignition voltage [34]. The measuring spark plug must be designed to be as close as possible to the original equipment with respect to heat range and position of the electrical arc (spark position is taken to mean the position of the electrode in with respect to the plug face). The displacement of spark position can have a deteriorating effect on the ignition quality particularly in gasoline direct injection (GDI) engines. In GDI engines, the stratified mixture needs to be ignited in the area where the mixture strength is sufficiently rich to initiate combustion. Thus, spark position is more critical in this type of combustion process.

An additional technical requirement related to sensor working environment other than heat range is a separation of high electrical voltage. The spark plug adaptor must separate the high voltage and charge signal, preventing cross-talk and interference of the signals. Additionally, it need to maintain the required level of insulation resistance to ensure no leakage of the high-voltage spark that could cause a misfire, even under the most extreme engine operating conditions [15].

Glow plug adaptors are the nonintrusive access technique for cylinder pressure measurement in compression ignition engines. Typically, most compression ignition engines are equipped with preheating probes (glow plugs) to preheat the cylinder gases, particularly during cold start conditions. For cylinder pressure measurement in diesel engines, glow plug can be removed and replaced with a measuring glow plug (Fig. 2.41) or glow plug adaptor (Fig. 2.42) which contains a suitable, uncooled miniature pressure transducer. The measuring glow plug or glow plug adapters are screwed into the mounting bore of the original glow plug and do not require separate measuring bore. Figure 2.41 presents the section view of measuring glow plug. The figure shows that a measuring probe with particularly small mounting and connection dimensions is screwed in a specially designed glow plug. The heating coil at the end of measuring glow plug necessitates an indication passage between the combustion chamber and measuring probe [18]. The arrangement can result in pipe oscillations, and it limits the frequency range for pressure measurement with good signal quality. The measuring glow plug is suitable for cylinder pressure measurement without dispense with functional glow plugs. The

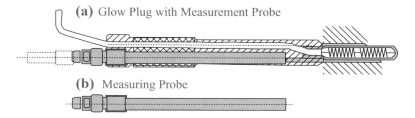

Fig. 2.41 Measuring glow plug with a miniature measuring probe for cylinder pressure measurement (Courtesy of Kistler)

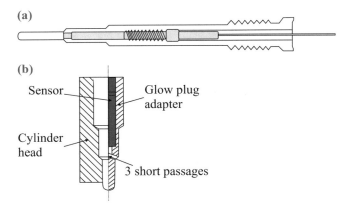

Fig. 2.42 Glow plug adapter with a miniature transducer for cylinder pressure measurement (Adapted from and courtesy of Kistler)

measuring glow plug is beneficial for cold start test without separate indicating bores, on-road measurement applications with glow function, and combustion diagnostics with lower speed and load range (as combustion temperature is lower) with glow function.

Glow plug adapters are also used for the cylinder pressure measurement test, where the engine can be readily started without original glow plugs. Figure 2.42 shows a glow plug adapter for the pressure measurement in a diesel engine. An adapter fitted with a miniature pressure sensor can be mounted on the place of the original glow plug. The cylinder pressure sensor mounted in glow plug adapter is connected to the combustion chamber through several short passages (Fig. 2.42b). The advantages of this arrangement are (1) simplest method of mounting with additional measuring bore in the engine head, (2) design conforming to specified glow plug bore, and (3) high signal quality ensured by measurement close to combustion chamber [18].

Typically different mounting positions are chosen depending on the applications. Some applications (such as engine peak pressure monitoring for continuous operation) requires the selection of the durable measuring system. In such applications, the advantageous mounting position of the sensor (for low thermal load) and high level of robustness of sensor is required. Recessed position of sensor ensures long service life due to lower thermal load exposure with glow plug adapter. Figure 2.43 shows the two different arrangements of pressure sensor mounting in glow plug adapter. The larger recessed position mounting (Fig. 2.43b) has a longer life due to lower thermal load, but it can have the pipe oscillations, which restrict the measurement frequency. Thus, based on the application and measuring requirement, the suitable mounting position of the sensor can be selected in the glow plug adapter.

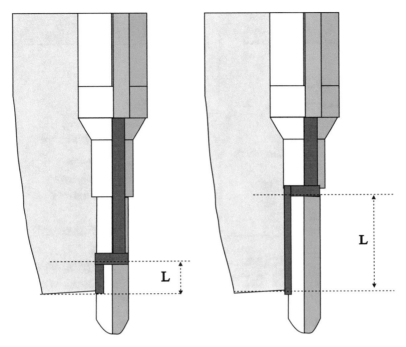

Fig. 2.43 Typical position of a miniature pressure sensor in glow plug adapter for development work

2.2.6   Transducer Selection

The cylinder pressure measurement is used for different tasks such as thermodynamic analysis, noise estimation, calibration, and monitoring of engine. During cylinder pressure measurement, the engine is also operated at different conditions such as knocking, cold start, misfire, etc., which affect the working environment of the pressure sensor. For the particular measurement, the selection of the correct type of transducer and the installation environment is a crucial task for the desired quality of pressure data. Performance of pressure sensor can be maximized with respect to measurement quality and reliability by appropriate selection of sensor and mounting position and type. The final choice will be a compromise among factors and boundary conditions that include (1) measurement task, target, and focus, (2) installation location, (3) installation space, (4) access pathways into the cylinder, (5) cost of sensor and installation, (6) installation effort and time, and (7) permanency of the installation [15]. The measurement task/objective has a significant role on the choice of the transducer. Presently, a wide variety of transducers are available, and many of them can be used for a particular application.

Figure 2.44 shows the relationship between the measuring application, the transducer choice, and the installation method. The measurement task defines the

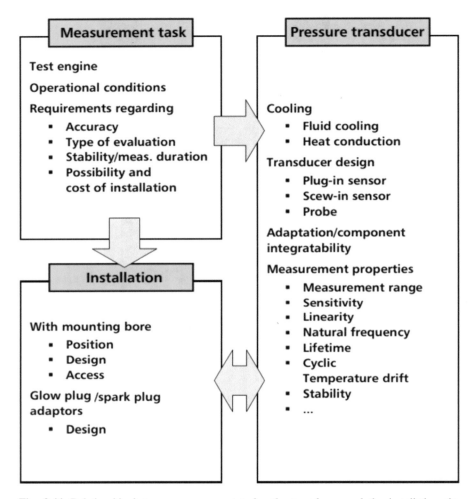

Fig. 2.44 Relationship between measurement tasks, the transducer, and the installation site (Courtesy of AVL)

operational conditions of the engine and the sensor. The measurement application also specifies the requirements regarding accuracy, stability, measurement duration, installation costs, etc., which governs the type/design of sensor to be used and its installation methods (Fig. 2.44). For the maximum accuracy of sensors, specifications of thermal shock value (Δp, ΔIMEP), linearity, and sensitivity are important parameters. The thermal shock specifications are important if the friction loss or gas exchange work is evaluated from firing engine. Pressure sensor must have high sensitivity in order to be able to effectively resolve the relatively small pressure differences (particularly during gas exchange). For the accuracy of data, the sensitivity of sensor needs to be high. The lower-sensitivity sensor requires a higher gain

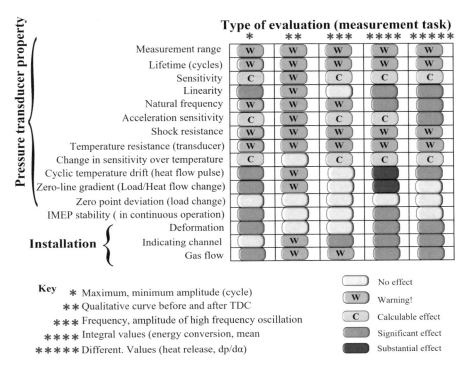

Fig. 2.45 Effect of transducer installation and properties on the measured signal (Courtesy of AVL)

of the amplifier, where noise will also get amplified. Thus, signal quality gets affected.

For thermodynamic analysis, accurate determination of TDC and specification of thermal shock (Δp, ΔIMEP) are important for the pressure sensor. Recommended limit values are 0.5 bar for Δp and $\pm 2\%$ for ΔIMEP [18]. Figure 2.45 presents the summary of the effect of transducer installation and properties on the measured pressure signal and its analysis. The figure shows an evaluation of the metrological properties of pressure transducers and how their measuring position influences the measurement task, which helps in the selection of pressure sensor by looking at the datasheet of a sensor provided by the manufacturer. Thermal drift (short term and long term) has a substantial effect on the evaluation of combustion parameters that are evaluated by integral of measured pressure signal (Fig. 2.45). The other important variables are linearity, sensitivity, IMEP stability, and installation parameter, which affect the quality of measurement and analysis results.

Figure 2.46 presents a simple flowchart for the selection of pressure transducer to be used for engine combustion measurement. After defining the basic requirement of measurement task such as accuracy, stability, cost, etc., the major decision is required whether sensor installation is intrusive or nonintrusive. In case of

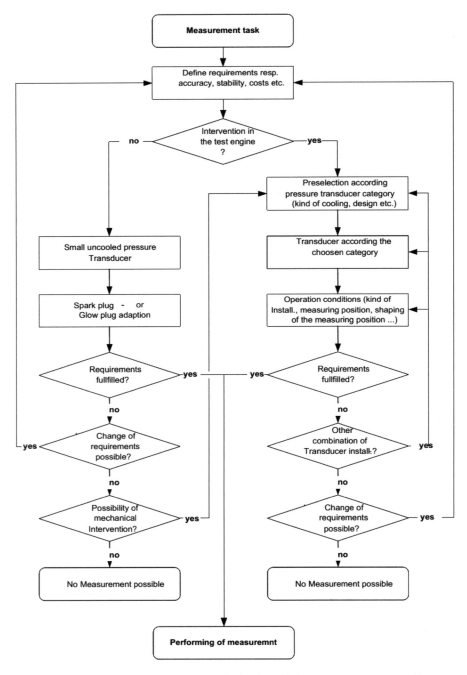

Fig. 2.46 Flowchart for pressure transducer selection for cylinder pressure measurement (Courtesy of AVL)

nonintrusive installation, the miniature uncooled pressure sensor is typically selected and mounted in glow plug or spark plug adapters depending on the engine. With intrusive installation, there is the choice of installation position. The design of pressure sensor is also affected by the available space for mounting and working environment of the sensor during an engine test. The information about the extent of the effects and the metrological properties of the selected transducer should be used to check whether the requirements can be fulfilled and, if so, the measurement can be started.

2.3 Alternatives to Piezoelectric Pressure Transducer

Typically, an ideal transducer for cylinder pressure measurement should have the following features: (1) high sensitivity, accuracy, and linearity; (2) absolute pressure measurement; (3) small size; (4) high natural frequency; (5) stable properties, irrespective of temperature, heat flow, and deformation; and (6) minimal number of components and interfaces in the measurement chain [15]. Presently, there is no single transducer or technology which fulfills all these requirements for all applications. Piezoelectric transducers are most commonly used for combustion pressure measurement in reciprocating engines, which is discussed in Sects. 2.1 and 2.2 in detail. Cost-effectiveness and demand for real-time pressure monitoring on production engines govern the development of alternative sensing technology. Several alternatives of the piezoelectric sensor are investigated over a long period for determination of combustion parameters such as ion current sensor [35–49], optical sensors [49–53], strain gauge-based sensor [54], piezoresistive sensor [17], angular speed [55], etc. Some of the important alternatives are briefly discussed in the following subsections.

2.3.1 Ion Current Sensor

The most practical realization of an ion current sensor is by use of the spark plug particularly in spark ignition (SI) engines. Using the ionization current method in conjunction with the spark plug is an alternative to the cylinder pressure sensor measurement. The ionization current contains information about the combustion process, and it reflects many parameters of the combustion process. The main challenge is to determine the ionization current properties that are useful for electronic engine control and how to extract them [41]. The combustion of fuel inside the engine cylinder produces ions and free electrons. A current can be generated (by applying electric field) and detected by locating two electrodes in the combustion chamber and applying a low DC potential difference between them. The benefit with ionization currents is the possibility of using a conventional spark plug as a sensor. The ignition system must only be slightly modified to allow electrode polarization

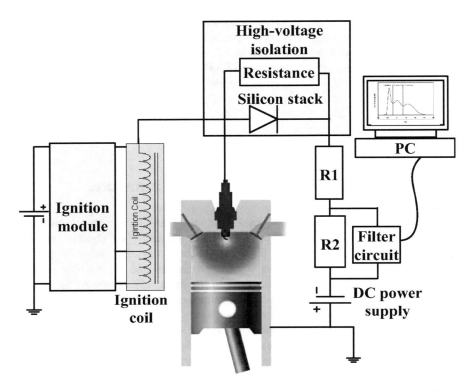

Fig. 2.47 Schematic diagram of ion current measurement setup on a reciprocating engine [35]

and current measurement. Figure 2.47 presents a schematic diagram of typical ion current measurement setup in an engine. In this arrangement, the spark plug acts as an ion current sensor (Fig. 2.47). The central electrode of the spark plug is used as the positive bias for the measurement of the ion signal produced in the cylinder. A bias voltage of 237 V across the spark plug electrode and a resistance of 241 kΩ are applied to obtain high strength ion signals. The ion current signal is acquired by dividing the 241 kΩ into the voltage drop across the resistor [35]. The ion current signal from spark plug has typically three main phases as depicted in Fig. 2.48. The first phase is called ignition phase, which initiates with the coil loading and finishes at the end of the spark. The flame front phase (second phase) appears during the displacement of the flame front in the spark plug electrode, and the post-flame phase (third phase) reveals the burnt gases behind the flame front [36]. The peak of the curve in the flame front phase is related to the combustion process in the flame kernel. It depends on the front flame propagation and ion probe position within the cylinder. The post-flame phase is related to the thermodynamic conditions in the burned gas. The ionization during this phase is due to the high temperature inside the engine combustion chamber. It is generally assumed that the dominating source for the formation of free electrons and the positives ions during post-flame phase is nitric oxide (NO) [39].

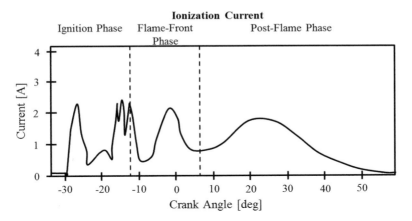

Fig. 2.48 Typical ion current curve with its three characteristic phases [36]

The investigation of ion species responsible for producing the ion current has been conducted [39]. The following chemi-ionization formation and recombination reactions are mainly responsible for ion current during combustion of hydrocarbon fuel. The ion formation reactions are presented as R1 to R3.

$$CH + O \rightarrow CHO^+ + e^- \tag{R1}$$

$$CHO^+ + H_2O \rightarrow CO + H_3O^+ \tag{R2}$$

$$CHO^+ + NO \rightarrow NO^+ + HCO \tag{R3}$$

The initiation of ion formation by CH radical (R1) has been experimentally established on a gasoline engine [42], and no ion current was detected for engine operation on hydrogen. The reaction (R1) is slow compared to the charge transfer reactions (R2) and (R3). The reaction (R2) is dominant in lean and slightly rich hydrocarbon-air flames and produces H_3O^+ ions. In rich and sooty hydrocarbon-air flames, reaction (R4) is dominant and produces $C_3H_3^+$ ions.

$$CH^+ + C_2H_2 \rightarrow C_3H_3^+ + e^- \tag{R4}$$

In sooty flames, charge transfer may occur between small ions such as $C_3H_3^+$ and H_3O^+ and large polynuclear aromatic hydrocarbons. The ionization potential of these large molecules is small enough so that the charge transfer is thermodynamically favorable [39]. In another study, the reactions of formation of positive ions of $C_2H_3O^+$, CH_3^+, and $C_3H_3^+$ are found in hydrocarbon flames along with H_3O^+ and CHO^+ [43]. By comparing the concentration and the mobility of these ions, the hydronium ion (H_3O^+) is recognized as the main source for producing the ion current signal during combustion measurement [37]. In addition, negative ions can be formed and contribute to the ion current. The ion recombination reactions are presented by reaction (R5).

$$H_3O^+ + e \rightarrow 2H + OH \tag{R5}$$

During post-flame period, thermal ionization takes places. The formation of ions at high temperatures in close to stoichiometric hydrocarbon-air flames is mainly by the reaction (R6). The destruction of NO^+ ions is by reaction (R7). In fuel-rich hydrocarbon-air flames, a fraction of the large hydrocarbon molecules and soot particles, both presented by R, can be ionized directly by reaction (R8) [39].

$$NO + M \rightarrow NO^+ + e + M \tag{R6}$$

$$NO^+ + e \rightarrow N + O \tag{R7}$$

$$R \rightarrow R^+ + e \tag{R8}$$

The fuel properties and engine operating conditions also affect the ion current signal. Typically, local information is obtained using the ion current method, but local measurement can be sufficient when the charge is homogeneous [10]. - Figure 2.49 depicts the numerically predicted distribution of hydronium (H_3O^+) ion in homogeneous charge compression ignition (HCCI) combustion chamber. In this case, the charge is prepared using double fuel injection, and injection ratio (Inj_{ratio}) of fuel injected in both injection events is varied. The figure shows the ion distribution for two injection ratios, and in the injection ratio 0.2, higher quantity fuel is injected during the compression stroke. For CA10 (crank angle corresponding to 10% heat release) position, the ion concentration is more heterogeneous for injection

Fig. 2.49 In-cylinder H_3O^+ ion distribution under different injection ratio conditions in HCCI combustion [37]

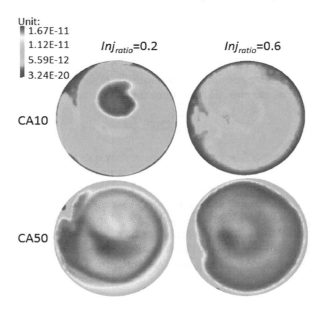

ratio 0.2 in comparison to injection ratio 0.6 (more premixed case). At the CA50 position, the ion concentration is distributed significantly for both operating conditions (Fig. 2.49). The numerically obtained data are in good agreement with the signal amplitude variation experimental engines. Therefore, ion-sensing system can be optimized by this tool [37] for better prediction of combustion parameters for effective combustion control and diagnostics of the engine.

The often cited advantages of ion sensing are that no additional sensor and no additional bore are required on engine head. No additional sensor requirement advantages are negated to some extent by the practical difficulty of packaging the current sensing registers since the ignition system itself has to be modified. The main disadvantage of the ion current sensor is that it measures local properties of the combustion products, rather than the global properties of the combustion gases in the entire cylinder [39, 41]. Additionally, both the ion current shape and magnitude depend on the (1) design of the sensor, (2) its location in the combustion chamber, and many (3) engine operating parameters such as A/F ratio, speed, load, and EGR. Since these parameters vary from one engine to another, strategies developed for one engine cannot be used for other engines particularly if the strategies are based on absolute values [41]. Another demerit of the ion current sensors is their inability to detect small cylinder-to-cylinder variations in multicylinder engines (particularly when the engine is operated with EGR) [39].

Ion current sensors appear to be quite resilient to electrode soiling. However, caution should be exercised since any electrically conducting deposits will permit current leakage. Operation of the spark plug is, on occasion, also compromised by fouling. Current leakage measurement shows that plug fouling by soot and/or water condensation takes place following a cold start [7]. The ion current signal typically appears in three parts (the interference of the ignition energy of the charge, interference of the spark plug ignition energy release, and combustion-related ion current) due to the inability to isolate the signal interference caused by ignition (Fig. 2.50). Sometimes a third peak also appears (Fig. 2.50), which is also created by spark plug ignition energy release. This is discovered by the nonidealities of components in the circuit that can be optimized by using more ideal components. The effective ion current signal from the combustion-related part can be used for combustion analysis. Figure 2.50 also depicts that the ion current signal strength is quite low during misfire condition relative to normal combustion conditions. However, the signal can be used for determination of engine combustion by careful calibration and validation [38].

The ion current sensing spark plug frequently serves as both sensor and igniter; this double duty is purely for convenience. It is not essential, and, where appropriate, the igniter function can be discarded, or an additional electrode can be provided in spark plug purely for ion sensing. Figure 2.51 shows the different methods for ion current sensing in reciprocating engines. The figure shows that ion current can be measured either by stand-alone ionization sensor or use the standard spark plugs. The spark plug can either have the dual function of sensor and igniter or additional electrode for ion current sensing. Using spark plug as a sensor, the position of sensing is fixed, while with stand-alone ionization sensor, the sensing position in the combustion chamber can be selected/defined by the user.

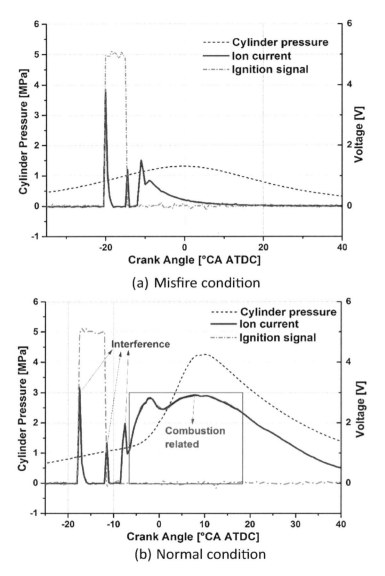

Fig. 2.50 Typical ion current signal in misfire and normal operating conditions in spark ignition engine [38]

Ion current sensing based on the conventional spark plug has certain additional limitations. For example, the electrode is optimized for spark discharge, rather than ion current measurement. The ion current measurement is limited to a single spatial location unless more sensors are installed. Moreover, the ion current signal is affected by the spark current during the sparking event. Use of a multielectrode spark plug which has three independent central electrodes is proposed to address

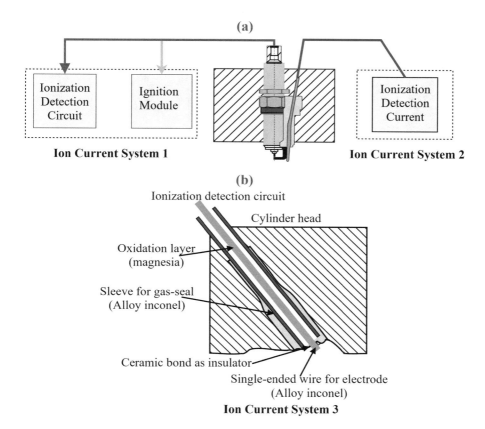

Fig. 2.51 Different methods of ion current sensing (Adapted from [41, 44])

some of the limitations of conventional spark plug [40]. Two of the electrodes are dedicated for ion current measurement, while the third electrode is used for spark discharge. Compared to the conventional spark plug, the ion signal COV of CA5 and CA50 for the multielectrode plug is a closer match to the overall trend and magnitude determined from the cylinder pressure signal. For measuring the combustion quality in the production SI engine, ring shaped gasket ion sensor are also proposed [45].

The ion current signal is normally too weak particularly at low engine speed and lower engine load. The stability of ion current signal under a wide range of engine speed and load should be solved before its industrial application. The ion current signal shape is also dependent on the location of ion current sensor because it provides the local information. Figure 2.52 illustrates the effect of ion current sensor location on the shape of the signal in SI engine. In the ion current measurement with a spark plug or its surrounding, there are two peaks in the signal waveform. It is also reported by several studies that the ion current traces in the gasoline engine have two peaks, under most operating conditions. It is well accepted that the first ion peak is

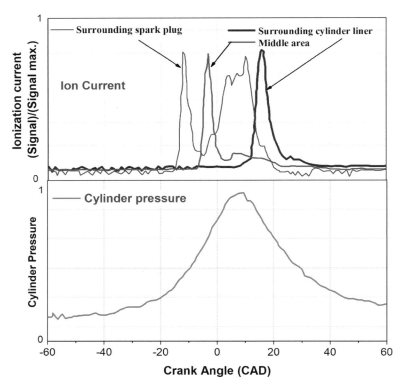

Fig. 2.52 Typical ionization current signal with the location of the sensor (Adapted from [44])

the result of chemi-ionization produced in the reaction zone of the flame kernel that is present in and around the electrode gap at that time. The amplitude of the first peak depends mainly on the equivalence ratio (Φ) of the mixture, which affects the concentration of the ion-producing species (CH and O) and the temperature of the combustion products. The gas temperature also depends on the thermal capacity of the charge immediately after the discharge of the electric spark in addition to Φ. The thermal capacity of the charge depends on its density, composition, and specific heat of its components, all of which depend on the operating variable of the engine, such as throttle valve opening, engine speed, and EGR rate [39]. As the flame front moves away from the spark plug zone, the combination reactions cause the ion current to drop and reach a minimum, after which it increases again toward the second peak. The second peak (post-flame zone) is due to thermal ionization (breaking of N_2) caused by the increase in the temperature of the gases around the spark plug.

In the ion-sensing position away from the spark plug, the two ionization timings (flame front and post-flame zones) are superimposed on each other and not separable (Fig. 2.52). This superimposition is due to the delay added by the front flame propagation. The second peak, mainly related to the cylinder pressure, is unchanged, while the first peak, related to the presence of the flame, is shifted toward the second peak [41]. The remote sensing plug detects the flame front as ion current, caused by ionization within the gap of its electrodes, yielding this single sharp spike. Under light loads, the second ion current peak disappears, because of low NO^+ concentration. The first ion peak is much higher than the second ion peak and is most suitable for the feedback control of the engine [39].

Figure 2.53 illustrates the ion current traces in HCCI and CI engines. Figure 2.53a shows that the ion current in HCCI engine has only one peak, due to the homogeneity of the charge. It can be used to detect the start of combustion under the different

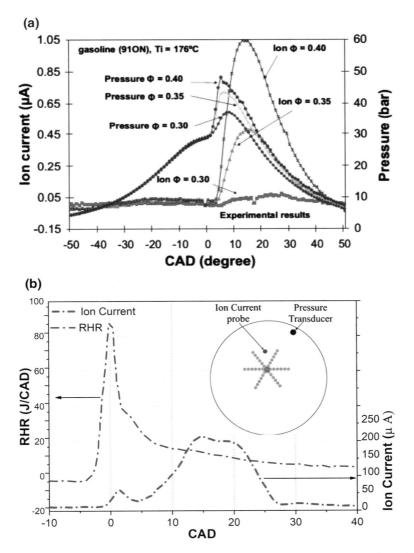

Fig. 2.53 (**a**) Ion current traces in HCCI engine at three equivalence ratios [46] and (**b**) heat release rate and ion current traces in compression ignition engine (Adapted from [39])

engine operating conditions. The strength of ion current also decreases as charge becomes leaner (Fig. 2.53a). Another study [47] also showed that the ion current amplitude is affected by both equivalence ratio and combustion phasing, but the ion current timing is only affected by the combustion phasing in HCCI engine.

Figure 2.53b shows the ion current trace in a diesel engine, which showed four peaks under medium and heavy loads. The possible sources of ionization are explained by considering the shape of the ion current trace, the location of the ion current probe relative to the fuel sprays, and the effect of the swirling motion in the original study [39] that is as follows. The first ion peak follows the peak of the premixed combustion fraction under different speeds and loads. The phase shift between the two peaks depends on the geometry of the ion current probe, its location in the combustion chamber particularly relative to the closest fuel spray, and the spray characteristics. The second ion current peak is caused by the turbulent premixed and diffusion flames carried to the probe by the swirling motion. In addition to gaseous components, these flames carry soot particles that contribute to the ion current. The third peak is due to flames and combustion products reflected on the combustion chamber walls. The fourth peak appears only under heavy loads and is considered to be caused by the third-in-line fuel spray that reaches the ion probe by the swirling motion. The ion current peaks in diesel engines may merge and produce one or more peaks under different probe locations and engine design and operating conditions. A more complete detail can be found in the original study [39].

Ion current sensors have been widely used in SI engines for in-cylinder combustion diagnostics; to determine the flame speed, A/F ratio, and mass fraction burned; to detect knocking, misfire, and partial burn; and to control the combustion stability. The use of ion current sensors in diesel engines has been limited compared to their use in spark ignition and HCCI engines [39]. Ion current sensors are found to be prone to soot deposition in diesel engines, where conducting tracks introduce offset to the signal [48]. The conductivity of soot deposits is not constant and changes during the engine cycle. This may be due to compression of soot by overlying gas. The short circuit can appear and then disappear as soot is re-entrained [7]. Soot particles can themselves carry charge [49], although the implications of this phenomenon for ion current sensing need more investigation.

2.3.2   Optical Sensors

For the study of engine combustion, optical methods are used in two ways. In the first way, luminosity-based sensors are used [50], while in the second way, optical techniques are used for pressure measurement [51]. The present section is focused on measurement of cylinder pressure by optical methods. Optical sensors were developed as an alternative to a piezoelectric sensor for combustion pressure measurement. Figure 2.54 illustrates the typical construction of optic-based pressure sensor. In this technique, the deflection (due to combustion pressure) of a measuring diaphragm is measured using a fiber optic-based method. A light source is used to

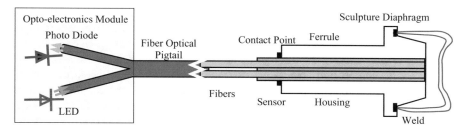

Fig. 2.54 Optical pressure sensor illustration (Adapted from [51])

send light to the metallic diaphragm. The light reflected from the deflected diaphragm is captured via an electronic optical receiver. The optical fiber system is
completely enclosed within the transducer body. The output from the transducer is
an electrical signal whose magnitude varies as a function of the change in cylinder
pressure. For a given diaphragm displacement caused by a pressure change, the
sensor response, i.e., sensitivity and linearity, can be adjusted by appropriate selection of optical fiber core diameters and layout [15]. The signal is nonmonotonic since
the deflection in the diaphragm changes, not just the total distance traveled by light
but also the area of intersection between the incident and received light as projected
onto the diaphragm [7]. The optical pressure sensors are reported in various designs
[52, 53], measuring the light reflected from a flexing hat-shaped metal diaphragm.

   To meet the requirements of high strength at combustion temperatures, the
diaphragm shape and material are selected having low creep and fatigue life of
well over a hundred million pressure cycles. To minimize the mounting space, the
measuring diaphragm should be as small as possible. The smallest sensor available
has a diaphragm diameter of approximately 1.7 mm [15]. The full-scale deflection of
the diaphragm is 10–15 μm. A critical aspect of the design is the hermetic seal since
the reflective properties of diaphragm surface are modified by oxidation [7]. The
advantage of the optical pressure-sensing technology is its compact size, which
permits easy integration into existing engine installations. Additionally, these sensors are highly immune to the effects of electromagnetic interference produced by
the engine's ignition and electrical systems.

2.3.3   Strain Gauges and Indirect Methods

Over the last few decades, advanced engine control systems have been developed
that use cylinder pressure signal as the primary feedback variable for closed-loop
combustion control. The cylinder pressure sensors have high degrees of accuracy
due to direct access to the combustion chamber. However, application of pressure
sensor on production vehicle is limited by cost, reliability, and packaging challenges
associated with the intrusive mounting of the sensor. The cylinder pressure sensors

operate in the harsh conditions (high temperature and pressure in the cylinder), which strongly limit the lifetime of the sensor. Several alternative sensor types have been proposed to address the cost, durability, and mounting issues of direct access pressure transducers through an indirect cylinder pressure estimation. No access to the combustion chamber is necessary along with the benefit of reduced engine machinery in the indirect cylinder pressure estimation methods. Additionally, cheaper measurement devices can be used, as they do not operate in high pressure and temperature environment. However, numerical modeling is necessary to correlate the measured quantity to the cylinder pressure trend [56]. The alternative indirect sensing-based method includes washer-based load cells under the head bolts, spark plugs, injectors, and head (head gasket or boss), crank angle velocity or torque changes, engine vibration, and acoustic emissions [56, 57]. Among these methods, the crankshaft speed fluctuations can be directly correlated to the in-cylinder pressure by a mathematical relation. This relation is achieved by calculating the equilibrium of the forces acting on the crank mechanism [58]. The other methods (vibrations, acoustic emissions, and forces acting on the head's components) are indeed related to the in-cylinder pressure, but it is very difficult to define a mathematical relation that expresses a direct relationship. In such cases, the pressure trend can be evaluated by using different approaches as frequency response functions (FRF), autoregressive moving average techniques (ARMA), and artificial neural networks (ANN) [56].

A study proposed a system for measuring the in-cylinder pressure using strain gauges by mounting strain gauges on external walls of the cylinder [54]. This strategy takes advantage of the fact that the cylinder block is usually made of aluminum or cast iron, hence metallic materials showing an elastic behavior. Unlike viscoelastic materials, such metallic materials are characterized by a linear relationship between stress and strain that does not depend on time; for this reason, any change in the in-cylinder pressure will cause an instantaneous change in the deformation of the cylinder wall which is not influenced by the time history of the deformation. Thus, by installing the strain gauges on cylinder wall at an appropriate position, combustion pressure information can be obtained. The strain gauge must be positioned near the combustion chamber, in correspondence of TDC. At TDC position of the piston, the component of the inertia forces acting perpendicularly on the internal surface of the cylinder is null because the piston roll is parallel to the cylinder axis. Consequently, the mechanical deformation of the cylinder surface is due only to the combustion pressure and is not influenced by the inertia forces when the piston is at TDC. Considering the high sensitivity of strain gauge to temperature variations, the researchers selected the placement of the sensor on the part of the cylinder wall that is in contact with the coolant flowing in the block. The study showed that it is possible to retrieve both the instant of time in which the maximum pressure occurs and the magnitude of the maximum pressure with high precision from the measurement of the maximum deformation of a strain gauge positioned along the circumferential direction in correspondence of TDC [54].

Another study used a strain washer placed on an engine stud for estimation of cylinder pressure [56]. It was possible to measure the stress due to the pressure variation into the engine cylinder during the combustion process. To correlate the in-cylinder pressure and the stress signal, a numerical model based on the neural network was developed. The results showed the good agreement between the direct and indirect pressure measurements [56].

The cylinder pressure can also be estimated back by measuring the engine vibrations with an accelerometer placed on the cylinder head or the engine block. The rapid pressure variations in the cylinder transfer part of the combustion energy to the engine block in the form of mechanical vibrations. This strategy of pressure estimation is intrinsically unable to reconstruct low amplitude stresses such as in the compression phase. Additionally, besides to the combustion information, the vibration signal contains also the effects of unwanted vibration sources as piston slap, mechanical unbalances, valves impacts, gear transmissions, and other stochastic forces. However, important excitation forces have their characteristic frequency. Therefore, it can be isolated by curtailing and filtering the signals, and several models are demonstrated to reconstruct the cylinder pressure by vibration measurements such as inverse filtering, time series model, and neural networks [59].

The acoustic emissions produced during the combustion phase can be used to reconstruct the pressure trend. Typically, a microphone is mounted on top of the cylinder block to acquire the acoustic data. During the engine operation, a wide range of noise sources is possible such as combustion, valve clatter, piston slap, turbulent gas flow, and many other fluids and mechanical events. Therefore, it is necessary to isolate the significant features and remove the events not related to the combustion for cylinder pressure estimation. The pressure signal reconstructed from the acoustic emissions is generally more accurate than that obtained from the vibrations because of high signal-to-noise ratio [60].

In-cylinder pressure can also be reconstructed from measured instantaneous crankshaft speed [55]. During engine combustion cycle, the crankshaft speed varies due to the variation of cylinder pressure, which varies the torque acting on the crankshaft and, thus, leads to acceleration. Therefore, it is possible to trace back the in-cylinder pressure by measuring the instantaneous crankshaft speed by phonic wheel or an optical encoder. The instantaneous variation in crankshaft speed depends on a number of cylinders and mean engine operating speed. The instantaneous variations in crankshaft speed decreases with the higher number of cylinders and higher mean engine speed, which limits this strategy of pressure estimation. However, it can be overcome by accurate modeling of the vehicle and the engine mechanism [55, 58].

2.4 Crank Angle Encoder

In modern engines, the cylinder pressure data is recorded using the high-speed data acquisition system. Any signal can be recorded with either internal time clock or external time clock (trigger). The internal time clock of data acquisition system can record the data at a fixed set time interval, which is defined by sampling frequency provided by the user. However, in external trigger mode, the user defines the points (instants) where data needs to be recorded by giving an external clock pulse. For the cylinder pressure measurement in reciprocating engines, typically external crank angle clock is used. The events occurring in the reciprocating engine are dependent on the crankshaft rotation, which governs the motion of piston and valves, fuel injection timings, spark timings, etc. It should be noted that these events are not dependent on time but depend on the angular position of crankshaft relative to a reference position (TDC or BDC). Since the instantaneous crankshaft speed varies with crank angle position, the time interval for traveling same crank angle duration varies depending on the position of the piston, even in one rotation of the crankshaft. Therefore, it is important to record the data on crank angle basis.

For thermodynamic analysis, precise cylinder volume as a function of crankshaft position is calculated from engine geometry. There is a high requirement for precise crank angle measurement along with obtaining high accuracy cylinder pressure data for combustion analysis. Two commonly used methods for the crank angle measurement are (1) a crank angle encoder and (2) a teethed/slotted wheel. The most accurate method is to use a crank angle encoder as a trigger source to guarantee that each pressure data is recorded at a predefined crank angle position. Although this solution shows high angular accuracy, it cannot be applied in production engine due to the practical and price restrictions and reliability reasons. Therefore, wheel-based solutions are commonly used in production applications [27].

2.4.1 Working Principle and Output Signal

The encoder senses mechanical motion of the shaft and translates this information (position) into useful electrical signals. Shaft encoders are used to determine the location of movable machine members such as crankshaft in the engine for accurate positioning. A rotary encoder is a transducer used for converting rotary motion or position into a series of electronic pulses. Figure 2.55 illustrates the different types of

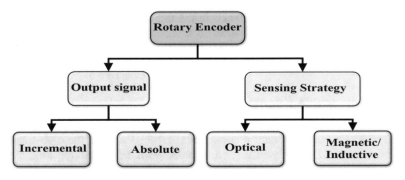

Fig. 2.55 Types of rotary shaft encoders

the rotary encoder based on their output signal and sensing strategy. Based on the output signal, the rotary encoders are of two types, namely, incremental encoders and absolute encoders. The incremental encoder has output as a series of square pulses, from which absolute position cannot be marked. On the basis of sensing strategy, rotary encoders can be divided as an optical encoder and magnetic/inductive encoders. Mostly, an incremental rotary encoder with optical sensing strategy is used for cylinder pressure measurement in reciprocating engines during laboratory tests. However, on production engines, inductive crank angle encoders are typically used.

Figure 2.56 shows the basic principle of incremental optical encoder along with its output signal. The basic components of a rotary encoder consist of light-emitting diodes (LED), a coded disk, and light detectors on the opposite side of the disk as illustrated in Fig. 2.56. The disk is mounted on the rotating crankshaft. Patterns of opaque and transparent sectors are coded into the disk. Incremental encoders work by rotating the coded disk in the path of a light source with the coded disk acting as a shutter to alternately shut off or transmit the light to a photodetector. This generates square wave pulses, which can then be interpreted into position or motion. The resolution of the encoder is same as the number of lines on the coded disk. Since the resolution is "hard coded" on the coded disk, optical encoders are inherently very repeatable and very accurate. There are two main signal outputs from encoder: the

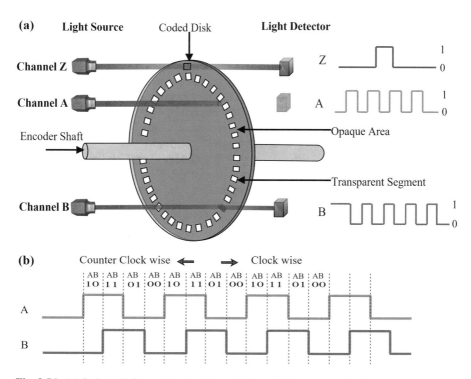

Fig. 2.56 (a) Optical shaft encoder principle and (b) quadrature output signal

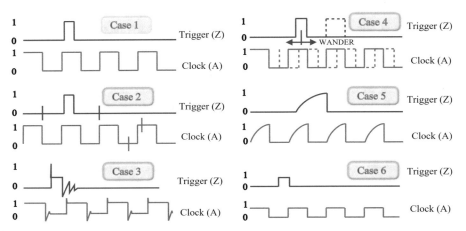

Fig. 2.57 Practically possible encoder signals during measurement (Adapted from [61])

first comprises a train of clock pulses (A or B); the second comprises just a single pulse (Z) per revolution of the crankshaft (Fig. 2.56a). An incremental rotary encoder, also known as a quadrature encoder, has two outputs called quadrature outputs. It consists of two tracks and two sensors whose outputs are called channels A and B. The signals from the A and B channels are a 1/4 cycle out of phase with each other (Fig. 2.56b). As the shaft rotates, pulse trains are generated on these channels at a frequency proportional to the shaft speed, and the phase relationship between the signals yields the direction of rotation. By counting the number of pulses and knowing the resolution of the disk, angular motion can be measured. The A and B channels are used to determine the direction of engine shaft rotation by assessing which channels "leads" the other. Third output channel, known as index (or reference) signal, yields one pulse per revolution, which is used in counting full revolutions and as a reference to define the zero position.

Figure 2.57 shows the different types of encoder signal encountered during cylinder pressure measurement [61]. The encoder signal presented in case 1 is the ideal situation for a good signal. The encoder signal can have noise spikes (case 2 in Fig. 2.57) due to poor signal connections and ground loop noise. The noise spikes can cause a false trigger, or poorly clocked data can be achieved. Ringing in the signal (case 3) can be due to high system inductance. This type of signal can also lead to false clocked data. As presented in case 4, wandering trigger and clock can be obtained due to loose encoder disk or faulty shaft coupling and encoder electrics. Rounding can also appear in signal (case 5) due to excessive system capacitance, which leads to inaccurate trigger and poor repeatability. The lower amplitude of trigger and clock signal (case 6) is typically achieved due to the poor power supply, bad connection, and lower impedance input device. This type of signal also results in improper triggering and clocking [61].

The data acquisition system must always record an engine cycle reference signal along with input from various engine sensors. This signal allows the transducer voltage to be related to the position of mechanical components of the engine. The cycle reference signal must be obtained from an engine crankshaft sensor. Typically, crank angle encoder produces one pulse (Z signal) in every rotation of encoder disk. However, in a four-stroke engine, the combustion cycle completes in two rotation of the crankshaft. Thus, the distinction is required for reference signal that in which stroke it occurs. Figure 2.58 presents different ways to generate the reference signal for four-stroke cycle. The first method (Fig. 2.58a) is to skip one of the two reference signals using a simple toggle-type flip-flop circuit. The reset switch is operated momentarily to select the desired output while observing the cylinder gas pressure [50].

The second approach (Fig. 2.58b) involves the cylinder gas pressure signal being supplied to a threshold detecting circuit which produces a digital "high" output when

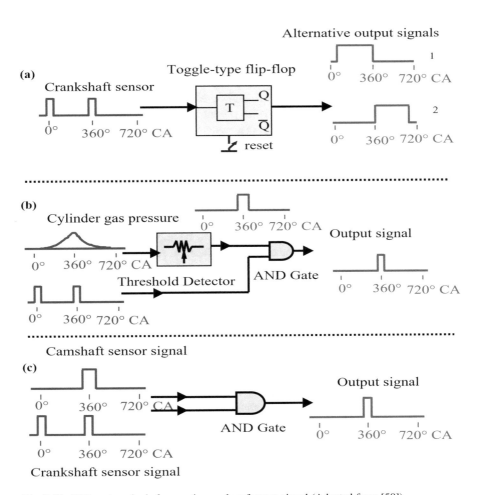

Fig. 2.58 Different methods for creating cycle reference signal (Adapted from [50])

gas pressure exceeds the threshold. The reference signal together with threshold detector output can then form the input of AND gate. This generates a "high" output at the TDC of the engine power revolution only. The third approach (Fig. 2.58c) is a similar strategy, where instead of threshold detector signal, camshaft sensor signal is provided. The cycle reference signal must be set accurately with respect to the position of the piston. The reference signal can also be dynamically set by using a proximity sensor [50].

Several automotive engines use 60-1 or 60-2 encoder wheels (60 teeth but missing 1 or 2 in a row) to create a reference position and control timing events. The signal is typically picked up by Hall effect sensor. Reference position can be determined by observing if the time from the last tooth is more than twice the time of the previous tooth. The wheel having the 60 teeth provides a resolution of 6 crank angle degree, and between the teeth, interpolation is used to find the exact position. Inductive crank angle measurement is one of the most common solutions for production applications because of its reliability, durability, and cost-effectiveness [27]. An inductive sensor consists of a permanent magnet with a ferromagnetic core surrounded by a coil, and it is usually mounted with an air gap to the ferromagnetic teethed or slotted flywheel. When teeth or slots pass through the sensor's magnetic source, the magnetic flow is changed, and an electric analog signal is generated. However, in practice, this system always produces considerable deviation from the actual crank angle position caused by mechanical and operating sources [27].

2.4.2   Resolution Requirement

For combustion pressure measurement and analysis, the crank angle resolution is one of the most important variables that needs to be adequately selected for better estimation of combustion parameter. The crank angle resolution is defined as the crank angle interval at which the pressure data are measured. Increasing the crank angle resolution (short CA interval) has three main advantages: (1) the bandwidth is increased allowing higher cylinder pressure variations to be detected and analyzed; (2) it increases the accuracy of identifying the crank angle position at which a certain absolute, or rate of change of parameter value occurs; and (3) the accuracy of the crank angle phasing (between pressure and volume curve) may be improved (if TDC determination using encoder), depending on the type of data acquisition and analysis system being used [62]. Figure 2.59 illustrates the effect of crank angle resolution on the measured cylinder pressure curve. The figure shows that the higher variations in pressure signal (also in terms of frequency response) can be detected by using higher-resolution crank angle encoder. Thus, higher resolution is preferred for frequency-based combustion analysis, and for lower encoder resolution, some of the frequencies present may not be detected in the pressure signal.

Typically, one of the two (A or B) clock signals from crank angle encoder are used for cylinder pressure measurement. The crank angle resolution defines the clock of data acquisition system. To increase the resolution of measurement, either

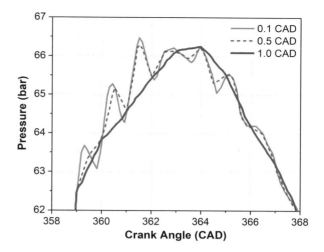

Fig. 2.59 Effect of crank angle encoder resolution on measured cylinder pressure (Adapted from [63])

both the clock signals can be used, or angle encoder of higher resolution needs to be selected/used. Figure 2.60 shows the method to increase the resolution of data acquisition system. Three different resolutions of the clock signal (R, 2R, and 4R) can be produced by decoding transitions of A and B by using sequential logic circuits. During acquisition of shaft encoder pulse used as an external trigger, the data can be acquired at different resolutions by selecting the position of data acquisition. When data acquisition is at rising or falling edge of the clock, then it is equal to the original resolution. The resolution of data acquisition is double with the acquisition at both rising and falling edges of the clock. Resolution can be further doubled (4R) by using both A and B signals and data acquisition at both the edges (Fig. 2.60). Higher resolution is required for the engine test conditions such as knocking, where higher data acquisition rate is required to understand the engine combustion phenomenon.

In contrast, increasing the crank angle resolution has a number of disadvantages: (1) it reduces the upper limit of the maximum engine speed and/or number of data channels due to sampling rate limit of data acquisition system, (2) it reduces the number of consecutive engine cycles that can be acquired due to system memory capacity limitations, (3) it reduces the speed of data handling and processing due to the increased volume of data, (4) it increases the data storage requirements, and (5) it increases the sensitivity to noise in some of the derived variable data [62]. A major problem with very high crank angle resolution is noise in some of the derived parameter data. This noise can be reduced by filtering, mathematical smoothing, and using a coarser calculation crank angle resolution. Additionally, there are certain calculations or applications that do not require such high crank angle resolution. Thus, adequate selection of crank angle resolution can be made based on application and measurement task. Typically, for IMEP calculation, friction measurement, heat release calculation, calculation of polytropic exponent, and direct analysis of pressure curve, the resolution of one crank angle degree (CAD) is suggested and

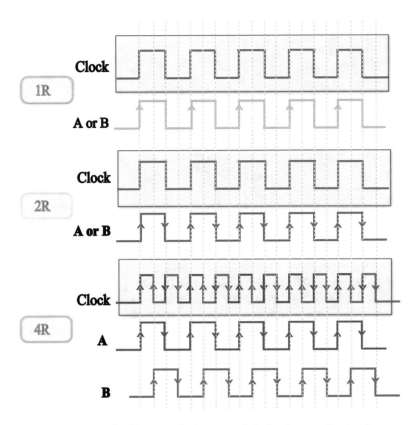

Fig. 2.60 Method for creating higher-resolution external clock using encoder signals

appropriate. However, crank angle resolution less than 0.2 CAD is suggested for knocking and combustion noise analysis [15]. A variable crank angle resolution technique which dynamically adjusts the crank angle interval over which the calculation is performed has been demonstrated to have good results [62].

Discussion/Investigation Questions

1. Discuss the applications where high accuracy and/or high repeatability of cylinder pressure measurement is required during engine development and testing process. Also, list the applications with moderate requirements on accuracy or repeatability for cylinder pressure measurement.
2. Discuss and list the design issues related to combustion sensors (piezoelectric pressure sensor, ion current sensors, and optical sensors) on automotive engine installed in laboratory test cell or a production vehicle.

3. Discuss the main advantages and disadvantages of piezoelectric sensors. List and explain the properties required of materials suitable for transduction elements in sensors. Fill the table showing the basic characteristics of sensing materials with notations low, very low, high, very high, and normal.

Transduction principle	Strain sensitivity	Threshold	Span-to-threshold ratio
Piezoelectric			
Piezoresistive			
Inductive			
Capacitive			

4. Discuss the advantages of direct pressure derivative measurement using current-to-voltage converter over charge amplifier-based measurement for cylinder pressure indicating by the piezoelectric transducer. Justify your answers.
5. Differentiate between piezoelectric and piezoresistive sensors. Explain why piezoelectric sensors are used for in-cylinder pressure measurement and piezoresistive sensors are used for manifold pressure and fuel line pressure measurement. Complete the following table showing different criteria for which a measurement technology is preferable to the others. Use the notation "↑" for preferred and "↓" for not preferred.

Criteria	Static measurement	Quasi-static measurement	Dynamic measurement	Pressure pulsation	Small sensor dimension	Wide temperature range
Piezoelectric						
Piezoresistive						

6. Explain the longitudinal, transverse, and shear cuts in piezoelectric elements and how it affects the charge output of the sensor.
7. Draw a typical cross-sectional diagram of a piezoelectric pressure transducer, and discuss all the major elements along with their functions. The piezoelectric sensor can be divided into two categories based on the output signal. Pressure sensors are available as charge output (PE) and voltage output (IEPE). Discuss the merits and demerits of PE and IEPE sensors.
8. Discuss the merits and demerits of pressure sensor mounting on engine cylinder head using direct installation and installation using adapters/sleeves.
9. Define the thermal drift during cylinder pressure measurement and write its causes. Discuss at least three methods to characterize the short-term thermal drift in measured in-cylinder pressure signal.
10. Discuss the methods for mitigating the short-term thermal drift from pressure sensors during in-cylinder pressure measurements in reciprocating engines.
11. Discuss the advantages of cooled pressure sensor over uncooled pressure sensor and uncooled pressure sensor over cooled pressure sensor designs. Write the applications when you will prefer cooled pressure sensor and applications when uncooled pressure sensor will be preferred.

12. Four possible positions (A, B, C, and D) for intrusive pressure sensor installation (red circle) are shown in Fig. P2.1 for a two-valve spark ignition engine with homogeneous combustion. Arrange four configurations (A, B, C, and D) in ascending order of preference for pressure sensor installation based on thermal as well as gas dynamics considerations. Justify your answer by discussing the merits and demerits of each sensor installation position.

Fig. P2.1 Engine head cross section with transducer installation at different locations

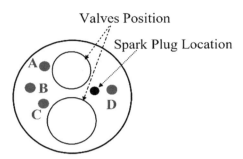

13. Two possible mounting positions (A & B) for intrusive pressure sensor installation (red circle) is shown in Fig. P2.2 for a two valve diesel engine with piston bowl combustion chamber. Write the preferred method for pressure sensor installation in this case and justify your answer by discussing the merits and demerits of both the positions. Discuss the effect of both measuring position by drawing a typical graph of the measured pressure signal.

Fig. P2.2 Engine head cross section with transducer installation at different locations along with piston bowl (Courtesy of Kistler)

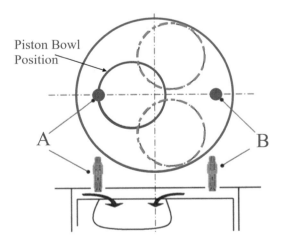

14. Figure P2.3 shows the typical variations of pressure difference due to thermal shock as a function of crankshaft position for different spark timings, engine loads (BMEP), engine speeds, and operating air-fuel ratios. Discuss the effect of

thermal shock on pressure measurement by explaining the trend observed with different engine operating parameters. Explain the reasons why maximum thermal shock occurs at 2000 engine speed instead of 1000 rpm (Fig. P2.3c). Thermal shock is also highest for stoichiometric ($\lambda = 1$) engine operations (Fig. P2.3d). Justify this observation with the explanation of combustion process.

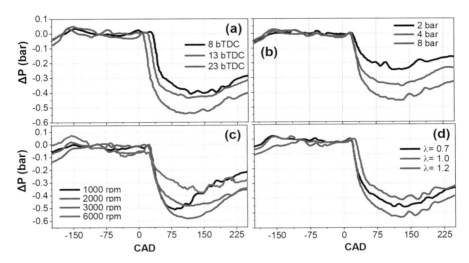

Fig. P2.3 Variation of pressure difference due to thermal shock at different engine operating conditions (adapted from [26])

15. Assume a recessed installation of the miniature pressure transducer on engine head (Fig. 2.35b). Calculate and plot the frequency of pipe oscillations in measured pressure signal for cylinder gas temperature of 500, 1000, and 2000 K as a function of measuring channel length. Measuring channel radius can be considered as 1.5 mm, and cavity volume (V_{cv}) ahead of the pressure transducer is 12 mm³. Discuss the effect of gas temperature on the frequency of oscillations at particular channel length.

16. Explain the trend in variation of the pipe oscillation frequency of the measured signal with advanced spark timing in a recessed installation of the pressure sensor. Explain how you will choose the channel length for pressure sensor mounting during knocking operating conditions.

17. Explain the working principle of ion current sensor. Write the merits and limitations of using the ion current sensor for combustion diagnostics and engine control. Discuss the advantages of using multielectrode spark plug over signal electrode spark plug for ion current sensing.

18. What are the typical ions formed during combustion, which are mainly responsible for ion current signal? Discuss the effect of air-fuel ratio, EGR, and

combustion temperature on the signal strength of ion current sensor in spark ignition engine.

19. Draw the typical shape of the ion current signal as a function of crank angle position for spark ignition (SI), compression ignition (CI), and homogeneous charge compression ignition (HCCI) engines. Justify the different peaks observed in the curve for different combustion modes.

20. Discuss the sources of error during cylinder pressure measurements. Specifically list the possible error sources at the level of the pressure sensor, crank angle encoders, TDC determination, transmission cables, signal conditioning, data acquisition, and data processing.

21. Discuss the reasons why crank angle encoders (not time-based signal recording) are used for cylinder pressure measurement in reciprocating engines. Define the resolution of crank angle encoder. Write the effect of crank angle resolution on cylinder pressure measurement and its further analysis for different combustion parameters.

22. Discuss the difference between the optical encoder and inductive-type encoder. In terms of accuracy, which one is preferred for cylinder pressure measurement? Write the sources of error in inductive crank angle measurement system.

23. Calculate the obtained sampling frequency for a four-stroke spark ignition engine that is running at 5000 rpm and uses the incremental crank angle encoder of two pulses per degree. Assume that you are using only "A" pulse of the encoder. Comment on whether this resolution is sufficient for knocking combustion analysis. Discuss the ways to increase the sampling frequency of the measurement.

References

1. Teichmann, R., Wimmer, A., Schwarz, C., & Winklhofer, E. (2012). Combustion diagnostics. In *Combustion engines development* (pp. 39–117). Berlin, Heidelberg: Springer.
2. Van Basshuysen, R., & Schäfer, F. (2016). *Internal combustion engine handbook-basics, components, systems and perspectives* (2nd ed.). Warrendale: SAE International. ISBN: 978-0-7680-8024-7.
3. Obert, E. F. (1968). *Internal combustion engines* (3rd ed.). Scranton, PA: International Textbook Co.
4. Amann, C. A. (1985). *Classical combustion diagnostics for engine research* (No. 850395). SAE Technical Paper.
5. Bueno, A. V., Velásquez, J. A., & Milanez, L. F. (2012). Internal combustion engine indicating measurements. In *Applied measurement systems*. London: InTech.
6. Bueno, A. V., Velásquez, J. A., & Milanez, L. F. (2009). A new engine indicating measurement procedure for combustion heat release analysis. *Applied Thermal Engineering, 29*(8–9), 1657–1675.
7. Eastwood, P. (2009). Combustion sensors. In *Automotive sensors* (pp. 115–157). New Jersey: Momentum Press.
8. Pischinger, R. (2002). *Engine indicating user handbook*. Graz: AVL List GmbH.
9. Gautschi, G. (2002). *Piezoelectric sensorics: Force, strain, pressure, acceleration and acoustic emission sensors, materials and amplifiers*. Berlin, Heidelberg: Springer.

10. Maurya, R. K. (2018). *Characteristics and control of low temperature combustion engines: Employing gasoline, ethanol and methanol.* Cham: Springer.
11. Luján, J. M., Bermúdez, V., Guardiola, C., & Abbad, A. (2010). A methodology for combustion detection in diesel engines through in-cylinder pressure derivative signal. *Mechanical Systems and Signal Processing, 24*(7), 2261–2275.
12. Shimasaki, Y., Kobayashi, M., Sakamoto, H., Ueno, M., Hasegawa, M., Yamaguchi, S., & Suzuki, T. (2004). *Study on engine management system using in-cylinder pressure sensor integrated with spark plug* (No. 2004-01-0519). SAE Technical Paper.
13. Maurya, R. K., Pal, D. D., & Agarwal, A. K. (2013). Digital signal processing of cylinder pressure data for combustion diagnostics of HCCI engine. *Mechanical Systems and Signal Processing, 36*(1), 95–109.
14. Kusakabe, H., Okauchi, T., & Takigawa, M. (1992). *A cylinder pressure sensor for internal combustion engine* (No. 920701). SAE Technical Paper.
15. Rogers, D. R. (2010). *Engine combustion: Pressure measurement and analysis.* Warrendale: Society of Automotive Engineers.
16. Claudio Cavalloni, C., & Sommer, R. (2003). *PiezoStar® crystals: A new dimension in sensor technology.* Winterthur: Kistler Instrumente AG.
17. Takeuchi, M., Tsukada, K., Nonomura, Y., Omura, Y., & Chujou, Y. (1993). *A combustion pressure sensor utilizing silicon piezoresistive effect* (No. 930351). SAE Technical Paper.
18. Gossweiler, C., Sailer, W., & Cater, C. (2006). *Sensors and amplifiers for combustion analysis, a guide for the user 100-403e-10.06.* © Kistler Instrumente AG, CH-8408 Winterthur, Okt.
19. Donovan, J. E. (1970). Triboelectric noise generation in some cables commonly used with underwater electroacoustic transducers. *The Journal of the Acoustical Society of America, 48* (3B), 714–724.
20. Rosseel, E., Sierens, R., & Baert, R. S. G. (1999). *Evaluating piezo-electric transducer response to thermal shock from in-cylinder pressure data* (No. 1999-01-0935). SAE Technical Paper.
21. Lee, S., Bae, C., Prucka, R., Fernandes, G., Filipi, Z., & Assanis, D. N. (2005). *Quantification of thermal shock in a piezoelectric pressure transducer* (No. 2005-01-2092). SAE Technical Paper.
22. Kuratle, R. H., & Märki, B. (1992). Influencing parameters and error sources during indication on internal combustion engines. *SAE Transactions, 101*, 295–303. Paper No-920233.
23. Randolph, A. L. (1990). *Cylinder-pressure-transducer mounting techniques to maximize data accuracy* (No. 900171). SAE Technical Paper.
24. Stein, R. A., Mencik, D. Z., & Warren, C. C. (1987). Effect of thermal strain on measurement of cylinder pressure. *SAE Transactions, 96*, 442–449. Paper No-870455.
25. Soltis, D. A. (2005). *Evaluation of cylinder pressure transducer accuracy based upon mounting style, heat shields, and water cooling* (No. 2005-01-3750). SAE Technical Paper.
26. Rai, H. S., Brunt, M. F., & Loader, C. P. (1999). *Quantification and reduction of IMEP errors resulting from pressure transducer thermal shock in an SI engine* (No. 1999-01-1329). SAE Technical Paper.
27. Storm, X., Salminen, H. J., Virrankoski, R., Niemi, S., & Hyvonen, J. (2017). *Analysis of cylinder pressure measurement accuracy for internal combustion engine control* (No. 2017-01-1067). SAE Technical Paper.
28. Davis, R. S., & Patterson, G. J. (2006). *Cylinder pressure data quality checks and procedures to maximize data accuracy* (No. 2006-01-1346). SAE Technical Paper.
29. Randolph, A. L. (1990). *Methods of processing cylinder-pressure transducer signals to maximize data accuracy* (No. 900170). SAE Technical Paper.
30. Higuma, A., Suzuki, T., Yoshida, M., Oguri, Y., & Minoyama, T. (1999). *Improvement of error in piezoelectric pressure transducer* (No. 1999-01-0207). SAE Technical Paper.
31. Roth, K. J., Sobiesiak, A., Robertson, L., & Yates, S. (2002). *In-cylinder pressure measurements with optical fiber and piezoelectric pressure transducers* (No. 2002-01-0745). SAE Technical Paper.

32. Hsu, B. D. (2002). *Practical diesel-engine combustion analysis*. Warrendale, PA: Society of Automotive Engineers.
33. Patterson, G. J., & Davis, R. S. (2009). Geometric and topological considerations to maximize remotely mounted cylinder pressure transducer data quality. *SAE International Journal of Engines, 2*(1), 414–420. https://doi.org/10.4271/2009-01-0644
34. Bertola, A., Dolt, R., & Höwing, J. (2015). *The use of spark plug pressure transducers for engine indication measurements-possibilities and limits* (No. 2015-26-0041). SAE Technical Paper.
35. Chen, Y., Dong, G., Mack, J. H., Butt, R. H., Chen, J. Y., & Dibble, R. W. (2016). Cyclic variations and prior-cycle effects of ion current sensing in an HCCI engine: A time-series analysis. *Applied Energy, 168*, 628–635.
36. Laganá, A. A., Lima, L. L., Justo, J. F., Arruda, B. A., & Santos, M. M. (2018). Identification of combustion and detonation in spark ignition engines using ion current signal. *Fuel, 227*, 469–477.
37. Dong, G., Chen, Y., Li, L., Wu, Z., & Dibble, R. (2017). A skeletal gasoline flame ionization mechanism for combustion timing prediction on HCCI engines. *Proceedings of the Combustion Institute, 36*(3), 3669–3676.
38. Chao, Y., Chen, X., Deng, J., Hu, Z., Wu, Z., & Li, L. (2018). Additional injection timing effects on first cycle during gasoline engine cold start based on ion current detection system. *Applied Energy, 221*, 55–66.
39. Henein, N. A., Bryzik, W., Abdel-Rehim, A., & Gupta, A. (2010). Characteristics of ion current signals in compression ignition and spark ignition engines. *SAE International Journal of Engines, 3*(1), 260–281. https://doi.org/10.4271/2010-01-0567
40. Dev, S., Sandhu, N. S., Ives, M., Yu, S., Zheng, M., & Tjong, J. (2018). *Ion current measurement of diluted combustion using a multi-electrode spark plug* (No. 2018-01-1134). SAE Technical Paper.
41. Peron, L., Charlet, A., Higelin, P., Moreau, B., & Burq, J. F. (2000). *Limitations of ionization current sensors and comparison with cylinder pressure sensors* (No. 2000-01-2830). SAE Technical Paper.
42. Reinmann, R., Saitzkoff, A., & Mauss, F. (1997). *Local air-fuel ratio measurements using the spark plug as an ionization sensor* (No. 970856). SAE Technical Paper.
43. Starik, A. M., & Titova, N. S. (2002). Kinetics of ion formation in the volumetric reaction of methane with air. *Combustion, Explosion and Shock Waves, 38*(3), 253–268.
44. Kato, T., Akiyama, K., Nakashima, T., & Shimizu, R. (2007). *Development of combustion behavior analysis techniques in the ultra high engine speed range* (No. 2007-01-0643). SAE Technical Paper.
45. Yoshiyama, S. (2010). *Detection of combustion quality in a production SI engine using ion sensor* (No. 2010-01-2255). SAE Technical Paper.
46. Bogin, G., Jr., Chen, J. Y., & Dibble, R. W. (2009). The effects of intake pressure, fuel concentration, and bias voltage on the detection of ions in a Homogeneous Charge Compression Ignition (HCCI) engine. *Proceedings of the Combustion Institute, 32*(2), 2877–2884.
47. Vressner, A., Hultqvist, A., Tunestål, P., Johansson, B., & Hasegawa, R. (2005). *Fuel effects on ion current in an HCCI engine* (No. 2005-01-2093). SAE Technical Paper.
48. Kubach, H., Velji, A., Spicher, U., & Fischer, W. (2004). *Ion current measurement in diesel engines* (No. 2004-01-2922). SAE Technical Paper.
49. Kittelson, D. B., & Collings, N. (1987). Origin of the response of electrostatic particle probes. *SAE Transactions, 96*, 353–363. Paper No 870476.
50. Zhao, H., & Ladommatos, N. (2001). *Engine combustion instrumentation and diagnostics*. Warrendale, PA: Society of Automotive Engineers. ISBN: 978-0768006650.
51. Ulrich, O., Wlodarczyk, R., & Wlodarczyk, M. T. (2001). *High-accuracy low-cost cylinder pressure sensor for advanced engine controls* (No. 2001-01-0991). SAE Technical Paper.
52. He, G., & Wlodarczyk, M. T. (1994). Evaluation of a spark-plug-integrated fiber-optic combustion pressure sensor. *SAE Transactions, 103*, 498–505. Paper No-940381.

53. Poorman, T. J., Xia, L., & Wlodarczyk, M. T. (1997). *Ignition system-embedded fiber-optic combustion pressure sensor for engine control and monitoring* (No. 970845). SAE Technical Paper.
54. Amirante, R., Casavola, C., Distaso, E., & Tamburrano, P. (2015). *Towards the development of the in-cylinder pressure measurement based on the strain gauge technique for internal combustion engines* (No. 2015-24-2419). SAE Technical Paper.
55. Moro, D., Cavina, N., & Ponti, F. (2002). In-cylinder pressure reconstruction based on instantaneous engine speed signal. *Journal of Engineering for Gas Turbines and Power, 124* (1), 220–225.
56. Romani, L., Lenzi, G., Ferrari, L., & Ferrara, G. (2016). *Indirect estimation of in-cylinder pressure through the stress analysis of an engine stud* (No. 2016-01-0814). SAE Technical Paper.
57. Andrie, M. J. (2009). *Non-intrusive low cost cylinder pressure transducer for internal combustion engine monitoring and control* (No. 2009-01-0245). SAE Technical Paper.
58. Liu, F., Amaratunga, G. A., Collings, N., & Soliman, A. (2012). *An experimental study on engine dynamics model based in-cylinder pressure estimation* (No. 2012-01-0896). SAE Technical Paper.
59. Jia, L., Naber, J., Blough, J., & Zekavat, S. A. (2014). Accelerometer-based combustion metrics reconstruction with radial basis function neural network for a 9 L diesel engine. *Journal of Engineering for Gas Turbines and Power, 136*(3), 031507.
60. El-Ghamry, M., Steel, J. A., Reuben, R. L., & Fog, T. L. (2005). Indirect measurement of cylinder pressure from diesel engines using acoustic emission. *Mechanical Systems and Signal Processing, 19*(4), 751–765.
61. Atkins, R. D. (2009). *An introduction to engine testing and development*. Warrendale: SAE International. ISBN 978-0-7680-2099-1.
62. Brunt, M. F., & Lucas, G. G. (1991). *The effect of crank angle resolution on cylinder pressure analysis* (No. 910041). SAE Technical Paper.
63. Katrašnik, T., Trenc, F., & Oprešnik, S. R. (2006). A new criterion to determine the start of combustion in diesel engines. *Journal of Engineering for Gas Turbines and Power, 128*(4), 928–933.

Chapter 3
Additional Sensors for Combustion Analysis

Abbreviations and Symbols

BMEP	Brake mean effective pressure
CFD	Computational fluid dynamics
CRDI	Common rail direct injection
ECU	Electronic control unit
EGPS	Exhaust gas pressure sensor
EGR	Exhaust gas recirculation
FM	Frequency modulation
LDV	Laser Doppler vibrometer
MR	Magnetoresistive
NTC	Negative temperature coefficient
PM	Particulate matter
SOI	Start of injection
TDC	Top dead center
UEGO	Universal exhaust gas oxygen sensor
A	Orifice area
d	Orifice plate diameter
I	Current flowing through hot wire
R	Electrical resistance of the wire
V	Airflow velocity past the wire
$\dot{m}$	Mass flow rate
C_d	Orifice discharge coefficient
n_c	Number of cylinders
N_m	Minimum engine speed
V_s	Swept volume
ρ	Density of air
ΔP	Pressure drop across the orifice

© Springer Nature Switzerland AG 2019
R. K. Maurya, *Reciprocating Engine Combustion Diagnostics*, Mechanical
Engineering Series, https://doi.org/10.1007/978-3-030-11954-6_3

3.1 Exhaust and Intake Pressure Sensors

Gas exchange process (gas flow into and out of the engine) affects the global engine operation, not only in the field of acoustic emissions but also in the domain of its performance and pollutant emissions. Increasingly stringent emission legislation also demands low-pressure measurement for analyzing and optimizing the gas exchange process [1]. The engine combustion process is directly affected by the gas flow dynamics in the engine combustion chamber, which is governed by the gas exchange process. Improved cylinder charging and charge motion can be achieved by optimizing the dynamic behavior of the gas flow, which is performed by low-pressure measurement and analysis [2]. A more detailed combustion analysis can be performed by understanding the charging and charge flow characteristics. The low-pressure measurement along with high-pressure measurement (in-cylinder) and valve lift can be used to determine accurate heat release during combustion. Additionally, measured exhaust and intake pressure data are typically used in the experimental analysis (e.g., calculation of residual gas fraction and trapping efficiency) and to experimentally validate the simulation data and results (1D or CFD models).

The inflow and exhaust of the engine combustion chamber (gas filling and residual gas fraction) are affected by the minor differences of pressures in the exhaust and intake manifold. Gas exchange analysis requires precise information of the manifold pressure (phase and absolute pressure value). Several factors govern the absolute pressure value and the dynamics of pressure signal in the exhaust and intake manifold, which include mounting of the sensor (flush mounting, fitting, directly or in adapter), type and performance of sensor, the radial position of the sensor and distance to cylinder head, setting of zero point (typically in warm conditions), measuring devices, and data acquisition [1]. Typically measuring inaccuracy less than 10 mbar is required for every engine test conditions. Inaccuracy of greater than 10 mbar leads to a significant error in estimation of the gas exchange phase because there exists a relationship between the gas mass flows and the absolute pressure values. Thus, the pressure must be measured by using high-precision sensors with highly sensitive measuring elements. To ensure the best possible accuracy of the measured data, any electrical (ignition, grounding, etc.) and physical (tension, temperature, gas dynamics, etc.) effects on the output signal need to be avoided.

The piezoresistive sensors are typically used for low-pressure measurement (or gas exchange). The absolute and dynamic pressure in the engine manifold is measured by piezoresistive pressure transducers. The piezoresistive pressure sensor is based on the measurement of variation in the electrical resistance of a semiconductor under mechanical load. This effect is about 70- to 150-fold greater than the strain-resistance effect exploited by metal strain gauges, which makes it possible to manufacture smaller sensors with high sensitivity and natural frequency [3]. The measuring element of a piezoresistive pressure sensor consists of a silicon load cell with a thin diaphragm, in which four resistors are implanted and connected to form a Wheatstone bridge. The registers are arranged in such a way that part of the resistors

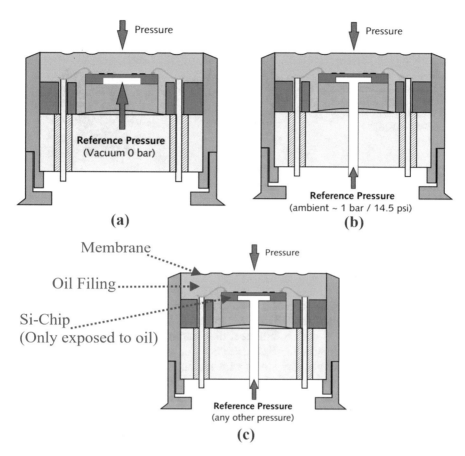

Fig. 3.1 Functional principle of the piezoresistive sensor (Courtesy of KISTLER). (**a**) Absolute pressure sensor, (**b**) gauge pressure sensor, (**c**) differential pressure sensor

is under tension and the rest are under pressure. The resistors are placed in the sensor chip that is integrated into the measuring cell as shown in Fig. 3.1. The pressure-dependent changes in the resistance are transformed into analog voltages by a separate electronic circuit. In recent versions, the pressure cell is integrated into the chip with "volume micromechanics" [4].

The piezoresistive pressure sensors measure the actual pressure relative to a reference pressure. The piezoresistive pressure sensors can be categorized as absolute, relative (gauge), and differential pressure sensors (Fig. 3.1). Absolute pressure sensors measure the pressure compared to a vacuum enclosed in the sensor element. Relative (gauge) pressure sensors measure the pressure in relation to the ambient air pressure. Differential pressure sensors measure the pressure difference between any two pressures. Thus, differential pressure sensors have two separate pressure connections. Depending on the application, absolute, relative (gauge), or differential pressure sensors can be suitably used.

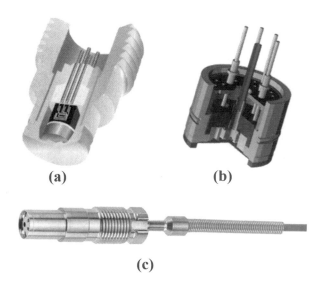

(a)   (b)

(c)

An involved packaging process is used to assemble the load cells into a pressure sensor. This involves attaching the load cell to a glass plate forming a base, placing it in a case, and sealing it with a thin steel diaphragm. The pressure is transmitted between the steel diaphragm and load cell via an oil cushion [3]. Figure 3.2 shows the actual piezoresistive sensors along with the cut sections with internal details. Piezoresistive sensors are suitable for measuring static pressures.

Instrumentation of the inlet and exhaust manifold has the specific challenges for achieving accuracy and durability. An appropriate sensor can be selected based on a number of application factors including thermal load, available space, soot content in gas, vibration, required accuracy, etc. Small pressure transducers are mostly appropriate for intake manifold pressure measurement, and it can be directly installed in the intake manifold. Exhaust gas pressure sensor (EGPS) essentially required to be actively cooled because exhaust gas temperatures can be as high as 1000 °C and even above depending on engine operating conditions. These sensors typically consist of steel diaphragm and filled with oil for transferring force to the measuring cell (media separation) because of harsh media (soot, combustion gases) exposure [1].

Exhaust gas pressure sensor also plays a key role in detecting the clogging degree of the particulate matter (PM) removal filter using a pressure signal and in sending feedback to the engine control system. The exhaust gas environment is extremely humid, and, thus, the condensed water of exhaust gas is supposed to be extremely acidic in nature. Additionally, soot can be easily accumulated under such a condition. Therefore, exhaust gas pressure sensors must possess three important factors: resistance to corrosion, resistance to icing, and resistance to clogging. A highly reliable (resistive to acid, icing, and clogging) against the exhaust gas environment and compact size sensor is developed by novel soft gel with high resistance to acid and applying the gel for the original back-surface sensing structure [5].

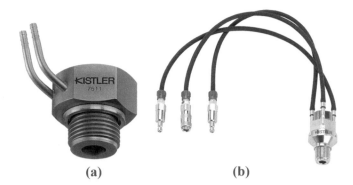

(a)   (b)

Fig. 3.3 Typical (**a**) cooling adapter and (**b**) cooled switching adapter for pressure measurement for gas exchange analysis (Courtesy of KISTLER)

The pressure sensor for gas exchange analysis has the following requirements: (1) sensor should not affect the gas exchange or combustion process; (2) sensor can be easily mounted; (3) sensor should be small size; (4) sensor should not be affected by mounting position, temperature, or vibration; and (5) sensor should have unlimited service life, and their characteristics should not change with time [3]. These factors can be considered for selection of the pressure sensor for gas exchange analysis.

Ideally during the measurement, the sensor temperature should be maintained to constant value for preventing shift in the thermal zero point. Additionally, to ensure no variation in zero-point value, the setting of zero point of the sensors can be done simultaneously. Particularly for exhaust sensors, cooling adapters are compulsory to protect the overheating of measuring element because of extremely high exhaust gas temperatures at some of the engine operating conditions. Figure 3.3a shows the typical cooling adapters for exhaust mounting of the pressure sensor. Adapters for mounting the sensor can be used with inlet and exhaust measurements, and it helps to improve the quality of measured data. A typical switching adaptor is shown in Fig. 3.3b, which fulfills some useful functions. Switching adapter acts as a damping adaptor and effectively isolates the sensor from structure-borne vibrations, improving the quality of the output signal. Additionally, it provides cooling water channels to maintain the sensor within operational temperature limits. This feature can be useful for pressure measurement in intake as well as for exhaust manifold. Modern engines with exhaust gas recirculation (EGR) and inlet charge boosting can operate with high charge temperatures in the intake manifold also. Thus, water cooling protects the sensor and stabilizes the signal, preventing drift and significant zero offset [2]. The switching adaptor also prevents the sensor exposure to hot gas for longer than the measurement time, which extends the life of the sensor.

The sensors should be located as close as possible to the engine valves for accurate measurement. This position of the pressure sensor prevents the superimposition of gas-dynamics effects (i.e., traveling waves) on the measured pressure data. Additionally, the measuring face of the sensor should not have any dead volume in

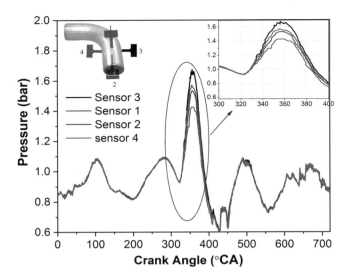

Fig. 3.4 Exhaust gas pressure signal for sensors placed on different planes at the same distance from the engine head (Adapted from [1])

front of it to avoid the pipe oscillations [2]. Flow effects in the exhaust manifold can influence the sensor signals due to the gas-dynamic oscillation in the exhaust. Hence, the installation of sensor must be selected at radial position with respect to the exhaust system geometry [1].

Figure 3.4 shows the exhaust gas pressure signal for sensors placed on different planes at the same distance from the engine head. Sensor output signals within the same measuring plane (1–2 or 3–4) show the obvious pressure differences due to the influences of the gas-dynamic effects. The differences in the pressure signals significantly occur only during the first blowoff wave [1]. The flow through the bend affected the pressure signal at position 3 (outer portion) to rise, while the pressure signal at position 4 (inner portion) is reduced considerably. The vertical position sensors 1 and 4 exhibit very small pressure differences, and, thus, no important effect is depicted due to the flow around the bend. The pressure difference between the inner and the outer walls results due to the curved flow in a bend. If required, the sensor needs to be mounted half way between the outside and inside radius for measuring the representative pressure curve [3]. The sensor should be mounted in a straight section of the manifold tube, where possible. The minimum pressure differences are obtained between the radially installed sensor and the mean pressure measured in the same plane in the case of sensors installed close to cylinder head. The sensor installed close to cylinder head has very low effect on signal differences due to the pipe geometry [1]. However, installation close to cylinder head could be critical due to availability of space. The mounting position close to the valve is no longer necessary in larger engines because of the low engine speed, no errors occur due to delays in traveling waves [3].

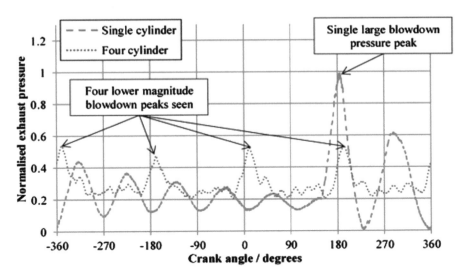

Fig. 3.5 Typical exhaust manifold pressure signal in a single- and four-cylinder engine at the same operating conditions [6]

The exhaust manifold pressure signals are significantly different for a single-cylinder engine and a typical multicylinder engine. Figure 3.5 shows the typical exhaust manifold pressure signal observed in a single-cylinder and four-cylinder engine for the same operating conditions. In the single-cylinder engine, a large single blowdown pulse is noticed at the exhaust valve opening position (Fig. 3.5). The four-cylinder engine shows four blowdown pulses which are at a much lower overall magnitude in comparison to the single-cylinder engine because of the cylinder-cylinder interactions. Optical exhaust pressure sensors are developed as an alternative to piezoresistive sensors to overcome the effect of operating temperature due to very high exhaust gas temperature [6].

Presently, tuned intake manifold is commonly used for naturally aspirated SI engines. Tuning of the manifold is performed to utilize such a manifold length which (together with intake valve opening and pressure wave exciting during the intake stroke) causes increase in the manifold pressure during intake stroke due to the interaction of pressure waves, thus improving the volumetric efficiency of the engine. In such cases, the manifold pressure signal curve will be dependent on the position of measurement and the engine operating conditions. Figure 3.6 presents the intake manifold pressure signal at different engine speeds. The lower harmonic orders have higher-pressure wave magnitude. It is also noted that not only is pressure wave amplitude magnitude important, the timing of the wave peak compared to intake valve timing is even more important, and it is better when the wave peak is located near the end of the intake stroke [7]. The impact of the distance between valves and measuring point is primarily evident in areas of rapid pressure changes. The bigger the distance, the bigger the time shift of measured signals [3].

Fig. 3.6 Comparison of
measured intake pressure
signal for different engine
speeds at 100 mm upstream
intake valve and inlet branch
length 500 mm (Adapted
from [7])

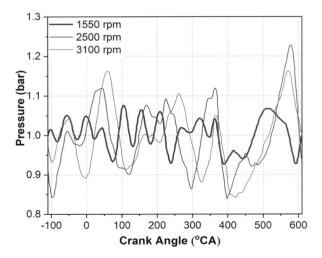

Accurate measurement of absolute pressure is one metrological requirement to enable correct calculation of the mass flow rates as an important part of a gas exchange analysis. The sensor must be highly sensitive to sufficiently resolve the relatively small pressure fluctuations in the inlet and exhaust manifolds. The gas fluctuations in the manifold are quite small in comparison to in-cylinder data, and, thus, high sensitivity is required to achieve a good signal-to-noise ratio.

Acceleration sensitivity is another important factor that needs to be considered for low-pressure measurement sensor. Particularly for applications in racing engines but also in production engines for vehicles, the low-pressure sensors must have the minimum possible acceleration sensitivity because the accelerations can be very substantial due to lower mass of inlet and exhaust manifolds. The accelerations of manifolds can add unwanted noise in the signal quality accordingly if it is not taken care by sensor [2, 3]. It is also advantageous to mechanically decouple the sensors from the engine with damping adapters, provided the sensors can nevertheless be mounted near the valves. Inlet pressure sensors are installed between the air cooler and cylinder in turbocharged engines. If charge temperature exceeds the sensor operating temperature or highest possible accuracy is required, sensors are installed in cooled mounting adapters (Fig. 3.3).

To lower the heating load from the sensor front, special protection screens are also used for reducing overheating. The main functions of a screen are (1) heat protection, (2) mechanical protection of steel diaphragm, (3) reduction of soot deposition in the sensor cavity, and (4) dissipation of pipe oscillation [1]. There is a distortion of the measured pressure signal depending on the design of screen and level of protection. The accuracy of low-pressure measurement also affects the cylinder pressure measurement and its further analysis, when pegging of the piezo-electric transducer is done by using a manifold pressure sensor [8].

3.2   Fuel Line Pressure Sensor

The dynamic behavior of the fuel injection system plays a key role in efficient engine operation. The fuel injection process affects the combustion process, which finally has an effect on overall fuel conversion efficiency. The time available for the fuel injection process is limited in modern high-speed diesel engines. The required fuel quantity needs to be precisely injected at the specified fuel injection timings for optimal engine performance. Thus, it is often necessary to measure the pressure in the fuel injection system of a diesel engine. Typically, fuel pressure is measured at three locations: the inlet and outlet position of the pipe conveying the fuel from the pump to injector and the fuel pressure in the injector itself [9]. The fuel injection pressure at these positions can often exceed 1000 bar in modern diesel engines. Additionally, the residual pressure between consecutive fuel injections at this position does not fall to zero (remains at a high level), due to nonreturn valve installed at the exit of the fuel pump.

Important governing factors for the selection of a fuel pressure sensor are as follows [9]. (1) Fuel pressure sensor must be small, which does not lead to significant cavities that can possibly interfere with the fuel pressure dynamics. (2) Installation of a fuel pressure sensor must not significantly alter the stiffness of the fuel line or injector. (3) Fuel pressure sensor must have sufficiently high-frequency response such that it can capture the rapid fuel pressure pulsation in the line. (4) Fuel pressure sensor must have a high natural frequency to avoid the resonant output oscillations.

Mechanical fuel injection systems (the generation before common rail fuel injection), the fuel pressure is measured between the cylinder head-mounted injector and the injector pump. Figure 3.7 presents ways for sensing the fuel line pressure by installing the sensor in the fuel line or in the injector body. A transducer boss (adapter) is welded onto the fuel pipe. A small-diameter communicating hole is then drilled through the adapter and the fuel pipe. An alternative boss (adapter)

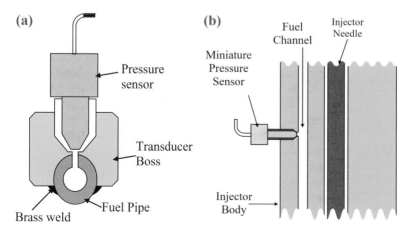

Fig. 3.7 Fuel pressure-sensing methods using sensor installation in (**a**) fuel line, and (**b**) injector body. (Adapted from [9])

design involves the clamping of the adapter onto the fuel pipe, and the fuel pressure
sealing is achieved by O-rings. Typical clamping-type adapter is shown in Fig. 3.8.

The fuel pressure measurement in mechanical fuel injection system technology is
straightforward. The fuel injection pressure is lower, and the line pressure is applied
only during the fuel injection process. Thus, the fuel pressure in the fuel line build
ups and decays in every engine cycle. Line pressure measurement can be significantly different from the combustion pressure measurement. Thermal loading is less
in fuel pressure measurement, but the hydraulic pressure applied is considerable
[2]. Piezoresistive sensor is typically used for measurement of fuel line pressure.
Strain gauge-based pressure transducer can be also be used for measurement of fuel
line pressure [9].

Figure 3.9 shows the variations of fuel line pressure at different engine load
(at constant speed 1800 rpm) and engine speed (at a constant load of 6.8 bar BMEP),
in an automotive engine with mechanical fuel injection system. Figure 3.9 depicts
that the fuel line pressure increases with increasing engine load as well engine speed.
At particular engine operating condition, the fuel line pressure builds and decays
after the injection duration (Fig. 3.9). Measured fuel line pressure can be used for the
evaluation of dynamic effects in the fuel injection system such as cavitation and
hydraulic pressure waves.

In modern diesel engines, the common rail direct injection (CRDI) technology
(latest generation) is employed, which poses new challenges for the measurement of
diesel fuel line pressure [2]. The CRDI technology affects the access point available
for measuring the fuel pressure. This technology is equipped with a common rail
system and electronically actuated injectors. The common rail maintains continuous
high-pressure fuel supply for the injectors which makes the fuel pressure independent of engine speed and load conditions. Thus, fuel quantity and fuel injection
timing can be independently controlled. This technology also minimizes the

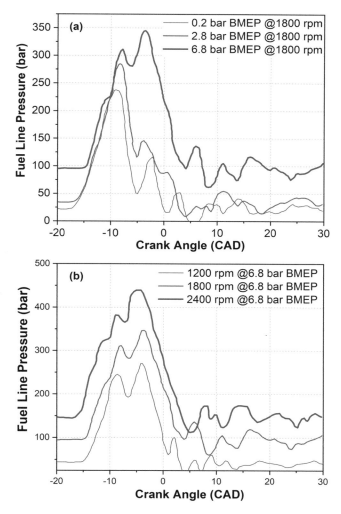

Fig. 3.9 Variations of fuel line pressure with (**a**) engine load and (**b**) engine speed, in an automotive engine with mechanical fuel injection system (Adapted from [10])

maximum pump torque requirement because high-pressure fuel delivery is achieved via common rail and peak flow rate does not require to coincide with the fuel injection event. Typically, peak pump torque is required at peak flow rate (high pressure) in a mechanical fuel injection system (Fig. 3.9). Additionally, complex injection strategy (multiple injections) can be achieved using CRDI technology to meet current and forthcoming emission legislation. The pilot fuel injection is typically used for improving the engine combustion noise. Thus, measurement of fuel pressure in CRDI is also important for understanding the pressure dynamics.

Fig. 3.10 (**a**) Rail pressure oscillations by needle opening and closing at different injection duration, (**b**) pressure drop due to fuel injection at different fuel injection pressures [11]

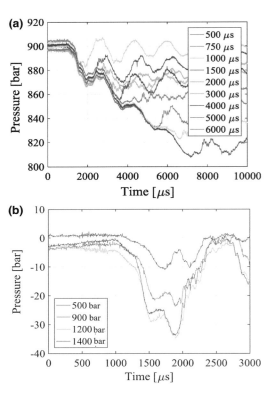

The quick closing and opening of the injector needle lead to high-frequency and high-amplitude fluid transients in the high-pressure common rail system [11]. While analyzing the measured fuel pressure data of CRDI system, understanding of the pressure wave propagation behavior is essential. The wave generated by one injector can affect other injectors and pipelines of the common rail system during pressure wave pressure propagation through a common rail circuit. In a high-pressure common rail system, the inertial effects of the fluid are typically more dominant than in the conventional hydraulic systems. Figure 3.10a shows the typical common rail pressure oscillations generated by fuel injections of different durations (500 μs–6 ms). The pressure oscillation either strengthens or attenuates at different fuel injection durations. Figure 3.10b shows the drop in common rail pressure by injection at different fuel injection pressures (500–1400 bar). The pressure drop is typically higher for higher fuel injection pressure (Fig. 3.10b).

The real CRDI system consists of a high-pressure pump, pressure-regulating valve, several injectors, and other components. The pressure oscillations to the fluid flow can be generated even at a distance from the original source. Thus, signal filtering is required to differentiate various sources of flow transients and to attenuate the pressure oscillation generated by injector needle closing and opening events. The detailed analysis of attenuated injection event can be used to estimate

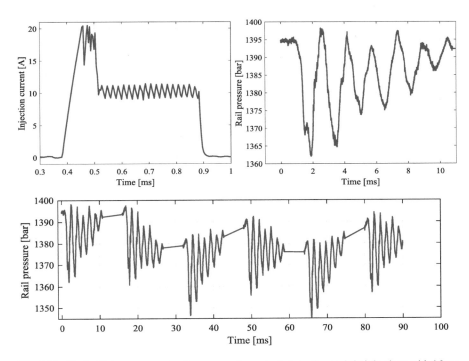

Fig. 3.11 Typical injector current, rail pressure after a single injection and six injections with 16 ms time between injections [11]

the fuel injection duration [11]. Figure 3.11 shows the typical pressure drop along with their oscillations because of the fuel injection event and corresponding control current of the injector. A six-injection event (corresponding to complete engine cycle) is also shown in lower part of Fig. 3.11. In every injection event, fuel from the common rail is injected into the combustion chamber by the injector, which suddenly drops the rail pressure (Figs. 3.10 and 3.11). This phenomenon as a feature of fuel injection and its pattern can be used to determine the duration of the injections.

3.3 Needle Lift Sensor

The movement of injector needle is frequently measured in a diesel engine to determine the start of fuel injection in the cylinder. Needle lift signal can also help in the identification of multiple injections in a particular engine cycle. The determination of the start of injection allows calculating the ignition delay, which plays an important role in diesel engine combustion. Additionally, needle lift and line pressure measurements are useful in the calculation of the fuel injection rate and identification of the dynamic behavior of fuel in the high-pressure fuel system at different engine operating conditions. The flow in the injector nozzle is strongly influenced by the internal geometrical factors of the nozzle and the motion of the injector needle. The needle motion inside diesel injector nozzle plays a decisive role for the spray dynamics [12–16]. The turbulence level in the nozzle flow affects the spray dynamics. The turbulence kinetic energy is mainly created (independent of presence or absence of cavitation) in the first stages of injector needle opening and in the last stage of closing due to the local acceleration of the flow at the nozzle inlet generated by the restricted passage. The turbulence level significantly decreases with the increase in inlet area and stabilizes at high needle lifts (over 150 μm) and starts increasing again when the injector needle descends below 150 μm, although to a lesser extent [13]. The lift profile of injectors affects the vapor penetration length in the chamber [12]. The injector needle lift profile has a significant effect on the amount of fuel injected and the momentum of the diesel jet.

Typically, the fuel injector is instrumented for the measurement of needle lift. Several production diesel engines operating on electronically controlled diesel pumps are equipped with instrumented injectors to determine the fuel injection timings [2].

Figure 3.12 shows the typical needle lift measurement method by means of variable inductance system, which operates on a frequency modulation (FM) principle. The inductive sensor coil is part of an electronic oscillator circuit that functions at 2 MHz frequency (approximately). The frequency of oscillator is partly determined by the inductance of the sensor coil. The armature rod penetrates further into the sensor coil when the injector needle rises, which leads to the variation in inductance of coils and its oscillation frequency. The variation in oscillation frequency is captured by an electronic circuit, which converts it into the variation of output voltage [9]. The electronics of the circuit is designed in such a way that output voltage varies linearly with the injector needle lift. Figure 3.12b shows the method for incorporating the sensor coil into the fuel injector. The frequency modulation method can also be used with a sensor that relies on the variation in the capacitance (Fig. 3.12a).

For the estimation of the injector needle lift position, other types of sensor including Hall-effect semiconductors and differential transformers are also used [2, 9]. The basic requirement is to generate a signal of sufficient accuracy and repeatability, with favorable signal-to-noise ratio and acceptable durability for the application.

Figure 3.13 illustrates the typical injector needle lift profiles at several injection-pulse durations for different fuel injection pressures. The start of injection (SOI) indicates the start of sending injection-triggering signal. The needle lift profile consists of two stages, namely, the opening stage and the closing stage. The needle lift profiles of different injection-pulse durations highly overlap in the opening stage of the injector needle. The period of needle opening stage nearly equals to the duration of injection pulse. The injector needle opening speed increases at a higher injection pressure, and the needle lift is also the higher. The needle dynamics is not altered by increasing the duration of injection pulse because duration only decides the timing of opening and closing of the fuel release valve. Figure 3.13b shows the

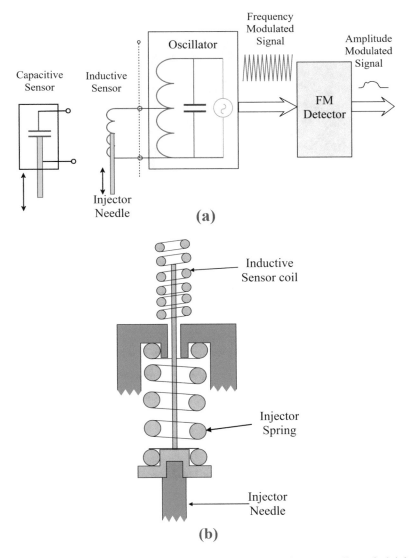

Fig. 3.12 (a) Typical needle lift-sensing methods, (b) installed sensor coil on fuel injector (Adapted from [9])

injector needle speeds for two fuel injection pressures. Abrupt velocity peaks appear at the beginning and closing of the needle movement (Fig. 3.13b). The injector needle opening speed is a function of the fuel injection pressure. The needle opening speed increases at the higher fuel injection pressure, and the needle closing speed is unaffected by fuel injection pressure (Fig. 3.13b) [15].

Figure 3.14 shows the variation of axial jet velocity (represents liquid-jet dynamics) with different needle lift profiles for two different fuel injection pressures. The

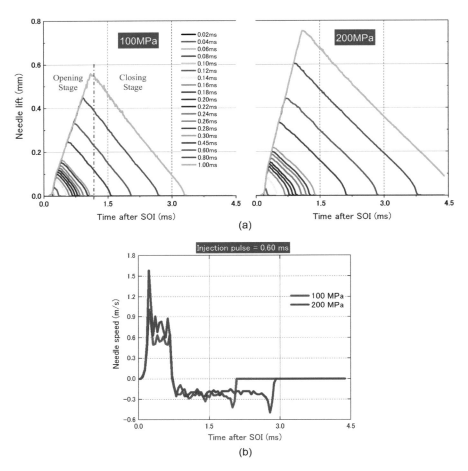

Fig. 3.13 Typical (**a**) injector needle lift profiles and (**b**) needle lift speed at different injection-pulse durations [15]

axial jet velocity profiles for different injection-pulse durations overlap with each other (Fig. 3.14). Axial jet velocity increases exponentially in the beginning at needle opening stage independent of the injection-pulse duration, and it reaches to a constant value after a particular needle lift value (Fig. 3.14b) for both the fuel injection pressures. Then after the needle reaches a certain critical height, the axial velocity becomes constant and steps into the steady state. The steady-state axial velocity increases with the increase in fuel injection pressure, but the trend of variation of axial velocity is similar (Fig. 3.14).

Typically, commercially available eddy-current sensors are used for measurement of injector needle lift position in modern electronic fuel injectors [17]. However, the repeatability and accuracy of the output signal are significantly affected by the presence of an electromagnetic disturbance. Modern electronic diesel injectors

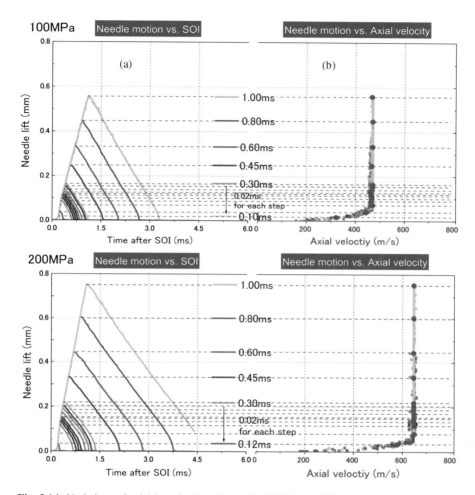

Fig. 3.14 Variations of axial jet velocity with needle lift for two different fuel injection pressure (**a**) needle motion with different start of injection (SOI), and (**b**) needle motion with axial jet velocity [15]

are typically actuated by providing high currents into a solenoid, which creates high electromagnetic fields. Additionally, the high levels of electromagnetic noise are produced by the several engine assistance devices; the measurement of injector needle lift can be very difficult with eddy-current type transducers. To overcome these issues, optical sensors are also developed for the measurement of needle lift of fuel injector [18, 19]. Figure 3.15 shows the typical optical transducer assembly on the test fuel injector. The optical sensor for needle lift measurement consists of a laser light emitter, a receiver, and a device used for modulating the intensity of the light reaching the receiver as a function of the position of moving element of the fuel injector. To get a clearance area that varied linearly with injector needle position, two rectangular windows are mounted on the injector body, while a third window is

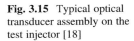

Fig. 3.15 Typical optical transducer assembly on the test injector [18]

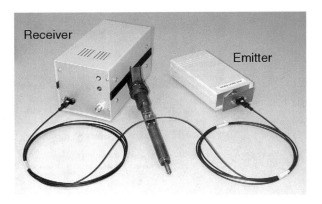

rigidly connected to the needle control piston and aligned with the fixed ones [18]. An optical fiber cable is used to transfer the laser lights between input and output. Light intensity is converted into an equivalent voltage signal that can be recorded and analyzed further for the determination of injector needle lift position.

3.4 Mass Flow Sensors

The mass flow measurement of air and fuel consumptions is an integral part of the combustion measurement and analysis of internal combustion engines. The measurement of air and fuel allows the calculation of various performance and combustion parameters such air-fuel ratio, combustion efficiency, fuel conversion efficiency, volumetric efficiency, residual gas fraction, etc. These parameters are helpful in analyzing the combustion quality and determination of optimal engine operations, and, thus, air and fuel mass flow rates are often measured during engine testing.

Figure 3.16 shows the different methods for airflow measurement in reciprocating engines. Typically, one of the two airflow rate measurements, i.e., instantaneous and quasi-steady, is required for engine research. The airflow is a highly unsteady process during the intake stroke of the engine. The airflow rate varies from zero to maximum value, and again comes back to zero within a combustion cycle (few milliseconds). Measurement of instantaneous airflow rate requires specialized fast-response measuring equipment. More often average airflow rate (quasi-steady flow rate) over the entire combustion cycle is required, which can be measured in a number of ways [9]. The systems for estimating the amount of air drawn in by the engine can be divided into measuring procedures based on either volume or mass basis.

The air-box method is the simplest method that is typically used for average airflow rate measurement (Fig. 3.16a). The air consumed by the engine from the air-box is filled by atmospheric air, which enters through a calibrated orifice meter or

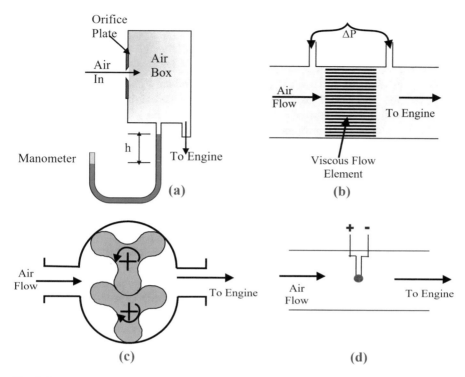

Fig. 3.16 Airflow measurement methods in reciprocating engines (**a**) air-box method, (**b**) viscous flow meter, (**c**) positive displacement flow meter, and (**d**) hot-wire flow meter (Adapted from [9])

a venturi. The volume of air-box needs to be sufficiently large to damp out the oscillating flow produced by the engine. The air mass flow rate ($\dot{m}$) by the orifice plate can be calculated using Eq. (3.1):

$$\dot{m} = C_{\mathrm{d}} A \sqrt{2\rho \Delta P} \qquad (3.1)$$

where C_{d} is the orifice discharge coefficient, A is the orifice area, ρ is the density of air, and ΔP is the pressure drop across the orifice. The Eq. (3.1) is applicable for incompressible flow, and it is better to limit the pressure drop (ΔP) below 1.2 kPa for good accuracy by selecting the appropriate design of orifice plate [9].

   If the pressure drop across the orifice plate is oscillating, then mass flow rate will also be oscillating and not steady. In such case, the accurate mass flow determination is difficult and leads to an error in air mass calculation. Therefore, a larger box volume in conjunction with orifice plate is required to eliminate the error due to the pulsating nature of the flow in the intake manifold of the reciprocating engines. The required volume of air-box for steady flow through the orifice can be estimated using Eq. (3.2) [9, 20]:

$$V_b = \frac{417 \times 10^6 n_s^2 d^4}{n_c V_s N_m} \tag{3.2}$$

where n_s is constant (with value 1 for two-stroke and 2 for four-stroke engine), d is orifice plate diameter (m), n_c is the number of cylinders, V_s is engine total displacement volume (m³), and N_m is the minimum engine speed of the measurement (rpm). It can noted that the volume of air-box is dependent on the orifice diameters, so both can be selected as per requirement. A significant disadvantage of air-box is a long tube that is often required to connect with the engine, which affects the breathing characteristics and performance of the engine [9, 21].

Viscous flowmeter is another meter that is another method of quasi-steady airflow measurement as illustrated in Fig. 3.16b. In this method, the air passes through a honeycomb of narrow parallel passages so dimensioned that the flow is viscous, the resistance of the element thus being directly proportional to the velocity within the working range. The meter requires calibration against a particular standard [20]. In principle, the flow-through instrument is laminar, and pressure drop across the meter is proportional to the airflow velocity or volumetric flow rate through meter. Typically viscous flowmeter is more suitable for use in pulsating flows than an orifice plate. The demerit of this method is its sensitivity to fouling of small flow area passages, which changes the calibration [9].

Airflow measurement using a positive displacement meter is illustrated in Fig. 3.16c. This method is largely suitable only for stationary engine operation because of their distinct inertia to change in airflow [4]. The accuracy of the positive displacement meters is not susceptible to pressure oscillations in the flow.

The most significant method of the air mass flow rate measurement is the hot-film anemometer, which can measure the instantaneous air mass flow rate. The functional principle of this method is illustrated in Fig. 3.16d. The wire is maintained at higher temperature by means of electrical supply. The heat loss rate can be calculated from energy supplied to the wire, which is also related to the air mass flow rate through the flowmeter by Eq. (3.3) [9]:

$$q = I^2 R = a + b(\rho V)^n \tag{3.3}$$

where q is the heat loss rate from the hot wire, I is the current flowing through hot wire, R is the electrical resistance of the wire, ρ is the air density, V is the airflow velocity past the wire, and a, b, and n are the constants estimated by calibration of the instrument. The hot-wire flowmeters can respond almost instantaneously to the variations in the airflow rate. The hot-wire thermal inertia should be very small for fast response, and, thus, the wire diameter must be very small (<100 μm) [9].

Continuous and discontinuous volumetric as well as gravimetric measurement methods are known for measuring the fuel consumption of internal combustion engines. The volume of fuel consumed by the engine is calculated in the volumetric measurement methods. The temperature-dependent fuel density must be taken into account for the determination of the mass of fuel consumed in case of volumetric measurement method. In contrast, the gravimetric measuring procedure directly calculates the mass of fuel consumed, and, thus, the additional uncertainty of determining the density is not required [4]. Several methods for fuel mass flow measurement are available, and the most suitable method for a particular application can be selected. The following factors can be considered for the selection of mass flowmeter: (1) volumetric or gravimetric, (2) the sensitivity to temperature and fuel viscosity, (3) absolute level of accuracy, (4) pressure difference required to operate, (5) wear resistance and tolerance of dirt and bubbles, (6) suitability for stationary/in-vehicle use, and (7) analog or impulse-counting readout [22].

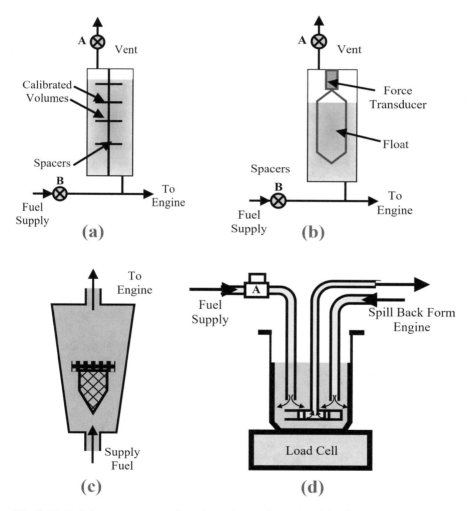

Fig. 3.17 Fuel flow measurement in reciprocating engines using (**a**) calibrated burette method, (**b**) cylindrical float based method, (**c**) rotameter based method, and (**d**) weighing (average gravimetric fuel consumption rate) method (Adapted from [9])

Figure 3.17 shows the different methods for average fuel flow rate measurement in reciprocating engines. The burette method of fuel flow rate for measuring the average fuel flow rate is illustrated in Fig. 3.17a. This method relies on the recording the time taken (by stopwatch) for consuming a particular amount of fuel by the engine. The burette is selected based on the choice of calibrated volumes in such a way that a larger calibrated volume be chosen at higher fuel consumption operating conditions. Figure 3.17b shows a method of measuring the average gravimetric fuel consumption rate. The advantage of this method is that it directly measures the mass of fuel consumption, not the volume. In this method, a cylindrical float is positioned inside a cylindrical vessel, and the buoyancy force is measured by a force transducer. The change in buoyancy force due to fuel consumption is directly proportional to the mass of consumed fuel [9]. A variable area flowmeter or "rotameter" can also be used (Fig. 3.17c), when great accuracy is not required. In this meter, a short conical float is free to move up or down a tapered transparent tube. The extent of rising of float provides an indication of the flow rate through the meter within an accuracy of a few percent. Figure 3.17d shows a weighing method for measuring the average gravimetric fuel consumption rate. The weight of fuel in the vessel at any instant can be measured using the load cell. The mass of fuel consumed by the engine over a period of time can be measured using this method.

Typical, gravimetric fuel flow measurement method includes the Coriolis and Wheatstone bridge-based devices according to the continuously measured flow principle. The method works according to the vessel principle, such as the drainage weight measurement and the weight measurement using burets and the weight principle. The continuously working mass flowmeters need the bubble separators to deaerate the approach flow and return flow. Gas bubbles in the fuel that have not been separated reduce the quality of the measuring signal and diminish the accuracy and dynamics of the system [4].

Figure 3.18 shows the schematic of the fuel mass flowmeter based on Coriolis effect. In this device, the fuel flows through the pipe sections that are electromagnetically stimulated to oscillate, with the result that the pipes are twisted because of

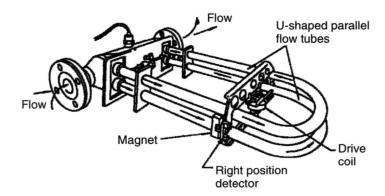

Fig. 3.18 Coriolis effect fuel flowmeter [22]

the Coriolis effect. The twist angle is calculated proportionally to the mass flow rate [4, 22]. Gravimetric methods of fuel mass determination are preferred over the volumetric methods for better accuracy of results.

3.5 Temperature Sensor

The engine management system uses several temperature inputs to improve the performance of the engine, control the emissions, and optimize the fuel conversion efficiency [23]. The most common applications for temperature sensing along with their typical measuring range are as follows: engine/coolant temperature $(-40...+130\,°C)$, intake air temperature $(-40...+120\,°C)$, engine oil temperature $(-40...+170\,°C)$, fuel temperature $(-40...+120\,°C)$, and exhaust gas temperature $(-40...+1000\,°C)$ [23, 24].

The coolant temperature sensor measures the coolant temperature, which is used by engine management to calculate the engine temperature for achieving the optimal engine operation. The intake air temperature sensor is installed in the intake tract. This signal is used for calculation of the intake air mass together with the signal from the boost pressure sensor. The intake air temperature is often integrated with the mass flow sensor or manifold absolute pressure sensor [23]. Additionally, the desired values for the various control loops (e.g., EGR, boost pressure control) can be adapted to the air temperature. Exhaust gas temperature measurement is an emerging application of temperature sensing due to aftertreatment technologies.

A wide range of temperature-sensing technologies is available such as resistive temperature transducers, thermistors, thermocouples, PN junction sensors, liquid crystal temperature sensor, heat flux gauges, etc. [23]. A temperature-sensing technology can be selected based on the application. Typically for engine temperature measurements, a temperature-dependent semiconductor measuring resistor of the NTC (negative temperature coefficient) is installed inside a housing, a sharp drop in resistance occurs when the temperature rises in NTC-type sensors. The measuring resistor is part of a voltage divider circuit to which 5 V is applied. Thus, the measured voltage through the measuring resistor is temperature-dependent. The engine management ECU estimates the particular temperature to every resistance or output voltage based on a calibration curve stored in it [24].

3.6 Valve Lift Sensor

The continuous development of internal combustion engines increasingly demands control on the gas exchange process. The gas exchange process is mainly governed by the valve train system, which influences the combustion behavior and power output of the engine. The necessary valve lift curve must be ensured within the whole operation range of the engine along with minimum friction [25]. The measurement of valve motion during the development of engine cylinder heads and in related component testing environments is typically required. Additionally, the fundamental research experiments may need the determination of valve motion in a running engine. The valve lift measurement is similar to the needle lift measurement, and,

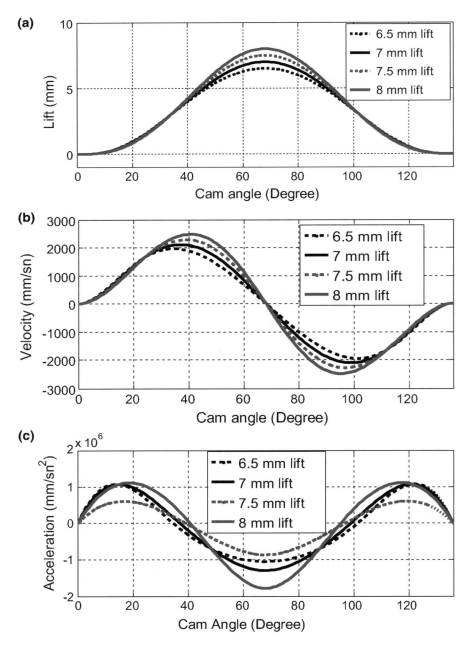

Fig. 3.19 Typical valve (**a**) displacement (lift), (**b**) velocity, and (**c**) acceleration curves as a function of cam position [26]

thus, similar technologies can be used for the measurement of valve displacement, namely, Hall effect, carrier frequency, and capacitive techniques with appropriate signal conditionings [2]. The recorded analog signals can be used to derive velocity and acceleration of reciprocating valve components. Figure 3.19 shows the typical valve lift, velocity, and acceleration profiles with respect to cam positions.

The analysis of the dynamic behavior of the valve train on motored test rigs has taken a firm position in the development of valve train components. However, influences caused by the gas exchange process, thermal effects, and additional issues of a complete engine under fired conditions are neglected with the motored approach. Thus, the differences between the ideal and real valve opening and closing together with real valve timing as well as detailed information about the gas exchange and the dynamic behavior of the valves need to be measured [25]. In order to measure the valve lift and velocity, typically a laser Doppler vibrometer (LDV) is used. The laser Doppler vibrometer is based on the principle of the detection of the Doppler shift of coherent laser light that is scattered from a small area of a test object. A study performed dynamic valve train investigations using LDV measurement on the motored test rig while on a fired engine using magneto-resistive (MR) sensors [25]. Magnetoresistive sensors are ideal for the position, speed, and angle detection. The signal of the sensors is generated by an external magnetic field with changing direction, which results into a change of the sensor's electrical resistance. If the sensor is mounted between a permanent magnet and a tooth structure of ferromagnetic material, the movement of this tooth structure deflects the magnetic field. The signal period characterizes the movement of the tooth structure. A linear relationship between the movement and the signal can be established for a particular tooth spacing.

3.7   Ignition Current Sensor

Ignition and combustion process in a gasoline or spark ignition engine are related by ignition signals. The timing of the ignition of the fuel-air mixture with respect to crankshaft position is critical for achieving the optimal fuel conversion efficiency. The ignition of the fuel-air mixture, the development of the flame front, and the combustion process must synchronize correctly with respect to the piston position and variations in-cylinder volume [2]. Therefore, the ignition timing has to be optimized and adjusted with respect to various engine operating parameters.

Ignition timing is defined as the exact instant a spark is sent to the spark plug. Spark timing is typically expressed in degrees of crankshaft rotation relative to TDC position. In most engines, the spark is designed to occur slightly before TDC due to the finite time that is taken for burning the fuel. If the ignition timing of an engine is not adjusted correctly, the spark may occur either before or after its best position. The early spark timings would result in difficult starting, poor performance, increased fuel consumption, and slow, rough idling. In later spark timing case, slow and jerky cranking with the warm engine may occur, and also some of the power strokes are wasted. Knocking may also be experienced while accelerating in

Fig. 3.20 Ignition time module and inductive pickup sensor (Courtesy of AVL)

early spark timings (depending on the amount of advanced spark timing). Therefore, it is very important to measure the exact ignition timing and calibrate according to system requirements [27]. Several factors affect the ignition timings including engine speed, type of fuel, mixture strength, and engine operating load. Direct measurement of ignition advance timing as well as absolute timing with respect to TDC is proposed for using directly on vehicles [27].

In a conventional ignition system, a high-tension voltage distribution to each cylinder via high-voltage cables and mechanical, high-voltage distributors is used. An inductive clamp can be used to capture the ignition timing signals. Figure 3.20 shows the commercially available ignition timing module along with an inductive pickup sensor. The module is available with an inductive current pickup that can be clamped onto the high-voltage cable of the particular cylinder for which ignition timing is measured.

The actual signal interface to the combustion measuring system depends on the type of ignition system. Typically, modern engines have a single ignition coil or one transformer per cylinder, and these are switched via a low-level on-off pulse. This pulse can be recorded on the combustion data acquisition system just by connecting the voltage signal through a suitable amplifier without any additional transducer [2]. The ignition time module can also detect and process lower-voltage signals coming, for instance, from the ECU, making the ignition time module suitable for any type of ignition systems (Fig. 3.20).

3.8   Sensors for Oxygen Detection and Air-Fuel Ratio Control

Air-fuel ratio plays an important role in the engine combustion process. Conventional spark ignition engine requires air-fuel ratio close to stoichiometric for sustaining the flame and complete flame propagation [8]. Typically, three-way catalytic converter is used as aftertreatment technology to meet the emission legislation. For the effective operation of the three-way catalytic converter, the engine must operate at the stoichiometric air-fuel ratio. For meeting this condition ($\Phi \sim 1$), a closed-loop control of air-fuel ratio is applied. Measurement of the oxygen concentration in the exhaust is used to determine the air-fuel ratio in the cylinder charge. Figure 3.21 shows the typical functional principle of zirconium oxide-based oxygen sensor and its characteristic output signal.

The oxygen sensor consists of a zirconia (ZrO_2) ceramic solid electrolyte. The electrolyte acts as a catalyst, and it is placed between two noble metal (platinum)

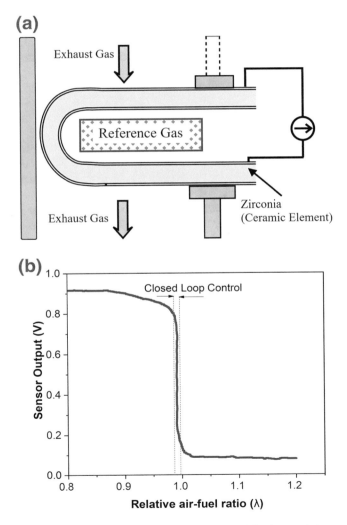

Fig. 3.21 (**a**) Typical functional principle of oxygen sensor and (**b**) characteristic output signal of oxygen sensor (Adapted from [9, 24, 28])

electrodes. The oxygen will flow from high partial pressure to low partial pressure when the partial pressure of oxygen is different at the two electrodes. The platinum electrode catalyzes the reversible reaction of oxygen disassociation and combination [28]. Oxygen from "air-side" (reference gas) of the sensor forms oxygen cations, which migrates across the zirconia to recombine and forming the oxygen. The potential difference created by ions is measured as an output voltage. The characteristic signal observed is shown in Fig. 3.21b. This sensor is only sensitive within a narrow λ range around $\lambda = 1$. Therefore, this signal is used as feedback for closed-loop combustion control of spark ignition engines.

The universal exhaust gas oxygen sensor (UEGO) is also developed to meet the wide range of air-fuel ratio control in lean-burn engines [9]. A UEGO sensor has a similar dimension to a lambda sensor. It is small and compact and can be easily installed in the engine exhaust. The UEGO sensor can be used to estimate the individual cylinder air-fuel ratio and closed-loop control of the engine [29, 30].

Discussion/Investigation Questions

1. Discuss the need for low-pressure (intake and exhaust manifold pressure) measurement for combustion and performance analysis of the engine. Write the characteristics required for the typical sensor used for the measurement of intake and exhaust pressure.
2. Discuss and list down the factor influencing the absolute pressure measurement and its dynamics in the intake and exhaust manifold in a modern engine.
3. Discuss the possible reasons for the use of adapters in inlet and outlet pressure measurement in modern engines. Comment on the best mounting position for the pressure sensor in the intake and exhaust manifold of the engine.
4. Discuss how low pressure (intake and exhaust) affects the combustion in SI, CI, and HCCI engines. Justify the importance of low-pressure measurement in three combustion modes.
5. Discuss the advantages provided by common rail direct injection (CRDI) technology over the distributor pump systems. Write the differences between the fuel pressure curves in the two fuel injection technologies.
6. Discuss the expected characteristics of a fuel pressure sensor for line pressure measurement in modern diesel engines.
7. Discuss the importance of the fuel line pressure and injector needle lift measurement. Write the information that can be generated by processing the signals acquired from line pressure and needle lift sensors. Draw the typical curves of these signals for CRDI and distributed pump engines.
8. Discuss how the opening phase of the injector needle (initial lift) affects the spray dynamics in modern diesel engines. Comment on the relation between needle lift and axial jet velocity of spray in CRDI engine.
9. Write the limitations of inductive (eddy-current)-based injector needle lift measurement in CRDI engines. Discuss one possible alternative for the needle lift

measurement with better accuracy and repeatability than the inductive-based method in CRDI engines.

10. Discuss the role of air and fuel mass flow measurement on performance and combustion analysis of an engine. Write the equations relating combustion/performance parameter involving air or fuel mass flow rates. Discuss the possible ways of calculating air-fuel ratio at particular engine operating condition.

11. Calculate the volume of air-box required for flow rate measurement in a single-cylinder four-stroke engine having a displacement volume of 500 cm^3 operated at 1500 rpm. Assume the diameter of orifice used measurement is 2.5 cm. Calculate the ratio of air-box volume and engine volume. What is the method you will suggest to reduce this ratio? Discuss the possible reasons why the higher volume of air-box is required at lower engine speeds and vice versa.

12. Discuss the limitations of airflow measurement with air-box and discuss flow-meter methods. Explain a method for instantaneous air mass flow rate measurement.

13. Discuss the reasons why gravimetric methods of fuel mass determination are preferred over the volumetric methods. Explain a volumetric and a gravimetric-based fuel mass flow rate determination in a diesel engine.

14. Explain why it is essential to operate a conventional spark ignition (SI) engine on the stoichiometric air-fuel ratio. Discuss the method to control the air-fuel ratio in SI engines.

15. Discuss the functional principle of the oxygen sensor (λ sensor) used in conventional SI engines. What are the possible methods for estimation of air-fuel ratio estimation in combustion engines? Comment on the merit of each method.

References

1. Czerwinski, J., Comte, P., Hilfiker, T., & Fürholz, A. (2012). *Research of techniques for low pressure indication in internal combustion engines* (No. 2012-01-0444). SAE Technical Paper.
2. Rogers, D. R. (2010). *Engine combustion: Pressure measurement and analysis.* Warrendale, PA: Society of Automotive Engineers.
3. Gossweiler, C., Sailer, W., & Cater, C. (2006). Sensors and amplifiers for combustion analysis, A guide for the User 100-403e-10.06, © Kistler Instrumente AG, CH-8408 Winterthur, Okt. 2006.
4. Van Basshuysen, R., & Schäfer, F. (2016). *Internal combustion engine handbook-basics, components, systems and perspectives* (2nd ed.). Warrendale, PA: SAE International.
5. Ueno, M., Izumi, T., Watanabe, Y., & Baba, H. (2008). *Exhaust gas pressure sensor* (No. 2008-01-0907). SAE Technical Paper.
6. Leach, F. C., Davy, M. H., Siskin, D., Pechstedt, R., & Richardson, D. (2017). An optical method for measuring exhaust gas pressure from an internal combustion engine at high speed. *Review of Scientific Instruments, 88*(12), 125004.
7. Vítek, O., & Polášek, M. (2002). *Tuned manifold systems-application of 1-D pipe model* (No. 2002-01-0004). SAE Technical Paper.
8. Maurya, R. K. (2018). *Characteristics and control of low temperature combustion engines: Employing gasoline, ethanol and methanol.* Cham: Springer.
9. Ladommatos, N., & Zhao, H. (2001). *Engine combustion instrumentation and diagnostics.* Warrendale, PA: SAE International.

10. Dhar, A. (2013). *Combustion, performance, emissions, durability and lubricating oil tribology investigations of biodiesel (karanja) fuelled compression ignition engine* (PhD thesis). IIT Kanpur.
11. Krogerus, T., Hyvönen, M., & Huhtala, K. (2018). Analysis of common rail pressure signal of dual-fuel large industrial engine for identification of injection duration of pilot diesfel injectors. *Fuel, 216*, 1–9.
12. Payri, R., Gimeno, J., Viera, J. P., & Plazas, A. H. (2013). Needle lift profile influence on the vapor phase penetration for a prototype diesel direct acting piezoelectric injector. *Fuel, 113*, 257–265.
13. Margot, X., Hoyas, S., Fajardo, P., & Patouna, S. (2011). CFD study of needle motion influence on the exit flow conditions of single-hole injectors. *Atomization and Sprays, 21*(1), 31–40.
14. Wang, T. C., Han, J. S., Xie, X. B., Lai, M. C., Henein, N. A., Schwarz, E., & Bryzik, W. (2003). Parametric characterization of high-pressure diesel fuel injection systems. *Journal of Engineering for Gas Turbines and Power, 125*(2), 412–426.
15. Huang, W., Moon, S., & Ohsawa, K. (2016). Near-nozzle dynamics of diesel spray under varied needle lifts and its prediction using analytical model. *Fuel, 180*, 292–300.
16. Payri, R., Gimeno, J., Venegas, O., & Plazas, A. H. (2011). *Effect of partial needle lift on the nozzle flow in diesel fuel injectors* (No. 2011-01-1827). SAE Technical Paper.
17. Coppo, M., Dongiovanni, C., & Negri, C. (2004). Numerical analysis and experimental investigation of a common rail type diesel injector. *Journal of Engineering for Gas Turbines and Power, 126*, 874–885.
18. Coppo, M., Dongiovanni, C., & Negri, C. (2007). A linear optical sensor for measuring needle displacement in common-rail diesel injectors. *Sensors and Actuators A: Physical, 134*(2), 366–373.
19. Amirante, R., Catalano, L. A., & Coratella, C. (2013). *A new optical sensor for the measurement of the displacement of the needle in a common rail injector* (No. 2013-24-0146). SAE Technical Paper.
20. Kastner, L. J. (1947). An investigation of the airbox method of measuring the air consumption of internal combustion engines. *Proceedings of the Institution of Mechanical Engineers, 157*(1), 387–404.
21. Stone, R. (1992). *Introduction to internal combustion engines* (2nd ed.). London: MacMillan.
22. Martyr, A. J., & Plint, M. A. (2007). *Engine testing* (3rd ed.). Oxford: Butterworth-Heinemann (Elsevier).
23. Turner, J. (2009). Temperature sensors. In J. Turner (Ed.), *Automotive sensors* (pp. 85–105). New Jersy: Momentum Press.
24. Reif, K. (2015). *Gasoline engine management*. Friedrichshafen: Springer Vieweg.
25. Kerres, R., Schwarz, D., Bach, M., Fuoss, K., Eichenberg, A., & Wüst, J. (2012). Overview of measurement technology for valve lift and rotation on motored and fired engines. *SAE International Journal of Engines, 5*(2), 197–206.
26. Çinar, C., Şahin, F., Can, Ö., & Uyumaz, A. (2016). A comparison of performance and exhaust emissions with different valve lift profiles between gasoline and LPG fuels in a SI engine. *Applied Thermal Engineering, 107*, 1261–1268.
27. Aggarwal, S., Subbu, R., & Gilotra, S. (2015). *A new approach in measurement of ignition timing directly on a two-wheeler using embedded system* (No. 2015-01-1642). SAE Technical Paper.
28. Swingler, J. (2009). Gas composition sensors. In J. Turner (Ed.), *Automotive sensors* (pp. 231–257). New Jersey: Momentum Press.
29. Benvenuti, L., Di Benedetto, M. D., Di Gennaro, S., & Sangiovanni-Vincentelli, A. (2003). Individual cylinder characteristic estimation for a spark injection engine. *Automatica, 39*(7), 1157–1169.
30. Cavina, N., Corti, E., & Moro, D. (2010). Closed-loop individual cylinder air–fuel ratio control via UEGO signal spectral analysis. *Control Engineering Practice, 18*(11), 1295–1306.

Chapter 4
Computer-Aided Data Acquisition

Abbreviations and Symbols

A/D	Analog to digital
AC	Alternating current
ADC	Analog-to-digital conversion
CA	Crank angle
CI	Compression ignition
D/A	Digital to analog
DAQ	Data acquisition card
DC	Direct current
F_s	Sampling frequency
IEPE	Integrated electronic piezoelectric
IMEP	Indicated mean effective pressure
N	Engine speed
p	Pressure
PC	Personal computer
PE	Piezoelectric
PXI	PCI eXtensions for Instrumentation
rpm	Revolutions per minute
S/H	Sample and hold
SI	Spark ignition
SNR	Signal-to-noise ratio
TDC	Top dead center
TTL	Transistor-transistor logic
V	Volume
θ	Crank angle position

© Springer Nature Switzerland AG 2019
R. K. Maurya, *Reciprocating Engine Combustion Diagnostics*, Mechanical
Engineering Series, https://doi.org/10.1007/978-3-030-11954-6_4

4.1   Introduction

In general, laboratory experiments are important for improving the understanding of
the physical phenomenon that occurs in nature as well as in engineering equipment
or machines such as internal combustion engines. The experiments can be conducted
for different purposes such as testing hypothesis, validating numerical or approximate solutions, determination of certain transport or material properties, exploratory
experiments for studying the process, or improving the design. Presently experiments become a design tool, especially for reciprocating internal combustion
engines. The experiments on reciprocating engines are conducted for improving
the design of engines for better performance or testing the effect of certain technologies on combustion and emission characteristics. Figure 4.1 illustrates the typical
steps involved in the experimental work. The first step is the designing of experiments, which is mainly the planning of experiments. The planning of experiments
includes recognizing the goal of the experiment, choice of input factors along with
their levels and ranges, selection of the response variable(s), and selecting the design
of experiments or selecting the test points before conducting the experiments.
Experimental setup fabrication along with the sensor or actuator installation is the
next step so that planned experiments can be performed to achieve the objectives of
experiments. While conducting the experiments, output (response) as well as
input signal data is recorded using data acquisition system, which is further analyzed
for obtaining meaningful conclusions.

In all the steps shown in Fig. 4.1, data acquisition is the topic of concern for the
present chapter. Signal processing of the measured data from combustion experiments (particularly cylinder pressure measurement) is presented in Chap. 5. Analysis
of measured data is discussed in Chaps. 6–10 for the different aspect of combustion
analysis in reciprocating engines.

Generally, in engine combustion experiments, a large amount of data needs to be
acquired within a fraction of seconds (in order of milliseconds). For such measurement requirements, high-speed data acquisition and processing system for continuous and automatic acquisition of data are essential. In case of modern reciprocating
engines with closed-loop control requirements, the measured data (often cylinder
pressure data) need to be processed in real time (online), and suitable action has to be

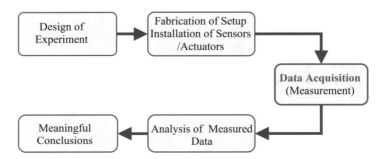

Fig. 4.1 Typical steps involved in the experimental work

taken based on the online data analysis. Thus, a data acquisition system having the essential characteristics of speed for fulfilling the demand of modern engine experiment is required.

Currently various systems are commercially available for combustion pressure measurement in modern reciprocating engines. These data acquisition systems are capable of logging a large amount of data at the required speed, processing it, and displaying the results in visual and printed forms. However, such commercially available data acquisition systems are expensive and have limited flexibility in their configurations. For particular laboratory experiments with a limited number of transducers and processing requirement, low-cost and flexible data acquisition systems can be configured. The PC-based or microprocessor-based data acquisition systems are highly flexible in their configurations, and thus, several options and expansion possibilities are allowed. Brief qualitative description of data acquisition systems and the functions of its elements which make up an overall data acquisition and processing are provided in this chapter.

4.2 Data Acquisition Principle

Data acquisition system quantifies and stores the experimental data in the required format. The main function of data acquisition system is to collect, transmit, and process the acquired data in the desired form such that it can be analyzed easily [1]. The essential sequence of operations involved in any data acquisition system includes the generation of input signals by sensors (transducers), signal conditioning, multiplexing, data conversion from analog-to-digital form and vice versa, and data storage and display [2]. Figure 4.2 illustrates the general data acquisition

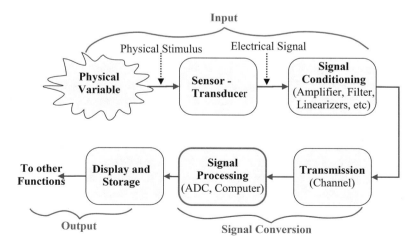

Fig. 4.2 Schematic diagram of typical data acquisition system (adapted from [1, 3])

system, which consists of a sensor-transducer, signal conditioning unit, signal conversion (analog to digital), data processing, data storage, and display units [1]. The essential component in a modern data acquisition system is the sensor or transducer that produces an electrical signal based on the changes in physical variable or property. Transducers are devices that are used to convert one type of physical phenomenon, such as temperature, strain, pressure, or light, into another type of signal. The most commonly used transducers convert physical quantities to electrical quantities, such as voltage or resistance. The characteristics of transducer typically defines many of the signal conditioning requirements of the data acquisition system.

Signal conditioning is basically the manipulation of generated signals to prepare them for digitization by data acquisition system. Typical signal conditioning may include one or more of the following operations: (1) amplification of signal produced, (2) linearization of signal, (3) filtering out the signal noise generated from the sensor and their circuitry, (4) transducer excitation, (5) correction of thermoelectric errors, (6) providing impedance matching, (7) compensation for the limitations of the sensor, and (8) arithmetic operations on the outputs of two or more sensors [2]. The data transmission unit allows the transmission of the data to data acquisition unit or display unit without any loss or contamination of data. The data needs to be transferred over a long distance or short distance depending on the experimental condition at a particular test setup in the laboratory. Strength of the generated signal and quality of transmission cable are important factors to be considered for data transmission. Typically, the transmission of an analog signal over a long distance is difficult, which may be converted into digital form using analog-to-digital (A/D) card. Transmission of digital data is relatively easier. The acquired data can be displayed and/or stored in an appropriate format that can use utilized for further analysis.

In typical engine combustion measurements, the number of measurands is more than one (such as cylinder pressure, ignition signal, needle lift signal, inlet, and exhaust pressure). For simultaneous data acquisition of various signals, several replicas of single-channel data acquisition systems will be required. The cost of this system will be quite exorbitant when the number of channels is higher than critical values. Additionally, handling and maintaining so many individual systems would be cumbersome. Therefore, single data acquisition system with the ability to acquire data from multichannel are essential. Typically, a scanner/programmer is employed which samples a data channel rapidly in sequence such that one conversion and output device is sufficient at any point of time [1]. Figure 4.3 schematically shows the typical multichannel data acquisition system. Multichannel data acquisition system contains additional data scanning system (data scanner/multiplexer), which is not present in signal channel system (Fig. 4.2). Multichannel data acquisition system can be programmed to acquire data or the desired range of data in a particular order. A multiplexer selects and routes one channel to the A/D converter for digitizing and then switches to another channel and repeats. Since the same A/D converter is used for sampling several channels, the effective rate of each channel is reduced in proportion to the number of channels sampled.

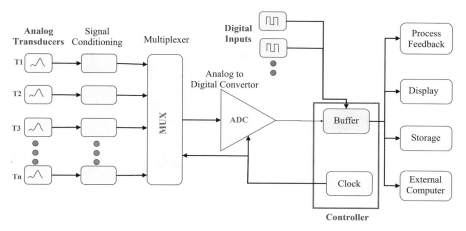

Fig. 4.3 Schematic diagram of a typical multichannel data acquisition system (adapted from [1])

Multiplexing system is typically required in data acquisition for engine research due to several rapidly changing variables including cylinder gas pressure, diesel injector needle lift, spark discharge transducer output, throttle angle during transient engine testing, instantaneous airflow into the engine, pressure pulsation in the inlet manifold, output from specialized instruments such as fast flame ionization detection, combustion optical sensors, fast-response surface thermocouples, etc. [4]. Switching between the channels in multiplexing generates a time skew between each channel sample. The time skew is dependent on the number of channel and sampling rate. This method is appropriate for applications where time relationship between sampled data over several input parameter is not important. Simultaneous sampling is required when time relationship between input parameter is important and needs to be analyzed. Simultaneously sampled multiplexer system is often used to monitor a large number of channels in a synchronized manner [2]. In this system, each channel of analog input is assigned to an individual sample and hold (S/H) device. Data updating in the sample mode is performed simultaneously for all the S/H devices using a clock. After receiving the data and locking in the hold mode, the multiplexer scans each S/H and feeds the output to the A/D converter.

Figure 4.4 illustrates the schematic of PC-based data acquisition system typically used for crank angle-based cylinder pressure measurement. For cylinder pressure measurement, a piezoelectric pressure transducer (Chap. 2) is typically used, which uses charge amplifier as signal condition unit. The data acquisition system digitizes and stores the input voltage from the charge amplifier when it is commanded to do so by an external clock. The clock output is a series of TTL pulses, and each pulse commands the acquisition system to digitize and record the voltages. The external clock pulse is typically generated from the incremental crank angle encoder. The data acquisition for engine combustion measurement uses the crankshaft encoder pulse because engine data needs to be presented in terms of crankshaft (piston)

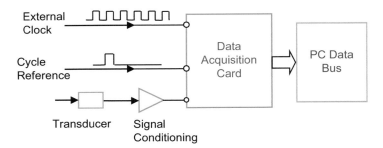

Fig. 4.4 Data acquisition principle using external clock and trigger signal generated from the crank angle encoder

position. For cycle reference, a trigger from TDC sensor or Z-pulse of angle encoder is typically used.

For selecting appropriate data acquisition system for particular application, the following important factors should be considered judiciously: (1) resolution and accuracy, (2) sampling rate per channel, (3) number of channels, (4) signal conditioning requirement of each channel, and (5) cost [1]. The number of channels can be decided on the basis of number of variables that need to be recorded during the experiment. The sampling rate is mainly governed by the frequency content of the signal and dynamic characteristics of the signal. Signal conditioning depends on the type of transducer used and data acquisition system.

4.3 Transducer and Signal Conditioning

4.3.1 Sensors and Transducers

Typically, a sensor is defined as a device which senses the process variable through its contact with the physical environment, and the transducer converts (transduces) the sensed information into a different form (typically electrical), yielding a detectable output [3]. The sensor contact always not need to be physical, and it can be located outside the environment of investigation (e.g., optical pyrometer) which is often known as a noninvasive sensor. An invasive sensor is located within the environment of the investigation. Generally, the signals between the sensor (transducer) and the detectable output are electrical, optical, or mechanical. Electrical sensors and transducers are often classified as active or passive. Active elements need an external power supply to generate a voltage or current output, while passive elements do not require an external power supply [3].

For engine research, several variables need to be measured, and various types of transducers are used depending on the measurand. The measurand can be categorized as static or dynamic measurand depending on its variation with time. The static measurand can be defined by a single value, while dynamic measurand is

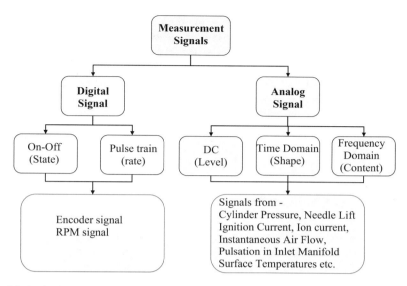

Fig. 4.5 Main signals involved in the data acquisition system (adapted from [5])

characterized by multiple values. Dynamic signals can be further characterized as a periodic or random signal. The measurement signal is a combination of single measurement values as a function of time, which describes the state of a system. A measurement signal is constituted of samples, and it is a tool to transfer the measurement information. Figure 4.5 summarizes the different types of measurement signals faced by a typical data acquisition system. The figure also depicts the typical signals acquired and used in engine combustion measurements. A measurement signal can be categorized as analog or digital by the way it conveys information. A digital signal has only two possible discrete levels—high level or low level. On the other hand, an analog signal contains information in the continuous variation of the signal with respect to time.

The primary characteristics of an analog signal are level, shape, and frequency (content) which are represented as DC, time-domain, and frequency-domain signals, respectively. The DC signals vary very slowly with time, and data handling hardware does not need to be of sophisticated speed. Time-domain (waveform shape) signals are required to be handled with precise timing, and for this type of signal, data acquisition hardware with onboard buffers as well as efficient data transfer capabilities is essential [6]. To estimate the frequency content, the system is called upon to do the heavy computation in calculating the Fourier transforms, etc. even though the signal itself is of the same complexity and speed as for time-domain signals. Thus, the computing system should also be powerful along with good data acquisition hardware. The digital signals are categorized as level signals (on-off/state) and pulse or wave-train signals. The state signals comprise of digital signal levels to be sensed or generated. Pulse and wave-train signals need precise timing, and therefore good onboard counters and timers are essential.

In engine combustion measurements, the digital signals are generated by crank angle encoder (Chap. 2), TDC sensor, or cam sensors. There are several transducers producing analog signals which need to be acquired by measurement system. Mainly for cylinder pressure measurement, the piezoelectric transducers are used along with needle lift signal, and fuel line pressure in diesel engines or ignition signal is SI engines. Few other analog signals are also typically measured such as instantaneous airflow into the engine, pressure pulsation in the inlet manifold, fast-response surface thermocouples (temperatures), and throttle position, especially during transient engine testing.

4.3.2   Signal Conditionings

Transducers convert the changes in physical variables into variations in electrical quantities. The electrical output of most transducers is not suitable as it stands for input to a data acquisition system. For example, cylinder pressure transducer converts in-cylinder pressure into charge (few pico-coulombs). A signal conditioning amplifier is used to convert this charge into a voltage and amplify it [4]. Signal conditioning forms the interface between the sensors and the data acquisition system that processes and stores the data. The signal conditioning amplifies the raw signal values from the sensors into a suitable, noise-free voltage signal, of appropriate amplitude, according to the measurement range of the physical value and input range of the data acquisition system. The signal amplification must be performed without adding any noise or frequency components to the actual experimental signal. The signal conditioner must also be capable of preventing and isolating unwanted noise from external sources [7].

For cylinder pressure measurement, piezoelectric (PE) and integrated electronics piezoelectric (IEPE) transducers (Chap. 2) are often used. The signal conditioning to be used is dependent on the type of sensor (PE or IEPE) and should be selected as charge amplifier for PE sensors and IEPE (Piezotron®) coupler for IEPE sensor. Figure 4.6 illustrates the circuit diagram of typical charge amplifier and IEPE coupler. A charge amplifier (also known as a charge-to-voltage converter) is the appropriate signal conditioning solution for piezoelectric (PE) sensors. This amplifier converts the charge signal of the sensor into a proportional voltage signal and, thus, makes the measurement available for further processing. An IEPE coupler (signal conditioning for IEPE sensors) supplies a constant current to power the sensor and decouples the measured AC signal from the DC power supply.

The circuit diagram of a charge amplifier has three essential components: (1) range capacitor (C_r), (2) time constant resistor (R_t), and (3) reset/measure switch (Fig. 4.6a). The range capacitor (C_r) is used to set the measurement range of the charge amplifier, which is performed by switching between different range capacitors. Switching over the measurement ranges makes it possible to measure across several decades with high signal-to-noise ratio (SNR) [8]. The time constant property of the charge amplifier is an important criterion that defines the performance of the amplifier for quasi-static measurements. The time constant of a piezoelectric

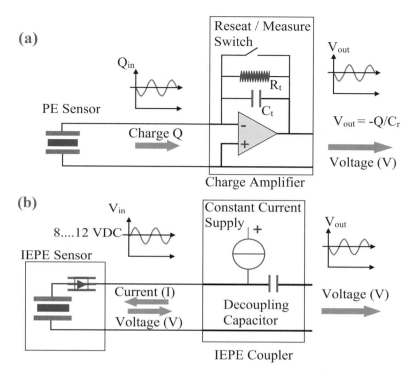

Fig. 4.6 Circuit diagram of (**a**) charge amplifier and (**b**) IEPE (Piezotron®) coupler (Courtesy of KISTLER)

system is a measure of the time it takes for a given signal to decay, not the time it takes the system to respond to an input. The time constant resistor (R_t) defines the low-frequency performance of the charge amplifier. In particular, the time constant determines the cutoff frequency for the high-pass characteristic of the charge amplifier. Switching between the different time constant resistors makes it possible to change the high-pass characteristic. The reset/measure switch is used to control the start of measurement or to set the zero point. The most important selection criteria for choosing a suitable charge amplifier are number of channels, measuring range, measurement type (quasi-static versus dynamic), and frequency range [8]. The frequency range of a charge amplifier is defined by the lower and upper cutoff frequencies. The lower cutoff frequency is determined by the measurement type (quasi-static or dynamic) and related high-pass characteristics.

Most modern amplifiers are equipped with internal drift-compensated circuits, which typically work by applying a compensating current to the charge amplifier input that is equivalent to the leakage current lost via the insulation resistance. Drift is a gradual change in the output signal that is not measured in the input signal. The drift of the output signal can be caused by a number of factors (e.g., temperature change, connecting cable properties), but a certain amount of electrical drift is always present because of the working principle of the charge amplifier electronics [7].

Often the measurement of additional phenomena around the engine subsystems during the combustion measurement can include the use of sensor technologies other than piezoelectric. Therefore, other signal conditioning units are also required in addition to the charge amplifiers. The types of signals most frequently found are analog signals from the various sensor technologies available such as piezo-resistive, Hall effect, strain gauge, etc. The suitable signal conditioning unit is required depending on the sensor technologies used for measurement.

4.4 Sampling and Digitization of Experimental Signal

To acquire data for physical variables, it is essential to convert them into voltage signals using transducers. These voltage signals are then converted into digital signals and stored using a data acquisition system. To be able to input the analog data to a microprocessor or digital circuit, the analog signal must be converted into coded digital values. The first step for converting analog signal to discrete values is to evaluate the signal at several discrete instants in time. This process is called sampling, and the result is a digitized signal composed of discrete values corresponding to each sample. This process is called analog-to-digital conversion (ADC) of signal. The ADC involves two steps: quantizing and coding. Quantizing is defined as the transformation of a continuous analog input into a set of discrete output states. Coding is the assignment of a digital code word or number to each output state [9]. Figure 4.7 illustrates the quantization and sampling of an analog signal. The quantization of voltages, i.e., the discretization in amplitude domain (Y-

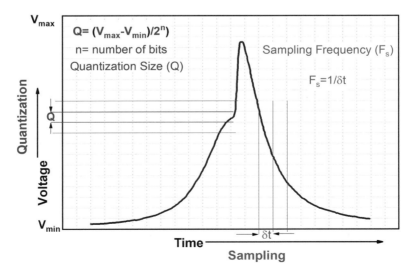

Fig. 4.7 Illustration of discretization of the measurement signal

axis), is done by analog-to-digital (A/D) converter. The minimum quantized voltage is defined as resolution, which is governed by the number of bits of A/D converter. Discretization in time (X-axis) is defined as sampling, which is characterized by the sampling rate.

One of the most important parameters of an analog input or output system is the rate at which the measurement device samples an incoming signal or produces the output signal. The sampling rate determines how often an analog-to-digital (A/D) or digital-to-analog (D/A) conversion takes place. To adequately reproduce the analog signal (i.e., to obtain a correct representation of signal with all important information), the input signal should be sampled at a sufficiently fast rate. The sampling theorem also called "Shannon's sampling theorem" states that it is required to sample a signal at a rate more than two times of the maximum frequency component in the signal to retain all frequency components. This means that for the correct representation of the analog signals, the digital samples must be taken at a frequency f_s such that $f_s > 2f_{max}$; f_{max} being the maximum frequency component in the input analog signal. The term f_s is known as the sampling rate, and the limit on the minimum required rate ($2f_{max}$) is called the Nyquist frequency. However, to retain the shape of the signal, the sampling frequency is typically more than ten times the frequency content. Very slow sampling results in a poor representation of the analog signal. Under-sampling causes the signal to appear as if it has a different frequency than it actually does. This misrepresentation of a signal is called aliasing. Figure 4.8 illustrates the aliasing due to under-sampling of analog signal by data acquisition system.

Typically, engine combustion measurements use an external clock for the sampling of the data. The TTL pulse from crank angle encoder is used as the clock. Thus, the sampling rate is governed by the resolution of the crank angle encoder. When sampling is made at either rising or falling edge of the pulse, the sampling rate can be calculated by Eq. (4.1).

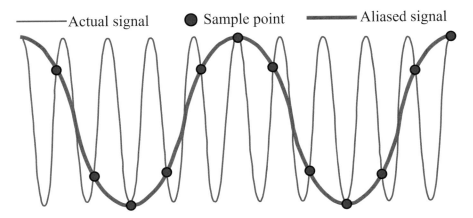

Fig. 4.8 Aliasing of the measurement signal (adapted from [9])

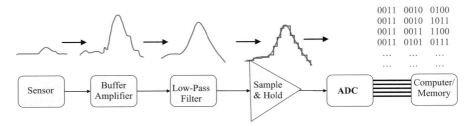

Fig. 4.9 Typical components used in analog-to-digital conversion (adapted from [9])

$$F_s(\text{Hz}) = \frac{360 \cdot x(\text{pulse}/^\circ\text{CA}) \cdot N(\text{rpm})}{60} \tag{4.1}$$

where F_s is the sampling frequency, x is the resolution of the encoder in pulse per crank angle degree, and N is the engine speed. Equation (4.1) clearly states that the sampling rate is dependent on engine speed, which means at lower engine speed, the sampling rate is also lower. Thus, appropriately higher resolution of the encoder is required, particularly when measurement is performed during knocking combustion.

The data conversion normally comprises the conversion of data from analog to digital (A/D) or from digital to analog (D/A). Conversion of signals from analog-to-digital form provides noise immunity to the data during its transmission [2]. In analog-to-digital conversion, the output will be a digital code that can be a straightforward binary representation of the signal, that is, a binary code (BC) or a binary-coded decimal (BCD). Figure 4.9 illustrates the different components used in A/D conversion process. The components used in the sequence include buffer amplifier, low-pass filter, sample and hold amplifier, analog-to-digital converter, and the computer (Fig. 4.9). These components must be properly selected and applied in a given sequence to properly acquire an analog voltage signal for digital processing. The resolution of an A/D converter is the number of bits used to digitally approximate the analog value of the input. The number of possible states N is equal to the number of bit combinations that can be output from the converter $N = 2^n$; n is the number of bits. For example, a 3 bit converter divides the analog range in $2^3 = 8$ divisions and similarly 12 bit converter in to 65,536 divisions, which provides an extremely accurate digital representation of the analog signal. The number of analog decision points that occur in the process of quantizing is $(N - 1)$. The analog quantization size (Q), occasionally called the code width, is defined as the full-scale range of the A/D converter divided by the number of output states (N). The process of A/D conversion requires a small but finite interval of time that must be taken into consideration when assessing the accuracy of the results. The conversion time, also called settling time, depends on the design of the converter, the method used for conversion, and the speed of the components used in the electronic design [9].

The analog-to-digital (A/D) system components shown in Fig. 4.9 are packaged in a variety of commercial products called data acquisition (DAC or DAQ) cards or modules. Figure 4.10 illustrates the typical DAQ products that are commercially

Fig. 4.10 Various commercial data acquisition products (Courtesy of National Instruments, USA)

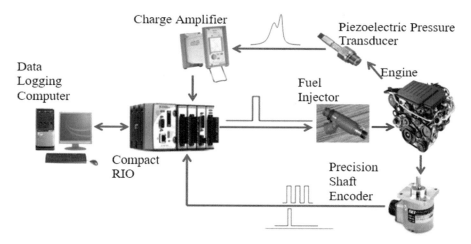

Fig. 4.11 Schematic diagram of data flow in the in-cylinder pressure measurement and fuel injection control [11]

available in a variety of form factors including PC and instrument panel plug-in cards, as well as stand-alone external units with standard interfaces (e.g., USB).

Presently, for closed-loop control of the engine, the cylinder pressure data acquisition and its online processing are required to take the decision for next combustion cycles. Based on the calculated combustion parameters from measured cylinder pressure data, the actuator signal (typically fuel injection) is generated. Real-time systems (such as PXI, CompactRIO) are required for engine control [10, 11]. In this kind of engine experiments, the data acquisition and engine control both can be performed with the same system. Figure 4.11 shows the simplified illustration of the cylinder pressure measurement along with fuel injection control. Engine combustion events are based on the piston position, and, thus, cylinder pressure measurement and fuel injection control require precise position, which is typically measured by crank angle encoder [10, 11].

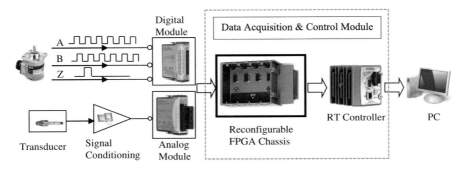

Fig. 4.12   Schematic diagram of data acquisition and control system [11]

Fig. 4.13 Timing of
cylinder pressure data
sampling by data acquisition
system

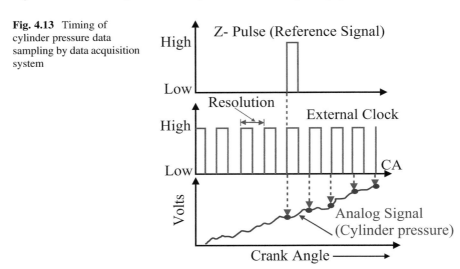

Figure 4.12 shows further details with the data acquisition module and encoder
signals. The crank angle encoder typically generates three pulses (A, B, and Z),
which is typically used for cycle referencing and crank angle-based data sampling
(see Chap. 2). The A and B pulses have the phase shift of one-fourth pulse
(quadrature encoder), and any one or both can be used for data acquisition depending
on the requirement and the capability of the data acquisition system. Signals of crank
angle encoder are connected to the appropriate module on the chassis. This signal is
used for the data acquisition of analog signal (cylinder pressure voltage). Additional
signals from the ignition, needle lift, airflow and pulsation, surface temperature, etc.
(not shown in Fig. 4.12) can also be used for data acquisition in the same system
using the adequate module.

   Figure 4.13 shows how the cylinder pressure data is acquired on the crank angle
basis. The Z-pulse of the encoder is typically used as a trigger to data acquisition
system and also acts as a reference for completion of the engine cycle. The A or B

pulse can be used as an external data acquisition clock, which decides when to actually sample the pressure signal. The data is typically acquired at the rise edge of the pulse. However, both rising and falling edge can be used for data acquisition of cylinder pressure, which will double the sampling rate. The sampling rate can be increased up to four times if both A and B pulses are used along with data acquisition at both rising and falling edges.

However, an increase in sampling rate increases the amount of data and, thus, also increases the memory requirements. Typically, the higher sampling rate is required for knocking conditions or at lower engine speeds, and data can be acquired at a lower sampling rate for regular performance or work analysis. Therefore, encoder pulses and data acquisition methods can be used accordingly based on the requirement.

4.5   Data Display and Storage

The computer-based data acquisition system consists of mainly two parts: a physical hardware system and data-processing software. A personal computer (PC) is typically used to store the data and perform calculations. The data acquisition cards and modules generally support several high-level language interfaces (e.g., C++, Visual Basic, FORTRAN, etc.) that give easy access to the product's features [9]. Easy-to-use software applications (e.g., National Instruments' LabVIEW software) are also available for graphical interfaces to program a data acquisition module to acquire and process the signals. Most modern systems employ standard PC technology as a platform for the user interface because this is familiar, generally utilized technology. The speed of this interface is a decisive factor in certain aspects of the system performance, for example, screen refresh rate and data-saving time.

In order to view the recorded and calculated data during a measuring sequence, display objects must be arranged according to the user requirements or preference. This must be a completely flexible part of the software user interface in case of commercial systems. Typical systems generally provide a number of standard graphical object types that users can select and integrate into their specific display configuration [7]. However, if software like LabVIEW is used for data acquisition and processing, the users can generate their own method of display and presentation of the data.

Some of the experiments are designed to conduct the experiments at constant combustion phasing (CA_{50}). In such experiments, online calculation and display of combustion phasing are required on the computer panel. Typically, the combustion phasing is estimated by integrating the heat release rate (Chap. 7) and computing the crank angle position of 50% heat release. The online calculation of combustion phasing is also required during closed-loop control experiments also. Typically, heat release calculation using measured cylinder pressure involves the derivative of pressure signal [10]. For real-time control application, it is difficult to include the pressure derivative in the computation due to its sensitivity to electrical noise in the signal especially in combination with a very high sampling rate [12]. A heat release

Eq. (4.2) is proposed for computation of net heat release, which does not involve pressure derivative term [12, 13]. The alternative method of heat release determination using experimental pressure signal is presented by Eq. (4.2) [13].

$$
\begin{aligned}
Q &= \frac{1}{\gamma - 1} \int_{\theta_{start}}^{\theta} d(pV) + \int_{\theta_{start}}^{\theta} p dV \\
&= \frac{1}{\gamma - 1} p(\theta) V(\theta) + \int_{\theta_{start}}^{\theta} p(\theta) \frac{dV}{d\theta} d\theta - \frac{1}{\gamma - 1} \{p(\theta_{start}) V(\theta_{start})\}
\end{aligned}
$$

$$\times \underbrace{\text{Calculated on Line}}\ \underbrace{\text{Estimated Constant, Added offline}} \tag{4.2}$$

Using this equation, combustion phasing can be easily calculated by assuming a constant value of the ratio of specific heat (γ), and it can be implemented for automatic engine control [12]. Thus, the heat release and combustion phasing can be displayed online during engine experiments.

Cylinder pressure analysis in reciprocating engines involves measuring very large quantities of data, typically several megabytes per test point depending on the crank angle resolution and number of cycles recorded. Derived results, such as peak cylinder pressure, IMEP, and burned angles can be calculated either in real time or by post-processing of the pressure data, and the need for the long-term saving of the raw data after processing is then optional. The demerit of saving the raw cylinder pressure data is mainly due to the large size of the files which means that the hard disk on the in-cell computer can only be used as temporary storage and larger-capacity storage systems are required for longer-term retention. A study proposed techniques for reducing the steady-state cylinder pressure data file size using variable crank angle resolution and assuming mean cycle characteristics are applicable over part of the engine cycle [14].

There are several possible options for reducing the amount of raw data stored such as (1) saving only mean (ensemble averaged) pressure data, (2) saving pressure envelope and standard deviation curves in addition to the mean pressure curve, (3) saving only part of the engine cycle rather than the whole measured crank angle range, and (4) varying the resolution over the engine cycle. There is a trade-off between the options which reduce the data storage most have the greatest restrictions (loss of information) and vice versa. The best approach for achieving a very large (~90%) reduction in the size of the raw pressure data is by employing a combination of variable crank angle resolution and mean cycle pressure data over part of the engine cycle [14]. It was found that the cylinder pressure curves for individual cycles for the exhaust induction and most of the compression stroke are well represented by a single mean cycle curve. Additionally, high crank angle resolution is only required over a small crank angle range near to TDC of the combustion phase; progressively coarser crank angle resolution can be used away from TDC without significant loss in accuracy. Thus, data storage can be optimized by proper use of crank angle resolution during a portion of the engine combustion cycle.

Discussion/Investigation Questions

1. Discuss the requirement of data acquisition systems for engine combustion analysis of reciprocating engines. List the signals which necessarily require high-speed data acquisition. Additionally, list down the signal that can be acquired with low-speed data acquisition systems.
2. Describe the constituents of a typical data acquisition systems used for cylinder pressure measurement in reciprocating engines.
3. Discuss the purpose of multiplexing in the data acquisition system. Explain why typically it is used in data acquisition of experiment test bench for engine research.
4. How you will differentiate between analog and digital signal. Describes the characteristics used to define analog and digital signal.
5. Define the static and dynamic measurand. Discuss the different types of dynamic signals. Write few examples of dynamic signals during engine combustion measurements.
6. Discuss why signal conditioning is required for various transducers. Write the typical process performed during the signal conditioning process. Describe typical sensors along with their signal conditioning required during engine combustion measurements.
7. Describe the important factors that need to be considered for selecting the appropriate data acquisition system for engine combustion measurements. Discuss the variable which is important for considering the data acquisition for typical CI or SI engines. Explain the parameter needs to be taken care for selecting the data acquisition for cylinder pressure measurement from the engine used for racing vehicles.
8. Define the sampling rate of the data acquisition system. Describe how sampling rate is provided in typical cylinder pressure measurement. How can you vary the sampling rate for the different portions of an engine cycle; for example, gas exchange and combustion period can have different sampling rates.
9. Assume a minimum 20 kHz sampling rate is required during knocking conditions of engines. Calculate the minimum resolution of crank angle encoder is required for engine operation at 1500 and 3000 rpm. Plot a graph of minimum resolution requirement as a function of engine speed in the range of 600–5000 rpm.
10. Define the resolution of A/D conversion of the signal. Write the typical resolution of a data acquisition system that is used for cylinder pressure measurement. Assuming pressure signal voltage is in the range of 0–10 V, calculate the minimum detectable voltage by data acquisition system.
11. Discuss how engine cycle reference is generated during cylinder pressure measurement. Describe the clock used for the measurement, and how you will calculate the sampling rate at a particular engine speed?
12. Define aliasing of the measured signal and describe when it can happen. Discuss the engine operating conditions at which a higher sampling rate is required and it

is a critical issue. For a fixed resolution of crank angle encoder, what is the major factor governing the sampling rate during cylinder pressure measurement on crank angle basis?

13. Calculate the sampling rate during cylinder pressure measurement at 1000 rpm with crank angle encoder of resolution six pulse per degree. Assuming only rising edge of "A" pulse of crank angle encoder is used, what will be the sampling rate when both (A and B) pulses of the encoder are used during data acquisition that is performed at rising as well as falling edge. Discuss the engine operating conditions, when you would like to use both pulses (A and B) of crank angle encoder.

14. Define the sampling rate and resolution of data acquisition card/system. How will you choose the data acquisition card for in-cylinder pressure measurement for a spark ignition engine? Discuss all the factors that need to be considered and equations involved in deciding the data acquisition card.

References

1. Mishra, D. P. (2014). *Experimental combustion: An introduction*. Boca Raton: CRC Press.
2. Rathakrishnan, E. (2007). *Instrumentation, measurements, and experiments in fluids*. Boca Raton: CRC Press.
3. Dunn, P. F. (2014). *Measurement and data analysis for engineering and science*. Boca Raton: CRC Press.
4. Ladommatos, N., & Zhao, H. (2001). *Engine combustion instrumentation and diagnostics*. Warrendale, PA: SAE International.
5. National Instruments. (1998). *LabVIEW-data acquisition basic manual*. Austin, TX: National Instruments Corporation.
6. Gupta, S., & John, J. (2010). *Virtual instrumentation using LabVIEW*. New Delhi: Tata McGraw-Hill.
7. Rogers, D. R. (2010). *Engine combustion: Pressure measurement and analysis*. Warrendale, PA: Society of Automotive Engineers.
8. Kistler. (2018). *Test & measurement pressure: Measurement equipment for demanding T&M applications*. Winterthur: Kistler Group.
9. Alciatore, D. G., & Histand, M. B. (2012). *Introduction to mechatronics and measurement systems* (4th ed.). New York: McGraw-Hill.
10. Maurya, R. K. (2018). *Characteristics and control of low temperature combustion engines: Employing gasoline, ethanol and methanol*. Cham: Springer.
11. Maurya, R. K. (2012). *Performance, emissions and combustion characterization and closed loop control of HCCI engine employing gasoline like fuels* (PhD thesis). Indian Institute of Technology Kanpur, India.
12. Wilhelmsson, C. (2007). *Field programmable gate arrays and reconfigurable computing in automatic control*. Lund: Lund University.
13. Tunestål, P. (2001). *Estimation of the in-cylinder air/fuel ratio of an internal combustion engine by the use of pressure sensors* (PhD thesis). Lund Institute of Technology, 1025.
14. Brunt, M. F., Huang, C. Q., Rai, H. S., & Cole, A. C. (2000). An improved approach to saving cylinder pressure data from steady-state dynamometer measurements. SAE transactions (2000-01-1211), 1381–1390.

Chapter 5
Digital Signal Processing of Experimental Pressure Signal

Abbreviations and Symbols

BDC	Bottom dead center
bTDC	Before top dead center
CA	Crank angle
CI	Compression ignition
COV	Coefficient of variation
CR	Compression ratio
CRR	Combustion reaction rate
DFT	Discrete Fourier transform
FIR	Finite impulse response
HCCI	Homogeneous charge compression ignition
IDI	Indirect injection
IIR	Infinite impulse response
IMEP	Indicated mean effective pressure
LPA	Least-square polynomial approximation
LSM	Least-square method
PSD	Power spectral density
SI	Spark ignition
TDC	Top dead center
VFO	Variable frequency oscillator
E_{bias}	Sensor offset voltage
K_{S}	Sensor gain
k	Harmonics
n	Polytrophic exponent
P	Pressure
P_{cyl}	True cylinder pressure
P_{offset}	Pressure offset
P_{peg}	Absolute pressure value at the pegging position

© Springer Nature Switzerland AG 2019
R. K. Maurya, *Reciprocating Engine Combustion Diagnostics*, Mechanical
Engineering Series, https://doi.org/10.1007/978-3-030-11954-6_5

ΔP	Pressure difference
$p(\theta_{\text{ref}})$	Cylinder pressure at a reference crank angle
V	Volume
$V(\theta_{\text{ref}})$	Cylinder volume at a reference crank angle
θ	Crank angle position
θ_{peg}	Crank angle position of pegging point
τ	Normalized time
ξ	Combustion reaction rate
σ	Standard deviation

5.1   Introduction

In-cylinder pressure measurement and analysis is the key tool for engine research and diagnosis since the advent of the reciprocating engine. In-cylinder pressure signal provides a large amount of information that can be used for analyzing combustion process and combustion quality. The accuracy of pressure measurements governs the quality of analysis of engine combustion process. In the last two decades, the real-time engine control and onboard supervision based on pressure signal have also become of particular interest to the automotive industry. Thus, measurement of the undistorted and unbiased cylinder pressure signal is essential for cylinder pressure-based combustion analysis and control. In-cylinder pressure is typically measured by piezoelectric transducer installed on the engine head (see Chap. 2). The experimental pressure signal typically requires four-step data processing before extracting the valuable information regarding the engine combustion process [1]. Four-step experimental signal data processing includes (1) crank angle phasing, (2) absolute pressure referencing, (3) cycle averaging, and (4) filtering (smoothening) [1, 2]. In-cylinder pressure data obtained from signal processing is used for further analysis to get information about engine combustion characteristics. The piezoelectric pressure transducer is based on the concept that measures only the variation in the combustion chamber pressure rather than the pressure itself. Hence, absolute pressure referencing (pegging) of the signal form piezoelectric transducer is required to get the absolute values of cylinder pressure. Different methods of pegging the pressure signal are discussed in Sect. 5.3. To perform an accurate combustion and performance analysis, a precise angle phasing between pressure and piston position (cylinder volume) is essential. The measured cylinder pressure is a function of piston position (cylinder volume), and thus, determination of exact crank angle for each pressure sample is of vital importance. Section 5.2 provides the various methods for crank angle phasing (TDC phasing) of experimental cylinder pressure signal. There exist various sources of error that affect the measured pressure signal even with the good performance of the piezoelectric sensor. Typical sources of error in measured pressure data include inaccurate calibration of the cylinder pressure measurement system, signal drift due to thermal shock, mechanical vibration noise and electrical noise along with the error in pressure referencing, and crank angle phasing [1]. The extent of error in combustion analysis by each of these sources of errors depends on the type of analysis to be conducted. The combustion parameters are typically calculated from the estimated heat release rate from measured cylinder pressure data. The heat release computation is typically affected by all these errors to some extent. Different averaging and filtering methods (Sects. 5.4 and 5.5) are used for minimizing some of the errors. Four-step digital signal processing is traditionally applied offline in such a way that allows a long signal processing time [3]. Recent developments in engine technologies demand advanced real-time diagnosis and control strategies, where long signal processing time cannot be affordable. Therefore, real-time signal processing methods with less processing time is required for engine control. The signal processing methods for offline as well as real-time application are discussed in the present chapter.

5.2 Crank Angle Phasing of Pressure Signal

Inaccurate determination of TDC, leading to the incorrect phasing of crank degree and measured cylinder pressure, is one of the most common errors that need to be eliminated for accurate calculation of combustion parameters. Incremental crank angle encoders are typically used for combustion pressure measurement. Encoders generate two signals in which one for crank angle mark (crankshaft position) and the other of one pulse per rotation for trigger signal. The trigger is used to determine the absolute position of the crankshaft. During the installation of crank angle encoder, the accurate determination of trigger mark relative to the true crank position is not always possible. Hence, a method is required to measure and establish the position of the trigger mark correctly with respect to the absolute position of the crankshaft, which is typically known as TDC determination [4]. After the determination of TDC offset, the measured crank angle marks can be shifted by the combustion measurement system to determine absolute crank position. Ideally, the peak pressure position should be at TDC position (minimum volume) during motoring conditions. However, in actual (real) motoring condition of the engine, the peak pressure precedes the TDC position due to wall heat transfer and mass losses. Figure 5.1 illustrates the TDC phase lag error with respect to motoring cylinder pressure. The difference between actual TDC position and the crank position corresponding to peak motoring pressure is defined as thermodynamic loss angle (Fig. 5.1). The TDC phase lag error ($\Delta\phi$) is defined as the angle between the peak pressure with the actual TDC and the peak pressure with a wrong TDC [5]. The thermodynamic loss angle creates the main difficulty in establishing the absolute position of the crankshaft using cylinder pressure measurement.

Determination of the absolute crankshaft position is imperative to the accuracy of subsequently calculated combustion analysis parameters. The slightest error in the measurement of crankshaft position is typically magnified by at least an order of magnitude with respect to calculations such as IMEP. Crankshaft position must be estimated accurately with respect to TDC position within $0.1°$ for calculation of

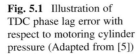
Fig. 5.1 Illustration of
TDC phase lag error with
respect to motoring cylinder
pressure (Adapted from [5])

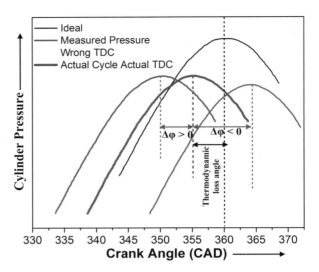

combustion parameters with sufficient accuracy [4]. The TDC offset determination
methods can be categorized mainly into two groups based on hardware requirements
for measurement [6]. The first group of methods requires additional hardware for
measurement to determine the accurate TDC position, and these methods include
dial gauge method (static determination), capacitive probes, microwave, etc. The
second group of methods do not require additional hardware but use in-cylinder
pressure signal, which is any way installed for combustion measurement. These
methods are based on algorithms which determines the TDC position from measured
in-cylinder pressure as a function of the crankshaft position. Various methods for
determination of TDC position for crank angle phasing are discussed in the following subsections.

5.2.1   Phasing Methods Using Additional Hardware

5.2.1.1   Static TDC Determination

The TDC position can be determined both statically (with a dial gauge) and
dynamically. In the static TDC determination method, measurement of TDC position is taken at the stationary crankshaft. Mechanical and manual intervention at the
engine is involved in this method. The static TDC position is determined using the
measurement of piston displacement. The engine flywheel or pulley can be marked
to show the correct TDC position [4]. Figure 5.2 schematically illustrates the static
TDC determination method. In this method, access to the piston with a dial gauge
precision measurement device is required as additional hardware. The dial gauge can
be installed through spark plug or fuel injection bore for the measurement of piston

Fig. 5.2 Illustration of static TDC determination method (Courtesy AVL)

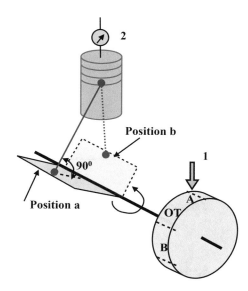

height. The crankshaft is rotated around 90° after TDC, and the depth of piston with dial gauge reading is recorded. Additionally, the pulley or flywheel is also are marked. The crankshaft is then further rotated manually with the effect of slowly lowering and then raising the piston until the exact same reading is obtained on the dial gauge. The pulley or engine flywheel is marked again relative to the static mark. Now, there are two marks on the flywheel, and two marks are exactly at the same distance from TDC. Thus, center of position between these two marks is the precise static TDC position.

In the static determination of TDC position, it is necessary to choose the position of marking on flywheel at an angle with sufficient distance from TDC. The reason for selecting the sufficient distance from TDC position is that piston movement around TDC is very small per degree of crankshaft movement [4]. However, the actual angle at which measurement is taken depends on the reach of the measuring equipment and the availability of access to the piston crown. The statically determined TDC position differs from the TDC that prevails during engine operation (i.e., dynamic TDC) due to the nonideal rigidness of the mechanical structure of reciprocating engines. Therefore, dynamic TDC determination methods are to be preferred [7].

5.2.1.2 Capacitive Probes (TDC Sensor)

The real dynamic TDC position can be determined using the TDC sensor with high levels of accuracy. In this method, the TDC sensor directly measures the piston displacement and can accommodate the elasticity of the crankshaft. This dynamic TDC determination method required additional hardware (capacitive probe), and it

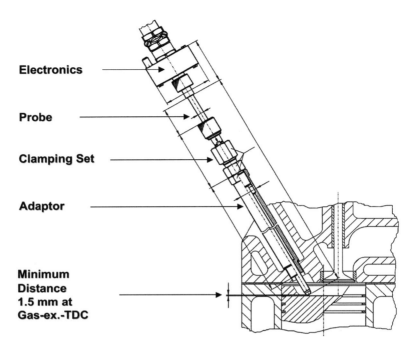

Fig. 5.3 Assembly and mounting of TDC sensor (Courtesy AVL)

can estimate the TDC position to an accuracy of 0.1° of crank angle [4]. The access of TDC sensor to the combustion chamber can be provided with an existing bore such as spark plug, glow plug, or fuel injector. Figure 5.3 presents the typical assembly and mounting of the TDC sensor on the engine head. The major components of the TDC sensor typically include the actual sensor, clamping piece, the adaptor, and evaluation electronics. The direct determination of TDC is beneficial in comparison to motoring cylinder pressure-based methods because this method does not require any correction by computing the degree of the thermodynamic loss angle. Additionally, no special machining or preparation of the engine is required, and thus, TDC sensor can be installed in a reasonably short time with high accuracy.

The functional principle of the TDC sensor is based on a capacitive measurement method. In this method, the changes in capacitance between the piston and the sensor head are measured by TDC sensor. The change in capacitance is inversely proportional to the gap between the piston and the sensor head. Figure 5.4 schematically presents the functional principle of a capacitive TDC sensor. Typically, a capacitive proximity sensor has two conductive plates separated by a dielectric material. An imbalance of electrical charges between the plates is created by applying a voltage difference. The capacitance determines the amount of current flow, which depends on the conductive plate proximity. In the capacitive TDC sensor, one conductive plate is the sensor probe, and another conductive plate is the piston (Fig. 5.4). The TDC sensor is installed in such a way that moving the piston does not touch the

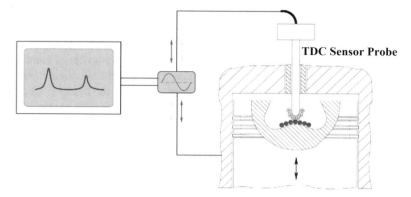

Fig. 5.4 Functional principle of capacitive TDC sensor [8]

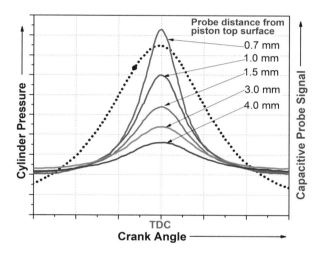

Fig. 5.5 TDC signal versus crank angle depending on the probe distance (Courtesy KISTLER)

sensor probe. The TDC sensor generates the signal with amplitude inversely proportional to the distance between the sensor tip and the piston top [8]. The maximum amplitude of the TDC sensor signal provides the exact location TDC. The output signal amplitude for compression conditions is lower in comparison to gas exchange conditions (Fig. 5.4). The lower amplitude during the compression conditions is caused by the cylinder pressure acting upon the engine components, which alters operating clearances [4]. Additionally, this phenomenon helps in discriminating compression TDC and gas exchange TDC positions.

Figure 5.5 presents the TDC signal as function of crank angle depending on the probe distance. The figure clearly illustrates that the signal amplitude depends on the distance between the probe and the piston. Minimal gap between the sensor probe and the piston is decisive for a good quality signal, but a certain gap must be present to prevent damage from touching of piston and sensor tip, which is typically between

0.5 and 1.5 mm [4]. The probe should be installed perpendicular to the piston crown in possible limits. This factor affects the choice of the measuring bore, and thus, if possible, the sensor should not be inclined more than 30° from the piston movement axis [4]. In order to determine the exact TDC position, the maximum amplitude of the TDC sensor signal must be evaluated. The TDC evaluation can be performed with great accuracy due to the high degree of symmetry of the signal (Fig. 5.5).

The TDC sensor is designed to measure the minimum clearance height, and thus, it is not always necessary to coincide with exact TDC position due to various factors such as bearing play and piston pin offset [4]. The output signal (analog voltage) of correctly installed TDC sensor is a smooth and symmetrical curve, which is typically used for TDC determination. Typically, the output signal is processed by indicating systems in connection with an angle encoder directly. The maximum position the TDC sensor's output signal cannot be used for the determination of the TDC location because the piston/crank movement is very small around TDC, which can lead to imprecise TDC sensing. Therefore, the "horizontal-cut principle" is used for calculation of TDC position, which is illustrated in Fig. 5.6. The output signal of TDC sensor is bisected at equidistant points before and after TDC, and the straight lines are connected from each symmetrical point. The center position is calculated from straight lines, and best fit line connecting to the multiple center points establishes the TDC position on the curve.

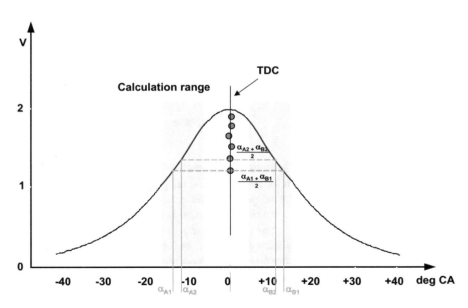

Fig. 5.6 Illustration of horizontal-cut principle for TDC determination in reciprocating engines (Courtesy AVL)

Fig. 5.7 Typical
microwave signal near TDC
(Adapted from [10])

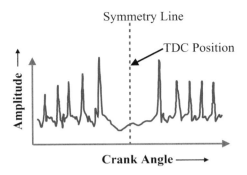

5.2.1.3   Microwave Probe

Microwave detection is another method for determination of dynamic TDC position
in a reciprocating engine via piston movement [9, 10]. In this method, low-power
microwaves are used to determine the piston proximity and movement. The measurement system measures the output signal with respect to crank position. A
computer-controlled variable frequency oscillator (VFO) enables the measurement
system to be utilized with a wide range of cylinder displacements. The uniqueness of
the microwave measurement process is the ability to accurately determine TDC
position in real time, with the engine running (or cold-motored) at virtually any
engine speed [4].

In this method, the combustion chamber of the engine is treated as a variable-
length microwave resonator. The TDC position is established by investigating a
series of resonance location data that is recorded as a function of crank angle
position. The probe couples the microwave with the engine combustion chamber
and determines the reflection coefficient of the microwave signal. The structure of
the probe is similar to a miniature whip antenna used on automotive vehicles [9]. The
probe, pre-chamber (in case of IDI engine), and cylinder comprise a microwave
cavity which is tuned by the piston position. Reflected signals from the cavity vary in
amplitude as the piston ascends in the compression stroke and descends in the power
stroke as shown in Fig. 5.7. Each peak position on the microwave signal corresponds
to the microwave resonance frequency for the mode. The detected microwave signal
shows a peak at every resonance dip because a detector with a negative output signal
was used. In principle output signal should be symmetrical with respect to TDC. The
TDC can be determined by calculating the center of symmetry (Fig. 5.7) [10].

5.2.2   Phasing Methods Using Measured Pressure Data

Measured in-cylinder pressure-based methods for dynamic TDC determination are
well established and typically supported by all cylinder pressure measurement
equipment. This method of TDC determination does not require any additional

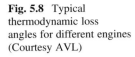

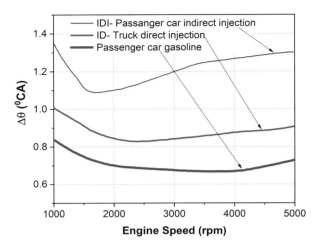

Fig. 5.8 Typical thermodynamic loss angles for different engines (Courtesy AVL)

hardware other than cylinder pressure sensor, which is anyway present for combustion analysis. Most of the methods use motored (unfired) cylinder pressure as a function of crank angle position for TDC determination. In ideal conditions (absence of heat transfer and mass losses), the peak pressure must occur at TDC position during unfired engine operating condition due to lowest cylinder volume at TDC position. Hence, the position of motoring peak cylinder pressure can be assumed as TDC position. However, in real conditions, the heat transfer and mass losses cannot be avoided. The peak pressure position occurs before real TDC position due to wall heat transfer and mass losses in real engine operating condition (Fig. 5.1). The difference between the peak pressure position and the actual TDC position is defined as the thermodynamic loss angle. The loss angle depends on a number of operating conditions (engine speed, temperature, blowby, etc.). Figure 5.8 depicts the thermodynamic loss angle as a function of engine speed for different engine combustion modes. The figure shows that the thermodynamic loss angle is higher at lower engine speeds because higher time is available for heat transfer at lower engine speeds.

Thermodynamic loss angle can be calculated by comparing the TDC sensor and the unfired cylinder pressure signals. However, this method of calculation of loss angle requires the additional hardware (TDC sensor) and its installation. Several algorithms are proposed in the published literature to calculate the thermodynamic loss angle only by measured pressure signal. The TDC determination methods based on polytropic exponent, symmetry of pressure curve, loss function, temperature-entropy diagram, inflection point, and IMEP-based calibration using measured cylinder pressure are presented in Sect. 10.1 of Chap. 10.

The accuracy of the TDC position depends on the estimation method. The application of particular method involves significant effort, cost, and time. Figure 5.9 shows the accuracy of different TDC determination methods with time and effort required. The static TDC determination method involves minimal expanse but considerable effort. This method is able to estimate reasonably accurate TDC

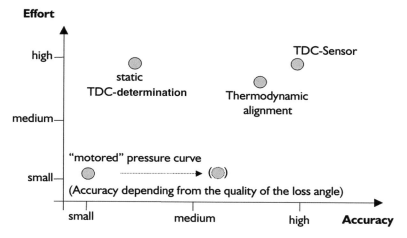

Fig. 5.9 Accuracy of different TDC determination methods with time and effort (Courtesy AVL)

position, but this does not account for dynamic effects. The TDC determination using the TDC sensor provides the most accurate TDC position estimation, but it also involves significant cost and effort. The motoring pressure-based estimation of TDC position has the accuracy depending on the algorithm used for calculation of thermodynamic loss angle.

5.3   Absolute Pressure Referencing (Pegging)

The quartz piezoelectric pressure transducers have been commonly used for the measurement of the cylinder pressure in reciprocating engines due to the advantages of good thermal performance and durability, high-frequency response, small size, light weight, large measuring range, etc. Inherent characteristics (working principle) of piezoelectric transducers require referencing of the output signal to absolute pressure (pegging). The piezoelectric pressure transducers can only measure the changing pressure content, i.e., only pressure variations in the combustion chamber, and not the physically correct absolute value of pressure in the combustion chamber. The charge output from the piezoelectric pressure sensor is supplied to the charge amplifier, which converts the charge output to a proportional voltage signal (see Chap. 2). The voltage signal is recorded by data acquisition system into digital format. The recorded voltage can be converted into absolute cylinder pressure data by Eq. (5.1) [11, 12]. This equation assigns a known absolute pressure value at particular pegging position in the engine combustion cycle:

$$P(\theta) = P_{\text{peg}} + \text{cal} \left[v_{\text{t}}(\theta) - v_{\text{t}}(\theta_{\text{peg}}) \right] \tag{5.1}$$

where P_{peg} is the known absolute pressure value at the pegging position, "cal" is calibration factor of the transducer (bar/Volt), v_t is the measured voltage, and θ_{peg} is the crank angle position of pegging point. In case of the noisy signal, it is important to use the average value of multiple points (10–15) at the pegging position to estimate $v_t(\theta_{peg})$, as presented by Eq. (5.2):

$$v_t\left(\theta_{peg}\right) = \frac{1}{w} \sum_{z=-k}^{k} v_t(\theta_z) \quad \text{where} \quad k = \frac{w-1}{2} \tag{5.2}$$

Pegging position depends on the point, where the user can provide a known absolute pressure value. Typically, two parameters, i.e., the intake manifold absolute pressure or exhaust backpressure and the polytropic exponent, are mainly used to correct the measured cylinder pressure signal. Absolute pressure referencing (correction) may also be required to compensate for inter-cycle and intra-cycle drift (long-term and short-term drift, respectively) which necessitates individual cycle referencing [13]. The cylinder pressure curve would drift more or less globally if it isn't corrected because the piezoelectric transducers can only measure the relative variations of cylinder pressure. Figure 5.10 shows the measured in-cylinder pressure curves of two cylinders from an eight-cylinder diesel engine, where one cylinder pressure is corrected using fixed polytropic exponent. The cylinder pressure curve for cylinder 5 is obviously unreasonable (Fig. 5.10) because both the two cylinders are in the normal working state and the curve of cylinder 1 has been already corrected. The reason is that the pressure curve of cylinder 5 drifts globally based on an arbitrary ground under the effect of the disadvantages of the quartz piezoelectric pressure transducers [14]. Therefore, the measured in-cylinder pressure signal by the piezoelectric transducers must be correctly referenced to get authentic combustion pressure data.

Fig. 5.10 Pressure curves of cylinder 1 and cylinder 5 from an eight-cylinder turbocharged diesel engine (Adapted from [14])

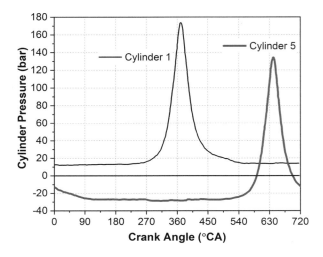

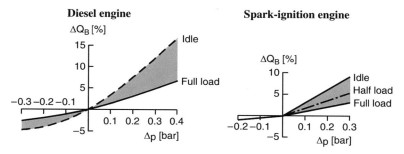

Fig. 5.11 Effect of referencing error in absolute pressure value on the energy balance [15]

Fig. 5.12 Effect of referencing error on cumulative heat release and peak pressure in a stoichiometric non-dilute operating condition at 2000 rpm (Adapted from [16])

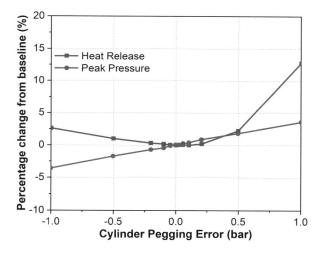

Incorrect pegging or absolute pressure referencing can affect calculated combustion parameters such as the heat release energy and heat release rate, mass fraction burned and burned angles, compression and expansion polytropic indices, estimated pressure drop across ports, estimated cylinder charge mass, estimated charge temperature and derived quantities, etc. Figure 5.11 shows the referencing error on energy balance in SI engine and diesel engine. A positive referencing error $(+\Delta p)$ leads excessive cylinder pressure (and vice versa), which results into smaller conversion rates before the TDC position and to larger ones after the TDC. Since most of the heat conversion takes place after the TDC, most of the changes occur in this part of the combustion sequence [15].

Most parameters are essentially unaffected for reasonably small errors in pressure referencing. Typically, IMEP is unaffected by absolute pressure corrections because it is a cycle-integrated parameter and shifting the pressure values does not change the area contained in the pumping or compression/expansion loops [16]. Figure 5.12 shows the effect of pegging error on cumulative heat release and peak pressure in a stoichiometric non-dilute operating condition at 2000 rpm. The figure shows that the

Fig. 5.13 Effect of
±0.1 bar pegging errors on
the calculated compression
polytropic exponent at part-
load and full-load operating
condition (Adapted from
[13])

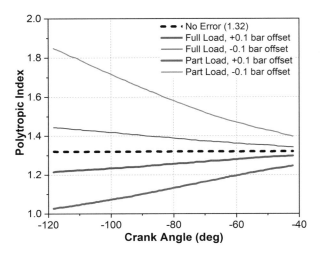

calculated heat release in the engien cycle is increased by over 10% for the case of
the pegging error of 100 kPa, but for all other cases, the effect is much more modest.
However, the peak pressure is shifted by exactly the amount of the pegging error, as
would be expected.

Figure 5.13 shows the effect of ±0.1 bar absolute pressure offsets on the
estimated compression polytropic exponent using simulated engine pressure data
at part-load and full-load operating conditions. The simulated data were generated by
assuming a fixed polytropic index of 1.32 throughout the compression process.
Figure 5.13 depicts that a low polytropic exponent is calculated when the pressure
is too high, and the error in the polytropic exponent is more sensitive to a specified
magnitude of pressure error at low engine load. Additionally, the largest polytropic
index errors are incurred early in the compression process (Fig. 5.13) and that the
erroneous polytropic indices vary greatly with crank angle.

Cylinder pressure-based charge temperature calculations are often used to calculate the mean gas temperature and derived parameters such as wall heat flux, gas
properties, and component temperatures. A study showed that the calculated temperature is very sensitive to pressure referencing, and with a −0.1 bar error at
low-load operating condition can lead to a change in peak temperature of 1000 K.
At full load, the same magnitude of pressure error would still cause a 250 K shift in
peak temperature [13].

For achieving good accuracy in the derived parameters, the accurate pressure
pegging is highly desirable for both mean cycle and individual cycles. A certain level
of referencing errors will always be present. However, in practice referencing errors
need to be reduced to an acceptable limit. The minimum pressure referencing
accuracy required depends on the type of analysis being performed and the
engine operating conditions [13]. Higher absolute pressure accuracy is typically
required for combustion analysis at low engine load and under slow burn conditions.
High referencing accuracy around the whole combustion cycle can only be attained

in the absence of all other sources of pressure measurement error. Thermal shock, long-term drift, and sensitivity errors mean that correct pressure referencing will only occur over a limited portion of the engine cycle. For example, thermal shock can distort the pressure by a variable amount over most of the engine cycle. In this condition, the accurate pressure referencing becomes difficult, and it is necessary to decide which part of the engine cycle needs to be most accurately referenced.

Several methods are proposed in published literature for absolute pressure correction including (1) pegging to intake manifold pressure near intake BDC using absolute intake manifold pressure sensor, (2) pegging to exhaust manifold pressure near exhaust TDC using absolute exhaust manifold pressure sensor, (3) referencing by an absolute cylinder pressure sensor exposed to the cylinder charge near to BDC, (4) installing a switching adapter to expose the piezoelectric sensor to a known pressure during part of the engine cycle, and (5) using a numerical referencing method based on polytropic index (constant or variable) [13]. Each of these approaches has advantages and disadvantages and an expected level of accuracy. An iterative method for determining the pressure offset is proposed using the tuning of heat release curve during motoring condition or compression phase of firing condition [17]. In this method, it is assumed that no heat is released during the compression phase or motoring condition, and the zero line can be corrected until this requirement is met. However, this method has the disadvantage of the enormous amount of computation, and it also requires corrected parameters used in heat release calculation. The commonly used methods for absolute pressure referencing are discussed in the following subsections.

5.3.1 Inlet and Outlet Manifold Pressure Referencing

Manifold pressure pegging has the advantage of relative simplicity but does require additional pressure transducer. Manifold pressure sensors are ideally required one per cylinder mounted in each runner, and water-cooled for exhaust pressure transducers [13]. Intake manifold pressure sensors are mostly absolute pressure sensors and have a high accuracy of approximately ± 10 mbar. The combustion chamber pressure can be adjusted to the intake manifold pressure if the flow in the cylinder is balanced. An appropriate time for this is the BDC position of the intake stroke because the piston speed is zero at this position. However, under highly tuned operating conditions, improper selection of the measurement point (or the use of a pressure transducer with inadequate frequency response) will introduce systematic errors [18]. Thus, intake manifold pressure referencing (IMPR) is an appropriate method for an untuned intake system or a tuned system which has a very low engine speed. Additionally, the measurement noise at intake BDC position can lead to incorrect referencing results for the all pressure data points of the engine cycle. Even if the average value of several points near intake BDC position is used for referencing, this method still has error with a tuned intake system or at high engine speed [19, 20].

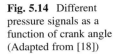

Fig. 5.14 Different
pressure signals as a
function of crank angle
(Adapted from [18])

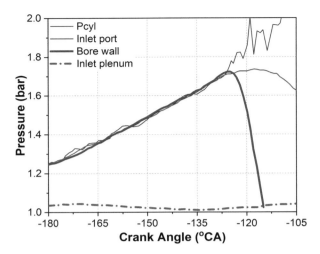

Intake manifold pressure sensor can be mounted centrally in the intake manifold
so that all cylinders use the same pressure value before the intake ports for correction, or the pressure sensor can be mounted in each inlet port of a cylinder near the
intake valve, which removes cylinder-specific differences. The most accurate but
also the most expensive method is to mount a sensor in each inlet runner. A study
mounted three absolute pressure sensors in (1) the inlet plenum, (2) the inlet port
7 mm above the valve seat, and (3) through the bore wall such that the transducer is
5 mm above the piston at BDC for pressure referencing [18]. Figure 5.14 shows the
measured pressure signal at each of these positions. The figure clearly depicts that
the common method of pegging to the pressure measured in the inlet plenum will
introduce systematic errors when manifold tuning is significant. Additionally, the
pressure measured in the inlet port close to the valve seat precisely follows the
pressure measured through the bore wall, up to the point where the piston covers the
transducer hole (at $-126°$ BTDC). Access to the inlet port is typically much more
convenient than through the bore wall.

In practice, the intake BDC is used for pressure referencing position. However
some dependence on the crank angle position used for the referencing would be
expected. Figure 5.15 shows the change in pressure referencing by varying the crank
angle over which the referencing is performed (datum value is BDC, ±10 crank
angle degrees average in all cases). The figure shows that changing the position for
intake manifold pressure referencing (IMPR) does produce a sizeable change in the
absolute pressure. A later referencing crank angle initially increases the absolute
pressure (Fig. 5.15). As expected, the variations with crank angle position are
greatest for the higher flow rate cases. The best location for pressure referencing is
reasonably the flat portion of the curve where the manifold and cylinder pressures are
the same or have a constant difference. The study concluded that pressure
referencing at 10–15° after BDC position is optimum for this engine [13].

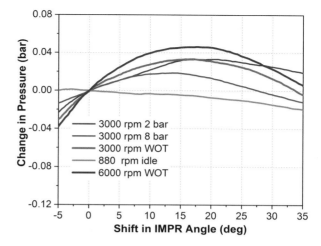

Fig. 5.15 Effect of manifold pressure sensor referencing crank angle position on referencing pressure (Adapted from [13])

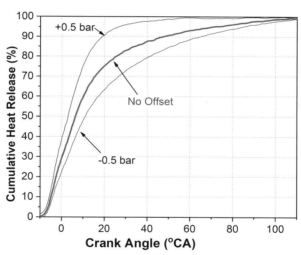

Fig. 5.16 Cumulative release curves with different pressure offset values (Adapted from [14])

The global drift of the cylinder pressure curve not only leads to wrong values of the whole engine cycle but also leads to the wrong calculation of combustion parameters. Figure 5.16 shows the cumulative heat release curves calculated from the measured cylinder pressure with different cylinder pressure offsets. The cumulative heat curves vary with the changes of cylinder pressure offsets, and in case of +0.5 bar offset, the heat release rate increases more rapidly with the increase of the crankshaft angle when compared with the curve with no offset (Fig. 5.16). The study showed that the ±0.5 bar offset could lead to up to 45% error in the calculation of combustion duration [14]. Thus, a proper correction method of cylinder pressure is the prerequisite for the analysis of the combustion process in reciprocating engines.

In the outlet pressure referencing method, the cylinder pressure during the exhaust stroke (typically TDC position or average pressure during exhaust stroke) is assumed to be equal to the exhaust backpressure. The pressure fluctuations of the exhaust manifold are more noticeable than those of the intake manifold [19]. Averaging the pressure data over several crank angle degrees can reduce the effect of this fluctuation and the measurement noise. However, an additional pressure sensor is required for exhaust backpressure measurement that can be used for pegging/referencing.

5.3.2   Two- and Three-Point Referencing

In two-point method, a fixed polytropic coefficient is assumed. In this method, cylinder pressure offset is calculated using measured cylinder pressure data at two points $\theta(i)$ and $\theta(i + 1)$ in the compression stroke (Fig. 5.17). The points are considered before the start of combustion and after the intake valve closing. The compression process in the engine is considered as polytropic compression process (PV^n = constant) before the start of combustion during the compression stroke. Thus, polytropic expression at two points can be written as Eq. (5.3) [21]:

$$P_{cyl}(i + 1) = P_{cyl}(i) \cdot \Omega(i)^n$$
$$\text{With}: \Omega(i) = \frac{V(i)}{V(i + 1)} \tag{5.3}$$

The true cylinder pressure (P_{cyl}) and measured cylinder pressure (P_m) are related to pressure offset (P_{offset}) by Eq. (5.4):

$$P_{cyl} = P_m + P_{offset} \tag{5.4}$$

In this method, the polytropic exponent is assumed to be known, and the pressure offset is constant. Thus, the pressure difference (ΔP) between two points of the

Fig. 5.17 Compression cylinder pressure of the reciprocating engine

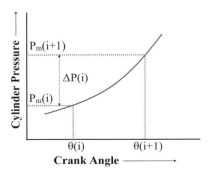

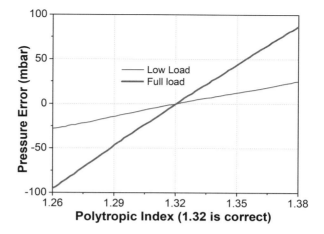

Fig. 5.18 Sensitivity of calculated absolute pressure to the assumed polytropic exponent (Adapted from [13])

cylinder pressure curve (Fig. 5.17) must be independent from the pressure offset value:

$$\Delta P(i) = P_m(i+1) - P_m(i) = P_{cyl}(i+1) - P_{cyl}(i) \quad (5.5)$$

By using Eqs. (5.3) and (5.5), the pressure offset can be calculated by Eq. (5.6):

$$P_{offset} = P_m(i) - \frac{\Delta P(i)}{\Omega(i)^n - 1} \quad (5.6)$$

Typical values of the polytropic exponent for CI and SI engines in motoring operation are $n = 1.37$–1.40, and for SI engines with a stoichiometric air-fuel ratio condition $n = 1.32$–1.33 [15].

Figure 5.18 shows the relationship between the assumed polytropic exponent and the absolute pressure referencing error for the case of simulated pressure data based on $n = 1.32$. The figure shows that errors in the fixed polytropic exponent of ± 0.05 cause referencing errors of typically ± 75 and ± 25 mbar for full-load and part-load conditions, respectively. The fact that the errors are proportional to load is favorable for combustion analysis because larger referencing errors can normally be tolerated at higher load [13].

Considering the error in the assumed polytropic exponent, it can be calculated as well from the measured pressure data. Thus, the pressure offset and polytropic exponent are the two unknown parameters which can be calculated from two independent equations. For generating two equations, an additional point can be considered in the compression stroke. This method is called three-point referencing method. Two equations are created in the same way as Eq. (5.6) with three points from the measured cylinder pressure signal in compression stroke as Eq. (5.7) [21, 22]:

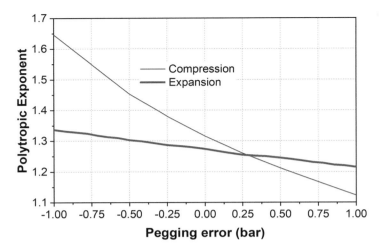

Fig. 5.19 Effect of pegging errors on compression and expansion polytropic coefficients (Adapted from [18])

$$\frac{P_m(i+1) - P_m(i)}{\left(\frac{V(i)}{V(i+1)}\right)^n - 1} = \frac{P_m(i+2) - P_m(i)}{\left(\frac{V(i)}{V(i+2)}\right)^n - 1} \tag{5.7}$$

Equation (5.7) cannot be solved analytically to determine the polytropic exponent "n." Therefore, it is solved iteratively by writing in the form $n = f(n)$. Successively, the cylinder pressure offset can be calculated by polytropic exponent using Eq. (5.6). The study used state space formulation to determine the pressure offset and polytropic exponent using tow extended Kalman filters [21].

Figure 5.19 shows the effect of pegging errors on the calculated values of compression and expansion polytropic coefficients. The effect of pegging errors is more pronounced on the compression metric than the expansion (Fig. 5.19) due to the expansion occurring at a much higher pressure, making a given pegging error a smaller percentage of the values used to calculate the coefficient. Figure 5.20 shows the effect of pegging errors on burn locations (10%, 50%, and 90%) using actual calculated as well as forced (assumed) polytropic coefficients. The figure depicts that a $-2°$ error in CA50 per bar of pegging error when the heat release algorithm uses the actual calculated polytropic coefficients. Figure 5.20 also depicts a $+2.3°$ error in CA50 per bar of pegging error when the using assumed (forced) polytropic coefficients [18].

In this method, referencing errors mainly produced from disturbed measured values and an incorrect determination of the polytropic exponent. Moreover, this method can only give correct results if the thermal shock remains constant between the two sampling points.

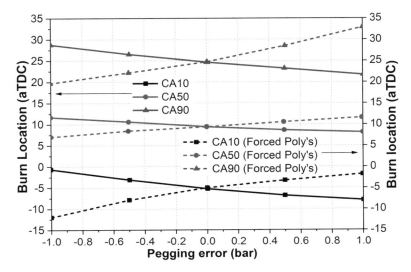

Fig. 5.20 Effect of pegging errors on burn locations using calculated and assumed polytropic coefficients (Adapted from [18])

5.3.3   *Referencing Using Least-Square Methods*

In order to reduce the sensitivity of measurement, the pressure sensor offset voltage is determined by evaluating a number of measurement samples (instead of two or three) and applying regression calculations. Typically, more than two measurement samples of the cylinder pressure signal are available during the compression process. The measured voltage during the compression phase or motoring engine operation can be written as Eq. (5.8) [19, 20]:

$$E(\theta) = K_S \cdot c(\theta) \cdot p(\theta_{\text{ref}}) + E_{\text{bias}}$$
$$c(\theta) = \left[\frac{V(\theta_{\text{ref}})}{V(\theta)}\right]^k \tag{5.8}$$

where E is voltage; K_S and E_{bias} refer to the sensor gain and sensor offset voltage, respectively; and $V(\theta_{\text{ref}})$ and $p(\theta_{\text{ref}})$ represent cylinder volume and pressure at a reference crank angle.

The Eq. (5.8) can be written in matrix form as Eq. (5.9):

$$y = X \cdot w \tag{5.9}$$

where

$$
y = \begin{bmatrix} E(\theta_1) \\ E(\theta_2) \\ \vdots \\ \vdots \\ E(\theta_N) \end{bmatrix}, \quad X = \begin{bmatrix} 1 & c(\theta_1) \\ 1 & c(\theta_2) \\ \vdots & \vdots \\ \vdots & \vdots \\ 1 & c(\theta_N) \end{bmatrix}, \quad w = \begin{bmatrix} E_{\text{bias}} \\ K_S \cdot p(\theta_{\text{ref}}) \end{bmatrix} \tag{5.10}
$$

Applying the linear least-square method, the parameter vector can be calculated as Eq. (5.11):

$$
w = \left(X^T X\right)^{-1} X^T y \tag{5.11}
$$

Since the sensor gain is already known, the sensor offset voltage and the reference cylinder pressure can be calculated simultaneously by Eq. (5.11).

The least-square method (LSM) is considered the best method for pegging the measured cylinder pressure [13]. This method assumes a polytropic coefficient, and it is becomes unsuitable when the polytropic coefficient is unknown. Similar to the two-point referencing method, this method also has a drawback of the assumption of a fixed polytropic coefficient. Therefore, an erroneous choice or changes of the polytropic coefficient can lead to pegging error [19]. Therefore, sensor offset voltage estimation by using the least-square method with the variable polytropic coefficient is proposed [19, 20]. This method has the assumption that the polytropic coefficient is slowly varying cycle-by-cycle and fixed during one cycle. The estimation of the polytropic coefficient on the ith cycle is derived using Eq. (5.12):

$$
\tilde{k}_i = \frac{\ln \left[p(\theta)/p(\theta_{\text{ref}}) \right]}{\ln \left[V(\theta_{\text{ref}})/V(\theta) \right]} \tag{5.12}
$$

In order to achieve robustness of estimation, a first-order auto-regressive filter is applied to the estimation result:

$$
\hat{k}_i = a \cdot \hat{k}_{i-1} + (1 - a) \cdot \tilde{k}_i, \quad 0 < a < 1 \tag{5.13}
$$

This approach alleviates the computational complexity and the sensitivity to polytropic coefficient error [19].

Least-square method (LSM) and two-point referencing methods have the demerit of assuming a fixed polytropic exponent, while three-point referencing uses calculated exponent, but it suffers from noise sensitivity. Figure 5.21 depicts the calculated polytropic exponent using three-point method and lease square with a variable polytropic exponent (modified LSM). The figure shows that both methods trace the tendency of a transition, and modified LSM method effectively determines the polytropic exponent. The modified LSM methods demonstrated the least sensitivity

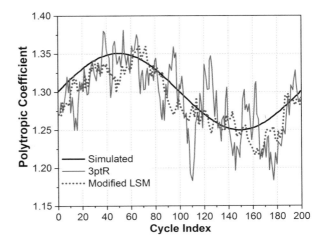

Fig. 5.21 Calculated polytropic exponents using three different methods [20]

to random noise [20]. Another model-based least-square method is proposed, which is computationally inexpensive, and well suited for real-time control applications [23].

5.3.4 Referencing Using Polytropic Coefficient Estimation

This method of pressure referencing also treats the compression stroke as a polytropic process and estimates the polytropic exponent. The polytropic exponent is a fixed value in the specified state, such as the same working condition in the particular engine, so that whole compression stroke seems to be adiabatic. However, in practical engine operating conditions, the adiabatic condition can only reach when cylinder charge temperature is close to the cylinder wall, which leads to the heat exchange close to zero. This condition occurs only in a certain crankshaft angle range during the compression stroke. Additionally, the loss of the charge through blowby can be neglected at the same time. Thus, crankshaft angle in this range can be used to analyze the curve of the polytropic exponent. A study found that the crank angle range for analyzing the polytropic exponent curve is between 80 °CA bTDC and 40 °CA bTDC for the turbocharged eight-cylinder diesel engine [14].

The polytropic exponent curve should be a horizontal line in the section of the compression stroke corresponding to the angle which can be regarded as the adiabatic process. However, if the curve is not a horizontal line, the cylinder pressure curve would experience global drift. In this case, the cylinder pressure curve needs to be added or subtracted by a fixed offset until the polytropic exponent curve becomes a horizontal line. In this process, the value of the polytropic exponent is calculated, and the cylinder pressure curve can be corrected simultaneously [14].

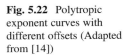

Fig. 5.22 Polytropic
exponent curves with
different offsets (Adapted
from [14])

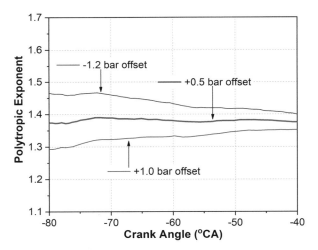

Figure 5.22 depicts the polytropic exponent curves calculated by the cylinder pressure curves with different offsets. The bigger the slope of the polytropic exponent curve is obtained for, the higher the offset of cylinder pressure (Fig. 5.22). Typically, the polytropic exponent should not be bigger than 1.4 in the adiabatic process. For the cylinder pressure offset of +0.5 bar, the polytropic exponent curve is a horizontal line, and the value of the polytropic exponent is not bigger than 1.4 in the analysis section. Therefore, the cylinder pressure curve corrected by +0.5 bar offset is the correct one, and the average value of the polytropic exponent curve in the angle analysis section is a correct polytropic exponent value. Thus, absolute pressure referencing of the measured pressure signal can be achieved using this method also.

5.4   Smoothing/Filtering of Experimental Data

Accurate measurement of in-cylinder pressure is clearly a prerequisite for good data and subsequent analysis. Typically, a large quantity of information related to the combustion process (heat release, combustion phasing, reaction rate, etc.) is generated by post-processing of the cylinder pressure signal. Therefore, it is essential to obtain actual physical information (cylinder pressure), which is free from signal noise. To conserve the useful physical information in the in-cylinder pressure signal, filtering (removal of high-frequency noise) or smoothing of the signal is necessary. Different filtering and averaging methods are generally used for smoothing the cylinder pressure signal, which leads to a precise combustion diagnosis [1, 2]. It is well known that differentiation (derivative) of signal leads to increase (amplification) of signal noise. Therefore, filtering the cylinder pressure signal is important because heat release calculation uses pressure derivatives. Signal noise present in the cylinder pressure data may lead to large error in heat release calculation. Figure 5.23 shows

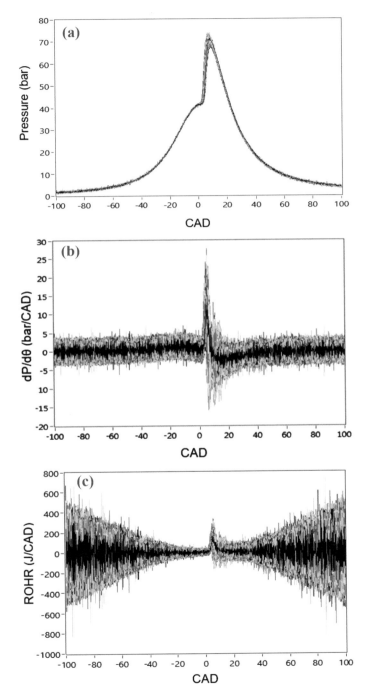

Fig. 5.23 (**a**) Measured cylinder pressure, (**b**) calculated pressure derivative, and (**c**) calculated heat release rate for 3000 consecutive cycles at 1200 rpm and $\lambda = 2.1$ in HCCI combustion engine [2]

the measured cylinder pressure signal and calculated pressure rise rate and heat release rate from the pressure signal for consecutive 3000 engine cycles in HCCI combustion engine at 1200 rpm. Cylinder pressure variations are maximum near TDC position (during combustion), and significant variation can be observed in peak pressure (Fig. 5.23a). Oscillations (variations) in pressure curves are relatively very small during the compression stroke in comparison to the variations during the combustion process in the engine cycle. Signal noise is not prominently perceived in the pressure curves. However, signal noise and cyclic variations are clearly perceived in calculated pressure derivative and heat release rate curves (Fig. 5.23a, b). Signal noise becomes amplified in the pressure derivative signal. Since heat release rate calculation involves pressure derivative, the heat release rate curves (Fig. 5.23c) has very high amplitude than the amplitude in the pressure curves (Fig. 5.23a). When the piston is away from TDC position, the heat release rate (Fig. 5.23c) is observed very high (where actually no heat release) due to signal noise in measured pressure signal. The possible reason for high heat release is that amplified signal noise gets multiplied by cylinder volume in heat release calculation. The cylinder volume as well as rate of change of cylinder volume is higher when piston is away from TDC position. This leads to a very large error in heat release rate calculation using noisy cylinder pressure signal.

There are several sources of signal noise in the measured pressure signal. Signal noise sources include the pressure conversion (non-flush sensor mounting, thermal effects, sensor resonance, lack of linearity in the sensor, vibrations, etc.), signal transmission (electrical effects, bad connections, etc.), and analog-digital conversion [1, 2]. Pressure waves caused by fuel injection or the rapid rate of premixed combustion or combustion chamber resonance are also recorded by the pressure transducer, which can cause errors in the calculation of heat release rate using first law of thermodynamics [24]. Variation in engine input parameters and variations in engine operating conditions also affect the measurement of the pressure signal and signal noise. Cyclic variations can even occur at steady-state engine operating conditions. The significance of this effect depends on the combustion modes such as SI, CI, or HCCI combustion.

Averaging of many engine cycles is suggested to reduce the errors in the processing of cylinder pressure due to signal noise. Averaging of cylinder pressure data for several cycles can only remove the random noise in pressure signal, and it cannot remove the systematic errors. Additionally, averaging several cycles is not suitable for the engine running under transient operating modes [24]. Averaging of cycles is also not possible when cyclic variations in the combustion parameters need to be analyzed. Averaging of pressure data can remove the random high-frequency noise from cylinder pressure signal. Figure 5.24 illustrates the removal of high-frequency noise by averaging different numbers of cycles in HCCI combustion data. The amplitude of power spectrum signal reduces with increase in number of averaging cycles (Fig. 5.24), which suggests that the averaging has reduced the random high-frequency noise.

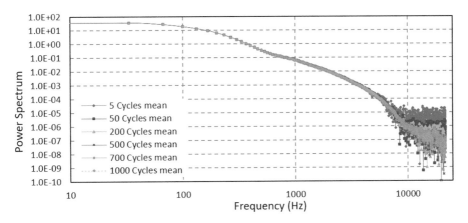

Fig. 5.24 Power spectrum of pressure signal for different numbers of cycles selected for averaging [2]

Filtering of signal noise is the only option in case of systematic errors and when averaging of the signal cannot be done. Even after averaging the pressure trace, it still requires filtering or smoothing after the premixed burn spike and the early pressure rise because of the effect of pressure waves initiated by combustion flame. Various filtering methods, such as moving average algorithms and low-pass FIR (finite impulse response) and IIR (infinite impulse response) filters, are discussed in next subsections of this section.

Another option to reduce the errors in heat release calculations is to directly measure the pressure derivative from the piezoelectric pressure sensor (see Chap. 2). This method is called current-to-voltage conversion, as rate of pressure change produces current that is converted into a voltage for data logging. Typically, cylinder pressure transducers, the charge is measured converted into voltage by the charge amplifier. This signal is used for computation of pressure derivative, where introduced signal noise gets amplified. Figure 5.25 presents the normalized heat release rate and mass fraction burned as a function of the crank angle measured by the current-to-voltage converter and charge amplifier. Figure 5.25a depicts that the heat release calculation has higher noise level in case of charge amplifier. Additionally, mass fraction burn calculation also affected by signal noise contains in the pressure signal measured using charge amplifier. It is shown that conversion of current produced from the piezoelectric transducer into an analog voltage signal reduces the quantization noise of pressure derivative data by about 70 times [25]. Various filters have been developed and recommended for filtering the signal noise to fulfill different requirements of processing the cylinder pressure for combustion diagnostics.

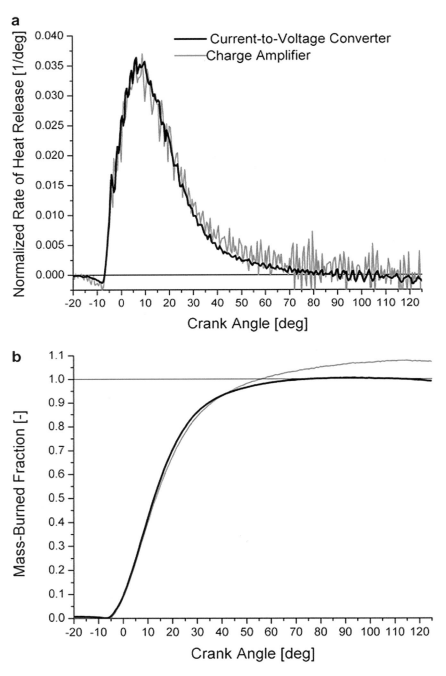

Fig. 5.25 (**a**) Normalized heat release rate, and (**b**) mass fraction burned as function of crank angle at full load [25]

5.4.1   Moving Average Filters

Cylinder pressure signal quality can be further improved by filtering and smoothing the experimental raw data. In some applications, filtering is targeted at much higher frequencies than the smoothing. For studies in which the frequencies of interest are the primary knock frequency and below, the data should be low-pass filtered at 30–50 kHz before digitization by high speed data acquisition system. Most charge amplifiers are equipped with an integral filter in this range. Smoothing is performed on the digitized data and can be successfully achieved using either the 3-point, 5-point, or 7-point least-square smoothing algorithms [24, 26]. Simple smoothing filters (such as a moving average or median filter) are frequently applied for decreasing the short-term signal fluctuations and estimating the long-term trend of the measured signal. The unweighted moving average filter can serve as a low-pass filter, and it uses the simplest convolution operation. More complex smoothing of measured signal with a weighted moving average is accomplished by applying the Savitzky-Golay convolution coefficients that can be computed from the least-square fit of subsets of adjacent data points with a low-degree polynomial [27].

Figure 5.26 presents the pressure signal filtering using four simple filters. Moving average filter with 5 and 10 data points spans, third-order polynomials for 1D median filter, and Savitzky-Golay filter with second-order polynomial with 7 points are used for smoothing (Fig. 5.26). Savitzky-Golay and the median filters show very good agreement with the raw data during compression stroke up until the start of combustion. The pressure curves with moving average filters deviated from measured raw data shortly before the initial pressure rise because of calculated average pressure in the window spread the effect of the quick pressure rise. Larger points moving average filter have the larger deviation as expected (Fig. 5.26). The high-frequency pressure signal noise generated during combustion is not eliminated effectively by any of the four filters. The worst performance is shown by median

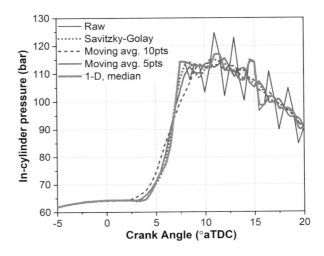

Fig. 5.26 Illustration of cylinder pressure signal smoothing using simple filters (Adapted from [27])

filter. Savitzky-Golay filter also exhibited higher oscillation in comparison to the moving average filters. Figure 5.26 demonstrates a trade-off between accuracy near the start of combustion and pressure oscillations during combustion with the simple filters [27].

A study used five-point moving average filter for filtering the signal noise only before the start of combustion and low-pass digital filter used for another portion of engine cycle because high-amplitude pressure oscillations are produced after starting of combustion [28]. A simple moving average filter can eliminate any background system noise or random electrical noise in the cylinder pressure signal before the initiation of combustion process. The moving average filter is not able to distinguish between the bands of frequencies. The performance of moving average filter is excellent in the time domain for smoothing a measured signal. However, moving average filter performs very badly in the frequency domain for smoothing measured signal as low-pass filter [28].

The moving average filter has an extra ability to remove any five to ten point noise spike present in the cylinder pressure signal (regardless of its amplitude). The spikes are typically picked up from the interference from other devices, etc. (such as spark plug). Figure 5.27 illustrates the elimination of the abrupt background noise (spike) from the cylinder pressure signal. The filtering algorithm of moving average filter recognizes the start and end of the sudden noise spike and replaces the points with the mean value of the noise start and endpoint. The smoothing ability of the moving average filter is typically dependent on the sampling interval.

Another smoothing algorithm $(2b + 1)$ points shown by Eq. (5.14) is proposed in [29]. The smoothing equation can be applied recursively (i.e., more than once), and it shows the significant removal of noise errors.

$$a_n = \frac{1}{b^2}\big[a_{n-(b-1)} + 2a_{n-(b-2)} + 3a_{n-(b-3)} + \cdots + ba_n + \cdots + 3a_{n+(b-3)}$$
$$+ 2a_{n+(b-2)} + a_{n+(b-1)}\big] \qquad (5.14)$$

Fig. 5.27 Random noise spike filtering using moving average filter from cylinder pressure signal (Adapted from [28])

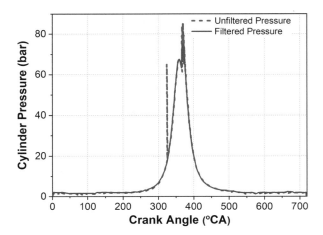

Least-square polynomial approximation (LPA) over a set of points can also be used for smoothing the cylinder pressure trace. Polynomials are the approximating functions of choice when a smooth function is to be approximated locally. However, if a function is to be approximated on a larger interval, the degree of the approximation polynomial can be unacceptably large. To solve this issue, the interval of approximation is subdivided into sufficiently small intervals in such a way that a polynomial of relatively low degree can offer a good approximation to the function interval. This polynomial approximation is performed such that the polynomial pieces blend smoothly and such smooth piecewise polynomial function is called a spline. The spline function is found to be very effective for smoothing both steady and transient cylinder pressure traces [24]. The spline function applied in the study is the cubic smoothing spline. The starting smoothing point has a significant effect on the accuracy for analyzing heat release rate and to eliminate systematic error.

Designing of filter for the combustion pressure signal faced several challenges to meet the requirement of signal processing over a wide range of operating condition (normal and abnormal). Typically, all filters introduce phase shift to a certain amount in the filtered output signal. A simple zero-phase (phase-less) filter is not an optimal smoothing method to this particular signal of combustion pressure, where a rapid increase in the signal occurs after the initiation of combustion. Even using a zero-phase digital filter for smoothing cylinder pressure signal leads to shift in the signal, and it is not able to track the abrupt pressure trace. Figure 5.28 illustrates the phase shift by zero-phase filter in the cylinder pressure signal and its comparison with five-point moving average filter. The zero-phase Butterworth filter applied to an oscillatory pressure signal is able to reduce the pressure oscillation but also adds the shift in the signal near the combustion starting point.

Identification of crucial filter parameters such as cutoff frequency, filter order, passband characteristics, etc. is an important issue particularly when the filter is applied online. Variation in engine operating conditions (speed and load) changes

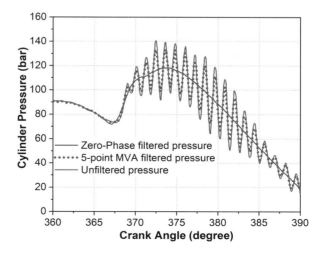

Fig. 5.28 Illustration of phase shift in filtered pressure signal with zero-phase Butterworth filter in pressure signal (Adapted from [28])

the frequency spectrum of the pressure curve. Thus, the predetermined/predefined cutoff frequency for all the engine operating condition is not a good strategy. Therefore, an adaptive method is a more suitable strategy for choosing the cutoff frequency for the cylinder pressure signal filter (particularly for online application) [28].

5.4.2 Low-Pass FIR Filter

Two types of digital filters are used for filtering signal noise from acquired cylinder pressure data: finite impulse response (FIR) and infinite impulse response (IIR). The impulse response (or response to any finite length input) of an FIR filter has finite duration, i.e., it settles down to zero in the finite amount of time. The phase shift in FIR filter is linear because it does not use feedback, and thus, it depends only on the input. Linear phase characteristics of FIR filter are their most significant advantage, but it needs more computation capacity in comparison to an IIR filter. The FIR filters are stable in comparison to IIR filters [28].

Payri et al. [1] developed low-pass FIR filters for offline and online filtering of the experimental cylinder pressure signal. Identification of the optimum cutoff frequency is the major problem with low-pass filter [1, 30]. The noise-to-signal ratio becomes important above the cutoff frequency. Additionally, the direct removal of the high-frequency band can lead to overshooting of the pressure signal (the Gibbs effect) which results in significant error in the heat release computation. This can be eliminated by smoothing the transition with a Hanning window [30, 31], which is defined between two cutoff frequencies: the stopband initial frequency and the stopband final frequency.

The offline FIR filter proposed is presented by Eq. (5.15) [1]:

$$P_k^{\text{filt}} = P_k \cdot \theta_k \tag{5.15}$$

$$\theta_k = \begin{cases} 1 & \text{if } k < k_c - \dfrac{k_{\text{stop}}}{2} \\[3mm] \dfrac{1}{2} \cdot \left[\cos \left(\dfrac{k - \left(k_c - \dfrac{k_{\text{stop}}}{2} \right)}{k_{\text{stop}}} \cdot \pi \right) + 1 \right] & \text{if } k_c - \dfrac{k_{\text{stop}}}{2} < k < k_c + \dfrac{k_{\text{stop}}}{2} \\[3mm] 0 & \text{if } k < k_c + \dfrac{k_{\text{stop}}}{2} \end{cases} \tag{5.16}$$

where P_k is the content of the averaged pressure signal at harmonic k and P_k^{filt} is the filtered value of the spectrum, k_c is cutoff harmonic, and k_{stop} is stopband edge harmonic. To get the averaged and filtered pressure in the temporal domain, inverse

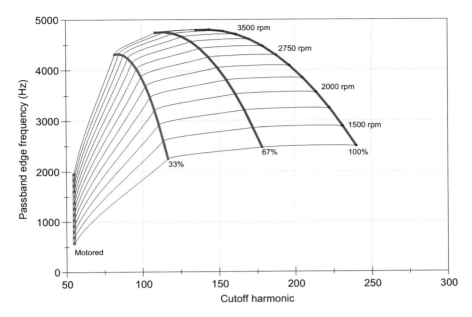

Fig. 5.29 Cutoff frequency map [1]

discrete Fourier transform (DFT) is used. The cutoff harmonic is estimated as the point where the average cycle harmonics meets with the noncyclic harmonics, attributable to signal noise and cycle-to-cycle variations.

The DFT-based filter calculation is quite a time-consuming, and acquiring several cycles before processing causes a significant delay from the data acquisition. Therefore, this method is not suitable for online filtering of the measured pressure signal. An online filter is proposed, where the cutoff frequency is estimated based on a cutoff harmonic map and fixed 1 kHz stopband is used. Figure 5.29 illustrates the cutoff frequency map for the test engine. The cutoff frequency depends on engine speed and load conditions.

In another study, Payri et al. [30] proposed an adaptive method for automatic determination of cutoff frequency for smoothing the cylinder pressure. This method is based on the statistical analysis of the DFT representation of the signal: signal-to-noise ratio is identified and used for detecting the frequency where the contribution of the noise equals that of the signal. Figure 5.30 compares the heat release rate calculated using different adaptive filters (statistical and map-based) at different engine loads. The figure suggests that map-based adaptive filter has more noise in heat release curve, and thus, statistical analysis-based adaptive filter (proposed) is better method for processing cylinder pressure signal. High repeatability is shown by this method, and it is able to adapt suitably the cutoff frequencies to the pressure signal bandwidth. Hence, this method can be used for the full engine operating range, without additional manual settings requirement.

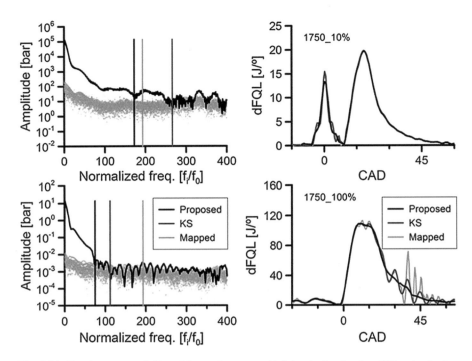

Fig. 5.30 Fourier spectra (left) and heat release rate (right) calculated using different adaptive filters at different engine loads [30]

Another study [32] proposed a procedure for designing an optimum equiripple FIR filter for filtering the combustion pressure signal of a diesel engine. A novel method of estimating the transition band frequencies and optimum filter order is presented. This method is based on discrete Fourier transform (DFT) analysis, which is the first step to determine the position of the passband and stopband frequencies. These passband and stopband frequencies are further used to estimate the most suitable FIR filter order.

5.4.3 Low-Pass IIR (Butterworth) Filters

The infinite impulse response (IIR) filter has an impulse response function, which is non-zero over an infinite length of time. The IIR filters use feedback, and thus, the phase shift is a nonlinear function of frequency. An IIR filter is also known as a recursive digital filter because its output is a function of previous outputs as well as the input. If $x[n]$ represents the nth input to the filter and $y[n]$ is the nth output of the filter, then a general IIR filter is implemented as Eq. (5.17):

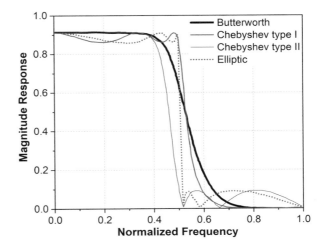

Fig. 5.31 Frequency responses of fifth-order low-pass IIR filters (Adapted from [28])

$$y[n] = a_0{}^*x[n] + a_1x[n-1] + \cdots + a_Mx[n-M]$$
$$- (b_1y[n-1] + b_2y[n-2] + \cdots + b_Ny[n-N]) \qquad (5.17)$$

Equation (5.17) depicts that the nth output is a linear function of the nth input, the previous M inputs, and the previous N outputs. The coefficients "a" and "b" are computed to give the IIR filter a specific frequency response and depending on the type of filter the number of coefficients, M and N vary. The IIR filters have sharper roll-off in comparison to FIR filters of the same order, and IIR filters demand lower computation capacity in comparison to FIR filters [28].

Several type of IIR filters are available such as Chebyshev filter, Elliptic filter, Butterworth filter, Bessel filter, etc. Figure 5.31 shows the frequency responses of 5th-order low-pass IIR filters with a cutoff frequency of 0.5 normalized units. The order of the filter determines the transition from passband to stopband. Among all these filters, the Butterworth filter has the flattest passband response and poor roll-off rate. Roll-off is defined as the steepness of a response function with frequency in the transition between a passband and a stopband. Chebyshev filter has a sharper roll-off and more passband ripple (Type 1) or stopband ripple (Type 2) in comparison to a Butterworth filter. The error between the idealized and the actual filter characteristics is minimized by Chebyshev filters over the range of the filter, but filtered signal has inherent passband ripples [28]. A study [27] conducted on aviation diesel engine, showed that the Butterworth filter calculates filtered pressure with reasonable accuracy for normal combustion. However, in the case of an abnormal (erratic) combustion, the Butterworth filter leads to biases, particularly at the start of combustion with a high-pressure rise rate. The Chebyshev Type 1 filter performed well, filtering both normal and erratic combustion in-cylinder pressure data. The Chebyshev Type 1 filter with optimal parameters (fifth-order polynomial and 0.001% allowed ripples) was found to be an optimal filter for analysis of an aviation diesel engine in-cylinder data [27].

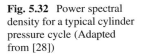

Fig. 5.32 Power spectral density for a typical cylinder pressure cycle (Adapted from [28])

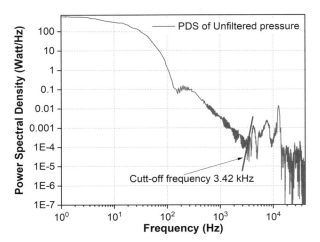

Low-pass filters are suitable for removing high-frequency noise. However determination of cutoff frequency to discriminate signal and noise is the key issue. The adaptive estimation of cutoff frequencies is required for filtering the in-cylinder pressure signal because it significantly varies with engine operating conditions. A study proposed the adaptive determination of cutoff frequency by spectral analysis of cylinder pressure trace [28]. The power spectrum analysis is used to analyze the oscillations in the pressure signal in the frequency domain. Figure 5.32 shows the power spectrum a measured combustion pressure signal of a particular engine cycle. The frequency content in the power spectral density (PSD) curve is typically dependent on the running conditions of the engine (engine load, engine speed, boost pressure, EGR, etc.). The adaptive determination of cutoff frequency is required for online application of filters. The cutoff frequency needs to separate the actual signal component and the signal noise. For adaptive determination of cutoff frequency, the slope of the PSD curve (Fig. 6.32) is calculated on a point-by-point basis. To find a strong trend in the slope of the PSD curve, a ten-point moving average is computed. The frequency component of the cylinder pressure signal progressively loses power with increasing frequency (Fig. 6.32). An abrupt increasing trend (positive slope) in the slope on a decreasing (positive slope) PSD curve suggests that noise content of the signal begins to dominate over the combustion frequency content. Frequency corresponding to the slope change point is considered as the cutoff frequency. The cutoff frequency can vary on a cyclic basis as well as engine operating conditions.

The pressure trace can be divided into motoring, combustion only, and noise signal by a decomposition method proposed in [33]. The PSD curve of these three components shows that low-frequency content up to 200 Hz dominated by the motoring curve, where piston movement affects the pressure curve. The mid-frequency content from 200 Hz to 3.42 kHz (for pressure curve in Fig. 5.32) is dominated by the combustion-only PSD curve. The unaccounted noise is

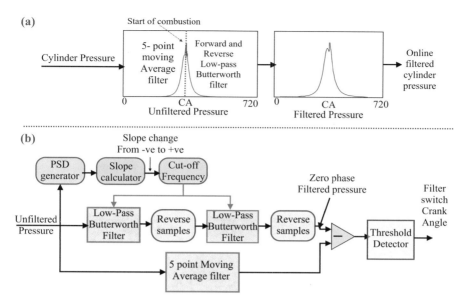

Fig. 5.33 Schematic diagram of a novel filter for cylinder pressure trace, (**a**) graphical illustration, and (**b**) flow-chart (Adapted from [28])

dominated by frequency content higher than 3.42 kHz [28]. This observation also validates the adaptive cutoff frequency determination method.

The actual information content of the cylinder pressure signal lies in frequency components in the low and mid-frequency region, and unaccounted noise is dominated in the high-frequency region. All digital filters (FIR or IIR type) lead to an inherent shift in the signal near the beginning of combustion because of an abrupt increase in the cylinder pressure signal. Therefore, a combination of two filters (five-point moving average and low-pass Butterworth filter) is used to avoid the phase shift. The schematic diagram of the proposed filter is presented in Fig. 5.33. The five-point moving average filter is applied till the start of combustion, and the Butterworth filter is used during combustion oscillations. The filter switch crank angle (start of abrupt combustion pressure) is determined by a threshold detector, which takes the difference of filtered signal of both the filters. When the difference crosses a particular threshold value, the filter switches. The cutoff frequency for the Butterworth filter is identified by the method described in Fig. 5.32. A fifth-order forward and reverse Butterworth filter (having a moderate roll-off factor) is used to filter the measured raw pressure signal. A forward and reverse Butterworth filter act as a zero-phase filter.

Figure 5.34 shows the comparison of different filter responses on the cylinder pressure signal. The five-point moving average filter works better during compression, and it is not able to remove the oscillations during combustion. The digital Butterworth filter creates a shift at the start of combustion (Fig. 6.34). However, the combined filter avoids the shift in the filtered pressure trace due to an intelligent switching between the filters just after the beginning of combustion in the cylinder.

Fig. 5.34 Comparison of
different filter responses on
the cylinder pressure signal
(Adapted from [28])

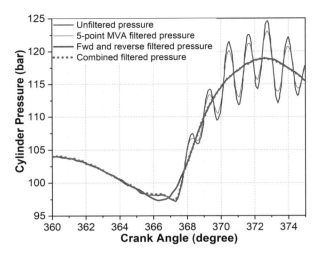

5.4.4 Thermodynamic Method-Based Filter

The measured cylinder pressure signal contains noise because of the natural characteristics of the combustion process and measurement system. Filtering/smoothing the pressure signal affects the successive heat release analysis. Several mathematical algorithms can directly smooth the cylinder pressure trace, but these methods are dependent only on mathematical algorithms. Thus, the loss of actual signal information is easier during the filtering/smoothing process. An improved methodology presented to smooth the pressure trace, which uses not only mathematic calculation but also thermodynamic knowledge [34].

In this method, average but not yet smoothed signal is used for heat release calculations. The cumulative heat release (reaction coordinate) is fitted using a series of Vibe functions. Successively, multiple Vibe functions obtained are further used as the heat release input in a single-zone simulation model to predict the pressure trace. The simulated pressure trace can be considered as a smooth characterization of the measured raw signal [34].

The combustion reaction rate (CRR) is calculated from the energy balance equation, which is presented in Eq. (5.18):

$$\mathrm{CRR} = \xi = \frac{mc_v\left(dT/dt\right) + p\left(dV/dt\right) + \dot{Q}_{\mathrm{loss}}}{u_{\mathrm{comb}}} \qquad (5.18)$$

The average measured pressure curve is used as the main input to the model. The Vibe combustion curve was originally proposed in [35]. The Vibe function is based on the first principle of chain reactions, where the radical formation is proportional to the quantity of fuel in the combustion chamber and the increase of radicals is proportional to the reduction of fuel [34, 36] as represented by Eq. (5.19):

$$\frac{dm_f^+}{dt} = km_f \quad \text{and} \quad dm_f^+ = -\mu dm_f \tag{5.19}$$

Then, the combustion reaction rate is calculated as Eq. (5.20):

$$\xi = \frac{dm_f}{dt} = -\frac{k}{\mu}m_f \tag{5.20}$$

The normalized combustion rate (Z), which is related to the normalized reaction rate (X), can be defined as Eq. (5.21):

$$z = \frac{dX}{d\tau} = \xi\frac{t_{comb}}{m_{f,0}} \quad \text{and} \quad X = \frac{m_f}{m_{f,0}} \tag{5.21}$$

where τ is normalized time $(\tau = t/t_{comb})$. Vibe proposed a model assuming a nonlinear time dependency of the reaction constant $(k \propto t^m)$. The normalized reaction rate and reaction coordinate can be represented as Eq. (5.22):

$$Z = a(m+1)\tau^m e^{-a\tau^{m+1}} \quad \text{and} \quad X = 1 - e^{-a\tau^{m+1}} \tag{5.22}$$

A series progression of Vibe functions was proposed for heat release calculation [37] as presented in Eq. (5.23):

$$X = \sum_{k=1}^{n} b_k X_k = \sum_{k=1}^{n} b_k \left(1 - e^{-a\tau^{m_k+1}}\right) \tag{5.23}$$

The normalized reaction rate (X) can be fitted using a nonlinear least-square fitting theory with different orders of Vibe function, assuming $m_{f,0}$ is known. The heat release characteristics of the cylinder are captured by fitting the parameters of multiple Vibe functions. The fitted multiple Vibe functions are used as an input for the estimation of the smoothed pressure signal using "cylinder process simulation model" [34]. It demonstrated that this method is able to remove the pressure oscillations. Additionally, three Vibe functions are the minimum, and four or five Vibe fitting functions are sufficient to get the heat release data with sufficient accuracy for that particular engine [34].

5.4.5   Wavelet Filtering

Computation of combustion metrics without careful noise rejection techniques affects the feedback control by bringing inaccurate or time-delayed signals to the feedback control system. A study developed a technique for de-noising pressure signals for the same cycle feedback control using a wavelet filter [38]. Wavelet filters allowed tuning of de-noising characteristics as a function of sampling crank angle resolution and desired noise elimination capability.

The wavelet-based method is known for effectively removing noise from a signal by identifying which component or components (of the signal) contain the noise and then reconstructing the signal without those parasitic components [39]. The wavelet de-noising process involves down-sampling of the convoluted signal, which in general leads to signal aliasing except when the convolution process involves a wavelet acting as a transfer function [38]. The convolution of the pressure signal with a transfer function is a desirable filtering option because the pressure signal being filtered consists of a fixed number of samples in an array consisting of a predefined positioning of the samples on the angular scale. To facilitate efficient filtration of noise affected pressure signals, wavelets are considered as the transfer function. The wavelet filter does not cause lag in signals, allowing for accurate estimation of the signal mean in the presence of noise without distorting the signal response with respect to crank angle. Wavelet-based filtration are fully immune to engine speed changes as the shape of the signal does not change with the engine speed and wavelets respond only to the pattern of sampling (number and distance between samples) [38].

5.5   Cycle Averaging of Measured Data

For automotive engine research and development, the cylinder pressure signal is always an important experimental diagnostic parameter, and it can provide the large amount of information by correct processing such as combustion phasing, thermal efficiency, knocking, cyclic torque variability, intake and exhaust tuning, cylinder balance, structural loading, and cyclic fueling variability [40]. Measured cylinder pressure data is also used for validation of various engine combustion models. For these purposes, typically ensemble averaged cylinder pressure (as a function of crank angle) is used to get the mean at desired accuracy, which further used to find average performance and combustion parameters. Thus, the averaged cylinder pressure must be robust to cycle-to-cycle variations in the signal. Additionally, mean average variables (fuel mass flow, air mass flow, engine speed, etc.) are used generally for an accurate heat release rate estimation. Therefore, it is important to average the measured cylinder pressure signals so that a representative (close to actual) thermodynamic cycle can be analyzed [1].

The cylinder pressure signal is typically oscillating and varying on cycle-to-cycle basis. The oscillation in the cylinder pressure signal can be partly due to the combustion process. Both chemical and physical phenomena are responsible for cycle-to-cycle variations. Various factors can be considered for combustion variations such as the variations in the fuel-air ratio, the residual gas fraction, the fuel composition, and the motion of unburned gas in the combustion chamber [40]. Cyclic variations can easily be recognized by plotting a number of engine cycle pressure data on one figure (Fig. 5.23). The stochastic (random) fluctuations in pressure signal can be removed by averaging the number of cycles. However, the systematic errors cannot be eliminated by averaging. Other important factors for pressure signal fluctuations are signal conversion, signal transmission, analog-to-digital conversion, etc.

Numerous consecutive engine pressure cycles are typically measured to minimize the cyclic variations. Any randomly selected engine combustion cycle may not be representative of the steady-state operation of the engine. The optimal number of cycles to be averaged for representing a steady-state operation of the engine depends on the type of engine combustion mode, the data acquisition system, and the engine operating conditions [1, 41, 42]. Generally, the engine stability is the major issue in the determination the number of cycles to be considered for averaging. In diesel engines, a lower number of cycles are typically required because of its relatively lower cyclic variation in comparison to corresponding spark ignition engine. The thermo-fluid dynamic processes are more stable in a diesel engine. Even for the same engine, different engine operating conditions have different levels of cyclic variations due to engine stability [1]. Typically, engines are more steady and stable at higher speed and load conditions, and lower load conditions (particularly idle operating conditions) have higher variations. Cylinder pressure data acquisition system can also affect the optimal number of cycle to record, in addition to the engine and operating point stability [22]. Additionally, the number of cycles to be recorded also depends on the most critical or demanding application for which data is recorded [43]. The optimal number of cycles to analyze the pressure rise rate or the heat release rate is not necessarily the same [2].

In most of the published study, experience-based rules are mostly used depending on the application. Two statistical methods are discussed for determination of optimal cycle number in the following subsections.

5.5.1   Method Based on Standard Deviation Variations

In this method, the variations in the standard deviation of the measured data at each crank angle position are used to determine the optimal number of cycle to measure. The optimal number of cycles is considered as the number of cycles where a further increase in the number of cycles will not improve the precision of the estimated results. To estimate the optimal number of cycles, the first large number of cycles (M) of pressure data is recorded using the high-speed data acquisition system. Next, a set of mean engine cycle (j) is calculated for a different number of selected cycles (m) using the Eq. (5.24):

$$\bar{Y}_{m,j}(\theta) = \frac{1}{m} \sum_{i=j}^{j+m} Y_i(\theta) \qquad (5.24)$$

where Y_i is a particular signal during ith cycle and θ is the crank angle (CA) position. Y_i can be a pressure or pressure rise rate or heat release rate signal. For every selected number of cycles, a set of $M - m + 1$ average signal $Y_{m,j}$ can be determined. The standard deviation as a function of crank angle position is calculated using Eq. (5.25) for each set:

$$\sigma_m(\theta) = \sigma_m\{Y_{m,j}(\theta)\} \qquad (5.25)$$

There exists a variation of standard deviation in all set $(M - m + 1)$ at every crank angle position, and using a range of variation in standard deviation at each position, a maximum and a minimum envelope curve can be derived. Figure 5.35 presents the variations of the standard deviation (for $m = 5$) as a function of crank angle position for the cylinder pressure traces recorded for 3000 consecutive engine cycles in HCCI engine. There is a 2996 set of data can be derived when five cycles are selected for 3000 consecutive engine cycles recorded. At each crank angle position, 2996 values of standard deviation are computed, and one such variation is shown with the thin

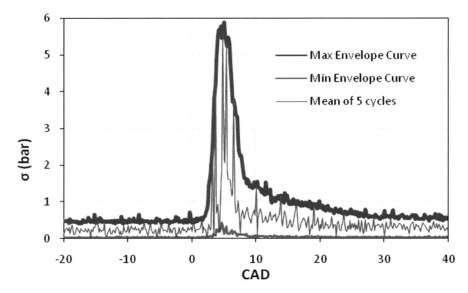

Fig. 5.35 Standard deviation and its maximum and minimum envelope curves of five averaged pressure cycles in HCCI engine [2]

blue line (Fig. 5.35). The upper and lower bold lines in Fig. 5.35 correspond to the maximum and minimum value of standard deviation at each crank angle position. The figure clearly shows that the variations in the standard deviation of the pressure signal are dependent on the crank angle position, and it is higher during the combustion period. The large increase in the standard deviation is because of higher cyclic variations in the combustion process.

Figure 5.36 shows the variations in standard deviation as a function of crank angle position with different number of cycles selected for averaging for cylinder pressure signal (with and without filter) in HCCI engine. The maximum and minimum envelope curve (as derived in Fig. 5.35) is presented with a different color for different numbers of cycles. Standard deviation value decreases with higher in number of cycles selected, and this trend is the same for both filter and non-filter pressure trace. The higher value of the standard deviation is lower for filtered pressure trace (Fig. 5.36b). The area between the maximum and minimum envelope curve also shrinks with the increasing number of cycles for both without filter and filtered pressure trace.

The maximum difference between the maximum and minimum standard deviation curves $\{(\sigma_{max} - \sigma_{min})_{max}\}$ in whole engine cycle is calculated for all the test conditions without filter and with different filters. Figure 5.37 presents the variations of $(\sigma_{max} - \sigma_{min})_{max}$ with different numbers of cycles selected for averaging at different HCCI engine operating conditions for pressure signal without filter and with different filters. The values of $(\sigma_{max} - \sigma_{min})_{max}$ are higher for raw pressure signal (without filter) in comparison to filtered pressure signal. The figure also depicts that the $(\sigma_{max} - \sigma_{min})_{max}$ value decreases rapidly with higher number of selected cycles. The value of $(\sigma_{max} - \sigma_{min})_{max}$ does not reduce after particular number for selected cycles in all the four cases, which means adding further number of cycles will not decrease the standard deviation and also will not improve the precision of the mean value. This observation suggests that this is the optimal point of number of cycles and the additional cycle will not increase the accuracy of results too much. Figure 5.37 also depicts that the minimum number of cycles (the point where additional cycle has no improvement in standard deviation) is highly dependent on engine operating condition and the type of filter used to smooth the pressure data. Single optimal number of cycles for all the operating conditions can be determined on the basis of the allowed threshold value of $(\sigma_{max} - \sigma_{min})_{max}$. Value of number of cycles at which $(\sigma_{max} - \sigma_{min})_{max}$ values are less than a threshold value can be considered as the optimal required number of cycles for analysis [2, 41]. The similar analysis is also conducted for calculated pressure rise rate and heat release rate; the optimal number of cycles is different depending on the application. A more complete detail can be found in the original study [2].

The value of $(\sigma_{max} - \sigma_{min})_{max}$ also depends on the engine combustion mode. Figure 5.38 shows the variations of $(\sigma_{max} - \sigma_{min})_{max}$ with different number of cycles selected for averaging cylinder pressure trace in a diesel engine. It can be noticed that for this particular operating condition, the optimal number of cycle is 25, which is very less in comparison to the HCCI engine (Fig. 5.37).

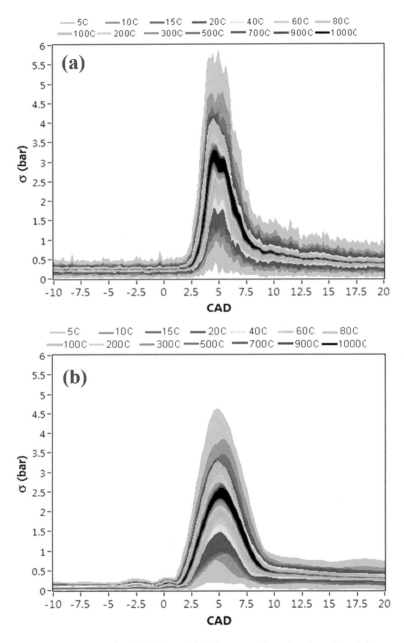

Fig. 5.36 Variations in standard deviation with different number of cycles selected for averaging for cylinder pressure in HCCI engine at 1200 rpm and $\lambda = 2.1$. (**a**) No filter. (**b**) Savitzky-Golay filter [2]

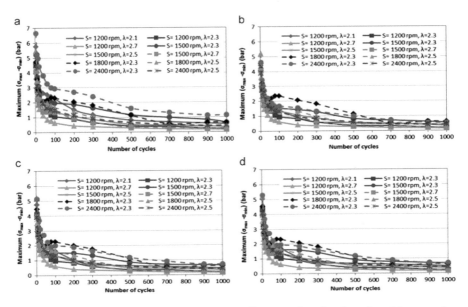

Fig. 5.37 Variation of the $(\sigma_{max} - \sigma_{min})_{max}$ with different number of cycles selected for averaging the pressure signal at different operating conditions of HCCI engine. (**a**) No filter, (**b**) Savitzky-Golay filter, (**c**) zero-phase filter, and (**d**) Butterworth filter [2]

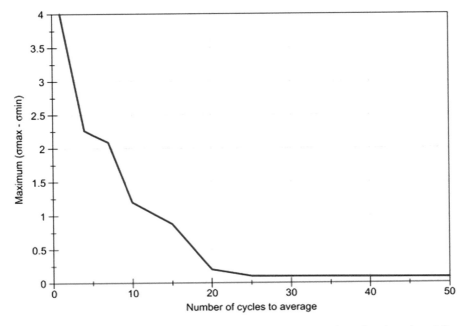

Fig. 5.38 Evolution of the $(\sigma_{max} - \sigma_{min})_{max}$ with different the number of cycles selected for averaging cylinder pressure in a diesel engine [1]

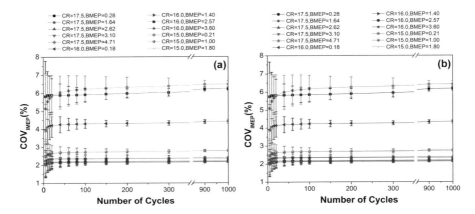

Fig. 5.39 The variation of COV$_{\text{IMEP}}$ with different number of cycles in a convention diesel engine for (**a**) no filter and (**b**) Butterworth filter [41]

To estimate the indicated engine performance, typically indicated mean effective pressure (IMEP) is used. Noise in the cylinder pressure signal and cycle-to-cycle variation in pressure signal result in the significant variation in IMEP value. The coefficient of variation (COV) of IMEP is typically used to characterize engine stability at particular engine operating point. To determine the optimum number of cycles for lower variation in the IMEP calculation, the COV$_{\text{IMEP}}$ is evaluated at various operating conditions for different number of cycles using the data of 2500 consecutive cycles (Fig. 5.39). For each number of cycles selected for calculation, the variation of COV$_{\text{IMEP}}$ is determined, the average value is shown by symbol, and the standard deviation is represented as error bars (Fig. 5.39). The figure depicts that average values and standard deviation of COV$_{\text{IMEP}}$ reduces with higher engine load. The variation of IMEP is also higher at lower compression ratio (CR). Figure 5.39 also illustrates that filtering the pressure trace does not lead to a significant reduction in standard deviation of COV$_{\text{IMEP}}$ (error bars). To eliminate the cyclic variations, the number of cycles that can be used depends on engine operating conditions. For standard deviation of COV$_{\text{IMEP}}$ less than 0.5, lower numbers of cycles (up to 60 cycles) are needed at higher engine load conditions in comparison to idle conditions (up to 200 cycles) [41].

5.5.2 Method Based on Statistical Levene's Test

In this method, the optimal number of cycles is determined using statistical Levene's test. After recording the sufficient number for pressure cycles, the data is divided into bundles of first 10, 20, 30, and so on and number of engine cycles. The standard deviation of IMEP (σ_{IMEP}) data in all the groups is calculated. In order to determine the optimal number of cycles, it is essential to show that the additional data after a certain point do not significantly vary the statistics of σ_{IMEP}. It means there is no need for an additional number of cycles for achieving a correct value of COV_{IMEP} if standard deviation remains the same as with the increase in cycle number [40].

The Levene's test is used because it provides better results under non-normality conditions [44], and the IMEP data may not be necessarily normally distributed. In this test, the test value M calculated by Eq. (5.26) is usually compared with the critical value of F-distribution for the particular degree of freedom and significance level (p-value). The null hypothesis that is "variances of the groups are equal" is rejected if the p-value is lower than the determined significance level (usually 0.05) otherwise accepted. The p-value calculated by Eq. (5.27) and is defined as the probability that a randomly drawn number from the F-distribution is greater than or equal to the test value achieved. Thus, the p-value depicts how strong evidence is for null hypothesis. The higher the p-value provides, the stronger evidence for the null hypothesis to be true [45]. This suggests that there is no requirement of an additional number of cycles to obtain true variance value if the p-value is large enough [40]. Therefore, the cycle number at which variance continuously remains constant afterward can be an optimal number of cycle. A more complete detail can be found in the original study [40]:

$$ M = \frac{(N-k)\sum_{i=1}^{k} N_i \left(\bar{Z}_{i.} - \bar{Z}_{i..}\right)^2}{(k-1)\sum_{i=1}^{k}\sum_{j=1}^{N_i}\left(\bar{Z}_{ij} - \bar{Z}_{i.}\right)^2} \qquad (5.26) $$

where $Z_{i.}$ is group mean and $Z_{i..}$ is overall mean, $Z_{ij} = \left|Y_{ij} - Y_{i.}\right|$, Y_{ij} is i^{th} data of jth group, and $Y_{i.}$ is mean of the ith group.

$$ P\text{-value} = p\left(M \le F_{v1;v2}\right) \text{ is the probability that } M \text{ is smaller than } F_{v1;v2} \qquad (5.27) $$

Figure 5.40 presents the variations of p-values with the relative air-fuel ratio and engine operating speed with different number of selected cycles. The p-value reaches near the value 1 for 50 cycles for both engine operating conditions (Fig. 5.40). After 50 engine cycles, the p-value remains high between 0.9 and 1 value which is the important factor for the null hypothesis test. Thus, it is concluded that 50 cycles are enough to estimate the true covariance value and to achieve the average pressure cycle at various engine operating conditions.

Discussion/Investigation Questions

1. Differentiate between static and dynamic TDC determination methods, and discuss their merits and demerits.
2. Why two points of measurement at sufficiently away from TDC is selected during static TDC determination using dial gauge instead of taking the measurement exactly at TDC position? What is the optimum position for dial gauge measurement for static TDC determination? Justify your answer.

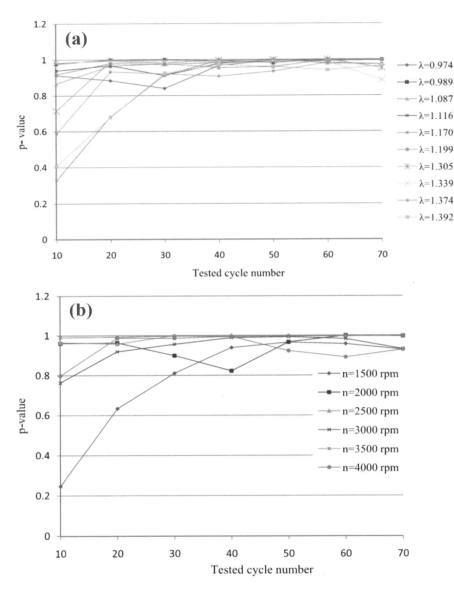

Fig. 5.40 The *p*-value variations with different cycle numbers at different (**a**) air-fuel ratios and (**b**) engine speeds [40]

3. Discuss the working principle of capacitive TDC sensor used for dynamic determination of TDC position. Write the advantages and disadvantages of TDC estimation using this method.

4. Draw typical output signal from capacitive TDC sensor for whole engine cycle of the four-stroke engine. Discuss the method for determination of compression TDC and gas exchange TDC positions.

5. Discuss the effect of the shape of the piston's upper surface on TDC sensor mounting and its signal output.

6. What is the thermodynamic loss angle? Explain the factors affecting the thermodynamic loss angle. Discuss the dependency of thermodynamic loss angle with engine speed.

7. Explain two thermodynamic methodologies of TDC determination in IC engines based on a motoring in-cylinder pressure data.

8. Make a list of the important combustion parameters affected by absolute pressure referencing error, and discuss whether the referencing error is affected by engine operating load. Write a performance or combustion parameter which is not affected by absolute pressure referencing, and justify your answer.

9. Why accurate pressure referencing is essential for mean and individual engine combustion cycle when combustion pressure is measured using a quartz piezo-electric transducer?

10. Discuss the different (at least three) methods used for pegging in-cylinder pressure data. How will you choose the pegging method for noisy in-cylinder pressure data? Discuss the possible error introduced in the cylinder pressure data due to pegging method and how these errors can be eliminated?

11. Discuss the best crank angle position for absolute pressure referencing using inlet and outlet manifold pressure sensor. Comment on the characteristics of pressure sensors required for both intake and exhaust manifold.

12. Discuss the limitations of absolute pressure referencing using intake or exhaust manifold pressure sensor.

13. Explain an efficient method of pegging calculation for real-time in-cylinder pressure offset compensation. Discuss why onboard cylinder pressure signal analysis is important.

14. Discuss the reasons for the higher preference for the polytropic exponent method for pressure correction over the manifold pressure referencing. Discuss the demerits of cylinder pressure correction using the polytropic exponent method.

15. Why filtering of in-cylinder pressure signal is required? Discuss the different methods of filtering/smoothening of the cylinder pressure signal. How optimal filtering frequency can be calculated for a given cylinder pressure signal?

16. Explain a method for adaptive determination of cutoff frequencies for filtering the cylinder pressure signal.

17. What are the sources of noise in the cylinder pressure signal of internal combustion engines? Discuss the different ways to remove the signal noise from the measured cylinder pressure signal by a piezoelectric pressure signal.

18. Differentiate between two types of digital filters, i.e., finite impulse response (FIR) and infinite impulse response (IIR), applied for filtering noise from measured cylinder pressure data.

19. Why averaging of cylinder pressure is done over a different number of cycles? Discuss the case when averaging cannot be done over cycles. Discuss a method for deciding the number of cycles sufficient for averaging.
20. Discuss the typical number of cycles that can be used in averaging for heat release analysis in CI, SI, and HCCI engines. Justify your answer based on the combustion process in three modes of engine combustion.

References

1. Payri, F., Luján, J. M., Martín, J., & Abbad, A. (2010). Digital signal processing of in-cylinder pressure for combustion diagnosis of internal combustion engines. *Mechanical Systems and Signal Processing, 24*(6), 1767–1784.
2. Maurya, R. K., Pal, D. D., & Agarwal, A. K. (2013). Digital signal processing of cylinder pressure data for combustion diagnostics of HCCI engine. *Mechanical Systems and Signal Processing, 36*(1), 95–109.
3. Luján, J. M., Bermúdez, V., Guardiola, C., & Abbad, A. (2010). A methodology for combustion detection in diesel engines through in-cylinder pressure derivative signal. *Mechanical Systems and Signal Processing, 24*(7), 2261–2275.
4. Rogers, D. R. (2010). *Engine combustion: Pressure measurement and analysis.* Warrendale: Society of Automotive Engineers.
5. Tazerout, M., Le Corre, O., & Rousseau, S. T. D. C. (1999). *TDC determination in IC engines based on the thermodynamic analysis of the temperature-entropy diagram* (No. 1999-01-1489). SAE Technical Paper.
6. Žvar Baškovič, U., Vihar, R., Mele, I., & Katrašnik, T. (2017). A new method for simultaneous determination of the TDC offset and the pressure offset in fired cylinders of an internal combustion engine. *Energies, 10*(1), 143.
7. Pischinger, R. (2002). *Engine indicating user handbook.* Graz: AVL List GmbH.
8. Bueno, A. V., Velásquez, J. A., & Milanez, L. F. (2012). Internal combustion engine indicating measurements. In Md. Z. Haq (Eds.), *Applied measurement systems.* InTech.
9. Yamanaka, T., & Kinoshita, M. (1991). Optimum probe design for precise TDC measurement using a microwave technique. *Journal of Engineering for Gas Turbines and Power, 113*(3), 406–412.
10. Yamanaka, T., Esaki, M. I. C. H. I. R. U., & Kinoshita, M. A. S. A. O. (1985). Measurement of TDC in engine by microwave technique. *IEEE Transactions on Microwave Theory and Techniques, 33*(12), 1489–1494.
11. REVelation operator reference manual. (2004). Madison: Hi-Techniques.
12. Maurya, R. K. (2018). *Characteristics and control of low temperature combustion engines: Employing gasoline, ethanol and methanol.* Cham: Springer.
13. Brunt, M. F., & Pond, C. R. (1997). *Evaluation of techniques for absolute cylinder pressure correction* (No. 970036). SAE Technical Paper.
14. Sun, W., Du, W., Dai, X., Bai, X., & Wu, Z. (2017). *A cylinder pressure correction method based on calculated polytropic exponent* (No. 2017-01-2252). SAE Technical Paper.
15. Teichmann, R., Wimmer, A., Schwarz, C., & Winklhofer, E. (2012). Combustion diagnostics. In G. P. Merker, C. Schwarz, & R. Teichmann (Eds.), *Combustion engines development* (pp. 39–117). Berlin: Springer.
16. Kaul, B. C., Lawler, B. J., Finney, C. E., Edwards, M. L., & Wagner, R. M. (2014). *Effects of data quality reduction on feedback metrics for advanced combustion control* (No. 2014-01-2707). SAE Technical Paper.

17. Maurya, R. K., & Agarwal, A. K. (2013). Investigations on the effect of measurement errors on estimated combustion and performance parameters in HCCI combustion engine. *Measurement, 46*(1), 80–88.
18. Davis, R. S., & Patterson, G. J. (2006). *Cylinder pressure data quality checks and procedures to maximize data accuracy* (No. 2006-01-1346). SAE Technical Paper.
19. Lee, K., Kwon, M., Sunwoo, M., & Yoon, M. (2007). *An in-cylinder pressure referencing method based on a variable polytropic coefficient* (No. 2007-01-3535). SAE Technical Paper.
20. Lee, K., Yoon, M., & Sunwoo, M. (2008). A study on pegging methods for noisy cylinder pressure signal. *Control Engineering Practice, 16*(8), 922–929.
21. Klein, P., Schmidt, M., & Loffeld, O. (2007). Estimation of the cylinder pressure offset and polytropic exponent using extended Kalman filter. *IFAC Proceedings Volumes, 40*(10), 175–182.
22. Randolph, A. L. (1990). Methods of processing cylinder-pressure transducer signals to maximize data accuracy. *SAE Transactions,* (900170), 191–200.
23. Tunestal, P., Hedrick, J. K., & Johansson, R. (2001). Model-based estimation of cylinder pressure sensor offset using least-squares methods. In *Proceedings of the 40th IEEE Conference on Decision and Control, 2001* (Vol. 4, pp. 3740–3745).
24. Zhong, L., Henein, N. A., & Bryzik, W. (2004). *Effect of smoothing the pressure trace on the interpretation of experimental data for combustion in diesel engines* (No. 2004-01-0931). SAE Technical Paper.
25. Bueno, A. V., Velásquez, J. A., & Milanez, L. F. (2009). A new engine indicating measurement procedure for combustion heat release analysis. *Applied Thermal Engineering, 29*(8–9), 1657–1675.
26. Randolph, A. L. (1994). *Cylinder-pressure-based combustion analysis in race engines* (No. 942487). SAE Technical Paper.
27. Kim, K. S., Szedlmayer, M. T., Kruger, K. M., & Kweon, C. B. M. (2017). *Optimization of in-cylinder pressure filter for engine research* (No. ARL-TR-8034). US Army Research Laboratory Aberdeen Proving Ground United States.
28. Dey, K. (2012). *Characterization and rejection of noise from in-cylinder pressure traces in a diesel engine* (Master's Thesis). University of Windsor.
29. Stone, R. (1999). *Introduction to internal combustion engines*. London: Macmillan Press Ltd.
30. Payri, F., Olmeda, P., Guardiola, C., & Martín, J. (2011). Adaptive determination of cut-off frequencies for filtering the in-cylinder pressure in diesel engines combustion analysis. *Applied Thermal Engineering, 31*(14–15), 2869–2876.
31. Shi, S. X., & Sheng, H. Z. (1987). Numerical simulation and digital signal processing in measurements of cylinder pressure of internal combustion engines. In *Proceedings of the Institution of Mechanical Engineers International Conference on Computers in Engine Technology (C345) C20/87* (pp. 211–218).
32. Rašić, D., Baškovič, U. Ž., & Katrašnik, T. (2017). Methodology for processing pressure traces used as inputs for combustion analyses in diesel engines. *Measurement Science and Technology, 28*(5), 055002.
33. Payri, F., Broatch, A., Tormos, B., & Marant, V. (2005). New methodology for in-cylinder pressure analysis in direct injection diesel engines—Application to combustion noise. *Measurement Science and Technology, 16*(2), 540.
34. Ding, Y., Stapersma, D., Knoll, H., & Grimmelius, H. T. (2011). A new method to smooth the in-cylinder pressure signal for combustion analysis in diesel engines. *Proceedings of the Institution of Mechanical Engineers, Part A: Journal of Power and Energy, 225*(3), 309–318.
35. Vibe, I. I., & Meißner, F. (1970). *Brennverlauf und kreisprozess von verbrennungsmotoren*. Berlin: Verlag Technik.
36. Stapersma, D. (2009). *Diesel engines, volume 3: Combustion* (5th print, pp. 636–660). Den Helder: Royal Netherlands Naval College.
37. Knobbe, E., & Stapersma, D. (2001, May). Some new ideas for performing heat release analysis. In *23rd CIMAC Conference*, Hamburg, Germany.

38. Malaczynski, G. & Foster, M. (2018) *Wavelet filtering of cylinder pressure signal for improved polytropic exponents, reduced variation in heat release calculations and improved prediction of motoring pressure & temperature* (SAE Technical Paper 2018-01-1150).

39. Donoho, D. L., & Johnstone, I. M. (1995). Adapting to unknown smoothness via wavelet shrinkage. *Journal of the American Statistical Association, 90*(432), 1200–1224.

40. Ceviz, M. A., Çavuşoğlu, B., Kaya, F., & Öner, İ. V. (2011). Determination of cycle number for real in-cylinder pressure cycle analysis in internal combustion engines. *Energy, 36*(5), 2465–2472.

41. Maurya, R. K. (2016). Estimation of optimum number of cycles for combustion analysis using measured in-cylinder pressure signal in conventional CI engine. *Measurement, 94*, 19–25.

42. Brunt, M. F., & Emtage, A. L. (1996). Evaluation of IMEP routines and analysis errors. *SAE Transactions*, 105(960609), 749–763.

43. Lancaster, D. R., Krieger, R. B., & Lienesch, J. H. (1975). Measurement and analysis of engine pressure data. *SAE Transactions, 84*, 155–172.

44. Levene, H. (Ed.). (1960). *Contributions to probability and statistics: essays in honor of Harold Hotelling*. Stanford: Stanford University Press.

45. Vardeman, S. B. (1994). *Statistics for engineering problem solving*. Duxbury Press.

Chapter 6
Engine Performance Analysis

Abbreviations and Symbols

ADC	Analog-to-digital converter
BDC	Bottom dead center
BMEP	Brake mean effective pressure
BSFC	Brake-specific fuel consumption
CCVL	Continuous variable valve lift
CDC	Conventional diesel combustion
CR	Compression ratio
CW	Counter weight
DPI	Difference pressure integral
EGR	Exhaust gas recirculation
F/A	Fuel-air ratio
GDI	Gasoline direct injection
HCCI	Homogeneous charge compression ignition
IMEP	Indicated mean effective pressure
MEP	Mean effective pressure
NMEP	Net mean effective pressure
PFI	Port fuel injection
PMEP	Pumping mean effective pressure
RCCI	Reactivity controlled compression ignition
SI	Spark ignition
SOE	Start of energizing
TDC	Top dead center
VVL	Variable valve lift
VVT	Variable valve timing
D	Diameter of the cylinder
P	Cylinder pressure
R	Crankshaft radius

© Springer Nature Switzerland AG 2019
R. K. Maurya, *Reciprocating Engine Combustion Diagnostics*, Mechanical
Engineering Series, https://doi.org/10.1007/978-3-030-11954-6_6

V	Volume of cylinder
X	Number of revolution per engine cycle
a	Acceleration of piston
A_p	Piston area
F_e	Engine flexibility
F_{IP}	Inertial force on piston
F_N	Engine speed flexibility
F_P	Piston force
F_T	Thrust force
I_S	Moment of inertia of crankshaft
m_f	Mass of fuel burned per cycle
m_p	Mass of piston
N_P	Engine speed at maximum power
N_T	Engine speed at maximum torque
P_b	Brake work
P_i	Indicated power
P_{in}	Intake pressure
Q_{LHV}	Lower heating value of the fuel
R_{bs}	Ratio of cylinder bore to stroke
r_c	Compression ratio
T_e	Engine torque
T_{IS}	Inertial torque on the crankshaft
T_{max}	Maximum engine torque
V_c	Clearance volume
V_d	Displacement volume
V_p	Piston velocity
W_b	Brake work
W	Indicated work
W_{pump}	Pumping work
α_c	Angular acceleration of connecting rod
α_e	Angular acceleration of engine
η_b	Brake efficiency
η_c	Combustion efficiency
$\eta_{g,i}$	Gross indicated efficiency
η_{ge}	Gas exchange efficiency
η_m	Mechanical efficiency
$\eta_{n,t}$	Net indicated efficiency
η_{tm}	Thermodynamic efficiency
ω_c	Angular velocity of connecting rod
ω_e	Angular velocity of engine
θ	Crankshaft position
ϕ	Equivalence ratio

6.1   Indicating Diagram Analysis

Engine performance generally means how well an engine is producing power (output) with respect to energy input or how effectively it provides useful energy with respect to some other comparable engine. Several measurable parameters are developed to determine and compare the engine performance for a particular engine as well as engines of different sizes. To maximize the performance of a particular engine, the developer must understand the combustion process and identify the inadequately designed components. Cylinder pressure-based combustion diagnostics can be used to understand the engine combustion process and to set the optimal engine operating conditions. It can also provide the direction to design or modify a particular engine component for better engine performance.

Accurate cylinder pressure can be acquired by instrumenting the engine (Chap. 2) and appropriately processing the recorded signal (Chap. 5). The measured pressure data can be graphically presented in typically three ways as shown in Fig. 6.1. The measured in-cylinder pressure data can be presented as a function of crankshaft position (P-θ) or combustion chamber volume with linear (P-V) as well as log scale ($\log P - \log V$), and each representation has its own particular characteristics

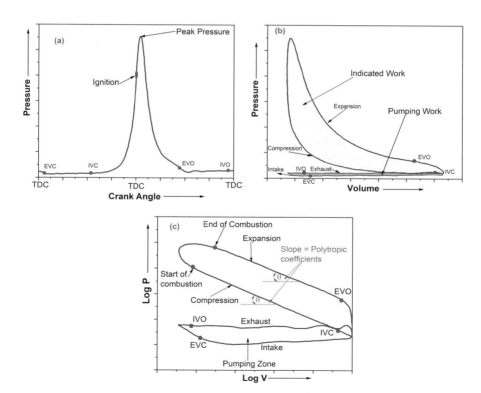

Fig. 6.1 Graphical representation of measured cylinder pressure data of one engine cycle

[1]. Crank angle-based events such as valve openings and closings, ignition and injection timings, etc. can be clearly identified in the P-θ diagram (Fig. 6.1a). In the P-V diagram (Fig. 6.1b), areas of loops represent the work produced by the engine. Area of the upper loop consisting compression and expansion strokes shows the indicated work produced in the cycle, and area of the lower loop consisting intake and exhaust strokes shows the pumping work required in the engine cycle (Fig. 6.1b). In case of the log P − log V diagram (Fig. 6.1c), the compression and expansion strokes are linear (straight line) before the start of combustion and after the end of combustion due to polytropic (adiabatic) relation between pressure and volume curve. The slopes of the straight lines in compression and expansion stroke are equal to the polytropic compression and expansion coefficients, respectively (Fig. 6.1c). Additionally, in this curve, the pumping loop is considerably expanded.

The cylinder pressure measurements and its adequate analysis can provide information regarding cylinder balance, combustion phasing, knocking, intake and exhaust tuning, cyclic torque variability, structural loading, thermal efficiency, and cyclic fueling variability. Effective analysis of these combustion and performance parameters often depends on the accuracy of the measured cylinder pressure data. Typical sources of error in the cylinder pressure measurement include absolute pressure offset, inaccurate measurement system calibration, incorrect crank angle phasing, sensitivity changes due to temperature variations, long-term drift, thermal shock distortion/short-term drift, effects due to transducer mounting, coarse analog-to-digital conversion and crank angle resolutions, mechanical vibration noise, and electrical noise [2]. The significance of each of these sources of errors in the cylinder pressure measurement is dependent on the analysis being performed using the measured pressure data.

Figure 6.2 illustrates the how different types of measurement error manifests in the cylinder pressure signal. The errors due to incorrect pegging (referencing) with cylinder volume leads to shift in the cylinder pressure signal (Fig. 6.2) as these errors typically change/shift the absolute value of pressure at the particular crank position. Thus, work calculation is not affected by this error because the area of the loop remains the same. Valve-closing noise occurs at the time of valve closure and appears in the cylinder pressure data as shown in Fig. 6.2. Valve-closing noise is a result of (1) overly accelerating the transducer (high-amplitude vibration) which generates an acoustic wave at valve impact and (2) from the fast changing stress distribution in the engine structure (specifically in the transducer mount) [1]. The amplitude of the valve noise rapidly declines after the first period (highest amplitude). Typically, valve closure noise is not an issue in production engines because the valve-seating velocities are considerably lower. Thermal shock is one of the major errors in the cylinder pressure measurement. The thermal shock causes the measured pressure early in the exhaust stroke to be considerably below the actual value (Fig. 6.2). Measurement error due to thermal shock is highest when the flame from the combustion process reaches the pressure transducer. Thermal shock gradually decreases after this point until the next combustion event. Several solutions are proposed to thermal shock problem such as water cooling, heat shield, appropriate sensor mounting position, etc. (Chap. 2). Electrical noise can appear on the cylinder

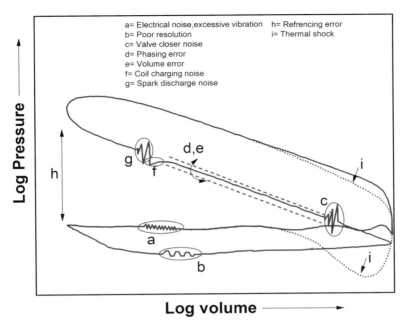

Fig. 6.2 Illustration of different types of cylinder pressure measurement error (adapted from [1])

pressure data in many different forms. One common source is conducted noise produced by the formation of a ground potential during coil charging. Another form of noise, namely, radiated noise which occurs at the time of spark discharge, may become prevalent in race engine applications. The electrical noise from each source has a distinct appearance [1]. To avoid some of the electrical noise, all instrumentation attached to the engine must be properly grounded and preferably shielded. Pressure signal needs to be recorded at sufficient resolution, and poor resolution leads to an error in representing the pressure signal (Fig. 6.2). However, a very high-resolution recording of cylinder pressure data leads to a large amount of data that can create data handling issues during post-processing.

   Cylinder pressure signal measured with sufficient accuracy can be used for further processing and analysis. Presentation of the measured pressure signal with crank angle position or cylinder volume also provides various information regarding the performance, combustion, different losses, and design of components. Figure 6.3 illustrates the variations between thermodynamic air cycle and the actual cycle in a P-V and $\log P - \log V$ diagram. The various losses can be identified by comparing the measured cylinder pressure and the cylinder pressure of an idealized cycle (Fig. 6.3). The addition of heat is typically assumed to be instantaneous, but in the actual engine cycle, the combustion of air-fuel mixture takes a finite amount of time. The non-instantaneous burning causes a reduction in the net area of the P-V diagram, which indicates a loss in work. This phenomenon is typically termed as time loss [3]. Typically, in the ideal cycle, adiabatic processes (no heat loss) are considered

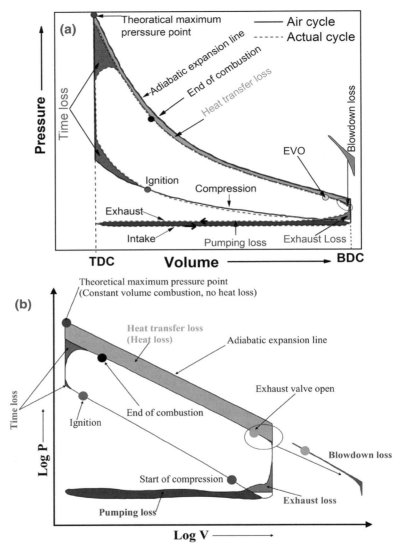

Fig. 6.3 Comparison of thermodynamic air cycle and actual cycle indicating various losses in the actual engine cycle (adapted from [3, 4])

during compression and expansion. However, the actual cycle has heat losses in the combustion chamber area, which reduces the work produced by the engine. In the ideal cycle, the instantaneous heat rejection is assumed, which is not possible in the actual cycle due to the inertia of the exhaust valve and exhaust gases. Thus, the exhaust valve opens before the piston reaches BDC position, leading to the blow-down losses. All the combustion products are not exhausted during exhaust stroke due to clearance volume. The residual gases mix with fresh charge inducted in the cycle leading to dilution and heating it, which lowers the work produced in the cycle, and termed as exhaust loss. The pressure in the cylinder is different during intake and exhaust stroke, which leads to the pumping losses. Pumping work (lower loop) is considered as loss because piston has to do the work on the gases in the cylinder. Typically, at part-load condition in SI engine, the area of the lower loop (pumping loss) increases due to restriction in air flow by the throttle.

Comparing measured in-cylinder pressure curve at different engine operating conditions also provide useful information to optimize several parameters such as the ignition timing, injection timings, valve timings, etc. for maximum thermal efficiency and lower exhaust emissions. The shape of indicator diagram provides an indication of how various processes are affecting the combustion and work produced in the cycle. Figure 6.4 shows the effect of advanced and retarded (late)

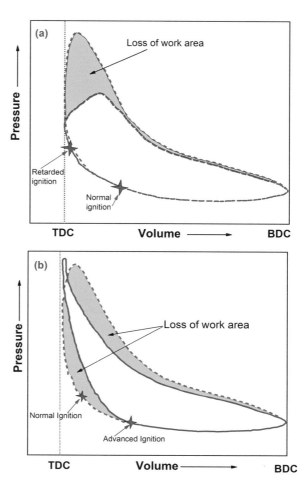

Fig. 6.4 Effect of (**a**) retarded spark timings and (**b**) advanced spark timing on work produced in the cycle (adapted from [3])

Fig. 6.5 Effect of fuel injection timing on cylinder pressure in a diesel engine (adapted from [3])

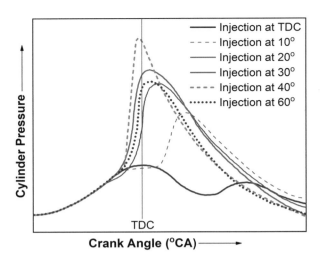

ignition timing on the work produced in the cycle. The figure clearly depicts the work loss (reduced loop area) in both advanced and retarded ignition timings. In case of retarded ignition timings, the peak pressure occurs when the piston is significantly away from TDC position due to finite time required for burning fuel-air mixture in the combustion chamber. Thus, the indicated work area is reduced, and lower work is obtained in this condition (Fig. 6.4a). When the spark timing is advanced, the combustion starts early, and pressure in the cylinder starts rising, which works against the piston motion. Thus, the indicated work area will be smaller than normal spark timings (Fig.6.4b) leading to lower work production in the cycle.

Similarly, cylinder pressure versus crank position (P-θ) can also be used for combustion analysis. The evolution of pressure as a function of crankshaft position and pressure rise rate is well presented in this curve. Figure 6.5 shows the *P-θ* diagram illustrating the effect of fuel injection timing in a compression ignition engine. The figure shows the how peak pressure and pressure rise rate are affected by fuel injection timings. For very late injection timings, the peak pressure is significantly lower because pressure rise due to combustion is utilized by overcoming the expansion of cylinder volume. Fuel injection 40° before TDC produces considerably higher peak pressure as well as higher pressure rise rate. For the injection timing much earlier (60° before TDC), the peak pressure again decreases (Fig. 6.5) due to sluggish burning [3]. For too early injection timing, the fuel is not properly prepared for rapid combustion due to lower air temperature at the timing of fuel injection. Thus, fuel injection timing can be adjusted to produce maximum power without engine roughness or emissions with the help of cylinder pressure measurement and its analysis.

The indicating diagram can be used to analyze the intake or exhaust manifold design as well the valve timings with respect to pumping loss evaluation. Figure 6.6 illustrates the effect of the improper design of intake/exhaust and improper valve timings on *P-V* diagram. If restrictions are imposed on the flow of charge through

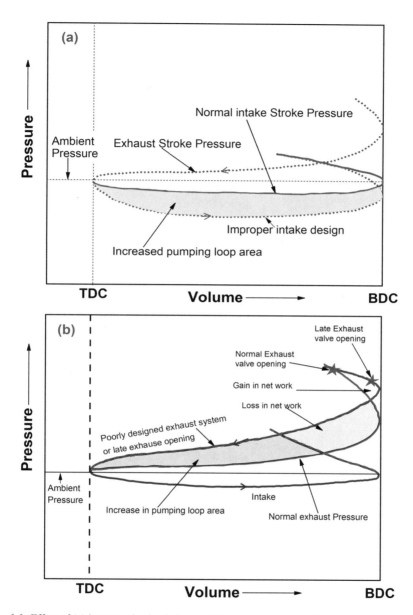

Fig. 6.6 Effect of (**a**) improper intake design and (**b**) exhaust valve opening on the pressure curve (adapted from [3])

intake system by undersized valves or other impediments, the reduction in pressure occurs, which reflect as pumping work requirement by increasing the area of pumping loop (Fig. 6.6a). Similarly, improper exhaust design or late valve opening timings leads to higher pressure in the exhaust stroke which leads to increasing the area of pumping loop. Thus, pumping work is increased due to improper design or valve timings, which can be optimized based on cylinder pressure measurement.

The measured cylinder pressure can be further processed to calculate other combustion and performance parameters. Combustion parameters and heat release calculation based on measured cylinder pressure are discussed in Chap. 7. Performance parameters such as indicated work, indicated torque, engine efficiency, etc. are discussed in the next section of the present chapter.

6.2   Engine Geometry and Kinematics

The kinematics of a reciprocating engine is similar to a slider-crank mechanism. Figure 6.7 shows a simplified reciprocating mechanism of the engine along with different terminologies used for characterizing engine geometry. The topmost piston position is defined as the top dead center (TDC), and the bottom most position is termed as a bottom dead center (BDC). The movement of the piston with respect to TDC position is denoted by variable $x(\theta)$, and the total distance traveled between

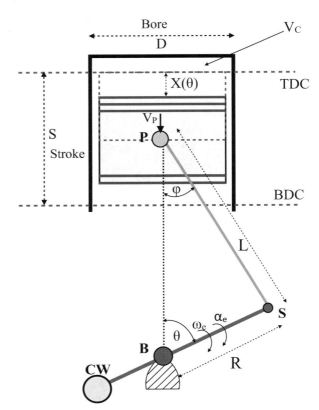

Fig. 6.7 Engine geometry and terminology

TDC and BDC position is defined as stroke (S). The diameter (D) of the cylinder is known as engine bore. The crankshaft radius is denoted as R, and connecting rod length is L. The connecting rod is joined to the piston via gudgeon pin or wrist pin (Point P) and to the crankshaft at point S using journal bearing. Crankshaft is supported on the main engine bearing (Point B). Typically, a counterweight (CW) is used for balancing the engine crankshaft.

The basic geometry of a reciprocating engine is defined by compression ratio, the ratio of cylinder bore to piston stroke (R_{bs}), and the ratio of crankshaft radius to connecting rod length (λ).

The compression ratio (r_c) is shown in Eq. (6.1).

$$\text{Compression ratio } (r_c) = \frac{\text{Volume at BDC position}}{\text{Volume at TDC position}} = \frac{V_d + V_c}{V_c} \tag{6.1}$$

where V_c is the clearance volume and V_d is swept or displacement volume. The relation between total volume (V), clearance volume, and swept volume are shown in Eqs. (6.2)–(6.4).

$$V(\theta) = V_c + x(\theta)\frac{\pi D^2}{4} \tag{6.2}$$

$$V_c = \frac{\pi D^2 R}{2(r_c - 1)} \tag{6.3}$$

$$V_d = \frac{\pi D^2 S}{4} \tag{6.4}$$

The ratio of the cylinder bore to piston stroke (D/S) is typically 0.8–1.2 for small- and medium-size engines and about 0.5 for large slow speed engines [5]. The ratio of crank radius (R) to connecting rod length (L) is denoted as λ ($=R/L$) typically less than one for all practical purpose, and its value ranges from 0.33 to 0.25 for small- to medium-size engine and 0.2 to 0.11 for larger slow speed engines.

The displacement of the piston as a function of crankshaft position (θ) can be calculated from engine geometry by Eq. (6.5).

$$x(\theta) = (L + R) - (R\cos\theta + L\cos\phi) \tag{6.5}$$

The sine law of trigonometry defines the relationship between θ and ϕ by Eq. (6.6).

$$L\sin\phi = R\sin\theta \tag{6.6}$$

Thus,

$$\sin\phi = \frac{R}{L}\sin\theta = \lambda\sin\theta \tag{6.7}$$

$$\cos \phi = \sqrt{1 - \lambda^2 \sin^2 \theta} \tag{6.8}$$

The $x(\theta)$ can be written in terms of engine geometry as Eq. (6.9).

$$x(\theta) = (L + R) - \left(R \cos \theta + L\sqrt{1 - \lambda^2 \sin^2 \theta} \right) \tag{6.9}$$

Differentiation of Eq. (6.9) provides the piston velocity (V_p).

$$\frac{dx}{d\theta} = R(-\sin \theta) + \frac{L}{2} \times \frac{-\lambda^2 2 \sin \theta \cos \theta}{\sqrt{1 - \lambda^2 \sin^2 \theta}} = -R\left[(\sin \theta) + \frac{\lambda}{2} \frac{\sin 2\theta}{\sqrt{1 - \lambda^2 \sin^2 \theta}} \right] \tag{6.10}$$

Piston velocity in terms of engine speed (ω_e) can be calculated as shown in Eqs. (6.11) and (6.12).

$$\frac{dx}{d\theta} = \frac{dx}{dt} \cdot \frac{dt}{d\theta} = \frac{dx}{dt} \frac{1}{\omega} = \frac{V_p}{\omega} \tag{6.11}$$

$$V_p = -\omega_e R\left[\sin \theta + \frac{\lambda}{2} \frac{\sin 2\theta}{\sqrt{1 - \lambda^2 \sin^2 \theta}} \right] = -\omega_e R\left[\sin \theta + \frac{\lambda}{2} \frac{\sin 2\theta}{\cos \phi} \right] \tag{6.12}$$

The acceleration of piston (a_p) can be obtained by differentiating the piston velocity (V_p), and the result is shown in Eq. (6.13) at constant engine speed (ω_e).

$$\frac{V_p}{dt} = a_p = -\omega_e^2 R\left[\cos \theta + \frac{\lambda \cos 2\theta}{\sqrt{1 - \lambda^2 \sin^2 \theta}} + \lambda^3 \frac{\sin^2 \theta \cos^2 \theta}{\left(\sqrt{1 - \lambda^2 \sin^2 \theta} \right)^3} \right]$$

$$= -\omega_e^2 R\left[\cos \theta + \frac{\lambda \cos 2\theta}{\cos \phi} + \lambda^3 \frac{\sin^2 \theta \cos^2 \theta}{(\cos \phi)^3} \right] \tag{6.13}$$

When the engine is accelerating with angular acceleration (α_e), then piston acceleration (a_p) can be represented by Eq. (6.14).

$$a_p = -\omega_e^2 R\left[\cos \theta + \frac{\lambda \cos 2\theta}{\cos \phi} + \lambda^3 \frac{\sin^2 \theta \cos^2 \theta}{(\cos \phi)^3} \right] - \alpha_e R\left[\sin \theta + \frac{\lambda}{2} \frac{\sin 2\theta}{\cos \phi} \right] \tag{6.14}$$

However, the additional second term (due to angular acceleration of engine) in Eq. (6.14) is typically very small and negligible in comparison to the first term, particularly at top engine speeds. Therefore, it is reasonable to always use Eq. (6.13) for calculation of piston acceleration even in accelerating condition of the recipro-cating engine [6].

The equation of piston speed and acceleration are highly nonlinear in nature, and they can be simplified by using Taylor series expansion as $\lambda^2 \sin^2\theta \ll 1$. The Eqs. (6.12) and (6.13) can be written as Eqs. (6.15) and (6.16) [7].

$$V_p = -\omega_e R \left[\sin\theta + \frac{A_2}{2} \sin 2\theta - \frac{A_4}{4} \sin 4\theta + \cdots \right] \tag{6.15}$$

$$a_p = -\omega_e^2 R [\cos\theta + A_2 \cos 2\theta - A_4 \cos 4\theta + \cdots] \tag{6.16}$$

where

$$A_2 = \lambda + \frac{\lambda^3}{4} + \frac{15\lambda^5}{128} + \cdots \tag{6.17}$$

$$A_4 = \frac{\lambda^3}{4} + \frac{3\lambda^5}{16} + \cdots \tag{6.18}$$

when $\lambda \ll 1$, neglecting the λ^3 or higher powers of λ, the piston velocity and acceleration can be easily calculated by Eqs. (6.19) and (6.20).

$$V_p = -\omega_e R \left(\sin\theta + \frac{\lambda}{2} \sin 2\theta \right) \tag{6.19}$$

$$a_p = -\omega_e^2 R (\cos\theta + \lambda \cos 2\theta) \tag{6.20}$$

After calculating the kinematics of piston, the kinematics of the connecting rod including the rotation speed and acceleration of its center of gravity position (Point G) can be estimated using classical dynamics. Figure 6.8 shows the kinematics terminology of the connecting rod.

The velocity and acceleration vectors of point S (where the shaft is joined with connecting rod) can be written in terms of angular velocity and acceleration of engine crankshaft. The velocity and acceleration of point S on the crankshaft can be written as Eqs. (6.21) and (6.22). Radius vector $\overrightarrow{R}$ is from the main bearing (point B) to point S. Perpendicular unit vectors are shown in Fig. 6.8.

$$\overrightarrow{V_s} = \overrightarrow{\omega_e} \times \overrightarrow{R} \tag{6.21}$$

$$\overrightarrow{a_S} = \overrightarrow{\alpha_e} \times \overrightarrow{R} + \overrightarrow{\omega_e} \times \left(\overrightarrow{\omega_e} \times \overrightarrow{R} \right) \tag{6.22}$$

$$\overrightarrow{\omega_e} = -\omega_e \hat{k} \tag{6.23}$$

$$\overrightarrow{\alpha_e} = -\alpha_e \hat{k} \tag{6.24}$$

$$\overrightarrow{R} = R \sin\theta \hat{i} + R \cos\theta \hat{j} \tag{6.25}$$

Fig. 6.8 Kinematics
terminology of connecting
road and crankshaft

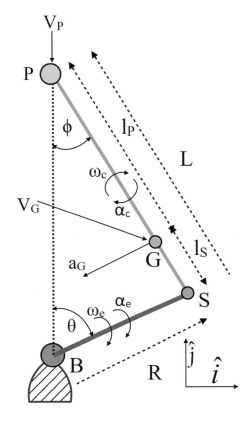

Fig. 6.8 Kinematics terminology of connecting road and crankshaft

The solution of Eqs. (6.21) and (6.22) provides the velocity and acceleration of point S.

$$\vec{V}_S = R\omega_e \left(\cos\theta\hat{i} - \sin\theta\hat{j}\right) \tag{6.26}$$

$$\vec{a}_S = R\left(\alpha_e\cos\theta - \omega^2_e\sin\theta\right)\hat{i} - R\left(\alpha_e\sin\theta + \omega^2_e\cos\theta\right)\hat{j} \tag{6.27}$$

The angular velocity (ω_c) and angular acceleration of the connecting rod can be easily calculated by knowing the velocity and acceleration of two points (S and P) on the connecting rod (rigid body). The angular velocity (ω_c) and acceleration (α_c) of connecting rod are related with two points by Eqs. (6.28) and (6.29) [6].

$$\vec{V}_S = \vec{V}_P + \vec{\omega}_c \times \vec{L} \tag{6.28}$$

$$\vec{a}_S = \vec{a}_P + \vec{\alpha}_C \times \vec{L} + \omega_c\left(\vec{\omega}_c \times \vec{L}\right) \tag{6.29}$$

where:

$$\vec{\omega}_c = -\omega_c \hat{k} \tag{6.30}$$

$$\vec{a}_c = -a_c \hat{k} \tag{6.31}$$

A solution of Eqs. (6.28) and (6.29) provides the angular velocity and acceleration of the connecting rod of the engine [6].

$$\omega_C = -\frac{R}{L} \cdot \frac{\cos \theta}{\cos \phi} \omega_e \tag{6.32}$$

$$\alpha_C = -\frac{R}{L} \cdot \frac{\cos \theta}{\cos \phi} \alpha_e + \frac{R}{L} \cdot \frac{\sin \theta}{\cos \phi} \omega_e^2 \tag{6.33}$$

The velocity and acceleration of the center of gravity of connecting rod (point G) can be estimated by considering the angular velocity and acceleration of rod and velocity and acceleration of any other point on the rod (point P or point S) using the equations similar to Eqs. (6.28) and (6.29). The velocity and acceleration of point G can be written as Eqs. (6.34) and (6.35).

$$\vec{V}_G = R\omega_e \left(C_1 \hat{i} + C_2 \hat{j} \right) \tag{6.34}$$

$$\vec{a}_G = R\alpha_e \left(C_1 \hat{i} + C_2 \hat{j} \right) - R\omega_e^2 \left(C_3 \hat{i} + C_4 \hat{j} \right) \tag{6.35}$$

where:

$$C_1 = \frac{l_A}{L} \cos \theta \tag{6.36}$$

$$C_2 = -\left(1 + \lambda \cdot \frac{\cos \theta}{\cos \phi} \cdot \frac{l_B}{L} \right) \sin \theta \tag{6.37}$$

$$C_3 = \frac{l_A}{L} \sin \theta \tag{6.38}$$

$$C_4 = \cos \theta + \lambda \cdot \frac{\cos 2\theta}{\cos \phi} \cdot \frac{l_B}{L} \tag{6.39}$$

Kinematic parameters of the connecting rod can be used in the determination of engine torque. Most important kinematic parameters are piston speed and piston acceleration. Typical variations of piston speed and acceleration as a function of crankshaft position at different engine speed are shown in Fig. 6.9. The peak piston speed and acceleration increase with engine speed. Piston speed attains zero value at both TDC and BDC positions, and reaches its peak value in the between TDC and BDC position. The variation in piston speed leads to differences in time travel in

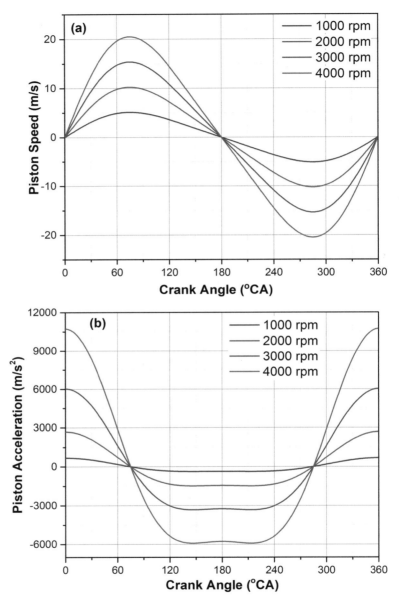

Fig. 6.9 Variations of piston speed and acceleration with crank angle position at different engine speeds of a typical engine

terms of crank angle depending on the position of the piston. Figure 6.9 also shows that the peak piston speed is not at exactly in the middle of the stroke, which can also be confirmed by the position where acceleration is zero. The peak position of piston speed depends on the rod ratio (λ) of the engine.

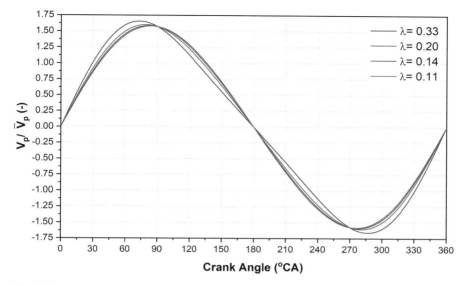

Fig. 6.10 Variations of piston speed as a function of crank angle at different rod ratios for a typical engine

Figure 6.10 shows the variation of piston speed as a function of crank angle at different rod ratios. Piston speed is normalized by dividing with mean piston speed. For relatively shorter connecting rod, the peak piston speed occurs early before the middle position of stroke (Fig. 6.10). It can be observed from Fig. 6.10 that relatively shorter connecting rod is slower at BDC range and faster at TDC range, and longer connecting rod is faster at BDC range and slower at TDC range.

In case of longer connecting rod, the piston moves slower during the middle of stroke to TDC position which may lead to a change in ignition timing because more time is spent due to lower piston speed. In expansion stroke, longer connecting rod spends less time for the travel from the middle of stroke to BDC, which makes less time availability for the exhaust to escape at the time of exhaust valve opening. The longer connecting rod engines could have relatively higher pumping losses. Shorter connecting rod spends less time near TDC position. Thus, it will suck air harder in the early part of the intake stroke and exerts more force to the crank pin before the middle of the power stroke. Thus, shorter connecting rod engines may require stronger wrist pins, piston pin bosses, and connecting rods than a relatively longer connecting rod. Shorter connecting rod spends more time at the bottom which can be beneficial in reducing intake charge being pumped back out in the intake manifold as the valve closes. Thus, varying the connecting rod ratio (λ) changes the instantaneous piston speed near TDC for a particular engine.

The mean piston speed is a measure for comparing the drives of various engines. Mean piston speed provides information on the engine load on the sliding partners and indications of the power density of the engine [8]. Typically, mean piston speed is between 8 and 15 m/s for typical automotive engines, and it is higher for racing engine up to 33 m/s [5, 9]. The mean piston speed can be calculated using Eq. (6.40).

$$\overline{V}_p = \frac{1}{\pi} \int_{180}^{360} V_p(\theta)d\theta = 2SN \tag{6.40}$$

where S is the stroke length and N is the engine speed.

The mean piston speed increases with an increase in the rotational engine speed. The mean piston speed is comparable among different types of reciprocating engines, and it is a better metric than engine speed (rpm). The mean piston velocity also defines the engine design limits. The mean piston speed affects the frictional losses and the gas exchange process significantly. Gas flow velocities in the intake manifold and gas flow velocities in the combustion chamber scales with mean piston speed. Thus, gas flow through valves and turbulent velocity of flow in the cylinder are related with mean piston speed. Typically, maximum gas flow is limited by the occurrence of sonic flow (Mach number equal to one) in the valve aperture [10]. Frictional power loss is related to engine speed via second-order polynomial. Thus, mechanical efficiency will be dependent on mean piston speed independent of engine size. The mechanical efficiency at lower speed is around 0.85 at low engine speed and drops to 0.6 at a mean piston speed of 20 m/s. This is almost independent of engine size because piston rings and cylinder are made of roughly same material, and lubricating oils are also used of similar viscosity [10]. Temperature rise of engine components such as exhaust valves and piston crowns is typically proportional to the turbulent velocity that is roughly proportional to the piston speed. A high mean piston speed is one of the structural limits imposed on any engine design [9]. The inertial forces, frictional wear, gas flow resistance during intake, and engine noise also increase at higher mean piston speed. Typically, inertial forces on connecting rods (and other reciprocating components) are proportional to the square of the piston speed (see Eqs. 6.20, 6.27, 6.33, and 6.35). Maximum allowed inertial forces limit the mean piston speed and, thus, engine speed. The engine speed is also limited by another constraint of time required for fuel-air mixture formation in the combustion chamber for reciprocating engines with internal charge preparation (direct fuel injection) such as diesel engine.

6.3 Indicated Torque Calculation and Analysis

To characterize the operating status and performance of an engine, torque is one of the key parameters. The engine torque is directly related with the drivability and the responsiveness of the automotive vehicle. The driver of the vehicle can feel the torque produced by the engine, and it is also a contributing factor in customer satisfaction. Thus, the sophisticated control of torque is required on an automotive engine [11]. Cylinder pressure measurement provides new opportunities for advanced closed-loop control, which leads to a reduction in engine emissions and improved fuel conversion efficiency [12].

Indicated torque (gas torque) is one of the parameters that can be calculated from the measured cylinder pressure data. The indicated torque is the main contributor to the output torque from the engine crankshaft. The other contributing factors are friction torque and mass torque imposed by the piston assembly. Various ways of engine torque estimation are explored because direct measurement of shaft torque using torque sensor encounters problems in cost, reliability, and vibration/noise immunity [11]. Engine torque estimation using instantaneous crank angle fluctuation is proposed by several studies [13–16]. This method of torque estimation employs a dynamic model of the engine shaft and the frequency spectrum of the measured engine speed. This method is simple and easy to implement, but it has inaccuracy [15, 17]. Engine torque can also be calculated from indicated mean effective pressure (IMEP) which is estimated from measured cylinder pressure data [5, 11].

The estimation of the indicated torque is possible using cylinder pressure measurement, which eliminates the requirement of an indicated torque model. The combustion process in the cylinder leads to high cylinder pressure that exerts a force on the top surface of the piston. This force drives the piston downward that pushes the connecting rod which forces the crankshaft to produce the engine torque (T_e), which is also known as indicated torque. The forces acting on the piston, connecting rod, and crankshaft are shown in Fig. 6.11. Inertial forces are shown to deal with this problem as static instead of dynamic using D'Alembert's principle [6]. The piston force (F_P) is created by gas pressure in the cylinder, and it can be calculated by Eq. (6.41).

$$F_P = PA_P \tag{6.41}$$

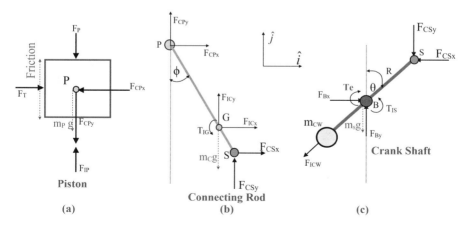

Fig. 6.11 Free body diagram of the piston, connecting rod and crankshaft of a reciprocating engine (adapted from [6])

To compute the gas torque (indicated torque), the frictional forces and gravitational forces due to the mass of components are not considered (indicated as red in Fig. 6.11). The forces acting on the piston are thrust force (F_T) from cylinder wall, inertial force due to piston acceleration (F_{IP}), and horizontal and vertical component of forces from connecting rod (F_{CPx} and F_{CPy}). The inertial force on the piston can be calculated using Eq. (6.42), where the acceleration of piston (a_P) calculated from kinematics using Eq. (6.14) and mass of piston (m_P) is known.

$$F_{IP} = m_P a_P \qquad (6.42)$$

The forces on connecting rod endpoints (P and S) as well as its center of gravity (G) are shown in Fig. 6.11b. At the piston side of connecting rod, the forces are in opposite direction as of piston (F_{CPx} and F_{CPy}). The horizontal and vertical components of forces that push the crankshaft at point S are denoted as (F_{CSx} and F_{CSy}). Forces and torque on point G are inertial forces and torque that can be calculated using Eqs. (6.43) to (6.45), where mass of connecting rod (m_C) and moment of inertia about G (I_G) are known and horizontal and vertical components of acceleration (a_{Gx} and a_{Gy}) as well as angular acceleration of connecting rod (α_C) can be calculated from Eqs. (6.35) and (6.33).

$$F_{ICx} = m_C a_{Gx} \qquad (6.43)$$

$$F_{ICy} = m_C a_{Gy} \qquad (6.44)$$

$$T_{IG} = I_G \alpha_C \qquad (6.45)$$

The forces and torque acting on crankshaft are shown in Fig. 6.11c. The forces acting on the crankshaft are the main bearing forces (F_{Bx} and F_{By}) and the forces exerted from the connecting rod at point S. The inertial torque on the crankshaft can be calculated using Eq. (6.46), where moment of inertia of crankshaft (I_s) and angular acceleration of engine/crankshaft are known.

$$T_{IS} = I_s \alpha_e \qquad (6.46)$$

Engine torque can be calculated by Eq. (6.47) using torque balance around the main bearing B.

$$T_e = R\left(F_{CSy} \sin\theta - F_{CSx} \cos\theta\right) - T_{IS} \qquad (6.47)$$

The Eq. (6.47) has two unknowns (F_{CSx} and F_{CSy}), which can be calculated by force and moment balance on the connecting rod (Fig. 6.11b). The force and moment balance equations are written as Eqs. (6.48)–(6.50). These three equations have four unknowns, which require an additional equation for solving. Additional Eq. (6.51) can be derived by vertical force balance on the piston (Fig. 6.11a).

$$F_{CSx} + F_{CPx} + F_{ICx} = 0 \tag{6.48}$$

$$F_{CSy} + F_{CPy} + F_{ICy} = 0 \tag{6.49}$$

$$F_{CSx}L\cos\phi + F_{CSy}L\sin\phi + T_{IG} + F_{ICx}l_A\cos\phi + F_{ICy}l_A\sin\phi = 0 \tag{6.50}$$

$$F_{IP} - F_{CPy} - F_P = 0 \tag{6.51}$$

Two unknowns (F_{CSx} and F_{CSy}) in the engine torque calculation equation can be computed from Eqs. (6.48) to (6.51). The engine torque can be calculated using the Eqs. (6.52) and (6.53).

$$F_{CSx} = \left(F_{IP} - F_P + \frac{l_B}{L}F_{ICy}\right)\tan\phi - \frac{l_A}{L}F_{ICx} - \frac{T_{IG}}{L\cos\phi} \tag{6.52}$$

$$F_{CSy} = F_P - F_{IP} - F_{ICy} \tag{6.53}$$

The thrust force acting on the piston by cylinder wall can also be calculated by using Eq. (6.54), which can be derived by using horizontal force balance on the piston and Eq. (6.48).

$$F_T = -F_{ICx} - F_{CSx} \tag{6.54}$$

It is clear from Eq. (6.54) that the thrust force varies with crank angle position, piston mass, piston acceleration, and cylinder pressure. The thrust force acts on one side of the cylinder in the plane of connecting rod during intake and expansion stroke of the cycle. It acts on the left side (as shown in Fig. 6.11a) in clockwise rotation of the engine shaft. This is known as the major thrust side of the cylinder because of higher cylinder pressure in power stroke leading to higher thrust force. During compression and exhaust stroke, the thrust force occurs on the other side of the cylinder, which is known as the minor thrust side (anti-thrust side) [18]. Thrust force on minor thrust side is relatively lower due to lower cylinder pressure during compression and expansion strokes. The wear due to friction occurs more on the major thrust side of the cylinder wall in comparison to minor thrust side. Modern automotive engines use pistons with less mass and shorter skirts to reduce the friction.

There are simplified models of indicated torque calculations in the published literature [6, 7]. Typical indicated torque curve calculated from cylinder pressure in a single-cylinder four-stroke engine is shown in Fig. 6.12. The figure shows that the torque is higher in power stroke, and other strokes torque is lower, and this nonuniform torque requires heavier flywheel. Multicylinder engines are typically used to achieve relatively more uniform torque from the engine crankshaft. The actual variation in resultant torque depends on the number of cylinders and their firing order.

The average engine torque can be determined by taking the mean value of the torque fluctuations in one complete engine combustion cycle at particular engine

Fig. 6.12 Variation of indicated torque in a single-cylinder four-stroke engine (adapted from [6])

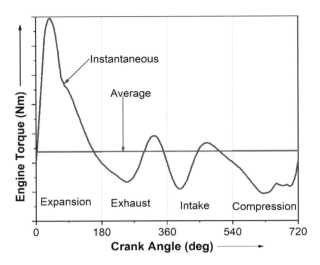

speed considering quasi-steady conditions. Typically, average brake torque on crankshaft is measured by engine dynamometer on a test cell. The torque and power characteristics of reciprocating engines are discussed in Sect. 6.7.

6.4   Work and Mean Effective Pressure Calculation and Analysis

In reciprocating combustion engines, work is produced by gases in the combustion chamber by moving the piston. Work is the result of a force acting through a distance, and it is typically calculated by Eq. (6.55).

$$W = \int F dx = \int P A_{p} dx = \int P dV \qquad (6.55)$$

where P is the cylinder pressure and A_p is the piston area, and V is the volume of cylinder.

Work done in a complete engine combustion cycle can be calculated by using measured cylinder pressure data and calculated cylinder volume (Eq. 6.2) by engine geometry. Net work done in a particular engine cycle is calculated by Eq. (6.56) using measured cylinder pressure data as a function of crank angle position [19].

$$W_{\text{net}} = \frac{2\pi}{360} \int_{-360}^{360} \left(P(\theta)\frac{dV}{d\theta} \right) d\theta \qquad (6.56)$$

Indicated work is typically defined as the mechanical work transferred from the gases to the piston during compression and expansion stroke, and sometimes called as gross indicated work. Indicated work can be calculated by Eq. (6.57). This work is equal to the area of the upper loop in a P-V diagram (Fig. 6.1b yellow region).

$$W_{\text{gross, ind}} = \frac{2\pi}{360} \int_{-180}^{180} \left(P(\theta)\frac{dV}{d\theta} \right) d\theta \tag{6.57}$$

Pumping work is the area of the lower loop in P-V diagram, which is a smaller area (Fig. 6.1b). Pumping work can be calculated during intake and exhaust stroke by Eq. (6.58).

$$W_{\text{pump}} = \frac{2\pi}{360} \left[\int_{-360}^{-180} \left(P(\theta)\frac{dV}{d\theta} \right) d\theta + \int_{180}^{360} \left(P(\theta)\frac{dV}{d\theta} \right) d\theta \right] \tag{6.58}$$

To compare the work output of engines of different sizes, mean effective pressure (MEP) is typically used. Mean effective pressure is same as the constant pressure, which acting on the piston is through the stroke would produce the same work per cycle as the work calculated by measured cylinder pressure by Eq. (6.57). The cylinder pressure is significant only during compression and expansion strokes in an engine combustion cycle. The work is done on the gases during compression stroke, while gases do work on the piston in the expansion stroke. Hence, net pressure would be the difference between these two. Therefore, the MEP is expected to be about the difference between the average pressures on the compression and power strokes [10]. By definition, the MEP is the ratio of work done per cycle and displacement volume (V_d) of the engine. The MEP has units of pressure, and it is a valuable measure of engine capacity to do work independent of the engine size. The MEP also depicts how well the available displacement volume of the engine is utilized for generating power.

Depending on the type of work used, different MEP parameters can be defined such as net mean effective pressure (NMEP), gross indicated mean effective pressure (IMEP), and pumping mean effective pressure (PMEP). Different parameters can be calculated by Eqs. (6.59)–(6.61).

$$\text{NMEP} = \frac{W_{\text{net}}}{V_d} \tag{6.59}$$

$$\text{IMEP} = \frac{W_{\text{ind}}}{V_d} \tag{6.60}$$

$$\text{PMEP} = \frac{W_{\text{pump}}}{V_d} \tag{6.61}$$

The IMEP is a very important and fundamental engine performance parameter that is used extensively in engine development work. The determination of IMEP has two key elements, namely, (1) the measurement of digital cylinder pressure as a function of crankshaft position and (2) the numerical integration of the cylinder pressure and cylinder volume for calculation of work produced (Eq. 6.57). The errors in the accurate cylinder pressure measurement and numerical integration (calculation) result into the error in IMEP determination. The IMEP can be calculated by different discretized Eqs. (6.62)–(6.64) or using Simpson's method of integration [20]. Equations (6.62) and (6.63) are most commonly used in practice for calculation of IMEP.

$$\text{IMEP} = \frac{\Delta\theta}{V_\text{d}} \cdot \sum_{i=n_1}^{n_2} P(i) \cdot \frac{dV(i)}{d\theta} \tag{6.62}$$

$$\text{IMEP} = \frac{1}{2V_\text{d}} \cdot \sum_{i=n_1}^{n_2} [P(i) + P(i+1)] \cdot [V(i+1) - V(i)] \tag{6.63}$$

$$\text{IMEP} = \frac{1}{2V_\text{d}} \cdot \sum_{i=n_1}^{n_2} P(i) \cdot [V(i+1) - V(i)] \tag{6.64}$$

where $P(i)$ is the cylinder pressure at crank angle position "i," $V(i)$ is cylinder volume at crank angle position "i," V_d is the displacement volume of cylinder, n_1 is the intake BDC integer crank angle position, and n_2 is the exhaust BDC integer crank angle position.

Crank angle resolution of cylinder pressure measurement can create the difference in computed IMEP values depending on the equation used for calculation. Figure 6.13 illustrates the error in IMEP calculation using Eqs. (6.62), (6.63), and (6.64) as a function of crank angle resolution of measured cylinder pressure data, and IMEP calculation methods are labeled as method1, method2 and method3

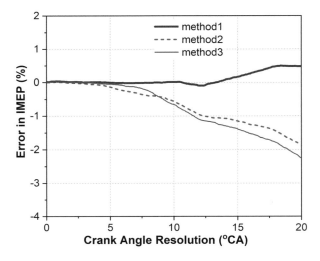

Fig. 6.13 Variation in error of IMEP calculation as a function of crank angle resolution (adapted from [20])

respectively. The figure depicts that all the equations produce similar results for crank angle resolutions up to about three degrees and significant differences are detected at higher resolutions. The differences in the accuracy are the result of the numerical integration process using different equations. Typically, encoders with crank angle resolution of one degree or less are used for the experiments, and in such case the differences in calculated IMEP are negligible (Fig. 6.13). Thus, in such cases, the equations for IMEP calculation can be selected based on computational effort and data processing time [20]. However, sometimes 60-2 teeth wheel is used for sensing the crank angle, which has the resolution of six crank angle degree, that may lead to error in IMEP calculation. The Eq. (6.62) seems to be the best IMEP calculation equation, which consistently produced the best accuracy at coarse crank angle resolutions.

Even using the best equations for IMEP calculation, the error can be introduced by random noise in pressure data or by engine operation at heavy knock conditions. The error is largest for the heavy knock and random noise cases but is always below $\pm2\%$, even at resolutions greater than $10°$ [20]. This degree of error is relatively small considering the extremely large fluctuations in the pressure data, and the effects which it would have on other parameters such as pressure rise rate and mass fraction burned. The relatively small errors appeared using coarse crank angle resolution, and noisy pressure data is due to (1) lower sensitivity to cylinder pressure fluctuations around TDC because of the low rate of change of volume, and (2) the errors created for individual steps and sections of the cycle tend to cancel out over the integration period.

The contribution to the IMEP is small near to TDC and reaches a maximum typically $40°$ after TDC position where the maxima of the product of pressure and volume change rate occur. Figure 6.14 illustrates the percentage contribution over 10-degree steps to the final gross IMEP for a typical full-load gasoline engine. The figure depicts that the peak negative contribution occurs at about $20°$ before TDC,

Fig. 6.14 The contribution to gross IMEP at every ten crank angle degree for full load in a typical gasoline engine (adapted from [20])

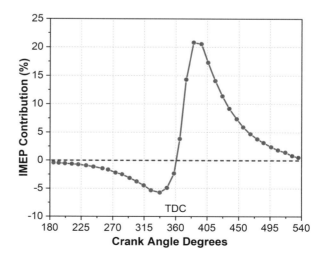

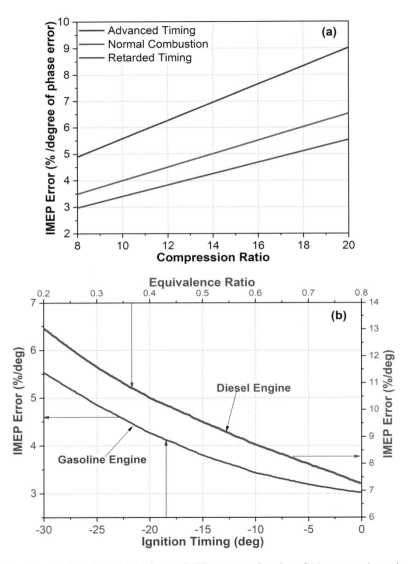

Fig. 6.15 Effect of crank angle phasing on IMEP error as a function of (**a**) compression ratio and (**b**) equivalence ratio and ignition timings (adapted from [20])

and the peak expansion stroke work occurs at about 35° after TDC position. Thus, the engine operation can be designed for maximum work output based on the contribution to IMEP on crank angle basis.

Figure 6.15 shows the effect of crank angle phasing on the error in IMEP calculations. The figure illustrates that the error of one crank angle degree in crank angle phasing can lead to a large error (up to 12%) in IMEP calculations depending on the engine operating conditions. The magnitude of the IMEP error percentage per degree of crank angle phase shift is mainly dependent on compression ratio, heat released per unit mass of charge, ignition/injection timing, and overall combustion duration [20]. Figure 6.15a depicts that IMEP error increases linearly with compression ratio for stoichiometric engine operation of gasoline engines for different ignition timings. Figure 6.15b confirms that diesel engines will have relatively larger sensitivity than gasoline engines. Advancing the injection timing (for diesel engines) or ignition timing (for spark ignition engines) and fast-burn combustion systems increases the IMEP error for a given amount of phasing error (Fig. 6.15b). Figure 6.15b also depicts that percentage of IMEP error increases with a leaner mixture (part-load operation of a diesel engine), and thus IMEP error can easily exceed 10% error at part-load condition. A similar effect is expected with an increase in exhaust gas recirculation.

Errors in the calibration of any part of the measurement system, i.e., pressure transducer, charge amplifier, and analog-to-digital converter (ADC) will produce proportional errors in the IMEP and PMEP. Thus, calibration of the complete measurement system is required to eliminate these errors. Cylinder volume calculations as a function of crank angle position require geometrical parameters such as bore, stroke, and connecting rod length as input in the measurement system. The manufacturer does not frequently provide the connecting rod length. In case of no availability of detailed geometric data for a particular test engine, the connecting rod length is estimated, which can lead to a relatively large error in IMEP calculation. Figure 6.16 shows the error in IMEP calculation due to an error in connecting rod length (up to ±10%). The figure shows that the IMEP error depends on the engine operating conditions, and it is a maximum of about 0.16% per 1% error in the assumed connecting rod length. For a normal range of expected errors in connecting rod length, this is a relatively small error [20].

In summary, crank angle phasing error, measurement system sensitivity, and thermal shock [21] errors are potentially the most influencing sources of error in IMEP calculation. Signal noise, connecting rod length, and crank angle resolution

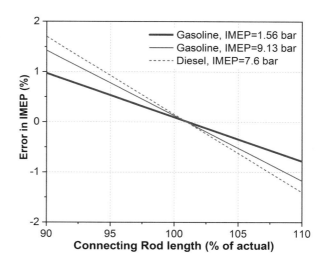

Fig. 6.16 Effect of connecting rod length on the gross IMEP calculations (adapted from [20])

errors are relatively smaller. The magnitude of IMEP errors for diesel engines is typically larger (typically twice) than the gasoline engines [20].

Determination of running quality of a reciprocating engine most significantly deals with the estimation of the IMEP, the detection of misfires and partial burns, and how the statistics of IMEP relate to customer satisfaction [22]. A poorly running engine can lead to driver discomfort because of the resulting vibrations being transmitted through the chassis and into the passenger compartment [23]. Misfires and partial burns will also dramatically affect emissions and, if left undetected, can result into catalyst damage which further worsens the situation [24]. The misfires and partial burns during engine operation also leads to drivability problems [5]. Typical methods of monitoring engine performance are based on the IMEP, and, thus, a number of ways are explored for IMEP estimation with easily measurable parameters [25–30].

To control engine torque, a conventional engine management system acquires feedback data such as air mass flow, crank position, air temperature, and air-fuel ratio. It is difficult to obtain accurate engine torque and it's precise control because these input data values are indirectly related to engine torque. Calculation of the IMEP using measured cylinder pressure data can be used to obtain more accurate torque data because IMEP is equivalent to indicated torque [28]. Thus, a real-time IMEP estimation algorithm is required for torque-based engine control.

A faster method of IMEP calculation is developed by using a very simple formula (Eq. 6.65) involving only two harmonic coefficients of the cylinder gas pressure. This equation avoids the necessity to calculate the variation of the cylinder volume, requiring less arithmetic operations than the formula based on the integration of the elemental work. The direct values of the sampled cylinder pressure are the only data required to calculate the IMEP and the harmonic coefficients of the tangential gas pressure [30].

$$\text{IMEP} = \frac{2\pi}{N}\left[\sum_{j=1}^{N} P_j \sin\theta_j + C\sum_{j=1}^{N} P_j \sin 2\theta_j\right] \tag{6.65}$$

where C is the constant determined by Eq. (6.66) using the connecting rod ratio ($\lambda = R/L$).

$$C = \frac{\lambda}{2} + \frac{\lambda^3}{8} \tag{6.66}$$

Another real-time IMEP estimation method is developed based on difference pressure integral (DPI) in the expansion stroke [28, 31]. The DPI is calculated from cylinder pressure by Eq. (6.67).

$$\text{DPI} = \sum_{k=0°\text{ aTDC}}^{180°\text{ aTDC}} \left[P_{\text{firing}}(k) - P_{\text{motoring}}(k)\right] \tag{6.67}$$

The IMEP is calculated by a linear relationship with DIP as represented by Eq. (6.68).

$$\text{IMEP} = a \cdot \text{DPI} + b \qquad (6.68)$$

where constants a and b are functions of engine speed (rpm) and start of energizing (SOE) the fuel injector. The constants are obtained by regression of the measured data, and obtained functions can be represented by Eqs. (6.69) and (6.70).

$$a = a_1 \text{SOE}^2 + a_2 \text{SOE} + a_3 \text{rpm} + a_4 \qquad (6.69)$$

$$b = b_1 \text{SOE} + b_2 \text{rpm} + b_3 \qquad (6.70)$$

At a particular engine operating condition, the work output of a reciprocating engine depends on a number of parameters, which affects the engine combustion process. Equation (6.71) presents the main parameters which affect the IMEP in a reciprocating engine [19, 32].

$$\text{IMEP} = f\left(\phi, P_{\text{in}}, \text{CA}_{50}, N, \dot{Q}_{\text{out}}, \text{fuel}, \%\text{EGR}, \text{CR}\right) \qquad (6.71)$$

where ϕ is the equivalence ratio, P_{in} is the boost pressure, CA_{50} is the crank angle corresponding to 50% heat release (combustion phasing), N is the engine speed, $\dot{Q}_{\text{out}}$ is heat transfer loss rate, fuel is the used fuel type (octane/cetane), %EGR is the fraction of exhaust gas recirculation (EGR) at particular operating condition, and CR is the compression ratio of the engine.

Figure 6.17 shows the variation of IMEP with EGR at different compression ratios (CR) and intake pressures. The figure shows that engine IMEP decreases with the increase of cooled EGR rate due to the displacement of fresh air at constant throttle position. The decrease in volumetric efficiency leads to deterioration in

Fig. 6.17 Variation of IMEP with EGR rate for engine operation at different compression ratios and boost pressure [33]

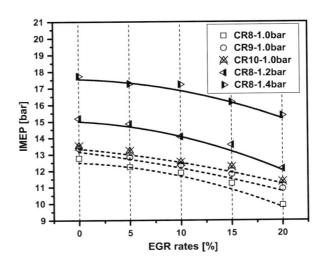

IMEP due to stoichiometric air-fuel mixture requirement in the SI engine. Increasing the compression ratio leads to a gain in IMEP. Thus, increasing the compression ratio can recover the loss of IMEP due to EGR. The IMEP at 10% EGR rate condition is restored almost identical to that with no EGR condition by increasing the compression ratio to 10:1 (Fig. 6.17). Much more significant improvement can be found with increasing intake pressure than in compression ratio as boost pressure increases the volumetric efficiency.

Figure 6.18 shows the variation of IMEP with engine speed at full engine load in conventional gasoline and diesel engine. In a gasoline engine, full-load IMEP initially increases with engine speed, and after reaching to a peak, it starts decreasing

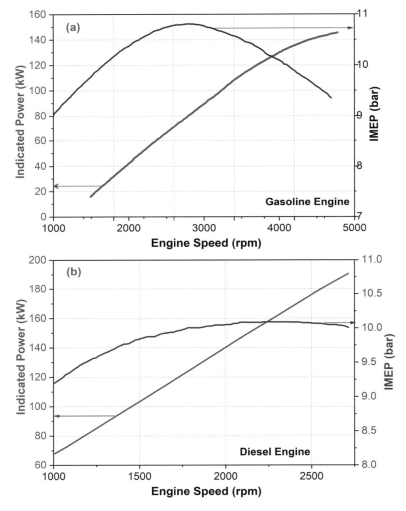

Fig. 6.18 Variation of IMEP with engine speed in a conventional (**a**) gasoline engine and (**b**) diesel engine (adapted from [5])

with engine speed (Fig. 6.18a). This trend follows the variation in volumetric efficiency, which initially increases with engine speed and decreases after a partic- ular speed due to higher flow friction and chocking. Due to stoichiometric air-fuel operation in SI engine, the IMEP is reduced as the lower amount of fuel can be burned to produce power during engine operation at lower volumetric efficiency. In a naturally aspirated diesel engine, the IMEP vary modestly with engine speed because intake system of a diesel engine can have large flow areas. There is no throttle in a diesel engine, and the load is mainly controlled by fuel quantity. Heat transfer loss is higher at lower engine speeds. Diesel engine typically operates at overall leaner mixtures, and, thus, at higher engine speed, the same amount of fuel can be burned which results in almost constant IMEP at higher speeds (Fig. 6.18b).

The work output is mainly dependent on the amount of fuel burned, which is governed by the amount of air intake (volumetric efficiency). The quantity of fuel burned in the combustion chamber controls the pressure generated and work out of the engine in the combustion chamber. The work is also governed by displacement in addition to the pressure/force (Eq. 6.55). Therefore, combustion phasing is another major factor, which governs the work produces as it controls the effective expansion ratio. Figure 6.19 shows the effect of combustion phasing (CA_{50}) on IMEP in an HCCI combustion engine. The figure shows that for a particular λ (amount of fuel), the IMEP decrease with retarded combustion phasing. Combustion temperature is lower at retarded combustion phasing (due to high cylinder volume expansion), which leads to the higher amount of unburned fuel and lower combus- tion efficiency. Combustion phasing is typically controlled by intake air

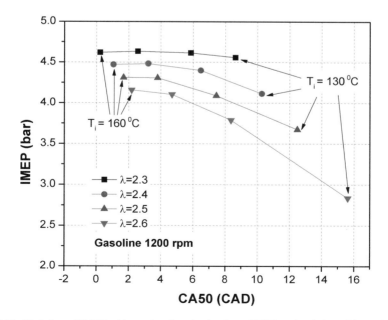

Fig. 6.19 Variation of IMEP with combustion phasing in an HCCI engine (adapted from [34, 35])

temperature in HCCI engine. Higher inlet temperature results into advanced combustion phasing, and advanced combustion phasing (close to TDC position) leads to higher IMEP (Fig. 6.19).

6.5 Engine Efficiency Analysis

Analysis of engine efficiency is an integral part of research and development work. Efficiency analysis helps to compare the improvement occurred by changes in engine design, fuel, combustion mode, and engine operating conditions. Understanding of complete energy flow through different engine components is required before performing the efficiency analysis. Figure 6.20 illustrates the energy transfer through different parts of the engine. Energy input in the engine is in the form of chemical energy stored in the fuel. Chemical energy is converted into heat energy after combustion, which is the total in-cylinder energy. A part of this energy is converted into useful work, which is the main output of the engine. Rest of the energy is utilized in the engine or comes out as heat energy, which is released in the form of heat rejected through radiator, heat rejection from the surface to ambient, and energy leaving through exhaust tail-pipe (Fig. 6.20). There are piston work transfers in and

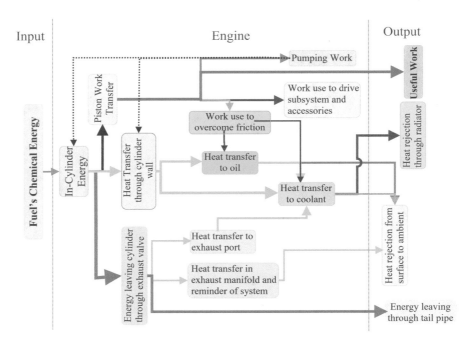

Fig. 6.20 Energy flow in through different engine components in a reciprocating internal combustion engine (adapted from [9])

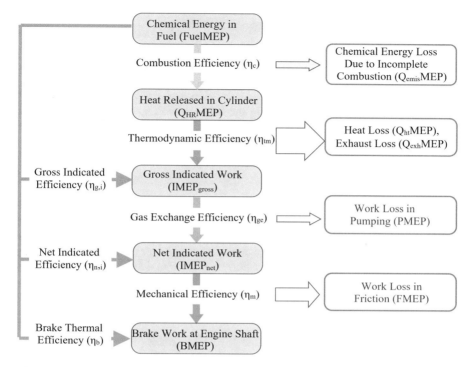

Fig. 6.21 Efficiency breakdown in a reciprocating internal combustion engine (adapted from [37, 38])

out of the engine cylinder, and net work transfer can be on the order of 40% of the fuel energy at full engine load. The net heat transfer out of the cylinder is on the order of 10–15% of the fuel energy, and the exhaust energy leaving the cylinder brings out on the order of 45–50% of the fuel energy at full-load engine operating conditions. These numbers vary significantly depending on engine operating conditions and the combustion modes [9]. Heat loss is the biggest contributor to an inefficient engine, and it occurs from several points in the engine between components. Some of the places and processes where heat loss occurs include cylinder bore to water jacket, combustion chamber deck to water jacket, piston crown to connecting rod and then to the oil, inlet valve radiated and convection, exhaust valve radiated and convection, fuel heat to fuel in evaporation, exhaust gases ejected, and blow-by losses [36].

Engine efficiency can be divided into different process efficiencies in order to find out the negatively contributing processes toward overall engine efficiency, which can be further improved. Figure 6.21 depicts the efficiency breakdown in a reciprocating internal combustion engine along with associated losses. The energy flow in the figure is presented as mean effective pressure (MEP) which is defined as the work or energy per cycle normalized by the engine's displacement volume (V_d). In a combustion engine, the chemical energy stored in the fuel is converted into heat energy during the combustion process. The energy contained by the fuel is shown as FuelMEP, which can be calculated by Eq. (6.72).

$$\text{FuelMEP} = \frac{m_f Q_{\text{LHV}}}{V_d} \tag{6.72}$$

Where m_f is the mass of fuel burned per cycle and Q_{LHV} is the lower heating value of the fuel.

The heat released (HR) during the combustion process is denoted as $Q_{\text{HR}}\text{MEP}$, which can be computed by integrating the heat release rate (dQ_{HR}) obtained from measured cylinder pressure (see Chap. 7).

$$Q_{\text{HR}}\text{MEP} = \frac{Q_{\text{HR}}}{V_d} = \frac{\int dQ_{\text{HR}}}{V_d} \tag{6.73}$$

The heat release during a cycle can also be estimated by measuring the unburned gas species in the exhaust and computing the chemical energy lost in unburned or partially burned species. Chemical energy lost due to incomplete combustion in the cylinder is denoted as $Q_{\text{emis}}\text{MEP}$. The combustion efficiency (η_c) can be defined as the ratio of heat released in the combustion chamber $(Q_{\text{HR}}\text{MEP})$ to total fuel energy injected (FuelMEP). The chemical energy lost is expelled out of the cylinder as partially, or completely unburnt fuel $(Q_{\text{emis}}\text{MEP})$ is contributing to a loss in combustion efficiency.

$$\eta_c = \frac{Q_{\text{HR}}\text{MEP}}{\text{FuelMEP}} = 1 - \frac{Q_{\text{emis}}\text{MEP}}{\text{FuelMEP}} \tag{6.74}$$

The heat released in the combustion chamber is converted to mechanical work transfer to the piston, which is denoted as gross indicated work (IMEP$_{\text{gross}}$). During the piston work conversion process, a fraction of the heat produced in the combustion chamber is lost in the form of heat transfer $(Q_{\text{ht}}\text{MEP})$ and exhaust heat through tailpipe $(Q_{\text{exh}}\text{MEP})$. Thermodynamic efficiency (η_{tm}) is defined as the ratio of gross indicated work to the total heat released in the cylinder as shown in Eq. (6.75). Thermodynamic efficiency is a measure of how efficiently engine can convert the heat energy of the combustion chamber into piston work.

$$\eta_{\text{tm}} = \frac{\text{IMEP}_g}{Q_{\text{HR}}\text{MEP}} = 1 - \frac{Q_{\text{ht}}\text{MEP} + Q_{\text{exh}}\text{MEP}}{Q_{\text{HR}}\text{MEP}} \tag{6.75}$$

The gross indicated efficiency $(\eta_{g,\,i})$ is defined as the ratio of gross indicated work to fuel's input energy. The relation to calculate the gross indicated efficiency is presented by Eq. (6.76).

$$\eta_{g,i} = \eta_c \cdot \eta_{\text{tm}} = \frac{\text{IMEP}_g}{\text{FuelMEP}} \tag{6.76}$$

Part of the produced indicated work is used in gas exchange process during intake and exhaust stroke. The work loss during intake and exhaust stroke is defined as pumping loss (PMEP). The gas exchange efficiency (η_{ge}) is defined as the ratio of net indicated work to the gross indicated work.

$$\eta_{ge} = \frac{\text{IMEP}_n}{\text{IMEP}_g} = 1 - \frac{\text{PMEP}}{\text{IMEP}_g} \tag{6.77}$$

The net indicated efficiency ($\eta_{n,i}$) is defined as the ratio of net indicated work to fuel's input energy. The relation to calculate the net indicated efficiency is presented by Eq. (6.78).

$$\eta_{n,i} = \eta_c \cdot \eta_{tm} \cdot \eta_{ge} = \frac{\text{IMEP}_n}{\text{FuelMEP}} \tag{6.78}$$

Different moving parts of the engine consume a fraction of the net work produced to overcome the friction, and lost work is denoted as frictional work (FMEP). The final remaining work available at the crankshaft is termed as brake work (BMEP). The measure of all the frictional processes in the engine is mechanical efficiency (η_m), which is calculated by Eq. (6.79).

$$\eta_m = \frac{\text{BMEP}}{\text{IMEP}_n} \tag{6.79}$$

The brake efficiency (η_b) is defined as the ratio of brake work available at the crankshaft to total fuel energy injected into the cylinder. The brake efficiency is calculated by Eq. (6.80), which accounts for all chemical, heat, and friction losses in the reciprocating engine.

$$\eta_b = \eta_c \cdot \eta_{tm} \cdot \eta_{ge} \cdot \eta_m = \frac{\text{BMEP}}{\text{FuelMEP}} \tag{6.80}$$

Engine efficiency depends on the various parameters such as compression ratio, combustion phasing, combustion duration, equivalence ratio of the fuel-air mixture, engine speed, engine combustion mode, etc., which governs the combustion efficiency, heat loss, and exhaust loss. Figure 6.22 shows the energy and exergy distribution in conventional diesel combustion (CDC), homogeneous charge compression ignition (HCCI), and reactivity controlled compression ignition (RCCI) engine at different combustion phasing (CA_{50}). Incomplete combustion loss is increasingly higher as combustion phasing is retarded in all the combustion modes (Fig. 6.22a). Diesel combustion shows the highest combustion efficiency (lowest incomplete combustion loss) due to heterogeneous high-temperature combustion. The HCCI and RCCI combustion have relatively lower combustion temperature and, thus, lower combustion efficiency than CDC [19]. The HCCI engine shows relatively higher combustion efficiency than RCCI engine because of total premixed fueling. The RCCI combustion is controlled by reactivity stratification by direct injection of high reactivity fuel along with premixed low reactivity fuel. Hence, the fuel-air mixture near cylinder wall is relatively leaner and less reactive in comparison to HCCI combustion conditions. The lower temperature near-wall boundary leads to higher incomplete combustion in RCCI engine than HCCI engine [39]. However,

Fig. 6.22 (**a**) Energy and (**b**) exergy distributions in CDC, HCCI, and RCCI combustion mode at different combustion phasing [39]

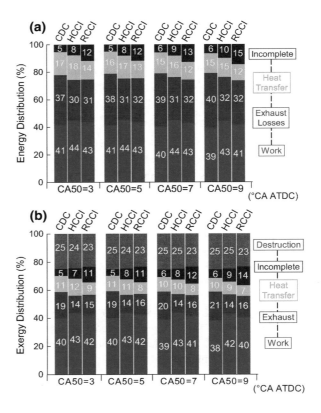

RCCI combustion has lower heat transfer loss in comparison to CDC and HCCI combustion at constant combustion phasing (Fig. 6.22a) due to lower temperature gradient near the cylinder wall in RCCI engine. For constant combustion phasing, the exhaust losses of HCCI and RCCI are much lower than that of CDC (Fig. 6.22a), which leads to the possibility of higher work extraction. The combustion duration is lowest in HCCI combustion, which leads to highest work extraction due to close to constant volume combustion.

In comparison to results obtained from the first law of thermodynamics, second law analysis provides an extra term, i.e., exergy destruction (Fig. 6.22b). The exergy destruction is an additional restriction to the maximum achievable work from the engine. The exergy fractions of incomplete combustion, heat transfer, exhaust losses, and net work are all smaller than their corresponding energy fractions, and the overall variation trends of each part are consistent with those from the energy distribution (Fig. 6.22b).

Analysis of energy fractions of incomplete combustion, heat transfer, exhaust losses, and net work can be helpful in identifying the areas of improvement to increase the overall engine efficiency. Figure 6.23 illustrates the typical variations in thermodynamic, brake, combustion, and gas exchange efficiencies in SI and

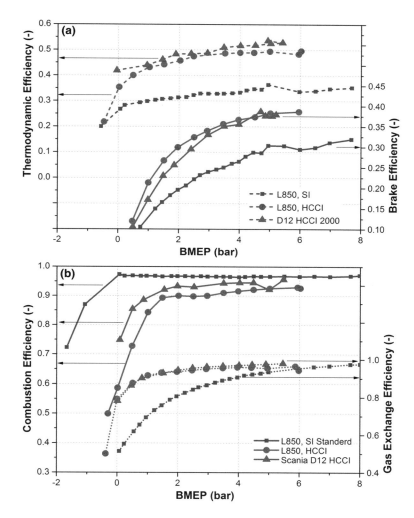

Fig. 6.23 (**a**) Thermodynamic and brake efficiency and (**b**) combustion and gas exchange efficiency as function of BMEP in HCCI engine (adapted from [40])

HCCI engine. The efficiencies can be calculated by the Eqs. (6.75), (6.80), (6.74), and (6.77), respectively. The HCCI engines show relatively higher thermodynamic efficiency as compared to SI engine due to higher compression ratio and leaner mixture operation (higher γ of working fluid). Higher thermodynamic efficiency leads to higher brake efficiency (Fig. 6.23a). Combustion efficiency increases with engine load and higher for SI engines due to high-temperature combustion (Fig. 6.23b). The gas exchange efficiency is higher in the HCCI engine due to unthrottled engine operation. The overall brake efficiency of HCCI engine is higher in comparison to SI engine. Figure 6.24 illustrates the dependency of brake

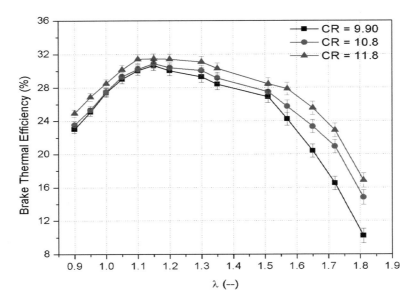

Fig. 6.24 Brake efficiency variation with relative air-fuel ratio in SI engine for different compression ratios [41]

efficiency on compression ratio and relative air-fuel ratio in SI engine. The brake efficiency is lower at richer mixture due to lower combustion efficiency. The brake efficiency reaches to a maximum and then starts decreasing with leaner mixtures (Fig. 6.24).

6.6   Gas Exchange Analysis

Cylinder pressure measurement along with intake pressure measurement can be jointly used for gas exchange analysis and the optimization of valve timings. The valve timings govern the pumping work and the residual gas fraction in the cylinder. It is well established that the gas exchange process is strongly influenced by variable valve timing (VVT), and it has the capacity to reduce the pumping loss and improve the engine performance [42, 43]. However, current VVT technologies have demerit of only changing the valve timings, i.e., valve opening (or closing) time, and cannot change the valve opening duration. Additionally, the current VVT is not able to vary the valve lift. These constraints limit the pumping loss reduction advantages using VVT technology. Another effective way to reduce the pumping loss of engine is the variable valve lift (VVL) [42]. The valve timing and valve lift both can be effectively varied as per the requirement of a particular engine operating conditions by using VVL technology. Thus, the VVL can improve the gas exchange

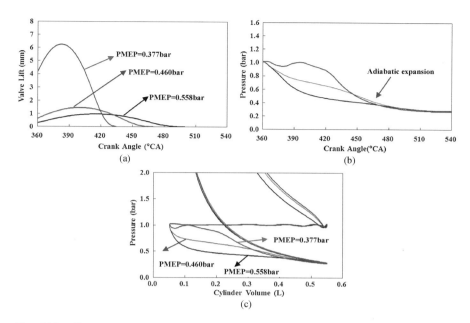

Fig. 6.25 Effects of valve lift and opening duration of the intake valve on the intake process of continuous variable valve lift engine at 2000 rpm and 2.3 bar IMEP in SI engine [42]. (**a**)Valve lift. (**b**) Cylinder pressure in intake process. (**c**) Cylinder pressure in gas exchange process

process and engine performance. The equations for computing the gas exchange efficiency and pumping work are described in Sect. 6.5.

Figure 6.25 shows the effect of intake valve lift and opening duration of continuous variable valve lift (CCVL) engine operated at 2000 rpm and 2.3 bar IMEP. The cylinder pressures at BDC position are very close to each other at all the three different valve lift conditions (Fig. 6.25) that leads to almost same engine loads. However, the intake processes for all the three lift conditions are different from each other, which leads to the difference in intake loss. Lowest valve lift and longest opening duration of the intake valve lead to large pressure loss due to the very little openness of the intake valve. This pressure drop causes the cylinder pressure much lower than intake manifold pressure which increases the pumping work (PMEP = 0.558 bar). With higher valve lift and shorter duration, the pressure loss decreases, which improves the gas exchange efficiency. At highest valve lift condition, the cylinder pressure during the intake process is approximately equal to intake manifold pressure, which leads to lowest PMEP (Fig. 6.25). Thus, cylinder pressure measurement can be used to improve and optimize the gas exchange performance of a reciprocating engine.

The variable valve lift technologies are widely used in the advanced engine combustion modes such as HCCI combustion. The VVL and VVT technologies affect the residual gas fraction in the cylinder, which affects the dilution of charge as well as the temperature of charge before combustion. In HCCI combustion,

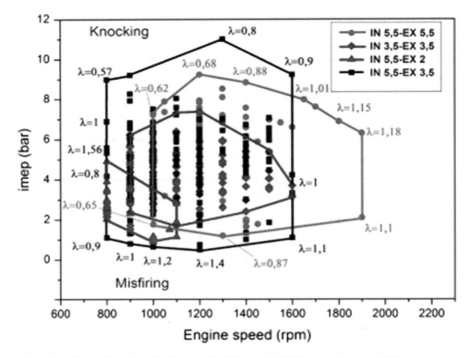

Fig. 6.26 Effects of intake and exhaust valve lifts on the HCCI operating range [44]

autoignition of leaner charge is problematic, which is improved by trapping the hot residual gases. Figure 6.26 shows the effect of intake and exhaust valve lift on the HCCI operating range. The number in the legend shows the lift in mm for intake (IN) and exhaust (EX) valves. Figure 6.26 depicts that the operating range of the HCCI engine can be extended using reduced valve lifts.

Thus, optimization of the gas exchange process helps to improve the performance of reciprocating internal combustion engines. Cylinder pressure-based combustion diagnostics of the gas exchange process is beneficial for optimizing the performance of engine.

6.7 Torque, Power and Engine Map

Torque is a good indicator of an engine's ability to do work. Torque is defined as the force acting at moment distance and has units of Nm. In other words, torque is a measure of how much a force acting on an object causes that object to rotate. Typically, torque is measured by dynamometers on an engine test cell. Different types of engine dynamometers including hydraulic, eddy current, direct current, alternating current, and friction are used for engine testing. The engine dynamometers measure quasi-steady engine torque or average torque, not the instantaneous torque (as calculated in Sect. 6.3). The torque is measured at engine crankshaft, and, thus, it is denoted as brake torque.

Power is defined as the work per unit time, which is the rate of work of an engine. There are different kinds of power depending on the type of work is used for calculation. Typically, indicated (P_i) and brake power (P_b) are used for engine performance analysis.

$$P_i = \frac{W_i}{\text{time/cycle}} = \frac{W_i \cdot N}{X} \tag{6.81}$$

$$P_b = \frac{W_b}{\text{time/cycle}} = \frac{W_b \cdot N}{X} \tag{6.82}$$

where W_i and W_b are the indicated and brake work per cycle, respectively, N is the engine speed in revolution per second, and X is the number of revolution per cycle. The $X = 1$ for two-stroke engine and $X = 2$ for four-stroke engine.

The power equations can be represented in terms of mean effective pressure (MEP) as Eq. (6.83).

$$P = \frac{dW}{dt} = \frac{W \cdot N}{X} = \frac{MEP \cdot V_d \cdot N}{X} = \frac{MEP \cdot n \cdot A_p \cdot S \cdot N}{X} \tag{6.83}$$

where V_d is the displacement volume of the engine, n is the number of cylinders, A_p is the area of the piston, and S is the stroke length.

Specific power equations can be derived using Eq. (6.83) in terms of mean piston speed ($\overline{V}_p$).

$$\frac{P_b}{nA_p} = BMEP\frac{\overline{V}_p}{2X} \tag{6.84}$$

$$\frac{P_b}{V_d} = BMEP\frac{\overline{V}_p}{2 \cdot S \cdot X} \tag{6.85}$$

Power produced per unit piston area, often called as specific power, is typically measure of utilization of available piston area irrespective of cylinder size. Typically, current engines have specific power around 0.38 kW/cm^2, which is at mean piston speed around 15 m/s. The BMEP of naturally aspirated gasoline and diesel engines are 11–15 bar and 10 bar, respectively, and in turbocharged conditions 15–20 bar and 15–22 bar for gasoline and diesel engines, respectively. Power per unit of displacement volume (power density) is another measure that is traditionally used, and it depends on the length of stroke [10]. The Eq. (6.85) shows that the shorter stroke will have higher power output per unit volume. Typically, the power density of naturally aspirated gasoline and diesel engines are 50–70 kW/L and 20 kW/L, respectively, and in turbocharged conditions 70–120 kW/L and 40–65 kW/L for gasoline and diesel engines, respectively. It can be noted from Eq. (6.85) that brake power can be increased in only three ways: (1) by increasing BMEP, (2) increasing total displacement volume, and (3) increasing engine speed, which is limited by material strength.

The power equations can be expressed in terms of different engine efficiencies, which provides more insight on the power generation from reciprocating engines.

$$P = \eta_b \cdot \dot{m}_f \cdot Q_{LHV} = \eta_c \cdot \eta_{tm} \cdot \eta_{ge} \cdot \eta_m \cdot \dot{m}_f \cdot Q_{LHV}$$
$$= \eta_c \cdot \eta_i \cdot \eta_m \cdot \dot{m}_f \cdot Q_{LHV} \tag{6.86}$$

where $\dot{m}_f$ is the mass flow rate of fuel and Q_{LHV} is the lower heating value of the fuel. The indicated thermal efficiency (η_i) can be defined in terms of thermal energy to work conversion, which is the product of thermodynamic efficiency and the gas exchange efficiency.

$$\eta_i = \eta_{tm} \cdot \eta_{ge} = \frac{IMEP_n}{Q_{HR}MEP} \tag{6.87}$$

The power term can also be represented in terms of air flow rate by Eq. (6.88).

$$P = \eta_c \cdot \eta_i \cdot \eta_m \cdot \dot{m}_a \cdot (F/A) \cdot Q_{LHV} \tag{6.88}$$

where $\dot{m}_a$ is the mass flow rate of air and (F/A) is the fuel-air ratio of engine operation.

Air flow rate is an important factor governing the power produced from the engine. The displacement volume of the engine would be filled with air truly if air is an incompressible and inviscid medium. Thus, mass entering is somewhat less depending on engine speed because air is compressible and viscous medium [10]. The viscous (turbulent) pressure drop occurs in the cylinder manifold, which leads to lower density in the cylinder than the air density at the inlet of the manifold. The other major factor is the occurrence of shock at the valve opening at very high piston speed, which chocks the air flow. Considering these factors, a volumetric efficiency term is introduced, which is defined as the ratio of the actual air mass flow and the theoretical air mass that can be achieved if air is inviscid and incompressible fluid.

$$\eta_v = \frac{\dot{m}_a}{\rho_i \cdot V_d \cdot \left(\frac{N}{X}\right)} = \frac{\dot{m}_a}{\rho_i \cdot A_p \cdot S \cdot n \cdot \left(\frac{N}{X}\right)} = \frac{m_a}{\rho_i \cdot V_d} \tag{6.89}$$

The equation of power can be written in terms of volumetric efficiency as Eq. (6.90).

$$P = \eta_c \cdot \eta_i \cdot \eta_m \cdot \eta_v \cdot \rho_i \cdot V_d \cdot \left(\frac{N}{X}\right) \cdot (F/A) \cdot Q_{LHV} \qquad (6.90)$$

The Eq. (6.90) is the main equation that can be very useful in engine design. Each term of the equation suggests the way for improving the performance of the engine. The volumetric efficiency is influenced by valve timings, cam modifications (including valve lift), tuning of intake and exhaust manifold, and number of valves per cylinder. Intake air density can be increased by turbocharging/supercharging, without throttle engine operation, inter-cooling, and use of fuel with a higher heat of vaporization (typically in a port injection system). Indicated efficiency can be improved by increasing compression ratio and reducing heat loss (piston coating, varying engine geometry, as well as number of cylinders) [10]. Mechanical efficiency can be improved by using low-tension piston rings, low friction coatings, and offset in piston pin and crankshaft.

Torque is related to the work and BMEP by Eq. (6.91).

$$2\pi \cdot T = W_b = \text{BMEP} \cdot \frac{V_d}{X} \qquad (6.91)$$

The Eq. (6.91) illustrates that for a particular engine, there are only two ways to increase torque: (1) by increasing the BMEP and (2) increasing the total displacement volume (increase bore/stroke or number of cylinders). For in-depth analysis, the engine torque can be written in terms of engine efficiencies using Eqs. (6.90) and (6.91).

$$T = \frac{P}{2\pi \cdot N} = \left(\frac{1}{2\pi \cdot X}\right) \cdot \eta_c \cdot \eta_i \cdot \eta_m \cdot \eta_v \cdot \rho_i \cdot V_d \cdot (F/A) \cdot Q_{LHV} \qquad (6.92)$$

Modern automotive engines have maximum torque per displacement volume in the range of 80–140 Nm/L [18]. Typically, power and torque curve for a turbocharged diesel engine with common rail direct injection system is shown in Fig. 6.27.

The torque curve of an engine is mainly dependent on volumetric efficiency, heat loss, and frictional loss. At higher engine speed, typically heat transfer decreases and frictional loss increases. Diesel engine operates without throttle with larger flow areas, and, thus, volumetric efficiency has a moderate effect on the torque curve. At higher engine speed, the torque decreases mainly because of increase in frictional loss.

Figure 6.27 also shows important points on the torque and power curves. The maximum engine power (P_{max}) is also known as rated power, i.e., the highest power an engine is allowed to develop in continuous operation. The engine speed (N_P) corresponding to maximum power is typically known as rated speed. The maximum engine torque is denoted as T_{max}, and its corresponding engine speed is (N_T). Engine torque at the operating point of maximum power is denoted as $T_{P_{max}}$, and engine

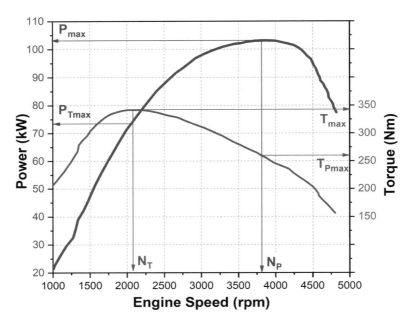

Fig. 6.27 Power curve of a modern turbocharged diesel engine (adapted from [45])

power at the operating point of maximum torque is denoted as $P_{T_{\max}}$ (Fig. 6.27). Engine torque flexibility (F_T) and engine speed flexibility (F_N) are defined by Eqs. (6.93) and (6.94) [6, 46].

$$F_T = \frac{T_{\max}}{T_{P_{\max}}} \tag{6.93}$$

$$F_N = \frac{N_P}{N_T} \tag{6.94}$$

The engine flexibility can be defined in terms of torque and speed flexibility by Eq. (6.95).

$$F_e = F_T \cdot F_N = \frac{T_{\max}}{T_{P_{\max}}} \cdot \frac{N_P}{N_T} \tag{6.95}$$

The engine with good flexibility can achieve maximum torque at lower engine speeds, or in other words, it produces higher torque at lower engine speeds. The higher engine flexibility engine will result in less frequent gear shifting [6, 46].

Figure 6.28 shows the engine torque curves for a different type of spark ignition engines. Engine torque increases as the engine is supercharged or turbocharged (Fig. 6.28a). Engine torque is proportional to the density of air (Eq. 6.92), which

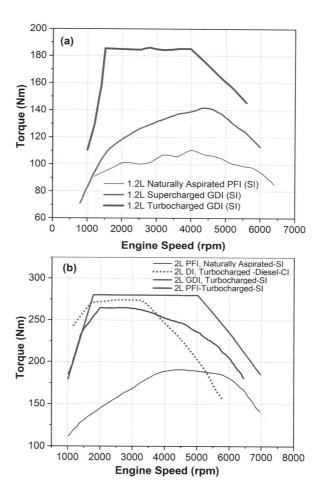

Fig. 6.28 Engine torques for various type of gasoline engines with different size (adapted from [47, 48])

increases with turbocharging that results into higher torque. Engine torque is dependent fuel injection strategy of spark ignition engine. The gasoline direct injection (GDI) engines have higher torque in comparison to the port fuel injection (PFI) engines with spark ignition (SI) as illustrated in Fig. 6.28b. It can also be observed by comparing Figure 6.28a, b that engine torque increases with an increase in displacement volume (Eq. 6.92).

The gasoline SI engine inherently operates at relatively higher engine speeds and over a wide speed range than diesel compression ignition engine (Figs. 6.27 and 6.28). Heavier components of diesel engines (due to higher peak pressures) limit their maximum engine speed. Additionally, the diesel combustion process also limits the maximum engine speed [9]. Diesel combustion occurs through a sequence of processes—atomization of liquid fuel injected into the cylinder, fuel droplets vaporization, mixing of the vaporized fuel with the surrounding air, and finally

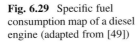

Fig. 6.29 Specific fuel consumption map of a diesel engine (adapted from [49])

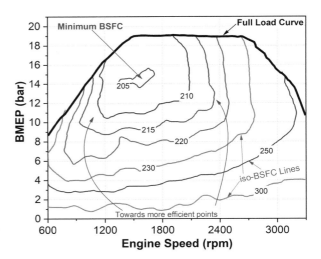

autoignition of the fuel-air mixture. These processes involved in diesel combustion do not scale with engine speed. Therefore, the maximum speed of a diesel engine is also constrained by the time needed for completing the combustion during an engine cycle.

The operating point of a reciprocating internal combustion engine is characterized by its speed and its torque. The full range of all possible engine operating points in a two-dimensional presentation is known as the "engine map." In the engine map, the operating range of engine is constrained by the full-load curve and by the minimum and the maximum engine operating speed. Figure 6.29 shows a typical engine map of a diesel engine. The brake-specific fuel consumption (BSFC) is shown on the engine map. For every engine operating point on the map, a corresponding BSFC value can be determined. Several engine operating points on the speed-torque map have the same BSFC value, and locus of these points is presented as iso-BSFC lines (Fig. 6.29). Minimum specific fuel consumption typically occurs at or near full-load curve and at relatively low engine speeds. With the increase in engine speed, the engine efficiency decreases because of the rapid rate of friction increase. At lower engine speeds, the BSFC again increases due to higher heat transfer losses. Minimum BSFC typically occurs between rated and peak torque speeds.

Figure 6.30 shows the typical engine map for a gasoline engine. The trend in BSFC lines is similar to the diesel engine. The minimum specific fuel consumption is found in the lower engine speed range in the range of high engine load. The BSFC gradient increases sharply toward the low engine load range. The main reasons behind this observation are the increasing throttle losses in the SI engine and the increasing proportion of friction in relation to the useful torque output. These two factors also lead to the significant increase in BSFC at constant engine load and increasing engine speeds. Toward the full-load range, the mixture has to be enriched for two reasons: (1) to counter the knock tendency of the engine and (2) to keep the

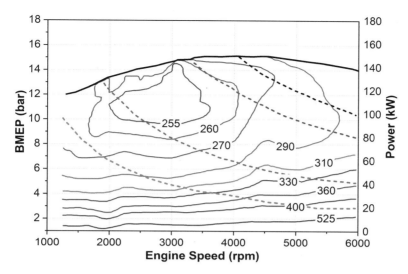

Fig. 6.30 Specific fuel consumption map of a 1.2 L gasoline SI engine (adapted from [50])

exhaust gas temperature below a critical limit temperature for catalytic converter aging. This leads to a sharper gradient of the BSFC increase in SI engine [8].

Discussion/Investigation Questions.

1. Discuss the information's revealed by the three types of presentation (P-θ, P-V, and $\log P - \log V$) of measured cylinder pressure data.
2. Discuss the sources of error in the measured pressure signal and how these sources manifested on the pressure trace. Write the possible solution for minimizing or eliminating the error from measured pressure trace.
3. Draw the P-V diagram of a naturally aspirated spark ignition engine at part-load and full-load operating conditions. Discuss how the indicating diagram will change when the engine is equipped with a turbocharger.
4. What is the difference you can observe in the P-V and P-θ diagram of conventional spark ignition and compression ignition engines?
5. Draw the cylinder pressure curve as a function of crank angle position for advanced and retarded ignition timings in a spark ignition engine. Justify the trends depicted in the graph.
6. Indicated mean effective pressure (IMEP) is typically used as indicated engine load. Write the equation for calculation of IMEP from the measured cylinder pressure sensor. Derive/define all the parameters used in the equation except measured cylinder pressure. Since pressure is measured as a digital value by the

data acquisition system, how can IMEP be calculated from digitized cylinder pressure data? Explain using digitized equations.

7. Explain why the IMEP and PMEP calculation is not affected by offset error (absolute pressure referencing) in the pressure signal measured by the piezo-electric pressure transducer.

8. Discuss and plot the differences between the ideal and actual valve timing diagram of a spark ignition engine at low and high speed. Justify the need for variable valve timings (VVT) in the engine.

9. Show by means of a diagram the energy flow in a reciprocating internal combustion engine and define combustion efficiency, indicated efficiency, mechanical efficiency, brake efficiency, and gas exchange efficiency.

10. Explain brake mean effective pressure (BMEP) and its significance. Write the typical values of BMEP for gasoline and diesel engines. Why BMEP of the naturally aspirated diesel engine is lower than a naturally aspirated spark ignition engines.

11. Suggest reasons for lower BMEP at maximum rated power as compared to BMEP at maximum torque for a given engine.

12. Why larger volume size engines (marine or locomotives) have generally higher thermal efficiency as compared to lower size engines (car or truck).

13. What are the possible ways to increase the power output of a production engine (CI and SI)?

14. Brake torque initially (at lower speed) increases with engine speed and after certain engine speed starts decreasing in engines. Suggest reasons for the given observation.

15. Engine "X" generates 45 kW of power by consuming 9 kg/h of diesel fuel. Engine "Y" generates 55 kW of power by consuming 10 kg/h which engine would one can use to do a job that required 35 kW of continuous power? Assume constant efficiency over the operating range of the both engines.

16. Mean piston speed is a parameter used for comparison of the engine of different sizes. Write the typical mean piston speed of a passenger automotive engine and a racing car engine. Discuss the major factors/parameters which are influenced by mean piston speed, which governs the engine design.

17. Typically rotation engine speed (rpm) of the diesel engine is lower in comparison to the corresponding spark ignition engines. Explain this observation and justify your answer.

18. Write the typical compression ratios for spark ignition and compression ignition engine. Discuss the typical effect of compression ratio on specific fuel consumption of an engine at particular engine operating condition.

19. Write the expression for BMEP in terms of indicated thermal efficiency, mechanical efficiency, combustion efficiency, fuel-air ratio, and intake air pressure. Discuss the ways to improve the BMEP of the engine by looking at the derived expression.

20. A designer wants to achieve 50 kW/L at 12 bar BEMP in an automotive spark ignition engine. What should stroke length be used for this particular engine? Discuss the effect of stroke length on the engine performance and design.

21. Discuss at least three thermodynamic levers that can be used for increasing the thermal efficiency of an engine.

Problems

Problem 1. A single-cylinder four-stroke engine with bore diameter "D," stroke "S," and connecting rod length "L" is operated at engine speed "N" rpm. The engine has a crank radius of "R" and piston pin offset "δ." Derive an expression for instantaneous cylinder volume, piston speed, and acceleration. Clearly state your assumptions if any.

(a) Calculate instantaneous piston speed and acceleration (without pin offset case) for $N = 1000$, 2000, and 3000 rpm for D and $S = 90$ mm and $L = 160$ mm. Plot the curves for 0–360 crank angle degree (CAD) and compare the results. Comment on the maximum piston speed position observed.

(b) Plot the ratio of instantaneous piston speed to mean piston speed for $L/R = 3$, 5, and 9 for without pin offset conditions for 0–360 CAD. Calculate the following parameter and fill the Table P6.1, and comment on your observations.

(c) Calculate instantaneous piston acceleration without pin offset for $N = 1000$, 2000, and 3000 rpm for B and $S = 90$ mm and $L = 160$ mm. Plot the curves for 0–360 crank angle degree (CAD) and compare the results.

Table P6.1 Determination of parameters related to engine geometry

Computing parameter	$L/R = 3$	$L/R = 5$	$L/R = 9$
Crank angle position of maximum piston speed			
Piston speed at 10 CAD after TDC at $N = 1500$ rpm			
Piston speed at 60 CAD after TDC at $N = 1500$ rpm			
Time duration for piston travel between 30° before and after TDC at $N = 1500$ rpm			
Time duration for piston travel between 30° after TDC to 90° after TDC at $N = 1500$ rpm			

Problem 2. A single-cylinder four-stroke engine has bore diameter 90 mm, stroke 100 mm, and connecting rod length 160 mm. Connecting rod is having CM (center of mass) 100 mm from the small end.

(a) Calculate and plot connecting rod angular speed and angular acceleration versus crank angle position of the crankshaft at engine speed 2000 and 4000 rpm.

(b) Calculate velocity and acceleration of connecting rod center of mass at engine speed 2000 rpm.

Problem 3. A four-cylinder engine produces 27 kW at engine speed of 1500 rpm. The engine generated the average torque of 120 Nm when fuel supply of one the cylinders was stopped. Assume BSFC of the engine is 340 g/kWh and the lower heating value of the fuel is 43 MJ/kg. Calculate the indicated thermal efficiency of the engine. Clearly state your assumptions in the calculation. Discuss the method of indicated thermal efficiency calculation if cutting fuel supply of one of the cylinders is not allowed or more accurate estimation of indicated thermal efficiency is required.

Problem 4. Calculate the following parameters shown in the table and fill empty values in Table P6.2. Compare and explain any significant difference between engines. Comment on the mean piston speed with respect to engine size for the given engines.

Table P6.2 Engine characteristics parameters for different size and type of engines

Engines	BMEP [bar]	Torque [nm]	Specific power output (P/A_p) [kW/m^2]	Mean piston speed [m/s]
SI; 4 S; 4 C; $D \times S = 87 \times 92$; $P_b = 65$ kW, $N = 5000$ rpm				
SI; 4 S; 6 C; $D \times S = 90 \times 80$; $P_b = 90$ kW, $N = 4800$ rpm				
SI; 4 S; 8 C; $D \times S = 85 \times 65$; $P_b = 410$ kW, $N = 8000$ rpm				
DI; 4 S; 12 C; $D \times S = 135 \times 125$; $P_b = 950$ kW, $N = 2200$ rpm				
DI; 2 S; 14 C; $D \times S = 965 \times 2490$; $P_b = 80,000$ kW, $N = 100$ rpm				

Problem 5. Assuming the engine specifications given below, answer the following questions.

Configuration: I6 (in-line 6 cylinder).
Displacement: 7.6 L
Peak brake power output: 156 kW @ 2400 rpm
   Peak brake torque: 705 Nm @ 1450 rpm
      Aspiration: Turbocharged
      Combustion system: Direct injection
      Compression ratio: 17:1

(a) Calculate the clearance volume (cubic centimeter, cc) for one cylinder.
(b) Assume the stroke/bore ratio is 1.2:1, what is the average piston speed at peak power output and peak torque output?
(c) If the volumetric efficiency is 1.5 due to turbo charging, what is the equivalence ratio at 2000 rpm if the fuel flow rate is 30 kg/h? Assume the stoichiometric Air-Fuel ratio is 15:1; ambient air density is 1.181 kg/m^3.

(d) Assume this engine burns diesel with a representative molecule of $C_{15}H_{32}$. Show the balanced stoichiometric chemical reaction, and calculate the stoichiometric air-fuel ratio. Assume this engine can also operate with a fuel mixture of 1/3 biodiesel ($C_{19}H_{36}O_2$), 1/3 ethanol (C_2H_5OH), and 1/3 of diesel ($C_{15}H_{32}$) by MOLE. Show the balanced stoichiometric chemical reaction and calculate the stoichiometric air-fuel ratio. Why stoichiometric air-fuel ratio decreases in the second case?.

(e) For a certain condition, the engine is operated with an air-fuel ratio of $AF = 24$, and the cycle can be modeled by an air-standard dual cycle. Assume 60% of the fuel burned at constant volume and 40% burned at constant pressure and the combustion efficiency is 100%. Cylinder conditions at the start of compression are $T_1 = 50\ °C$ and $P_1 = 150$ kPa. $C_v = 0.821$ kJ/kg K, $C_p = 1.108$ kJ/kg K, heating value of the fuel is $Q_{LHV} = 42{,}500$ kJ/kg. What will be the cycle thermal efficiency? If the air flow rate is 200 kg/h and the mechanical efficiency is 80%, what will be the brake power in kW for this condition? What will be the brake-specific fuel consumption (BSFC) of the engine?

(f) Assume the cycle thermal efficiency is the same as part (e) and the volumetric efficiency is 1.7 for a certain operating condition with engine speed at 1500 rpm. Each cylinder has one injector, and each injector has six nozzle holes with a diameter of 200 μm. The discharge coefficient of the injector nozzle is 0.75. Assume the fuel injection pressure is 1200 bar across the injector nozzle and the injection duration is 8 crank angle degrees. Calculate the air-fuel ratio and IMEP by assuming 98% combustion efficiency? Fuel density is 800 kg/m³. Heating value of the fuel is $Q_{LHV} = 42{,}500$ kJ/kg.

(g) Assume the injection timing is 20° before TDC. Ignition delay is 0.0012 s at 1500 rpm. Assume constant volume combustion. What will be the indicated thermal efficiency if the ratio of the connecting rod to the crank radius is 4? Draw the P-V diagram. What if we have two injections in one cycle each with constant volume combustion? Draw a schematic P-V diagram and write the equation to calculate indicated thermal efficiency.

Problem 6. A four-stroke, four-cylinder, DI diesel engine having displacement volume of 2.5 L produces 200 HP at 4000 rpm in naturally aspirated conditions. The volumetric efficiency of engine is increased from 90% (naturally aspirated) to 110% (turbocharged) by boosting intake pressure twice as that for the naturally aspirated design. The air-fuel ratio is 18:1 for both turbocharged and naturally aspirated designs. What is the displacement volume required for the turbo engine to produce the same power at same rated rpm? Determine the fuel flow rate (cc/min) per cylinder per cycle for the turbocharged engine given that the density of diesel fuel is 830 kg/m^3. For this engine at 4000 rpm, if the fuel injection started 12 CAD bTDC and lasted for 600 μs, at what crank angle did the injection end? Clearly write your assumptions, if any.

Problem 7. A fuel economy map of a reciprocating engine is shown in the Fig. P6.1. Answer the following questions based on the figure shown.

(a) Based on the engine map, identify whether this map is for a compression ignition engine or spark ignition engine and a naturally aspirated engine or a turbocharged engine. Justify your answer based on your observation in the engine map.

(b) Discuss the curve of maximum BMEP curve with increasing engine speed. Discuss the factors which affect the maximum BMEP curve.

(c) Best fuel economy point (minimum specific fuel consumption) is shown as point P. Discuss why specific fuel consumption increases with an increase in speed, increase in BMEP, and decrease in BMEP as indicated in the figure.

(d) Calculate the torque and power of the engine at minimum specific fuel consumption point assuming the displacement volume of the engine as 6.7 L.

(e) Draw the line at which the engine can operate and produce 150 kW of power. Assume continuous variable transmission is available for the engine which makes it possible to operate the engine at any speed by performing a torque conversion through variable gear ratios. Mark the best engine operating point for producing 150 kW from this engine.

(f) Assume the bore stroke ratio of 0.92 for this particular engine, and determine the best possible number of cylinder and bore diameters for this particular engine.

Fig. P6.1 Reciprocating engine map (adapted from [49])

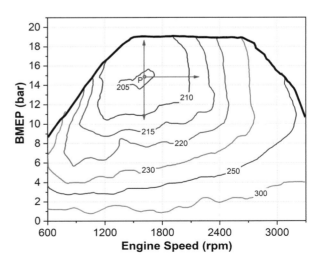

Problem 8. The variation of combustion efficiency (η_c), volumetric efficiency (η_v), mechanical efficiency (η_m), thermodynamic efficiency (η_{tm}), and gross indicated efficiency ($\eta_{g,i}$) with equivalence ratio for a four-stroke, single-cylinder SI engine at wide open throttle condition is shown in Fig. P6.2. Answer the following questions based on Fig. P6.2.

(a) Justify and discuss the trend observed for variation of all the efficiencies shown in the figure.
(b) Derive the equation which relates the brake power with all the efficiency terms shown in the figure.
(c) Draw the approximate variation (trend) of brake power and brake-specific fuel consumption with equivalence ratio for this particular engine. Explain the trend drawn for both curves.

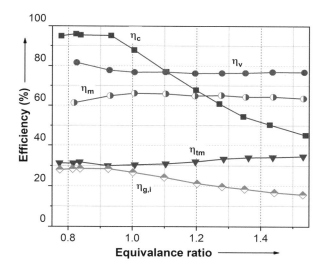

Fig. P6.2 Engine efficiency variations with equivalence ratio (adapted from [5])

Problem 9. A fuel consumption map of a four-cylinder spark ignition engine with bore/stroke of 90/80 mm is shown in Fig. P6.3. Answer the following questions based on the engine map.

(a) Fill the values of mean piston speed on the ticks shown in the figure on upper horizontal axis as shown in the figure.
(b) Draw the constant power lines on the engine map for the 10 kW, 30 kW, 50 kW, 80 kW, and 110 kW of power. Mark the most optimum point of operation for all the constant power engine operation.
(c) Draw an optimum operating line of the engine map by joining the all the fuel-efficient point on the different contours.

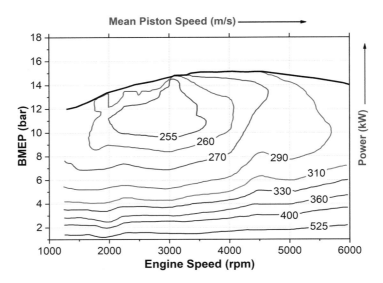

Fig. P6.3 Reciprocating engine map [50]

Problem 10. The cylinder pressure curve as a function of crank angle position for two different injection timings in an RCCI engine is shown in Fig. P6.4. Calculate the IMEP for the two conditions using Eqs. (6.62), (6.63), and (6.64) by taking 20 and 40 data points from the figure (or you can use your pressure data points for two different conditions). Write your observation and comment on the results. Discuss the role of crank angle resolution on IMEP calculation.

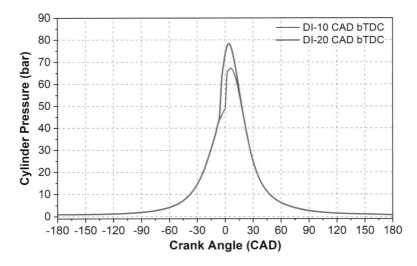

Fig. P6.4 Cylinder pressure trace at two different injection timings

Problem 11. The two engines are developed to produce the same maximum torque of 150 Nm. The torque and speed of the two engines can be assumed to be represented by equations given below. Assume the maximum engine speed of both the engines is 7000 rpm.

$$T_{e1} = 120 + 0.030(\text{rpm} - 1000) - 7.5 \times 10^{-6}(\text{rpm} - 1000)^2$$

$$T_{e2} = 120 + 0.020(\text{rpm} - 1000) - 3.33 \times 10^{-6}(\text{rpm} - 1000)^2$$

(a) Graphically plot the variation of engine torque and power with engine speed. Identify whether the engine is spark ignition or compression ignition.
(b) Calculate the engine flexibility of both the engine using Eq. (6.95). Comment in the engine flexibility of two engines.
(c) Discuss the design variations in these two engines based on the obtained torque curve.

Problem 12. Full-load torque curves for two different types of engines are given in Fig. P6.5. Assume that these engines are designed for use in the identical vehicle (same model of automotive vehicle). Identify the vehicle that provides the best acceleration from stop sign under normal conditions. Identify the vehicle which provides the maximum speed. Justify your answer with suitable reasoning. Comment on two types of engines shown in the figure.

Fig. P6.5 Full-load torque curve of two different engines

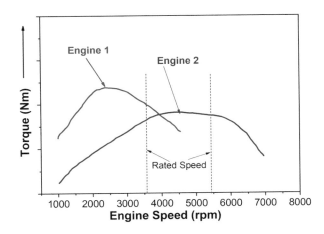

Problem 13. The typical brake-specific fuel consumption curve as a function engine speed, equivalence ratio, and displacement volume are shown in Fig. P6.6. Discuss the factors influencing the variation of BSFC with each parameter, i.e., engine speed, equivalence ratio, and displacement volume. On the same graph, draw the BSFC curve at a higher compression ratio, and justify the trend with suitable reasoning.

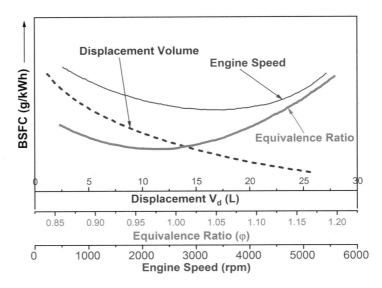

Fig. P6.6 Variation of BSFC of reciprocating engine with different engine parameters (adapted from [18])

References

1. Randolph, A. L. (1994). *Cylinder-pressure-based combustion analysis in race engines* (No. 942487). SAE Technical Paper.
2. Brunt, M. F., & Pond, C. R. (1997). *Evaluation of techniques for absolute cylinder pressure correction* (No. 970036). SAE Technical Paper.
3. Gill, P. W. S., & Ziurys, J. (1959). *Fundamentals of internal combustion engines*. New Delhi: Oxford and IBH Publishing.
4. Kar, K., Cheng, W., & Ishii, K. (2009). Effects of ethanol content on gasohol PFI engine wide-open-throttle operation. *SAE International Journal of Fuels and Lubricants, 2*(1), 895–901.
5. Heywood, J. B. (1988). *Internal combustion engine fundamentals*. New York: McGraw-Hill.
6. Crolla, D., & Mashadi, B. (2011). *Vehicle powertrain systems*. Chichester: Wiley.
7. Mallik, A. K., & Ghosh, A. (2004). *Theory of mechanism and machines*. New Delhi: Affiliated East-West Press (P) Ltd.
8. Van Basshuysen, R., & Schäfer, F. (2016). *Internal combustion engine handbook-basics, components, systems and perspectives* (2nd ed.). Warrendale: SAE International.
9. Hoag, K., & Dondlinger, B. (2016). *Vehicular engine design*. Vienna: Springer.
10. Lumley, J. L. (1999). *Engines: An introduction*. Cambridge: Cambridge University Press.
11. Park, S., & Sunwoo, M. (2003). Torque estimation of spark ignition engines via cylinder pressure measurement. *Proceedings of the Institution of Mechanical Engineers, Part D: Journal of Automobile Engineering, 217*(9), 809–817.
12. Pettersson, P. S., & Kjellin, A. (2017) *Torque estimation from in-cylinder pressure sensor for closed loop torque control* (Masters thesis). Chalmers University of Technology, Gothenburg, Sweden.
13. Rizzoni, G. (1989). Estimate of indicated torque from crankshaft speed fluctuations: A model for the dynamics of the IC engine. *IEEE Transactions on Vehicular Technology, 38*(3), 168–179.

14. Rosvall, T., & Stenlaas, O. (2016). *Torque estimation based virtual crank angle sensor* (No. 2016-01-1073). SAE Technical Paper.
15. Ginoux, S., & Champoussin, J. C. (1997). *Engine torque determination by crankangle measurements: State of the art, future prospects* (no. 970532). SAE Technical Paper.
16. Lee, B., Rizzoni, G., Guezennec, Y., Soliman, A., Cavalletti, M., & Waters, J. (2001). Engine control using torque estimation. *SAE Transactions,* Paper no - 2001-01-0995, 869–881.
17. Iida, K., Akishino, K., & Kido, K. (1990). IMEP estimation from instantaneous crankshaft torque variation. *SAE Transactions*, Paper no - 900617, 1374–1385.
18. Pulkrabek, W. W. (2004). *Engineering fundamentals of the internal combustion engine.* Upper Saddle River, NJ: Pearson Prentice Hall.
19. Maurya, R. K. (2018). *Characteristics and control of low temperature combustion engines: Employing gasoline, ethanol and methanol.* Cham: Springer.
20. Brunt, M. F., & Emtage, A. L. (1996). Evaluation of IMEP routines and analysis errors. *SAE Transactions, 960609,* 749–763.
21. Rai, H. S., Brunt, M. F., & Loader, C. P. (1999). *Quantification and reduction of IMEP errors resulting from pressure transducer thermal shock in an SI engine* (No. 1999-01-1329). SAE Technical Paper.
22. Arbuckle, J. S. (2006). *Indicated mean effective pressure estimation with applications to adaptive calibration* (PhD thesis). Michigan Technological University, Michigan, USA.
23. Hoard, J., & Rehagen, L. (1997). *Relating subjective idle quality to engine combustion* (No. 970035). SAE Technical Paper.
24. Hallgren, B. E., & Heywood, J. B. (2003). *Effects of substantial spark retard on SI engine combustion and hydrocarbon emissions* (No. 2003-01-3237). SAE Technical Paper.
25. Nishida, K., Kaneko, T., Takahashi, Y., & Aoki, K. (2011). *Estimation of indicated mean effective pressure using crankshaft angular velocity variation* (No. 2011-32-0510). SAE Technical Paper.
26. Hamedović, H., Raichle, F., Breuninger, J., Fischer, W., Fishcer, W., Dieterle, W., . . . Böhme, J. F. (2005). IMEP-estimation and in-cylinder pressure reconstruction for multicylinder SI-engine by combined processing of engine speed and one cylinder pressure. *SAE Transactions,* Paper no - 2005-01-0053, 135–142.
27. Jaine, T., Charlet, A., Higelin, P., & Chamaillard, Y. (2002). *High frequency IMEP estimation and filtering for torque based SI engine control* (No. 2002-01-1276). SAE Technical Paper.
28. Oh, S., Kim, D., Kim, J., Oh, B., Lee, K., & Sunwoo, M. (2009). *Real-time IMEP estimation for torque-based engine control using an in-cylinder pressure sensor* (No. 2009-01-0244). SAE Technical Paper.
29. Corti, E., Moro, D., & Solieri, L. (2008). *Measurement errors in real-time IMEP and ROHR evaluation* (No. 2008-01-0980). SAE Technical Paper.
30. Taraza, D. (2000). *A faster algorithm for the calculation of the IMEP* (No. 2000-01-2916). SAE Technical Paper.
31. Oh, S., Kim, J., Oh, B., Lee, K., & Sunwoo, M. (2011). Real-time IMEP estimation and control using an in-cylinder pressure sensor for a common-rail direct injection diesel engine. *Journal of Engineering for Gas Turbines and Power, 133*(6), 062801.
32. Saxena, S. (2011). *Maximizing power output in homogeneous charge compression ignition (HCCI) engines and enabling effective control of combustion timing* (PhD Thesis). University of California, Berkeley.
33. Feng, D., Wei, H., & Pan, M. (2018). Comparative study on combined effects of cooled EGR with intake boosting and variable compression ratios on combustion and emissions improvement in a SI engine. *Applied Thermal Engineering, 131,* 192–200.
34. Maurya, R. K. (2012). *Performance, emissions and combustion characterization and close loop control of HCCI engine employing gasoline like fuels* (PhD thesis). Indian Institute of Technology, Kanpur, India.

35. Maurya, R. K., & Agarwal, A. K. (2014). Experimental investigations of performance, combustion and emission characteristics of ethanol and methanol fueled HCCI engine. *Fuel Processing Technology, 126*, 30–48.
36. Atkins, R. D. (2009). *An introduction to engine testing and development.* Warrendale: SAE International.
37. Johansson, T. (2010). *Turbocharged HCCI engine, improving efficiency and operating range* (PhD thesis). Lund University, Sweden.
38. Shah, A. (2015). *Improving the efficiency of gas engines using pre-chamber ignition* (PhD thesis). Lund University, Sweden.
39. Li, Y., Jia, M., Chang, Y., Kokjohn, S. L., & Reitz, R. D. (2016). Thermodynamic energy and exergy analysis of three different engine combustion regimes. *Applied Energy, 180*, 849–858.
40. Hyvönen, J., Wilhelmsson, C., & Johansson, B. (2006). *The effect of displacement on air-diluted multi-cylinder HCCI engine performance* (No. 2006-01-0205). SAE Technical Paper.
41. Srivastava, D. K., & Agarwal, A. K. (2018). Combustion characteristics of a variable compression ratio laser-plasma ignited compressed natural gas engine. *Fuel, 214*, 322–329.
42. Li, Q., Liu, J., Fu, J., Zhou, X., & Liao, C. (2018). Comparative study on the pumping losses between continuous variable valve lift (CVVL) engine and variable valve timing (VVT) engine. *Applied Thermal Engineering, 137*, 710–720.
43. Basaran, H. U., & Ozsoysal, O. A. (2017). Effects of application of variable valve timing on the exhaust gas temperature improvement in a low-loaded diesel engine. *Applied Thermal Engineering, 122*, 758–767.
44. Cinar, C., Uyumaz, A., Solmaz, H., & Topgul, T. (2015). Effects of valve lift on the combustion and emissions of a HCCI gasoline engine. *Energy Conversion and Management, 94*, 159–168.
45. Abe, T., Nagahiro, K., Aoki, T., Minami, H., Kikuchi, M., & Hosogai, S. (2004). *Development of new 2.2-liter turbocharged diesel engine for the EURO-IV standards* (No. 2004-01-1316). SAE Technical Paper.
46. Lechner, G., & Naunheimer, H. (1999). *Automotive transmissions: Fundamentals, selection, design and application.* New York: Springer.
47. Kobayashi, A., Satou, T., Isaji, H., Takahashi, S., & Miyamoto, T. (2012). *Development of new I3 1.2 L supercharged gasoline engine* (No. 2012-01-0415). SAE Technical Paper.
48. Shinagawa, T., Kudo, M., Matsubara, W., & Kawai, T. (2015). *The new Toyota 1.2-liter ESTEC turbocharged direct injection gasoline engine* (No. 2015-01-1268). SAE Technical Paper.
49. DeRaad, S., Fulton, B., Gryglak, A., Hallgren, B., Hudson, A., Ives, D., . . . & Cattermole, I. (2010). *The new ford 6.7 L V-8 turbocharged diesel engine* (No. 2010-01-1101). SAE Technical Paper.
50. Fortnagel, M., Heil, B., Giese, J., Mürwald, M., Weining, H. K., & Lückert, P. (2000). Technischer Fortschritt durch Evolution—Neue Vierzylinder-Ottomotoren von Mercedes-Benz auf Basis des erfolgreichen M111 Teil 2. *MTZ-Motortechnische Zeitschrift, 61*(9), 582–590.

Chapter 7
Combustion Characteristic Analysis

Abbreviations and Symbols

ATDC	After top dead center
BMF	Burned mass fraction
BTDC	Before top dead center
CAD	Crank angle degree
CARS	Coherent anti-Stokes Raman scattering
CD	Combustion duration
CI	Compression ignition
CR	Compression ratio
DP	Difference pressure
ECU	Electronic control unit
EGR	Exhaust gas recirculation
EOC	End of combustion
EOI	End of injection
EVC	Exhaust valve closing
EVO	Exhaust valve opening
HCCI	Homogeneous charge compression ignition
HR	Heat release
HRR	Heat release rate
HT	Heat transfer
ID	Ignition delay
IGN	Ignition timing
IHR	Initial heat release
IHRR	Initial heat release rate
IMEP	Indicated mean effective pressure
IP	Intake pressure
IT	Ignition timings
IVC	Inlet valve closing

© Springer Nature Switzerland AG 2019
R. K. Maurya, *Reciprocating Engine Combustion Diagnostics*, Mechanical
Engineering Series, https://doi.org/10.1007/978-3-030-11954-6_7

LIF	Laser-induced fluorescence
LSR	Laser Rayleigh scattering
LTR	Low-temperature reaction
MP	Mixing period
NDP	Normalized difference pressure
°CA	Crank angle degree
PDF	Probability distribution function
PDR	Pressure departure ratio
PMEP	Pumping mean effective pressure
PMR	Pressure ratio management
PPC	Partially premixed combustion
PSD	Power spectral density
ROHR	Rate of heat release
RW	Rassweiler and Withrow
SI	Spark ignition
SOC	Start of combustion
SOV	Start of vaporization
SRS	Spontaneous Raman scattering
TDC	Top dead center
TP	Throttle position
TSA	Thermal stratification analysis
A	Area of the reaction front
B	Bore
D	Cylinder diameter
L	Characteristic length
P	Pressure
T	Temperature
u	Internal energy
V	Volume of cylinder
W	Work on the piston
Q_{ht}	Heat transfer
CA_{DD}	Detection delay crank angle
c_p	Specific heat capacity at constant pressure
c_v	Specific heat capacity at constant volume
$c_{v,b}$	Specific heat capacity of burn charge at constant volume
$c_{v,u}$	Specific heat capacity of unburnt charge at constant volume
HRR_{max}	Maximum heat release rate
L_c	Instantaneous combustion chamber height
n_a	Moles of air
n_c, n_e	Compression and expansion polytropic exponents
n_f	Moles of fuel
n_R	Moles of residual gases
Nu	Nusselt number
P_{atm}	Atmospheric pressure

P_{cyl}	Measured cylinder pressure
P_{diff}	Pressure difference
p_{EOC}	Pressure at end of combustion
P_{mot}	Motored cylinder pressure
Pr	Prandtl number
p_{SOC}	Pressure at start of combustion
r_{c}	Compression ratio
Re	Reynolds number
S_{p}	Mean piston speed
T_{atm}	Atmospheric temperature
T_{b}	Burned temperature
T_{EVC}	Temperature at EVC
T_{IVC}	Temperature at IVC
T_{U}	Burned zone temperature
T_{w}	Wall temperature
U_{s}	Sensible energy of the charge
V_{ch}	Characteristic velocity
V_{EOC}	Volume at end of combustion
V_{SOC}	Volume at start of combustion
V_{TDC}	Volume at TDC
x_{b}	Mass burn fraction
δQ_{hrr}	Elemental combustion heat released
$\Delta \theta_{\text{b}}$	Combustion duration
γ	Specific heat
θ	Crank angle position
λ	Excess air ratio (relative air-fuel ratio)

7.1 Introduction

In-cylinder pressure measurement and analysis is an essential tool for automotive engine research and diagnosis due to its direct relationship with the combustion process. The pressure developed in the combustion chamber at any crank angle position is the net result of the interplay of many overlapping phenomena such as combustion, change in volume, wall heat transfer, blowby, etc. The more that is known about the combustion process in the engine cylinder, the easier is to control it [1]. One approach for investigating the combustion process in an internal combustion engine is the heat release analysis. The heat release analysis provides valuable information on combustion behavior which is an important factor in fuel economy, engine performance, and emissions of harmful pollutants. Heat release analysis is a backward calculation process which requires pressure as input and analyzes the whole system within the framework consisting mainly of conservation equations for mass and energy and the equation of state. An insight into the engine combustion process, i.e., whether the burning rate is fast or slow, complete or incomplete, advanced or retard, and so on, can be obtained through the heat release analysis. Therefore, heat release analysis can be considered as a combustion diagnostic tool to gain an understanding about an existing engine which otherwise would be very difficult and would require highly advanced instrumentations and techniques, which may not even be possible to apply on production engines.

More stringent demands on emissions (by legislation) and efficiency (fuel economy) introduce new challenges for the design and control of modern automotive engines. Additionally, the complexity of the control system is increasing because of the added number of degrees of freedom available for engine control. This makes the task of calibrating a traditional open-loop engine control system more difficult and time-consuming. Thus, the interest in closed-loop engine control is growing as a solution to these problems. Combustion phasing (i.e., combustion event timing with respect to the position of the piston) is a central measure to control in an internal combustion engine as it affects both engine emissions and fuel conversion efficiency. The most common approach of measuring the combustion phasing is by means of heat release analysis [2]. Thus, online estimation of combustion phasing (or heat release) is required for condition monitoring and combustion management of reciprocating engines. Additionally, engine hardware development and control parameters tuning are usually supported by processing in-cylinder pressure data (heat release analysis) which allow examining a number of different solutions so that the combined effects of such solutions can be quickly tested experimentally.

Thermodynamic analysis of measured cylinder pressure data is a very powerful tool used for quantifying combustion parameters. There are two main approaches that are often referred to as "burn rate analysis" and "heat release analysis" [3]. Burn rate analysis is mainly used for determining burn crank angle positions (typically in gasoline engines), and it is used to obtain the mass fraction burned, which is a normalized quantity with a scale of zero to one. Heat release analysis is most commonly used for diesel engine combustion studies and produces absolute energy with units of Joules or Joules/degree. Different methods for the burn rate and heat release rate analysis (single/multi-zone model, Rassweiler and Withrow model, etc.) for offline and real-time estimation are discussed in this chapter (Sects. 7.2 and 7.5). Gas temperature and wall heat transfer (Sects. 7.3 and 7.4) are also required for performing accurate heat release analysis. Based on the calculation of heat release, different combustion parameters such as the start of combustion (SOC), end of combustion (EOC), combustion duration, etc. can be estimated. Typically, mass fraction burned is used for estimation of combustion phasing. Different methods that can be used for determination of mass fraction burned from measured cylinder pressure data are discussed in the next section.

7.2 Burned Mass Fraction Estimation and Analysis

Burned mass fraction (BMF) is a measure of the fraction of energy released from combustion of fuel to the total energy released at the end of the combustion process. Burned gas mass fraction is typically determined from the analysis of measured cylinder pressure [4]. The combustion rate is a very important parameter affecting fuel conversion efficiency, maximum temperature and pressure in a cycle, and exhaust emissions. The combustion rate is frequently quantified in gasoline engines by the estimation of burn angles which are the crank angles at which the BMF reaches a specified value. Conventional cylinder pressure-based techniques are able to predict the BMF as a function of crank angle position more accurately in SI engines than in diesel engines [5]. To calculate the burned mass fraction, cylinder pressure analysis can be performed using either relatively simple equations capable of real-time processing or by using more complicated post-processing thermodynamic heat release models. Several techniques have been proposed in the published literature for determination of burned mass fraction using measured cylinder pressure data.

7.2.1 Marvin's Graphical Method

Marvin proposed one of the first applications of experimental cylinder pressure data for direct diagnosis of engine combustion process [6]. Figure 7.1 illustrates the graphical method for calculation of burned mass fraction using measured cylinder pressure. In this method, the compression and expansion processes are considered as polytropic evolution (PV^n = constant) during closed valve conditions except for the period of combustion. The slopes of straight lines on the log P − log V curves

Fig. 7.1 Illustration of the graphical method of burned mass fraction calculation [7]

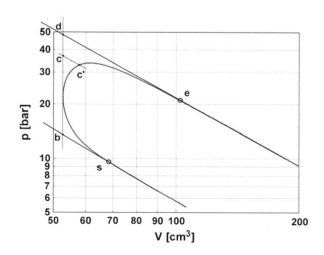

provide the compression and expansion polytropic index values. The positions of deviation of actual pressure from straight lines during compression and expansion phase are defined as the start of combustion (SOC) and end of combustion (EOC), respectively (Fig. 7.1), which are denoted by points "s" and "e," respectively. The points "b" and "d" on the curve are obtained by extrapolating the straight ideal compression and expansion lines.

The SOC and EOC points (s and e in Fig. 7.1) can be determined by considering the inflection points in the diagrams, which can be calculated using Eq. (7.1) [5, 7].

$$\frac{d^2(\log P)}{d(\log V)^2} = 0 \tag{7.1}$$

For a constant volume heat addition between points "b" and "d" (Fig. 7.1), ideally the chemical energy released during the combustion process generates a directly proportional pressure rise in the cylinder. Thus, burned mass fraction (x_b) at a particular point "c" is calculated by the ratio of the length b-c to length b-d, which is represented by Eq. (7.2).

$$x_b = \frac{P_c - P_b}{P_b - P_d} \tag{7.2}$$

For computation of burned mass fraction with experimental pressure data, a point corresponding to constant volume point (c) on the measured pressure data (point c') needs to be estimated. The point c' is determined as the intersection between constant volume combustion line and segment c'-c, which represents an equivalent and appropriate polytropic compression from point (c') to (c) [7]. It can be noticed that this method of burned mass fraction estimation is more valid for the engines which have a close to a constant volume heat release during combustion. Since conventional SI engine is close to Otto cycle (constant volume heat addition), thus, this method is more valid for SI engines. The BMF evaluated using Marvin's method is compared with theoretical two-zone model, and it is found that it has a difference of 2.5% between two curves and 1.5°CA difference in crank angle position corresponding to 50% BMF in both methods [5].

7.2.2 Rassweiler and Withrow Method

Rassweiler and Withrow [8] model is normally used for simple mass fraction burned calculations in spark ignition engines. The Rassweiler and Withrow (RW) method is a sort of analytical version of Marvin's graphical method of burned mass fraction evaluation. This method is established on the assumption that the variations in the cylinder pressure (due to piston movement and the gas-to-wall heat transfer) can be expressed by polytropic processes (PV^n = constant). The polytropic indices are based on calculated values obtained from the measured cylinder pressure data before and after combustion, the compression index (or exponent) being used up to TDC, and the expansion value thereafter [4]. The change in cylinder pressure consists of two components: (1) pressure rise due to combustion ($\Delta P_{comb,RW}$) and (2) pressure rise due to piston movement (volume change) and wall heat transfer. Pressure rise due to combustion is used for calculation of burned mass fraction, and it is calculated by Eq. (7.3).

$$\begin{cases} \Delta P_{comb,RW} = P(\theta_i) - P(\theta_{i-1})[V_{i-1}/V_i]^{n_c} & \text{if } \theta_i \le \theta_{TDC} \\ \Delta P_{comb,RW} = P(\theta_i) - P(\theta_{i-1})[V_{i-1}/V_i]^{n_e} & \text{if } \theta_i > \theta_{TDC} \end{cases} \quad (7.3)$$

where n_c and n_e are polytropic exponents during compression and expansion process.

The burned mass fraction (x_b) can be calculated using Eq. (7.4) in this method.

$$x_b = \frac{m_{b(i)}}{m_{b(total)}} = \frac{\sum_{j=0}^{i} \Delta P_{comb,RW}(\theta_j)}{\sum_{j=0}^{N} \Delta P_{comb,RW}(\theta_j)} \quad (7.4)$$

Here it is expected that the first point (sample 0) is between inlet valve closing (IVC) and the SOC and that last point (sample N) is after combustion has completed. The SOC and the EOC can be defined as the maximum (before TDC) and the minimum (after TDC) crank angle positions, respectively, where $\Delta P_{comb,RW}$ becomes virtually zero [5]. Figure 7.2 shows the variation of $\Delta P_{comb,RW}$ and

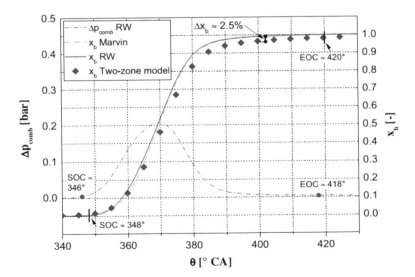

Fig. 7.2 Comparison of burned mass fraction calculated using the Marvin and RW methods in gasoline engine operating at 7.9 bar BMEP and 3300 rpm [5]

compares the burned mass fraction (BMF) calculation using Marvin's and RW methods in a gasoline SI engine. The figure shows that both methods have similar BMF using experimental data and a slight deviation from computed BMF using two-zone model.

The elemental combustion heat released (δQ_{hrr}) during the $\Delta\theta$ CA in SI engine can roughly be expressed as a heat released as constant volume process, as represented in Eq. (7.5).

$$\delta Q_{\mathrm{hrr}} = M c_{\mathrm{v}} \Delta T_{\mathrm{comb, RW}} \tag{7.5}$$

where M is the mass of the charge during combustion. The temperature increase ΔT_{comb} can be estimated as Eq. (7.6).

$$\Delta T_{\mathrm{comb, RW}} = \frac{\Delta p_{\mathrm{comb, RW}} V_{\mathrm{TDC}}}{MR} \tag{7.6}$$

where R is universal gas constant.

The elemental heat release can be expressed in terms of pressure rise due to combustion using Eqs. (7.5) and (7.6). The Eq. (7.7) shows that the heat release is proportional to the pressure rise due to combustion.

$$\delta Q_{\mathrm{hrr}} = \frac{C_{\mathrm{v}}}{R} V_{\mathrm{TDC}} \Delta p_{\mathrm{comb, RW}} \approx \frac{V_{\mathrm{TDC}}}{\gamma - 1} \{ p(\theta_i) - p_{i-1} [V_{i-1}/V(\theta_i)]^m \} \tag{7.7}$$

where V_{TDC} is the volume of the cylinder at TDC, m is the polytropic exponent, and γ is the ratio of specific heat of the charge. Since the combustion process does not occur at constant volume, the pressure rise due to combustion is not directly proportional to the mass of fuel burned. The pressure rise due to combustion has to be referenced to a datum volume (V_{TDC}).

McCuiston, Lavoie, and Kauffmann (MLK) suggested a similar relation for calculation of approximate burned mass fraction with the condition that the volume of the burning charge does not significantly changes during the combustion process [9]. The BMF equation proposed is shown by Eq. (7.8).

$$x_{\mathrm{b, MLK}} = \frac{p V^m - p_{\mathrm{soc}} V_{\mathrm{soc}}^m}{p_{\mathrm{EOC}} V_{\mathrm{EOC}}^m - p_{\mathrm{soc}} V_{\mathrm{SOC}}^m} \tag{7.8}$$

where p and V are the pressure and volume at a particular crank angle degree (CAD) and p_{SOC}, p_{EOC}, V_{SOC}, and V_{EOC} are the pressure values and volumes at SOC and EOC positions.

Figure 7.3 illustrates the BMF calculation using the MLK method in gasoline SI engine. The figure shows that calculated BMF using MLK method is virtually coincident with the BMF predicted using two-zone model and the difference between the MFB50 positions of the two methods is less than 0.2°CA [5].

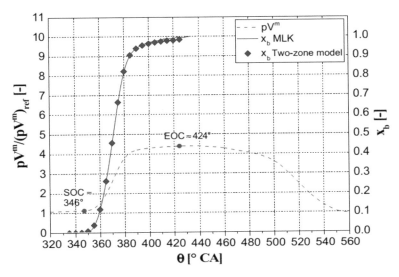

Fig. 7.3 Illustration of burned mass fraction calculation using the MLK method [5]

Development of combustion is often analyzed using BMF by estimating the flame development angle ($\Delta\theta_d$) and the rapid burning angle ($\Delta\theta_b$), which are defined as the crank angle from the spark timing position to 10% of BMF and from 10% to 90% of BMF, respectively. Figure 7.4 illustrates BMF as a function of crank angle in an SI engine operated at different fuels and ignition timings (IT). The BMF is calculated using Rassweiler and Withrow method using Eq. (7.4). Figure 7.4a shows that synthetic gases air-fuel mixture is burned faster than biogas, even though the ignition is more retarded [10]. The ignition timing of different fuel is selected corresponding to maximum thermal efficiency for a particular fuel. The evolution of the BMF at different ignition timing for syngas is presented in Fig. 7.4b. The figure depicts that the combustion tends to occur early as the spark timing is advanced.

7.2.3   *Pressure Ratio and Pressure Departure Ratio Method*

The pressure ratio management (PRM) is proposed by Matekunas [11] and is defined as the ratio of the cylinder pressure from a firing cycle $P(\theta)$ and the corresponding motored cylinder pressure $P_{mot}(\theta)$ at each crank angle position in the cycle. The motoring pressure can be obtained experimentally by measuring the pressure in the absence of combustion in the cylinder (running the engine by electrical means or combustion in other cylinder in case of multicylinder engine), or it can be obtained numerically by approximating the compression as a polytropic evolution. The pressure ratio can be represented by Eq. (7.9).

Fig. 7.4 Variations of BMF burned with CA position for (**a**) different fuels at ignition timing corresponding to the maximum thermal efficiency and (**b**) different ignition timing (IT) at syngas operation in an SI engine [10]

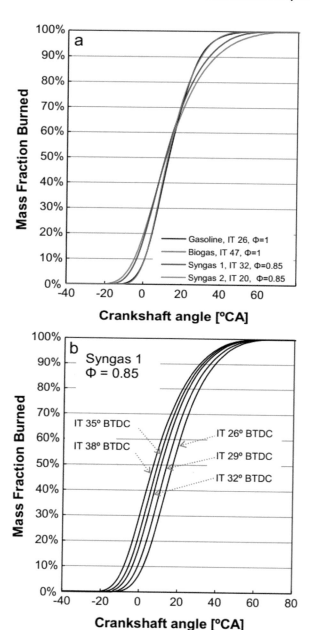

$$PR(\theta) = \frac{P(\theta)}{P_{mot}(\theta)} - 1 \qquad (7.9)$$

The pressure ratio (Eq. 7.9) is normalized by its maximum value, which produces traces that are similar to the burned mass fraction profiles. The BMF can be calculated by Eq. (7.10) using a pressure ratio management method.

$$x_{b,PRM} = \frac{PR(\theta)}{PR(\theta)_{max}} \qquad (7.10)$$

A similar approach for calculation of burned mass fraction in a diesel engine is also proposed [12]. The diesel pressure departure ratio (PDR) algorithm is proposed for fast and reliable determination of the mass fraction burnt. This algorithm is also based on the principle of the RW model which suggests that the fractional pressure rise due to combustion can be used for estimation of BMF. The PDR algorithm utilizes the fired and the motoring pressure signals for cyclic estimation of the CA50, and the mass fraction burned closely matches the normalized cumulative heat release curve [12] (Fig. 7.5). The pressure departure ratio is defined by Eq. (7.11).

$$PDR(\theta) = \frac{P(\theta) + C_1}{P_{mot}(\theta) + C_2} - 1 \qquad (7.11)$$

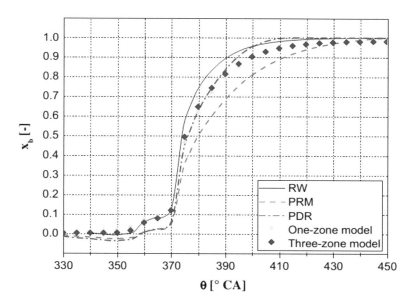

Fig. 7.5 Comparison of the mass fraction burnt calculation methods in a diesel engine using triple injection in the cycle [5]

where C_1 is the fired pressure characterization coefficient and C_2 is the motored pressure characterization coefficient. Constant C_2 is a correction for the polytropic relationship between the volume and the motoring pressure, while the C_1 is a correction for the difference between the pressure change due to the volume of the fired and the motored cycle. The coefficients C_1 and C_2 are constants for a particular engine configuration and are largely not affected by the boost pressure, EGR, etc. [12]. The PDR value is close to zero before SOC and reaches a peak value at the EOC.

The mass fraction burnt can be calculated by normalizing the PDR with its maximum value PDR_{max} as represented in Eq. (7.12).

$$x_{b,\,PDR} = \frac{PDR(\theta)}{PDR_{max}} \tag{7.12}$$

Figure 7.6 shows the comparison of the burned mass fraction in a diesel engine with multiple fuel injection in a cycle. The PDR method is found to be the best approximation of the mass burned fraction obtained by using single-zone and three-zone models for diesel engine [5]. Figure 7.6 illustrates that the RW method provides better performance than the PRM method. However, both RW and PRM methods seem inadequate for BMF estimation in diesel engines, and PDR is a relatively better method of BMF determination.

7.2.4   Vibe's Function-Based Method

The Vibe function [13] is often used as a parameterization of the burned mass fraction (x_b), and it is represented by Eq. (7.13).

$$x_b = 1 - \exp\left\{-a\left(\frac{\theta - \theta_0}{\Delta\theta_b}\right)^{m+1}\right\} \tag{7.13}$$

where θ is the crank angle, θ_0 is the crank angle at the start of combustion, $\Delta\theta_b$ is the combustion duration, a is duration parameter, and m is the shape factor. The shape parameter (m) governs the pressure rise rate and the location of the point of inflection of the BMF curve.

For a particular engine, the parameters "m" and "a" are estimated by trial and error [14]. The parameters are adjusted until the pressure-time curve produced by cycle simulation matches the experimental pressure-time signal. Typical values of the parameters "m" and "a" are 2 and 5, respectively.

7.2.5 *Apparent Heat Release-Based Method*

The burned mass fraction can also be estimated by computing heat release from measured cylinder pressure data using first law of thermodynamics with single-zone assumption. The detailed algorithm for calculation of heat release from experimental cylinder pressure data is presented in Sect. 7.5. The apparent heat release can be calculated using Eq. (7.14), and wall heat transfer as well as blowby losses are neglected in this equation.

$$\frac{\partial Q}{\partial \theta} = \frac{\gamma}{\gamma - 1} p \frac{dV}{d\theta} + \frac{1}{\gamma - 1} V \frac{\partial p}{\partial \theta} \qquad (7.14)$$

The cumulative heat release curve can be determined by integrating the heat release rate obtained using Eq. (7.14), starting from θ_{ivc} (crank angle of intake valve closing) to the current angle [15]. Starting point of integration can be any point between intake valve closing and start of combustion.

$$Q(\theta) = \int_{\theta_{ivc}}^{\theta} \frac{dQ(\alpha)}{d\alpha} d\alpha \qquad (7.15)$$

The mass fraction burned can be obtained by normalizing the heat release calculated using Eq. (7.15). The cumulative heat release profile is usually identified as a function of crank angle with the help of a normalized curve which starts with zero at SOC and ends with a value of one at EOC. Assuming that the BMF is directly proportional to the heat release, the burned mass fraction is calculated by Eq. (7.16).

$$x_{b,\,HR} = \frac{Q(\theta)}{Q(\theta)_{max}} \qquad (7.16)$$

The mass fraction burnt (MFB) calculated using this method is dependent on the values of the ratio of specific heat (γ) used in Eq. (7.14). The γ values are typically dependent on the gas temperature. Depending on the complexity, constant γ value as well as function of temperature can be used for calculation of BMF [4].

The sensitivity of MFB calculations to common sources of error in the measured pressure data shows that MFB errors produced as a result of pressure data errors should be relatively small for well-designed algorithms [4]. The highest sensitivity is to absolute pressure referencing errors (especially at part load), and this sensitivity can be made worse by resulting errors generated in the calculated compression and expansion exponents. Signal noise should not produce significant errors for normally defined burn angles but can cause problems with burn angles below 10% and above 90% and major problems with burn rates. A minimum of 150 cycles should be adequate for burn angle statistics which is less than would normally be required for IMEP calculations [4]. A crank angle resolution of 1.0° is recommended for most applications, but coarser resolutions of several degrees could be used if necessary with only a small loss in accuracy.

7.3 Combustion Gas Temperature Estimation

Thermodynamic state of the in-cylinder gases is defined by properties such as pressure, temperature, and density. Pressure is normally uniform in the combustion chamber, and it can be readily measured using a piezoelectric pressure transducer. Charge density can also be easily determined from cylinder volume and the amount of gases trapped in the cylinder. The gas temperature during compression can be considered uniform except for thermal boundary layers near the cylinder wall [16]. Temperature is normally not uniform during combustion in the conventional SI or CI engines because combustion is not occurring uniformly throughout the combustion chamber. However, combustion occurs almost homogeneously throughout the combustion chamber except in the proximity of combustion chamber walls in the HCCI engine. Thus, the assumption of uniform temperature during combustion is more closer to actual value is possibly in HCCI combustion. The bulk temperature of the charge can be estimated using ideal gas equation by assuming uniform temperature within the combustion chamber. The average gas temperature as a function of crank angle position can be calculated using Eq. (7.17) [1].

$$T(\theta) = \frac{P(\theta)V(\theta)n(\theta)}{P_{\mathrm{IVC}}V_{\mathrm{IVC}}n_{\mathrm{IVC}}}T_{\mathrm{IVC}} \tag{7.17}$$

where P, V, and T are the cylinder pressure, volume, and temperature of the cylinder gases, respectively. Temperature of gases at intake valve closing (IVC) position is an important variable, which is required for computation of bulk charge temperature.

One of the methods to calculate the temperature at IVC (T_{IVC}) is given by Eq. (7.18) [17].

$$T_{\mathrm{IVC}} = \frac{1}{(n_{\mathrm{a}} + n_{\mathrm{f}} + n_{\mathrm{R}})} \times \frac{P_{\mathrm{IVC}}V_{\mathrm{IVC}}}{R} \tag{7.18}$$

where n_{a}, n_{f}, and n_{R} are the moles of air, fuel, and residual gases, respectively, and R is the universal gas constant. Pressure at IVC position can be estimated by measured pressure data, and the volume at IVC position can be calculated by engine geometry. Moles of air and fuel can be estimated from the measured air and fuel values and their respective molecular weights. Moles of residual gas can be estimated by Eq. (7.19).

$$n_{\mathrm{R}} = \frac{P_{\mathrm{EVC}}V_{\mathrm{EVC}}}{T_{\mathrm{EVC}}R} \tag{7.19}$$

Pressure and volume known at exhaust valve closing (EVC) can be estimated similar to inlet valve closing (IVC). Temperature at exhaust valve closing is unknown in Eq. (7.19). The temperature measured close to the exhaust valve in the exhaust manifold can be used to estimate the T_{EVC}. The temperature measured is a multi-cycle-averaged exhaust temperature, and it is lower than the cylinder temperature at EVC due to heat transfer. However, it is demonstrated that if the exhaust manifold temperature and pressure are used to calculate the moles of residual gas instead of T_{EVC} and P_{EVC}, the trapped gas fraction is very close to the experimentally measured residual fraction using the in-cylinder CO_2 method [17]. Thus, exhaust manifold temperature and pressure can be used to estimate the moles of residual gases, and temperature at IVC can be computed.

The gas temperature at IVC is influenced by heat transfer and residual gases, which are normally difficult to measure and determine. The air in the intake manifold is heated from intake manifold temperature (T_{im}) to T_a by the hot valves and the locally high heat transfer coefficients in the cylinder. Additionally, inducted fuel undergoes evaporation that affects the charge temperature. Another method is proposed to determine the charge temperature at IVC in a gasoline engine by considering the energy equation with a lumped process for heating, evaporation, and mixing the initial air-fuel mixture temperature [15]. The resultant temperature of the air-fuel mixture is calculated by Eq. (7.20).

$$T_{af} = \frac{m_a c_{pa} T_a + m_f c_{pf} T_f - m_f h_{v,f} + Q}{m_a c_{pa} + m_f c_{pf}} \qquad (7.20)$$

where $h_{v,f}$ is the vaporization enthalpy for the fuel and Q is the heat added to the fresh mixture, which is difficult to estimate without detailed measurements. The residual gases and fresh charge are mixed in the combustion chamber, and temperature of the charge can be determined using Eq. (7.21). The residual gases have also been cooled down by heat transfer to the cylinder walls before mixing during the gas exchange process.

$$T_{ivc} = \frac{m_{af} c_{paf} T_{af} - m_r c_{pr} T_r}{m_{af} c_{paf} - m_r c_{pr}} \qquad (7.21)$$

In this method, several variables need to determine using complex models. Therefore, charge temperature can be calculated by the simplified model using Eq. (7.22). In this equation, heat transfer is neglected which provides $T_{af} = T_{im}$, and c_p is assumed the same for both residuals and fresh air and fuel mixtures.

$$T_{ivc} = T_{af}(1 - x_r) + x_r T_r \qquad (7.22)$$

where x_r is the residual gas fraction, which can be estimated by several models using cylinder pressure data (Chap. 10).

The combustion gas temperature plays a significant role in NO_x formation and its subsequent emissions. NOx formation is very sensitive to peak temperature encountered during the combustion. At temperatures above 1800 K, the NO_x formation rate increases rapidly. The maximum temperature in the combustion chamber also affects the combustion efficiency. At lower temperatures, combustion efficiency decreases,

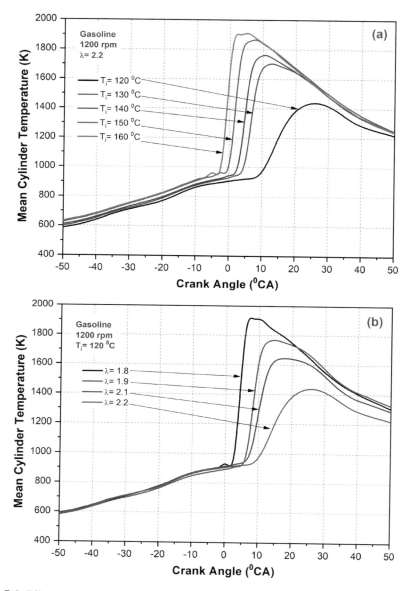

Fig. 7.6 Effect of λ and T_i on average cylinder temperature using gasoline at 1200 rpm in HCCI engine

which leads to higher unburned hydrocarbon emissions. The mean gas temperature history in the combustion chamber is shown in Fig. 7.6 at 1200 rpm in the HCCI engine using gasoline. Figure 7.6b shows that the mean gas temperature in the combustion chamber is highest for the richest mixture and decreases as the mixture becomes leaner. Mean gas temperature in the combustion chamber is also affected by the initial temperature of the charge (Fig. 7.6a). Mean gas temperature of the combustion chamber is higher for a charge with higher initial temperature due to advanced combustion phasing. When intake air temperature increases, from 120 to 130 °C, change in mean gas temperature is large as compared to other inlet air temperatures due to a substantial increase in combustion efficiency and advanced combustion phasing (Fig. 7.6a).

Figure 7.7 shows the variation for T_{max} in HCCI operating range for gasoline and ethanol fuel at 1200 rpm. The figure depicts that the contour lines of T_{max} are inclined, which means T_{max} is dependent on λ as well as intake air temperature (T_i). It can also be noticed that the value of T_{max} is highest closer to the knock boundary for richer mixture for both the fuels (Fig. 7.7). Maximum value of T_{max} in HCCI operating range is relatively higher for ethanol in comparison to gasoline due to advanced combustion phasing.

Temperature determined from Eq. (7.17) by assuming a single zone in the cylinder can be used to calculate the burned gas temperature by considering two zones (burned and unburned) in the combustion chamber. Two-zone model is divided into two zones: one containing the unburned gases and the other containing the burned gases, separated by an infinitesimal thin divider representing the flame front. Each zone is homogenous considering that temperature, thermodynamic properties, and the pressure are the same throughout the zones. A simple two-zone model is proposed to determine the burned zone temperature (T_b) and the unburned zone temperature (T_u) [18].

The unburned zone temperature (T_u) is assumed to be equal to the single-zone temperature (T) before the start of combustion (SOC). The unburned zone temperature (T_u) after SOC can be estimated assuming adiabatic compression of the unburned charge as shown in Eq. (7.23).

$$T_u = T_{u,SOC} \left(\frac{p}{p_{SOC}} \right)^{1-1/\gamma} = T_{SOC} \left(\frac{p}{p_{SOC}} \right)^{1-1/\gamma} \tag{7.23}$$

Energy balance between the single-zone and the two-zone models can be written by Eq. (7.24).

$$(m_b + m_u) c_v T = m_b c_{v,b} T_b + m_u c_{v,u} T_u \tag{7.24}$$

where $c_{v,b}$ and $c_{v,u}$ are the constant volume-specific heat of burned and unburned charge. Assuming $c_v = c_{v,b} = c_{v,u}$, i.e., a calorically perfect gas, the Eq. (7.24) can be written as Eq. (7.25).

$$T = \frac{m_b T_b + m_u T_u}{m_b + m_u} = x_b T_b + (1 - x_b) T_u \tag{7.25}$$

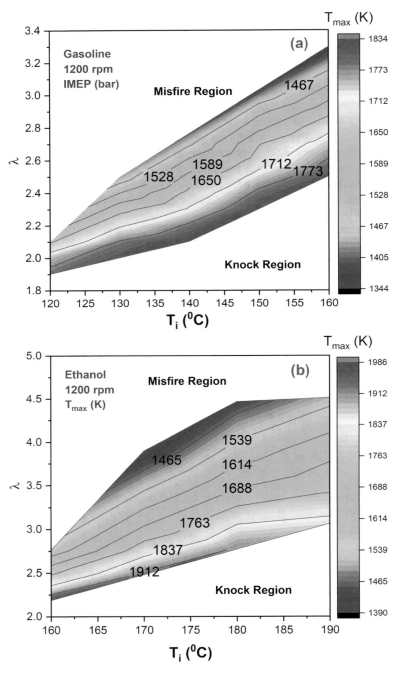

Fig. 7.7 Variation of T_{max} w.r.t. λ and T_i in HCCI operating range for gasoline and ethanol at 1200 rpm

where the single-zone temperature can be seen as the mass-weighted mean temperature of the two zones. The burned temperature (T_b) can be calculated by Eq. (7.26).

$$T_b = \frac{T - (1 - x_b)T_u}{x_b} \tag{7.26}$$

Heider [19] has defined an approach of burnt zone temperature calculation to determine the NO_x emissions more precisely. Thereby, the difference between the unburnt and burnt zone temperature is modeled by Eq. (7.27). As shown in Eq. (7.28), B describes the effect of the combustion using the cylinder pressure (p) minus the motored engine pressure (p_0), completed by the burned mass m_b.

$$T_b - T_{ub} = B \cdot A^* \tag{7.27}$$

$$B = 1 - \frac{\int_{\varphi_{SOC}}^{\varphi} (p - p_0) \cdot m_b \cdot d\varphi}{\int_{\varphi_{SOC}}^{\varphi_{EVO}} (p - p_0) \cdot m_b \cdot d\varphi} \tag{7.28}$$

A^* is an engine and combustion system-specific factor for the determination of the temperature level in the reaction (burnt) zone at the start of combustion. A method for estimation of A^* defined depending on engine stroke and global air-fuel ratio [20]. For engines with wide combustion bowl and without inlet swirl, the values of A^* are around 1600 K for a typical diesel engine. However, this model is not providing satisfactory results for constant volume combustion. The improved correlation for burnt zone temperature calculation based on combustion chamber thermodynamics and optical analysis is proposed in reference [21]. The found model in Eq. (7.29) calculates a temperature offset to the mass averaged temperature T which considers the combustion process by means of the normalized burn heat release (x) and the mixture air-fuel ratio λ.

$$\begin{aligned} T_b &= (A^* \cdot (1 - x) + x \cdot A) \cdot (1 - x) + T \\ A^* &= (0.9 - 2.26 \cdot \lambda + 1.9 \cdot \lambda^2 - 0.5 \cdot \lambda^3)^{-1} \end{aligned} \tag{7.29}$$

Further, an engine-specific factor A describes the temperature level in the burning zone at start of combustion. For engines up to 500 cc displacement per cylinder, good results can be reached with $A \approx 850$ K [21].

Combustion gas temperature plays a significant role in combustion and emission. For understanding autoignition kinetics of end-gas temperature is required with accuracy. Additionally, cylinder gas temperature is required for estimating the convective heat transfer and radiative loss to the wall and the thermal stress to wall material [16]. For measuring the gas temperature, the sensor used in the measurement system must be robust enough to deal with the environment inside the cylinder of a firing engine, and it should be capable of tracking the very rapid changes in charge temperature that occurs during the compression and expansion strokes.

The basic principle of thermometry is to bring a well-characterized material into thermal equilibrium with the body to be measured. The temperature measurement is made by measuring the change in some property of the well-characterized material. Thermocouples are the first choice for the measurement of gas temperature, but their response is too slow. The response of thermocouples is also very sensitive to the flow conditions surrounding the bead. Since the time lag is a function of the rate of heat transfer into the thermocouple, anything that affects the heat transfer, such as turbulence levels, affects the response time. This effect makes correction for the time lag difficult [22]. High-speed thin film resistance temperature sensor is also demonstrated for measurement of gas temperature. The in-cylinder gas temperature measurement has been dominated by optical and acoustic measurement [16, 22]. Optical temperature measurement techniques are based on the emission and absorption, either singly or in combination, of light in the infrared through an ultraviolet range of the spectrum. The use of spectroscopic techniques is limited to optically thin gases due to the self-absorption. On the other hand, radiation thermometers are suited for thick gases where emissivity or absorptivity of gas can be measured accurately. Temperature measurement by spontaneous Raman scattering (SRS), laser Rayleigh scattering (LRS), coherent anti-Stokes Raman scattering (CARS), and laser-induced fluorescence (LIF) has been developed [16]. Acoustic techniques make use of the variation in the speed of sound with the temperature of a gas. The time of flight of an acoustic pulse, as well as acoustic resonance, can be used for determination of gas temperature [23].

7.4 Wall Heat Transfer Estimation

In reciprocating combustion engines, the heat transfer from the combustion gases to the cylinder walls is an important factor in the engine's design because it affects fuel conversion efficiency, power, and emissions. Thus, a lot of studies are focused on investigating an accurate model of the heat transfer, for engine design and optimization purposes [24]. Several potential advantages can be gained by accurate thermal predictions such as improved cooling systems (lighter and small pumps as well as heat exchangers), lower thermal distortion (reduced friction and optimized piston ring assembly), and advances in engine simulation and modeling (by applying correct boundary conditions) [25]. These advantages definitely contribute to improving fuel economy and reducing engine emissions. The thermal condition of the combustion chamber is closely tied to heat transfer from the hot gas to the walls. Thus, a good understanding of the heat transfer process in the combustion chamber is prerequisite for developing advanced engine control strategies and thermal management scheme.

The heat transfer from combustion gases to the coolant in reciprocating internal combustion engines represents between one-fourth and one-third of the total energy released by the mixture of fuel and air during combustion. The entire heat rejection to the cooling fluid depends mainly on engine type and operating conditions. The modes of heat transfer are typically involved in the heat transfer process from combustion gases to the coolant through metallic components of the engine. Heat transfer from combustion gases to liner takes place mainly by forced convection with a contribution by radiation. Heat transfer by radiation mode is dominant in conventional diesel engines because of the generation of highly radiative soot particles. The radiative heat transfer accounts for only 5% of the total heat flux in an SI engine, and it is usually neglected [26]. Heat transfer from the liner to coolant takes place by conduction mode.

Heat transfer in combustion engines is a nonuniform and unsteady process because heat flux between combustion gases and metallic surface (liner) varies in time and space significantly during an engine cycle. In-cylinder heat fluxes oscillate between a few MW/m^2 during combustion and close to zero or even negative during intake stroke [25]. Heat flux measurement and analysis of heat transfer are typically conducted during engine operation in motored condition and fired conditions. In motoring operation, the engine is driven by external power (such as an electric motor). Measurements in motoring operating conditions are taken to investigate the effect of the gas properties and the gas flow (without the influence of combustion) on heat transfer. Figures 7.8 and 7.9 show the factors affecting the heat flux in motoring and firing operating conditions, respectively. The factors affecting the heat flux are categorized into three main groups, namely, (1) in-cylinder flow, (2) gas properties, and (3) temperature difference between the combustion gases and the cylinder wall [24].

In-cylinder flow and gas properties play role in convective heat transfer and affect the convective heat transfer coefficient. To determine the factors that can be actually varied on the engine, the three groups are further expanded. Various factors affecting the heat flux in reciprocating engines are divided into three categories, namely, (1) controllable factors, (2) uncontrollable factors (ambient conditions), and (3) unchangeable factors (fixed for a particular engine). The pathways from these three types of factors to the top can be used to describe the trends in the heat flux in motored and fired operating conditions (Figs. 7.8 and 7.9). The ambient conditions are typically fixed during experiments. In motoring operating condition, three independent parameters, i.e., (1) the throttle position (TP), (2) the compression ratio (CR), and (3) the injected gas (type and quantity), can be changed for varying the heat flux. In firing operating conditions, the ignition timing (IGN), excess air ratio (λ), fuel type, engine load, and EGR are the additional variables that can be controlled to affect the heat flux (Fig. 7.9). In motored conditions, no significant spatial variation is observed, and thus, measurement at one location can be conduction. However, significant spatial variation occurs in heat flux during firing operation [24].

Figure 7.10 depicts the effect of excess air ratio on the peak heat flux, which is presented as a normalized value along with the measurement location from three

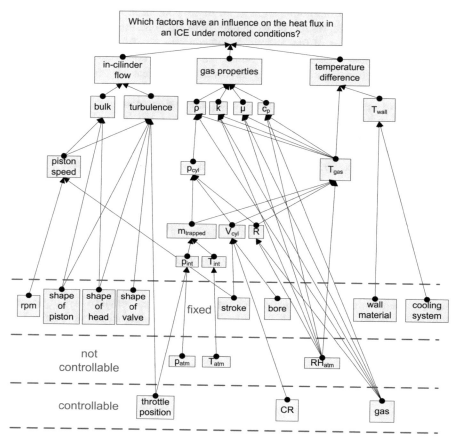

Fig. 7.8 Identification of the factors affecting the heat flux under motored engine operation [24]

studies. The figure depicts a maximum in the peak heat flux as with mixture richness. The maxima of peak heat flux around $\lambda = 1$ can be attributed to adiabatic temperature (highest for stoichiometric mixtures) and flame speed (the highest for richer mixtures) [24].

Engine speed significantly affects the heat transfer from the walls. The peak and total cycle heat flux increases with engine speed that is attributed to an increase in gas velocity and turbulence [24, 27, 28]. The increase in heat transfer happened mainly at the beginning of the expansion stroke, possibly due to the higher gas temperatures in that part of the engine cycle. The variation in spark timing around MBT affects both the rise and the peak value of the heat flux [27]. The operational properties are affected by increasing compression ratio such as (1) surface-to-volume ratio around TDC, (2) the gas pressure, (3) peak burned gas temperature, (4) gas motion increase, (5) faster combustion, and (6) the decrease in gas temperature late

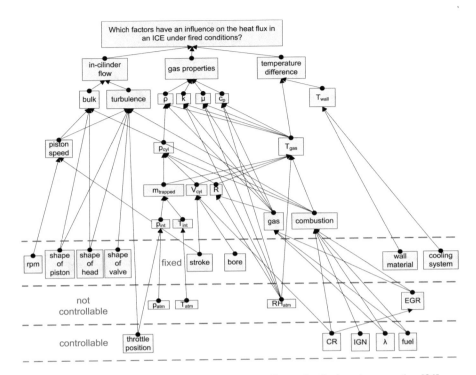

Fig. 7.9 Identification of the factors affecting the heat flux under fired engine operation [24]

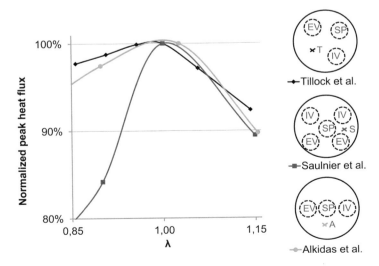

Fig. 7.10 Variation of normalized peak heat flux with excess air ratio (λ) [24]

Fig. 7.11 The variation of
convection coefficient for
different fuels [29]

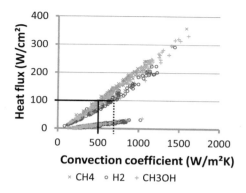

× CH4   ○ H2   + CH3OH

in the expansion stroke. The first five properties increase the heat transfer; the last
one decreases it [24].

Convective heat transfer is also affected by fuel used for operating the engine.
Figure 7.11 shows the heat flux is with the convection coefficient for the three fuels
at different engine operating conditions. Heat flux is measured for all the condition at
the same location. Figure 7.11 depicts that a certain heat flux level consistently
corresponds with a higher convection coefficient for hydrogen (illustrated in
Fig. 7.11 at a level of 100 W/cm^2). The convection coefficient of methane is
significantly lower than hydrogen at particular heat flux. It suggests that for the
same difference between the gas and wall temperature, the heat flux of hydrogen will
be higher in comparison to methane [29].

Figure 7.12 illustrates the spatial variation in heat flux in a particular combustion
cycle and cyclic variation in heat flux at a particular engine operating condition.
Figure 7.12a depicts two local maxima in the heat flux traces at P2 and P4 positions.
The P1 position corresponds to the spark plug installation. The first peak appears
when flame passes over the measurement position and the second peak occurs when
the temperature difference between combustion gases and wall reaches to maximum
[29]. The maximum of gas temperature occurs before the flame reaches to the P3
position, and thus, it has only one maximum point in the curve (Fig. 7.12a).

Figure 7.12b depicts the typical cyclic variations in the measured pressure and
heat flux. The cycle-averaged traces are shown with a solid red line; the cycles which
differ the most (max and min) and the least (best) from the mean cycle are plotted
with dashed black lines. The figure shows that there is a much steeper rise in the heat
flux trace at the instant of the flame arrival in the case of individual cycles. This rise
is not that steep in the case of the mean cycle due to the cyclic variations in the time
of the flame arrival [29].

The convective heat transfer is a transient process but can be assumed to be quasi-
steady. The rate of heat loss through the cylinder walls can be calculated using
Eq. (7.30).

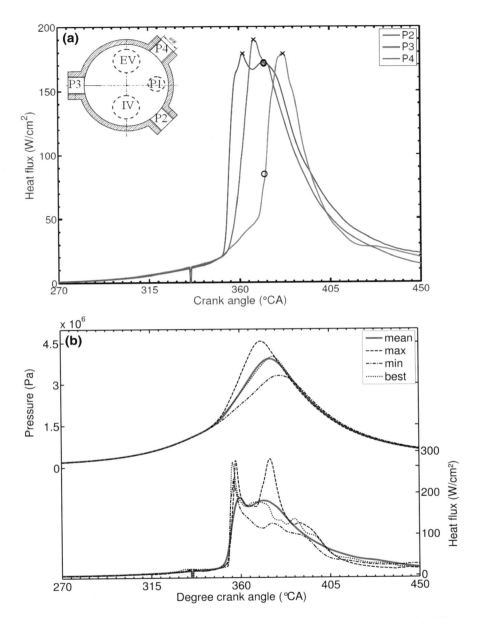

Fig. 7.12 (**a**) Spatial variation of heat flux at particular engine cycle and (**b**) cyclic variation in heat flux at particular engine operating condition [29]

$$\frac{\partial Q_{ht}}{\partial \theta} = \frac{h(T - T_w)A}{360N} \quad [\text{J/CAD}] \tag{7.30}$$

where h is heat transfer coefficient, T_w is the average cylinder wall temperature, A is the actual cylinder wall area, and N is the engine speed. It is normally assumed that T_w is constant over the entire cycle. h, T, and A are functions of crank angle position [1].

The most commonly used heat transfer models are empirical models that calculate the spatially averaged heat flux as a function of crank angle.

One of the early empirical models for the heat transfer coefficient is proposed by Eichelberg [30] which is given by Eq. (7.31).

$$h = \alpha_s \cdot S_p^{\,1/3} \cdot P^{1/2} \cdot T^{1/2} \tag{7.31}$$

where S_p is the mean piston speed, α_s is scaling factor, and P and T are pressure and temperature of the combustion chamber, respectively. The scaling factor depends on engine and combustion modes. The value of α_s is 0.25 in reference [31], and 7.67×10^{-3} is used in reference [32]. The scaling factors need to be calibrated for a particular engine. This model provides acceptable accuracy and easier to use and compute [32].

Heat transfer coefficient can be estimated by Hohenberg model which is shown in Eq. (7.32) [33].

$$h = \alpha_s \, V^{-0.06} P^{0.8} \, T^{-0.4} \left(S_p + 1.4\right)^{0.8} \tag{7.32}$$

Here S_p is the mean piston speed, α_s is scaling factor, and P, V, and T are pressure, volume, and temperature of combustion chamber, respectively. Another correlation that is normally used for computation of heat transfer coefficient by Woschni model [34] and presented by Eq. (7.33).

$$h = 131 \times (\text{Bore})^{-0.2} \times P^{0.8} \times T^{-0.55} W^{0.8} \tag{7.33}$$

where

$$W = C_1 \times 2.28 \times S_p + C_2 \times 3.24 \times 10^{-3} \times \frac{V_d}{V_{IVC}} \times \left(\frac{P - P_{mot}}{P_{IVC}}\right) \times T_{IVC} \tag{7.34}$$

where S_p is the mean piston speed and C_1 and C_2 are constants. Typically, heat loss through the crevices is assumed to be small and neglected.

Generally, all the models are based on the Pohlhausen equation (7.35) which characterizes the convective heat transfer on a flat plate using the boundary layer theory [35].

$$Nu = a \cdot Re^b \cdot Pr^c \tag{7.35}$$

where a, b, and c are model coefficients and Nu, Re, and Pr, are the Nusselt number, Reynolds number, and Prandtl number, respectively. Prandtl number is typically lumped with scaling constant "a" because it does not change significantly throughout the engine cycle for most fuels [29]. Thus, empirical correlations for the convective heat transfer coefficient can be presented as a function of a characteristic length L, a characteristic velocity V_{ch}, the thermal conductivity k, the dynamic viscosity, and the density of the combustion gases (Eq. (7.36)).

$$h = a \cdot V_{ch}{}^b \cdot L^{b-1} \cdot k \cdot \mu^{-b} \cdot \rho^b \tag{7.36}$$

Various proposed models differ with the selection of characteristics length and velocity and estimation of thermal properties of the charge. The model coefficient is typically computed by regression with experimental data.

The Eq. (7.36) can be approximated as Eq. (7.37) by certain assumptions about the thermal properties [35].

$$h = a \cdot V_{ch}{}^b \cdot L^d \cdot T^e \cdot P^b \tag{7.37}$$

The characteristics length (L) and velocity (V_{ch}) along with model coefficients for several models are proposed in reference [35].

Annand [36] proposed a correlation for both SI and CI engines, the constant b depends on the type of engine, and the heat transfer coefficient is given by Eq. (7.38).

$$h = a \cdot \frac{k}{B} Re^{0.7} + b \frac{(T^4 - T_w{}^4)}{(T - T_w)} \tag{7.38}$$

where T and T_w are the temperature of combustion gases and wall, k is thermal conductivity of gases, and B is the bore of the engine. For SI engines, the values of $a = 0.35$–0.8 and $b = 4.3 \times 10^{-9}$ W/m^2 K^{-4}.

Bargende [37] proposed another model for the heat transfer coefficient as shown by Eq. (7.39).

$$h = 3.5212 \cdot P^{0.78} \cdot \left(0.5\sqrt{8k_{spec} + S_p{}^2}\right) \cdot \left(V^{1/3}\right)^{-0.22} \cdot T^{-0.477} \tag{7.39}$$

where k_{spec} is the specific kinetic energy [m^2/s^2], S_p is mean piston speed, and P, V, and T are pressure, volume, and temperature, respectively.

Chang [38] proposed a new heat transfer coefficient for HCCI combustion by modifying the Woschni model. It was found that the original Woschni model cannot match well measurements in the HCCI engine since the unsteady gas velocity term causes overprediction of heat transfer during combustion. This in turn leads to underprediction during compression and hence undesirable consequences regarding predicting ignition. The proposed model is given in Eq. (7.40).

$$h = \alpha_s \cdot P^{0.8} \cdot \left(c_1 \cdot S_p + c_2 \cdot \frac{V_s \cdot T_r}{P_r \cdot V_r} \cdot (P - P_0) \right)^{0.8} \cdot L_c^{-0.2} \cdot T^{-0.73} \qquad (7.40)$$

where L_c is the instantaneous combustion chamber height.

A study applied the models of Chang, Hohenberg, and Woschni in a single-zone engine simulation of HCCI engine and showed that the shape of the heat flux curve predicted by Hohenberg and Chang model is very similar if the same length scale is used [39]. Figure 7.13 shows the variation of heat flux with a crank angle for different heat transfer coefficient model in motored and fired operation of HCCI engine. Figure 7.13a shows that the models of Woschni, Hohenberg, Chang, and Hensel underestimate the heat flux throughout the entire combustion cycle, except during the final part of the expansion phase. Subsequently, these models underestimate both the maximum heat flux and the total heat loss [35]. The Bargende model overestimates the heat flux in the whole combustion cycle in motoring conditions. Figure 7.13a clearly shows that none of the models can be used without calibrating the scaling coefficient first. After calibration in motoring conditions, the coefficient is used for heat flux estimation in fired engine operation (Fig. 7.13b). The models of Annand, Hohenberg, and Bargende are able to capture the effect of the combustion on the heat transfer. The models accurately predict the maximum heat flux during fired operation using the scaling coefficient found during motored operation [35].

A detailed assessment of the models of convective heat transfer coefficients can be conducted by presenting the Nusselt number as a function of the Reynolds number on a logarithmic scale. The relationship between the Nusselt and the Reynolds number must be a straight line on a logarithmic scale if the model perfectly predicts the convection coefficient because all the models are based on the power law as shown in Eq. (7.35) [35]. Figure 7.14 depicts the variation of the Nusselt with Reynolds number for the models of Annand, Woschni, and Bargende. Yellow square and red dot show the SOC and the EOC, respectively. The Annand and Woschni models show a distinct horseshoe shape with two straight legs around the top dead center position, representing the compression and the expansion phase (Fig. 7.14). Subsequently, a better model for heat transfer coefficient can be achieved if different coefficients a and b are used for the compression and expansion phase [35]. The linear relationship only occurs during the compression and expansion immediately before and after the combustion. The prediction of the heat flux cannot be improved in the other regions by using different values for the coefficients a and b (Eq. 7.35).

Knocking engine operating conditions affected the cylinder heat transfer, and it significantly increases. A study found that knock intensities above 0.2 MPa influenced the heat flux, and at knock intensities above 0.6 MPa, the peak heat flux was 2.5 times higher than for a non-knocking cycle [40]. The influence of knocking on heat flux is believed to be due to an increased charge motion as a result of the autoignition. This influences the heat transfer coefficient as the Reynolds

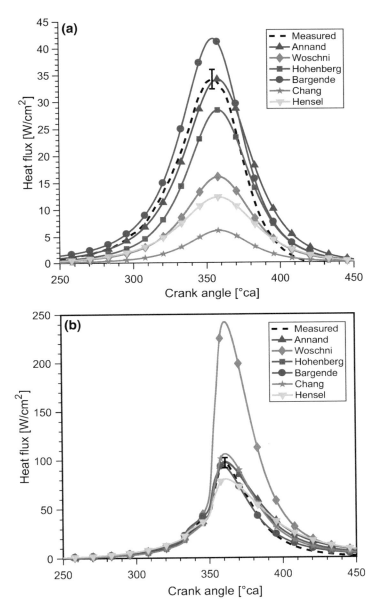

Fig. 7.13 Variation of heat flux during (**a**) motoring operation and (**b**) firing operating conditions with coefficients for motored operation [35]

number is increased. There were no evident indications that the heat flux was affected by the direction of the pressure waves in relation to the wall at which measurements were made.

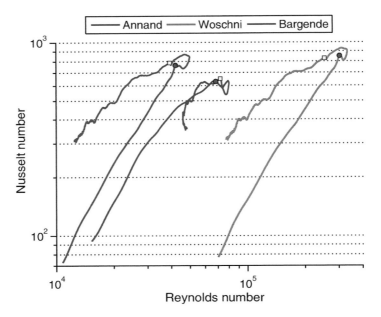

Fig. 7.14 Nusselt and Reynolds number on a logarithmic scale [35]

A study separated the heat loss into the loss due to normal heat transfer and into the loss due to knock in an HCCI engine [31]. Total heat transfer is separated into the normal convection heat transfer and the effect of knock by using the heat transfer coefficient (Eq. 7.41), by turn the term of dp/dt (pressure rise rate) in the coefficient on and off.

$$h = C_1 \cdot S_p^{1/3} \cdot P^{1/2} \cdot T^{1/2} \cdot \left(1 + \frac{dp/dt}{C_2}\right)$$

$$dp/dt = dp/dt \quad (dp/dt > 0)$$

$$dp/dt = 0 \quad (dp/dt \le 0)$$

$$(7.41)$$

It is demonstrated that changes in heat transfer rate can be expressed by knock in the HCCI engine with adding a term of *dp/dt* (pressure rise rate) to heat transfer coefficient by Eichelberg model [31].

Another study investigated the effect in-cylinder pressure oscillations have on piston heat transfer using two fast-response surface thermocouples embedded in the piston top for transient temperature measurement [41]. To quantify the magnitude of in-cylinder pressure oscillations, the metric of signal energy (SE) is introduced. The SE is calculated by integrating the power spectral density (PSD) over the first resonant mode. The limits of integration were selected to be 5.3 and 8.0 kHz, as denoted by a subscript designation specifying the frequency range, $SE_{5.3-8.0}$.

$$SE_{5.3-8.0} = \int_{5.3}^{8.0 \ \text{kHz}} PSD \cdot df \qquad (7.42)$$

Increase in the closed-cycle integrated heat flux is found with an increase in $SE_{5.3-8.0}$, which strongly suggests that the higher pressure oscillations are increasing the heat transfer rate [41]. It was found that before SOC, the heat transfer coefficient was effectively the same for cycles with high and low $SE_{5.3-8.0}$ at the same cylinder pressure. However, after combustion, the cycles with high $SE_{5.3-8.0}$ showed higher heat transfer coefficients than those with low $SE_{5.3-8.0}$ at the same pressure. This strongly suggests a correlation with the cylinder pressure oscillations.

7.5 Heat Release Rate Analysis

Cylinder pressure traces have been used for a long time to monitor the combustion event and extract information regarding burn rate, combustion phasing, and combustion duration. Heat release analysis by using a pressure sensor signal is a well-recognized technique for evaluation of the combustion event and also for combustion diagnostics. In this section, calculation of heat release from measured cylinder pressure data is discussed.

7.5.1 Rassweiler and Withrow Model

Rassweiler and Withrow (RW) heat release model is an approximate model that is loosely based on the Rassweiler and Withrow mass fraction burnt model (Sect. 7.2.2). The equation for heat release uses the same basic combustion pressure rise equation as the standard RW equation which is based on the assumption that the pressure changes due to (1) piston motion and (2) charge to wall heat transfer can be represented by polytropic processes [3]. However, the heat release equation does not assume that the proportion of fuel burned is proportional to the increase in the corrected combustion pressure. The main assumptions in RW heat release model are (1) the combustion can be idealized as being divided into a number of constant volume combustion processes and (2) the pressure change due to piston motion and charge to wall heat transfer can be modeled as a series of polytropic processes [3, 42].

The pressure rise between two positions 1 and 2 in the engine cycle due to combustion can be written as Eq. (7.43) by assuming pressure change due to compression can be represented by a polytropic process with polytropic index n [3].

$$\Delta P_2 = P_2 - P_1 \left(\frac{V_1}{V_2}\right)^n \tag{7.43}$$

Assuming ideal gas relationships, the incremental temperature rise (ΔT_2), due to combustion, can be approximated by Eq. (7.44).

$$\Delta T_2 = \frac{\Delta P_2 \cdot V_2}{m \cdot R} \tag{7.44}$$

Equation (7.44) is an approximation because it assumes R is a constant and the combustion occurs at a fixed volume. The incremental heat release energy ($\Delta Q_{hr,2}$), required to produce the temperature rise, can be achieved by using the first law for the assumed constant volume process as shown in Eq. (7.45) [3, 42].

$$Q_{hr,2} = m \cdot C_v \cdot \Delta T_2 = \frac{V_2}{\gamma - 1} \cdot \left\{ P_2 - P_1 \left(\frac{V_1}{V_2}\right)^n \right\} \tag{7.45}$$

Equation (7.45) provides the approximate incremental gross heat release energy for the crank angle interval from θ_1 to θ_2, and this can be integrated over the combustion period to get cumulative gross heat release. The net cumulative heat release can be obtained if the polytropic index (n) is replaced by the ratio of specific heats (γ) which ignores the heat transfer [3, 42].

A potential advantage of this equation is that charge to wall heat transfer is included for in the polytropic index term. To be most accurate, this term would need to be varied continuously during the heat release calculations, but it is assumed here that using the measured polytropic coefficients immediately prior to ignition and shortly after the end of combustion (EOC) can be used without significant loss of accuracy [3].

Figure 7.15 illustrates the effect of the polytropic index (assumed constant value over the entire cycle) on gross heat release curve which is computed using Eq. (7.45) with the calculated polytropic index with temperature-dependent γ. The figure depicts that the polytropic index variation considers the effect of the heat transfer, which has a large effect on the heat release rate after the EOC. However, the advantage of this model over the heat transfer term in the first law equation is that the polytropic index can be determined from the measured pressure-volume data, and thus, calculated values for each cycle can be used rather than assumed and probably erroneous values [3, 42].

Fig. 7.15 Effect of assumed fixed polytropic index on the cumulative gross heat release calculated by Eq. (7.45) with temperature-dependent γ (adapted from [3])

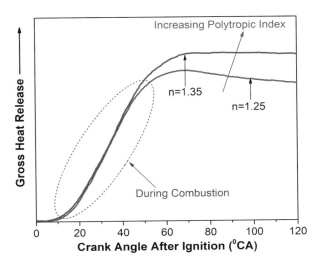

Fig. 7.16 Energy balance for engine combustion chamber with chemical heat release

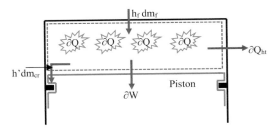

7.5.2   Single-Zone Thermodynamic Model

Simple single-zone models are usually employed in preference to the potentially more accurate multidimensional thermodynamic models due to their lower complexity and because they are numerically more efficient and normally yield similar results. The single-zone models do not include spatial variations and, thus, assume uniform charge temperature and composition in the entire combustion chamber. The first law of thermodynamics is employed to cylinder gases by assuming the combustion chamber as control volume as shown in Fig. 7.16. The state of the cylinder is defined in terms of average properties and does not differentiate the burned and unburned gases. The engine cylinder is assumed as a single zone, where no temperature gradient occurs and the reactants and products are completely mixed [2, 16, 43]. Single-zone model offers the advantage of including heat transfer and gas flow phenomena more simply. The combustion process is considered as separate heat addition process and contents of the cylinder as single fluid. The classical first law equation for the in-cylinder charge during the non-flow period between inlet valve closure (IVC) and exhaust valve closure (EVC) can be written for an incremental crank angle interval.

The gross chemical heat released (Q) due to combustion can be presented by Eq. (7.46) [16, 43].

$$\partial Q = dU_s + \partial W + \sum h_i \cdot dm_i + \partial Q_{ht} \tag{7.46}$$

Here U_s is the sensible energy of the charge (which means change in internal energy (u) or enthalpy (h) due to change in temperature alone and changes due to chemical reaction or phase change are ignored), ∂Q_{ht} is the heat transfer to the cylinder walls, and ∂W is the work on the piston, which is equal to pdV. The mass flux term indicates the flow across the system boundary, which includes crevice flows and injected fuel in case of direct fuel injection engines.

$$\sum h_i \cdot dm_i = \left(h' dm_{cr} - h_f dm_f \right) \tag{7.47}$$

Assuming U_s can be written as $m \cdot u(T)$, where T is the average charge temperature of the cylinder gases and m is the mass within the system boundary and u is internal energy.

$$dU_s = dm \cdot u + mC_v dT \tag{7.48}$$

For SI engines running on premixed charge, dm is caused by flows in and out of the crevices ($dm = -dm_{cr}$). The Eq. (7.46) can be written as Eq. (7.49).

$$\partial Q = mC_v dT + P \cdot dV + (h' - u)dm_{cr} + \partial Q_{ht} \tag{7.49}$$

By assuming that combustion gases follow the ideal gas law ($PV = mRT$), changes in temperature can be calculated by Eq. (7.50).

$$\frac{dT}{T} = \frac{dP}{P} + \frac{dV}{V} - \frac{dm}{m} \tag{7.50}$$

By substituting the Eq. (7.50) to Eq. (7.49), the heat release equation can be written as Eq. (7.51).

$$\partial Q = \left(1 + \frac{C_v}{R} \right) P \cdot dV + \frac{C_v}{R} dP \cdot V + \left(h' - u + C_v T \right) dm_{cr} + \partial Q_{ht} \tag{7.51}$$

Since gases are assumed as an ideal gas, then $C_p + C_v = R$, and $\gamma = C_p/C_v$. Substituting the values of γ, the final heat release rate equation is written as Eq. (7.52).

$$\frac{dQ}{d\theta} = \frac{\gamma}{\gamma - 1} P \cdot \frac{dV}{d\theta} + \frac{1}{\gamma - 1}\frac{dP}{d\theta} \cdot V + \left(h' - u + C_v T\right)\frac{dm_{cr}}{d\theta} + \frac{\partial Q_{ht}}{\partial \theta} \qquad (7.52)$$

The ratio of specific heat (γ) is the most important property used in the heat transfer calculation. The values of γ depend on temperature and the composition of charge (fuel, EGR, air-fuel ratio, ratio of reactant to combustion products). For spark ignition engine, the values of γ are obtained by matching single-zone model analysis to that of a two-zone model analysis for different fuels [44]. The calculation of heat transfer term in Eq. (7.52) is discussed in Sect. 7.4.

Crevice effect can be modeled by assuming the overall effect of crevice volume as single aggregate crevice volume, where the gas pressure is same as cylinder pressure, but the temperature is same as the wall temperature (T_w) [43].

$$m_{cr} = \frac{P \cdot V_{cr}}{R \cdot T_w} \quad \text{or} \quad dm_{cr} = \left(\frac{V_{cr}}{R \cdot T_w}\right) dP \qquad (7.53)$$

Assuming $\gamma(T) = a + bT$, the factor by which dm_{cr} is multiplied in Eq. (7.51) can be expressed as Eq. (7.54) [43].

$$\left(h' - u + C_v T\right) = T' + \frac{T}{\gamma - 1} - \frac{1}{b}\ln\left(\frac{\gamma - 1}{\gamma' - 1}\right) \qquad (7.54)$$

where the "primed" quantities are evaluated at cylinder conditions when mass is out of the control volume and evaluated at crevice condition when it re-enters the chamber.

Substituting the Eq. (7.54), the final gross heat release equation can be written as Eq. (7.55).

$$\frac{dQ}{d\theta} = \frac{\gamma}{\gamma - 1} P \cdot \frac{dV}{d\theta} + \frac{1}{\gamma - 1}\frac{dP}{d\theta} \cdot V + V_{cr}\left(T' + \frac{T}{\gamma - 1} - \frac{1}{b}\ln\left(\frac{\gamma - 1}{\gamma' - 1}\right)\right)\frac{dP}{d\theta} + \frac{\partial Q_{ht}}{\partial \theta}$$
$$(7.55)$$

The crevice volume is often ignored in the heat release analysis. The charge to wall heat transfer is often ignored for simplicity, and the heat release determined by ignoring crevice losses and wall heat transfer is referred to as "net" or "apparent" heat release. The net heat release equation can be represented by Eq. (7.56).

$$\frac{dQ_n}{d\theta} = \frac{\gamma}{\gamma - 1} P \cdot \frac{dV}{d\theta} + \frac{1}{\gamma - 1}\frac{dP}{d\theta} \cdot V \qquad (7.56)$$

The net heat release equation is the same for the direct injection engines also [16]. A zero-dimensional, three-zone heat release calculation utilizing thermodynamic equilibrium for direct injection (DI) compression ignition engine has been proposed [45]. In this method, an Arrhenius-based equation used to evaluate the rate

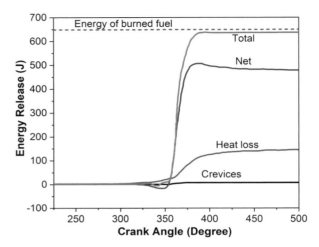

of fuel mass consumed by the combustion process, allowing the model to estimate
the rate of change of the chemical species within the cylinder, which provides the
properties of the bulk gas at a given crank angle for thermodynamic analysis.
Another study proposed a corrected temperature-based heat release analysis in
HCCI engine using equilibrium composition of species as a function of crank
angle position [46].

The cumulative heat release can be calculated by integrating the heat release rate.
Figure 7.17 illustrates the typical cumulative heat release as a function of crank angle
position in the HCCI engine. This figure also depicts the net heat release and the
losses due to heat transfer as well as crevices. Figure 7.17 shows that crevice loss can
be neglected due to very small contribution in the total heat release. The wall heat
transfer significantly contributes to the total heat release depending on the engine
operating conditions and combustion modes.

Specific heat ratio (γ) is another important factor that affects the heat release
calculation using in-cylinder pressure measurement. The ratio of specific heat for
single-zone model is computed by a more rigorous study using two-zone models
[18, 44]. However, the simpler linear model can be used for calculation of γ by
Eq. (7.57) [43].

$$\gamma = \gamma_0 - \frac{k}{100} \frac{T}{1000} \tag{7.57}$$

where γ_0 is the value of the ratio of specific heat at some reference temperature
(typically 300 K), and it depends on charge composition. For atmospheric air, γ_0 is
1.4, and for lean air-fuel mixtures, 1.38 is a usable value. The constant k is usually set
at 8.

Figure 7.18 illustrates the effect of γ without heat transfer consideration and the
effect of heat transfer at fixed γ (1.28) on the cumulative heat release calculated using

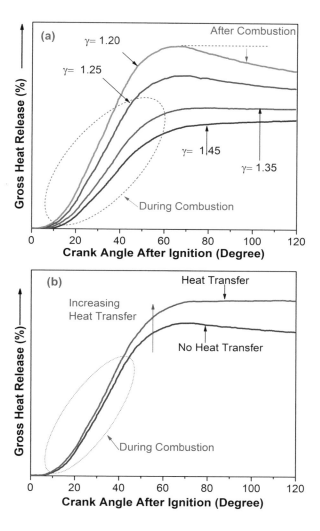

Fig. 7.18 (**a**) Effect of γ with no heat transfer and (**b**) effect of heat transfer at fixed γ (1.28) on the cumulative heat release calculated using single-zone model (adapted from [3])

single-zone thermodynamic model. Figure 7.18a shows that γ has a very large effect on both the magnitude of the heat release and the shape of the cumulative heat release curves. A low value of γ produces both (1) a heat release value which is too high and (2) a heat release rate which is negative after the completion of combustion. The opposite effect is present for high γ values although the effect is not linear with gamma, being inversely proportional to ($\gamma - 1$) [3]. Figure 7.18b depicts that heat transfer has a significant effect on both the peak cumulative gross heat release and the gradient of the curves following the completion of combustion. It suggests that heat transfer effects must be included if accurate gross heat release values are to be obtained.

Fig. 7.19 Effect of γ and
heat transfer rate on the
cumulative heat release
calculated using cylinder
pressure by single zone
model (adapted from [3])

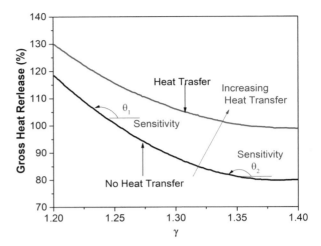

Figure 7.19 demonstrates the effects of the γ and heat transfer rate by presenting
the combined effects on the total (at $120°$ CA after ignition) cumulative gross heat
release energy. The figure shows that there is a strong interaction between the γ and
heat transfer with the calculated cumulative gross heat release. Sensitivity to γ is
highest for conditions where γ is at its lowest (Fig. 7.19). Engine operation at richer
mixture have lower γ values, and thus, relatively larger error is possible in heat
release calculation. Therefore, accurate and repeatable gross heat release data can
only be obtained if accurate values are used for both γ and heat transfer [3].

7.5.3 Real-Time Heat Release Estimation

The control system requirements of a modern engine are complicated because of the
wide variety of combustion modes and fuel injection strategies that can be utilized to
achieve high efficiency, low noise, or low emissions of oxides of nitrogen (NO_x) and
soot. Any means to control the phasing of the combustion process in a desired crank
angle window may help to alleviate the problems associated with the cyclic variation
and can stabilize these combustion regimes. Thus, online calculation of combustion
phasing is required, which can be calculated by calculation of real-time heat release.
There are several methods for the online calculation of combustion phasing
(discussed in Sect. 7.7) such as apparent heat release, RW methods, and pressure
departure ratio [48].

Present combustion measurement system provides the heat release curves on real-
time basis during engine experiments. Typically, apparent heat release is calculated
on a real-time basis. The basic algorithm for fast net heat release is proposed and
implemented by AVL based on single-zone model given by Eq. (7.58).

$$Q_i = \frac{K}{\gamma - 1} \{\gamma \cdot P_i \cdot (V_{i+n} - V_{i-n}) + V_i \cdot (P_{i+1} - P_{i-1})\} \qquad (7.58)$$

where K is the constant (based on unit conversion) and n is the crank angle interval.
The heat release calculation based on RW model based on Eq. (7.45) can also be used for online heat release computation. The γ values as a function of temperature can also be used for the calculation of heat release. For control purposes, combustion phasing is required which can be calculated by computing cumulative heat release or mass fraction burnt.

7.5.4 Characteristics of Heat Release Rate in Different Engines

The heat release rate can be calculated from measured cylinder pressure data using a single-zone model for different combustion modes such as SI, CI, HCCI, or dual-fuel combustion. The typical shape of heat release rate trace depends on the combustion mode as well as engine operating conditions. Important conclusions can be drawn based on analyzing the heat release rate traces at different engine operating conditions.

Figure 7.20 shows the measured cylinder pressure and calculated heat release rate for spark ignition (SI) and laser ignition (LI) of natural gas for different relative air-fuel ratios (λ). The figure shows that the maximum cylinder pressure and maximum heat release rate increase with relatively richer mixture (decreasing λ) up to stoichiometric mixture ($\lambda = 1$) for both SI and LI. For mixture richer than stoichiometric ($\lambda < 1$), maximum cylinder pressure and maximum heat release rate decreased. The typical heat release rate curve is almost symmetric in conventional spark ignition engine for normal combustion using homogeneous charge (Fig. 7.20). Initially, combustion of charge begins at a slower rate and increases rapidly to a very high combustion rate around halfway of the combustion process. After achieving the maximum point, the heat release rate starts decreasing toward zero in almost identical fashion as it increased. This almost symmetric characteristic heat release curve is obtained for SI engine at different load and speed conditions. The typical combustion duration obtained is around 50 CAD in SI and LI combustion (Fig. 7.20).

Figure 7.21 shows the heat release rate traces for conventional diesel combustion and modern diesel engine with multiple fuel injection in a cycle. Modern diesel combustion systems typically utilize common rail direct injection (CRDI), high-pressure fuel injection systems with multiple injections per cycle, variable geometry turbines, alternate combustion modes, and advanced exhaust gas recirculation (EGR) handling techniques to enable compliance with the diesel emission standards [48]. The present trend is to split the heat release into multi-events or even to shift the heat release away from the TDC in order to achieve lower the combustion

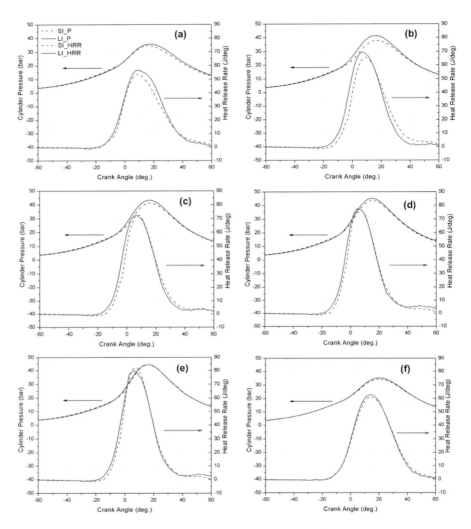

Fig. 7.20 Comparative variation of cylinder pressure and ROHR for SI and LI with crank angle position for (**a**) $\lambda = 1.35$, (**b**) $\lambda = 1.2$, (**c**) $\lambda = 1.15$, (**d**) $\lambda = 1.10$, (**e**) $\lambda = 1.00$, and (**f**) $\lambda = 0.90$ at ignition timing of $25°$ bTDC [49]

temperature to reduce NO_x emission and lower pressure rise rate [12]. The combustion phasing can be early or late depending on the boost pressure, EGR, and engine loads. Typical heat release rate curve in multiple injection strategy in diesel combustion is illustrated in Fig. 7.21b. In typical conventional diesel combustion with signal injection at a relatively lower injection pressure, the heat release consists of premixed combustion phase and diffusion-controlled combustion (Fig. 7.21). The peak of the premixed combustion phase depends on the ignition delay period, which depends on several factors such as fuel properties, injection timing, injection

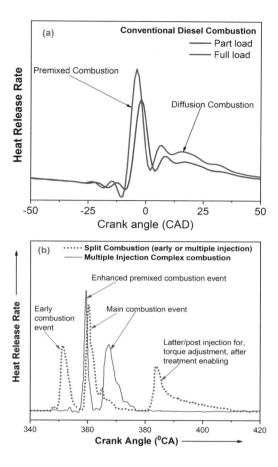

Fig. 7.21 Heat release rates for (**a**) conventional and (**b**) modern (multiple injection) diesel engines (adapted from [12])

pressure, engine load, etc. Thus, the combustion phasing and the combustion duration in modern diesel engine heavily depend on the mode of operation and fuel injection strategy. For example, the split injection can be used to control the combustion noise, while a retarded combustion phasing is essential to control the pressure rise rate and peak pressure at high boost pressure. Post-flame control can also be applied for torque modulation or the reduction of soot in certain cases (Fig. 7.21b) [12].

Figure 7.22 shows heat release rate traces at different λ in HCCI engine using ethanol and methanol. In the HCCI combustion, the charge is premixed, and combustion starts by autoignition. The typical shape of heat release curve in HCCI combustion is similar to SI combustion as both have almost symmetric characteristic heat release. However, in HCCI combustion the peak heat release rate is high, and the combustion duration is less than half (in terms of CA) of the combustion duration in SI engine. Figure 7.22 shows that the heat release rate and peak pressure rise rate increase as the mixture becomes richer (decrease in λ). Both fuels (ethanol and

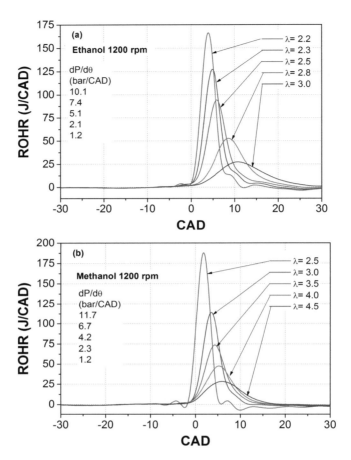

Fig. 7.22 Typical heat release rate trances for (**a**) ethanol and (**b**) methanol in HCCI combustion for different λ at 1200 rpm (adapted from [50])

methanol) show single-stage heat release process because these are high-octane fuels and do not show low-temperature heat release in naturally aspirated engine operation [1]. The highest heat release rate is achieved at richest (lowest λ) operating condition, and the lowest heat release rate is obtained for the leanest mixture (highest λ) at constant inlet air temperature in HCCI engine.

Figure 7.23 illustrates the typical heat release rate in a dual-fuel combustion engine with direct injection of high reactivity fuel and port injection of low reactivity fuel. Figure 7.23a depicts the heat release rate for three different conditions in dual-fuel combustion mode. The heat release rate shape of the first dual-fuel combustion mode is similar to the heat release rate shape of the conventional diesel combustion (Fig. 6.23a). However, the heat release rate shape of the second dual-fuel combustion mode looked like an "M shape," which has two maxima in the heat release rate. The third heat release rate shape of dual-fuel combustion mode 3 is similar to a "bell

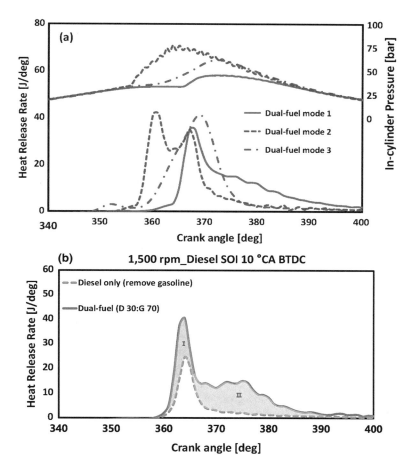

Fig. 7.23 (**a**) Heat release rate traces of dual-fuel combustion modes 1 (diesel SOI: 6°CA BTDC), 2 (diesel SOI: 13°CA BTDC), and 3 (diesel SOI: 46°CA BTDC) at 1500 rpm/Φ:0.44/gasoline ratio:70%, (**b**) comparison of dual-fuel combustion and diesel-only combustion [51]

shape" based on the premixed combustion. Figure 7.26b illustrates the difference in heat release rate when low reactivity fuel (gasoline) is added or not added at the constant diesel injection timing which was 10°CA bTDC. The shaded area (Fig. 7.23b) shows the increasing amount of heat release rate due to gasoline fuel during dual-fuel combustion. The "part I" area is originated by the combustion of gasoline (which is premixed with air where diesel is injected) as a premixed combustion phase. In "part II" area, the flame propagation of gasoline is proceeding, which was supported by the energy provided from the diesel mixing controlled and late combustion modes [51]. The gasoline fuel burned along with the burning of the diesel fuel, not after the end of diesel combustion.

7.6 Tuning of Heat Release

Cylinder pressure measurement and its analysis are routinely used for understanding the engine combustion because it provides a large amount of information such as IMEP, PMEP, peak in-cylinder pressure (critical structural constraint), combustion rate and phasing, etc. Several recent studies used cylinder pressure signal for online combustion diagnostics and control, trapped mass estimation, EGR control, emissions control, and noise control. The combustion diagnostics and control strategies require a reliable pressure data and error-free input parameters for accurate calculation of combustion parameters through heat release analysis. For better combustion diagnostics, understanding of the sensitivity of the combustion parameters to measurement error in the input parameters is required. The heat release analysis using single-zone model uses several input parameters, which can lead to error if input parameters have measurement error. Thus, tuning of the input parameters required for heat release computation is essential for the correct determination of heat release rate. In this section, tuning of input parameters based on measured motoring pressure and self-tuning methods are discussed in detail. A method for automatic identification of the heat release parameters is proposed in reference [52]. To have a well-determined reference for the heat release, the identification is performed for a motored cycle that is obtained by skipped firing. The parameters γ, crank angle offset, pressure sensor offset, and IVC cylinder pressure are identifiable without numerical difficulties. At least one of the parameters' initial temperature, wall temperature, and heat transfer coefficient, included in the heat transfer equations, must be fixed to a constant value or else the model is overparameterized. Another study proposed a method that the gross cumulative heat release should be constant following the end of combustion [3]. If the errors in the value of γ and measured pressure data are small, then any deviations from a constant heat release are due to heat transfer errors. Therefore, it is assumed that the heat transfer multiplier should be adjusted to give a cumulative gross heat release which is constant following the end of combustion. Potential problems with this method include the fact (1) that very slow burn cycles are difficult to cater for, (2) that pressure measurement errors will produce erroneous values of heat transfer rate and perhaps heat release rate, and (3) that "signal noise" effects may cause additional problems.

7.6.1 Motoring Pressure-Based Method

In this method, input parameters used for calculating heat release using a single-zone model are tuned using measured motoring pressure. The heat release calculated using cylinder pressure in motored operating condition must be zero because there is no combustion occurring during motoring operation. Figure 7.24 illustrates the measured cylinder pressure and the calculated heat release rate as a function of crank angle in motoring cycle with well-tuned parameters. The rate of heat release (ROHR)

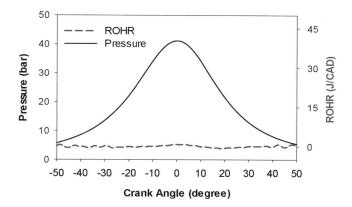

Fig. 7.24 Measured cylinder pressure and rate of heat release (ROHR) traces for motoring cycle [53]

trace is almost zero (Fig. 7.24). Thus, input parameters of gross heat release (inlet temperature and pressure, phasing between the pressure and volume trace, compression ratio and scaling factor of heat transfer coefficient) are considered to be well-tuned. Tuning of input parameter is performed iteratively with respect to calculated heat release, and the parameters giving almost zero cumulative heat release can be selected as a correct parameters.

Figure 7.25 illustrates the effect of incorrectly applied input parameters for calculation of heat release using a single-zone model. The figure shows that slight variations in some of the input parameters lead to the large and unacceptable errors in calculated cumulative heat release. Measurement errors in the phasing between pressure and volume, manifold pressure (for pegging of measured pressure signal), and compression ratio have significant error in cumulative heat release calculation.

The input parameters which provide almost zero cumulative heat release are used for calculation of heat release in the firing operating conditions. Figure 7.26 shows the measured cylinder pressure and calculated heat release rate using input parameters tuned in motoring conditions for an HCCI combustion. All the engine operating conditions were kept the same in the fired cycles except the amount of fuel injected (as compared to motoring cycle). The figure shows that the rate of heat release rate is almost zero at crank angle positions other than combustion duration.

The effect of measurement errors in the input parameters depends on the combustion parameter to be calculated using the measured cylinder pressure data. Table 7.1 illustrates the effect of measurement error in input parameters on the calculation of various combustion parameters in the HCCI engine. The calculation of errors is conducted over a wide range of load and speed conditions and summarized in Table 7.1. Table depicts that the maximum heat release rate is mainly affected by measurement errors in TDC location, compression ratio, and intake air pressure. The location corresponding to peak heat release rate is mainly affected by an error in TDC location and intake pressure. The combustion phasing (CA_{50}) is found to be a

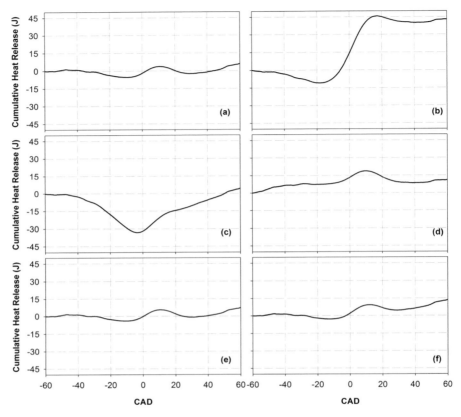

Fig. 7.25 Effect of incorrectly tuned input parameters on cumulative heat release during motoring (**a**) well-tuned parameters, (**b**) wrong phasing [*P* and *V* traces are phased apart by 1° CA], (**c**) wrong compression ratio [compression ratio is altered from 17 to 18], (**d**) wrong inlet air pressure [inlet air pressure altered from 0.99 to 0.89 bar], (**e**) wrong inlet air temperature [temperature is altered from 120 to 130 °C], and (**f**) wrong scaling of heat transfer coefficient [heat transfer increased 15%] [53]

very robust parameter for HCCI combustion control because it is mainly affected by measurement errors in TDC location. Cumulative heat release is strongly affected by measurement errors in TDC position, intake air pressure, and heat transfer determination. Calculation of maximum average gas temperature is affected mainly by measurement error in intake air pressure. The IMEP is mainly affected by phasing error of cylinder pressure and volume, and error of 1° CA leads to errors of up to 11.66%.

This method depends on the tuning parameters corrected in the motoring conditions. Thus, measurement of pressure data needs to be taken in similar operating conditions, which may be difficult in some of the engine test rigs. The other limitation is that some of heat transfer model may differ in motoring and firing conditions, and tuning parameters in motoring condition may give erroneous results

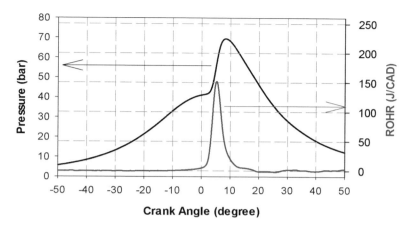

Fig. 7.26 Measured cylinder pressure and rate of heat release (ROHR) traces for a firing cycle in well-tuned parameters condition in HCCI engine [53]

Table 7.1 Effect measurement errors on HCCI engine combustion and performance parameters [53]

	IMEP (%)	PMEP (%)	HRR$_{max}$ (%)	CA$_{50}$ (CAD)	CD (CAD)	CAD HRR$_{max}$	Total HR (%)	T$_{max}$ (%)
TDC +1 (CAD) ↑	8.87– 11.66↑	1.23– 0.8↑	4.58– 7.80↑	1.0– 0.5↑	0.167– 0.5↑	1.166↑	9.88– 10.81 ↑	3.21– 3.94↑
r_c +1↑	NE	NE	6.90– 3.36 ↓	0.33– 1.0↑	5.16– 4.67↑	–	0.99– 1.10 ↓	4.51– 3.45↓
IP −0.1 (bar) ↓	–	0.02↓	4.48– 3.24↑	0.16– 0.33↓	0.50– 1.16↓	0.166↑	3.87– 2.53 ↑	10.68↑
IT +10 °C↑	NE	NE	1.02– 0.78↑	–	0.333↓	–	0.63– 0.67↑	2.54↑
HT+15% ↑	NE	NE	0.16– 0.41↑	–	0.5– 0.66↑	–	2.17– 2.52↑	–

NE no effect, r_c compression ratio, *IT* intake temperature, *IP* intake pressure, *HT* heat transfer, *TDC* top dead center, *HR* heat release, *CD* combustion duration, *PMEP* pumping mean effective pressure

(see Fig. 7.13). Additionally, this method is difficult to use online. The self-tuning method considers the heat transfer in a more sophisticated way, which may provide better results, and it can be used online.

7.6.2 Self-Tuning Method

Tunestal proposed an alternative method for self-tuning of the heat release using in-cylinder pressure [54]. This method is based on the least-square determination of the polytropic exponent immediately before and after the combustion event, respectively. The polytropic exponent cannot be estimated during the combustion event because it is not a polytropic process. A linear interpolation of the polytropic exponent with respect to crank angle is implemented during the combustion event between the compression and expansion values. This interpolated exponent values are used for computing the heat release. The interpolated exponents automatically incorporate an aggregate model for the heat losses, crevice losses, and blowby [54]. The interpolated polytropic exponent (n) trace is used as a replacement for the ratio of specific heat (γ) in the net heat release Eq. (7.56), which is derived using definition of specific heat ratio. The resultant heat release equation can be written as Eq. (7.59).

$$\frac{dQ_n}{d\theta} = \frac{n}{n-1} P \cdot \frac{dV}{d\theta} + \frac{1}{n-1} \frac{dP}{d\theta} \cdot V \qquad (7.59)$$

The crank angle intervals used for least-square fit to estimate polytropic exponent before and after combustion need to be selected based on the beforehand knowledge of the earliest and latest possible occurrence of combustion event, respectively. There is also a trade-off in choosing the interval lengths because a longer interval leads to improvement in statistics for the polytropic exponent estimation but also means more variation of the polytropic exponent over the interval [54]. Figure 7.27 illustrates the variations of polytropic exponents estimated before (compression) and

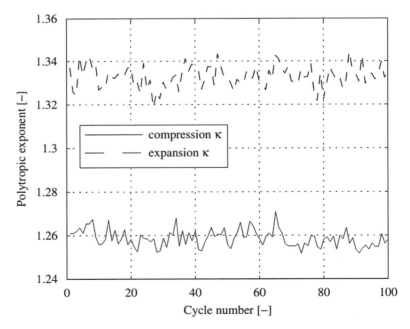

Fig. 7.27 Variations of polytropic exponents estimated before and after HCCI combustion ($\lambda = 2.5$) for 100 consecutive cycles [54]

after (expansion) HCCI combustion ($\lambda = 2.5$) for 100 consecutive cycles. The interpolated exponent between these two values is used during combustion process in Eq. (7.59).

The algorithm was proved to be fast and suitable for online calculation of the combustion phasing because in this model, there is no need to tune the heat transfer which allows a faster post-processing. The weak point of this model might be represented by the linear interpolation between the polytropic exponent before and after the combustion. A validation study of self-tuning heat release algorithm demonstrated that there is a very good match between the rate of heat release and the heat release computed with the modified Tunestal model and the Woschni model [55]. Additionally, the self-tuning heat release algorithm is user-independent which means that there are no errors introduced by the user; this allows a more realistic comparison between different engines and different operating points.

7.7   Estimation of Various Combustion Parameters

Combustion diagnosis and control have always been important in engine management systems. Presently, it is even more crucial due to the increasing demands to reduce pollutant emissions and engine noise. For this reason, a large number of parameters need to be evaluated in modern engine control systems. Engine combustion diagnosis and/or monitoring for closed-loop control has been implemented using the ion current sensor, optical combustion timing sensor, and cylinder pressure sensors. However, for combustion control in reciprocating engines, the in-cylinder pressure sensor is most widely used because ion current and optical sensors do not guarantee endurance during soot contamination (typically in a diesel engine). For combustion monitoring or diagnosis by cylinder pressure measurement, estimation of combustion indicators is required. Several combustion parameters such as start of combustion (SOC), ignition delay, the location of 50% fuel mass burned, combustion duration, etc. are estimated to obtain different kinds of information about the combustion process and combustion modes. This section describes different methods for evaluation of various combustion parameters in reciprocating engines using in-cylinder pressure measurement.

7.7.1   Start of Combustion

The start of combustion (SOC) is an important indicator due to its direct impact on heat release rate, which affects the pollutant formation and combustion noise. Thus, SOC directly affects the combustion as well as emission, and engine performance (torque and efficiency) typically increases with advanced SOC. The advanced SOC timings also lead to higher peak pressure and temperature in the cylinder, which increases the NOx emissions, and very advanced SOC may also hamper the engine efficiency. On the other hand, retarded combustion timing can result into incomplete

combustion leading higher unburned hydrocarbons. Therefore, optimal SOC position is essential at particular engine operating condition and combustion mode
[56, 57]. Typically, SOC is controlled by spark (in gasoline engines) or injection
(in diesel engines) timings, but ignition delay can vary during the lifetime of the
engine because of variation in fuel quality or gradual deterioration of the engine and
injection/spark system. Additionally, combustion in advanced premixed compression ignition engine is kinetically controlled, and ignition timing is governed by
pressure and temperature in the cylinder. Thus, closed-loop control of SOC is
required to correct unanticipated shifts in SOC timings. The requirements of the
pressure variable for engine control are (1) one-to-one correspondence to engine
operating conditions, (2) invariance of the target value (mapping of combustion
control is not required for constant pressure variable), (3) good signal noise immunity, (4) high contrast-noise ratio (CNR), and (5) low processor load (computational
effort) [58]. Moreover, early detection of multiple SOCs created by multiple fuel
injections is beneficial for not only next-cycle but also same-cycle combustion phase
control [57, 59].

For online (real-time) applications, the SOC estimation methods typically use the
cylinder pressure signal and its derivatives as well as the heat release rate (HRR),
such as (1) the crank angle position of abrupt deviation from motoring (without
combustion) pressure by measured pressure signal during combustion [58, 60, 61],
(2) the crank angle position of the first maximum value of the second derivative of
combustion pressure [62], (3) the crank angle position of the zero crossing (inflection
point) of the second derivative of cylinder pressure during combustion [63], (4) the
crank angle position of the first maximum value of the third derivative of cylinder
pressure [64], (5) the crank angle position of the minimum value of the heat release
rate before it starts increasing, (6) the crank angle position of the first zero crossing of
the heat release rate after fuel injection [48], and (7) the crank angle position of the x
% mass fraction burned ($x = 1\%$, 5%, or 10%) [60, 65]. Typically, for estimation for
rapid combustion period, the crank angle position corresponding to 10% heat release
(CA_{10}) is considered as SOC.

Figure 7.28 illustrates the different methods of start of combustion detection
based on measured cylinder pressure data and its analysis. The SOC is typically
characterized as the separation point between the measured cylinder pressure during
combustion and motored cylinder pressure trace (SOC_{SP}). This is the most direct
method of SOC detection. The SOC can also be defined by the location of the point
where compression curves start deviating from straight line in log P − log V curve
(Fig. 6.1).

To find the deviation of measured cylinder pressure from motoring curve (sharp
pressure rise is expected at the SOC), pressure derivatives with respect to crank angle
are used. The inflection point can be computed where the second derivative of
pressure has the value zero. This point can also be considered as SOC position.
Some of the studies have used crank angle corresponding to the maximum value of
the second derivative of cylinder pressure as SOC. Further research on detection of
SOC showed that the maximum value of the third derivative measured cylinder
pressure curve with respect to crank angle is closer to the actual SOC [64]. The

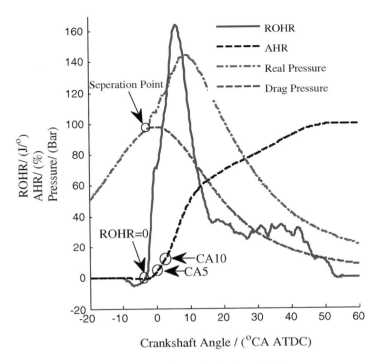

Fig. 7.28 Various start of combustion detection methods using cylinder pressure measurement [59]

physical meaning of the third derivative can be determined from the net heat release Eq. (7.56). From Eq. (7.56), it can be derived that near TDC or at TDC position (rate of change of volume is close to zero), the heat release rate can be written as Eq. (7.60).

$$\frac{dQ_n}{d\theta} \propto \frac{dP}{d\theta} \tag{7.60}$$

The Eq. (7.60) suggests that the peak heat release rate will coincide with the peak pressure rise rate. To define SOC, it is appropriate to denote the point where heat release rate significantly departs from zero (the point of maximum curvature of heat release rate) as the position of SOC. Mathematically, the maximum of second-order derivative of heat release rate can be designated as SOC position. Using Eq. (7.60), the maximum of the second derivative of pressure rise rate can be used as SOC.

$$SOC = \left(\frac{d^3 P}{d\theta^3}\right)_{max} \tag{7.61}$$

Since the cylinder pressure signal naturally contains electric noises, the accuracy of differential operations is very low, and it can lead to inaccurate results for real-time control systems [59, 60]. A threshold value of the third derivative of pressure is also used to calculate SOC as shown by Eq. (7.62) [66].

$$\frac{d^3P}{d\theta^3} > \frac{d^3P}{d\theta^3}\bigg|_{\lim} \qquad (7.62)$$

The threshold value used is 25 kPa/CA^3 for the detection of the main SOC in HCCI engine [66].

Heat release-based methods are also used for the detection of ignition timing or SOC in reciprocating engines. Typically, the SOC position is defined as the zero-crossing point of heat release rate (HRR) curve (SOC$_{HRR=0}$) [59]. The HRR is usually below zero before combustion due to vaporization of fuel in the cylinder. Thus, the first point when the HRR curve reached to zero value is considered as SOC. The calculation of HRR involves cylinder pressure derivative, and thus, numerical derivative operation on real-time system can be problematic due to electrical noise. Filtering operation can be performed on raw data to minimize the effect of noise, but it increases the computation time. Generally, CA5 or CA10 (5% and 10% heat release position, respectively) is used as SOC to obtain reliable results for control purposes. Ignition position is important in HCCI combustion as there is no direct control on ignition timings. However, accurate ignition timing detection is required for combustion diagnosis particularly in fully premixed combustion (HCCI) or partially premixed combustion (PPC). To determine the ignition timing in HCCI engine, a study used one crank angle before the 1% heat release position [67]. Sometimes crank angle position corresponding to 1% heat release is also used as ignition position.

Figure 7.29 illustrates the typical heat release rate trace for partially premixed combustion (PPC). The HRR curve can be divided into four distinct phases: (1) ignition delay (ID), (2) low-temperature reaction (LTR), (3) premixed combustion phase, and (4) late mixing controlled phase (Fig. 7.29). In this combustion, typically fuel injection ends before the SOC. Ignition delay (ID) phase is defined as the period between start of injection (SOI) and SOC. The duration between the end of injection (EOI) and SOC is denoted as mixing period (MP). The SOC is defined as the position at which HRR is zero (CA0). The HRR is negative in ID phase because most of the fuel is injected in this period and evaporation of droplets occurs. The LTR phase occurs as a small peak before the main heat release (Fig. 7.29). To determine the start of the main heat release position, the end of LTR (EoLTR) is defined. A Gaussian profile is used to distinguish the LTR and main heat release in a well-defined manner [68, 69]. Gaussian profile is fitted to the rising flank of the premixed peak, between the end of LTR and the actual peak, and the difference between the fitted curve and HRR curve is calculated. The position where the difference is zero is considered as EoLTR.

Another study used a threshold value of 0.2 J/CAD of HRR for estimation of ignition timing as well as ignition pressure and temperature in HCCI engine [70]. The conventional heat release method requires large computational effort in

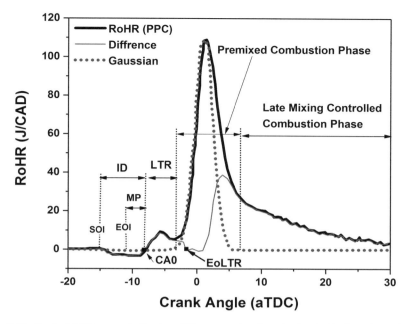

Fig. 7.29 Typical HRR curve indicating different combustion phasing in PPC engine (adapted from [68], Courtesy of Hadeel Solaka)

the calculation of heat release. Therefore, computationally efficient methods of SOC detection are proposed based on initial heat release [56].

The initial heat release (IHR) can be calculated from the heat release rate Eq. (7.56). The measured cylinder pressure is expressed as a summation of pressure difference due to combustion and motoring pressure ($P = P_{\text{diff}} + P_{\text{motoring}}$). The Eq. (7.56) can be written as Eq. (7.63) by substituting the pressure values [56].

$$\frac{dQ}{d\theta} = \frac{1}{\gamma - 1}\left(V\frac{dP_{\text{diff}}}{d\theta} + \gamma P_{\text{diff}}\frac{dV}{d\theta}\right) + \frac{1}{\gamma - 1}\left(V\frac{dP_{\text{motoring}}}{d\theta} + \gamma P_{\text{motoring}}\frac{dV}{d\theta}\right) \quad (7.63)$$

Typical variations of each term in Eq. (7.63) are presented in Fig. 7.30a. It can be noticed from Fig. 7.30a that the term (b) is zero near the early stage of combustion. Thus, it can be neglected from the Eq. (7.63). Thus, the initial heat release rate (IHRR) equation can be written as Eq. (7.64) [56].

$$\frac{dQ}{d\theta} \approx \frac{1}{\gamma - 1}\left(V\frac{dP}{d\theta} + \gamma P_{\text{motoring}}\frac{dV}{d\theta}\right) \quad (7.64)$$

The cumulative initial heat release (IHR) can be calculated from the Eq. (7.64) by integrating the equation. The variation of total heat release (HR) calculated using Eq. (7.56) and IHR calculated using Eq. (7.64) is shown in Fig. 6.30b.

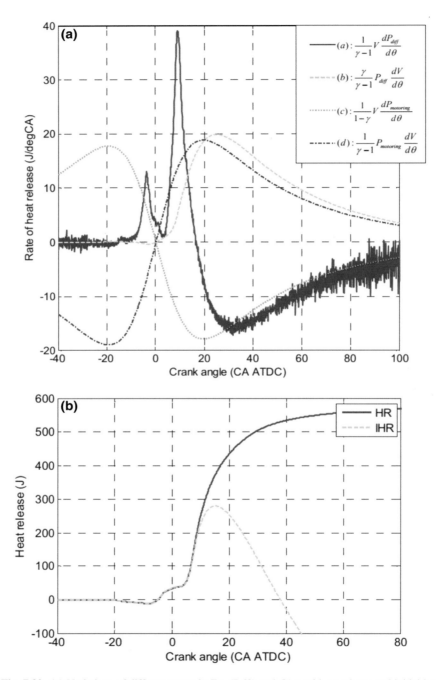

Fig. 7.30 (**a**) Variations of different terms in Eq. (7.63) and (**b**) total heat release and initial heat release variation [56]

The estimated SOC combustion is defined as a linear interpolation of 2% of IHR (IHR_2) as shown in Eq. (7.65).

$$SOC_{est} = a \times IHR_2 + b \tag{7.65}$$

This method can also be used for online estimation of SOC. It is demonstrated that this method reduces 49.8% of the cylinder pressure acquisitions and 44.8% of the calculation time compared to the conventional method [56]. However, this method still uses the pressure derivative, which is sensitive to electrical noise.

The difference between the measured cylinder pressure and the motoring pressure is also used for determination of SOC. The SOC estimated using this method is highly dependent on the accuracy of the determination of motoring pressure. The separating position of motoring and fired pressure is typically detected with a threshold, i.e., the crank angle position at which the pressure difference is greater than the threshold value is considered as the initial SOC (further correction also possible) [60]. A threshold of 10 bar is used in reference [58, 61], and 1.5 bar is used in [59]. However, these threshold values are too high for the detection of the start of low-temperature reactions in premixed compression ignition engines. The motoring pressure is typically predicted by assuming the polytropic process, and the determination of polytropic index is crucial for the accurate determination of motoring pressure. A fixed polytropic index limits prediction of the accurate motoring pressure because wall heat transfer strongly affects the polytropic process. A study [59] corrected the polytropic index by mean gas temperature, and the reference point is changed at every 10 °C, and the estimation error less than 1.5 bar in the motoring pressure is demonstrated. Another study used self-adaptive correction in the polytropic index for estimation of correct polytropic index [60]. The equivalent polytropic exponent can be calculated using Eq. (7.66).

$$k_e(i) = \frac{\log p(i) - \log p(i-1)}{\log V(i-1) - \log V(i)} \tag{7.66}$$

The $p(i)$ can be the same as or lower than $p(i-1)$ because of signal noise, and thus, a fixed reference is selected to avoid nonphysical values [57] as shown in Eq. (7.67).

$$k_e(i) = \frac{\log p(i) - \log p_{ref}}{\log V_{ref} - \log V(i)} \tag{7.67}$$

At the jth point in the mth engine cycle, the real value of the equivalent isentropic index $k_e(j, m)$ is calculated using Eq. (7.67). The predicted value of the equivalent isentropic index $k_{e,p}$ of the ith point is computed by Eq. (7.68).

$$k_{e,p}(i,m) = k_e(j,m) + C(i,j,m) \quad i \in (j, j+u) \tag{7.68}$$

where u is the prediction length and $C(i,j,m)$ is a self-adaptive correction factor for the prediction of the ith point at the jth point in the mth engine cycle. This factor is corrected once in the current engine cycle by Eq. (7.69).

$$C(i,j,m) = \alpha(k_e(i, m-1) - k_e(j, m-1)) + (1-\alpha)C(i,j,m-1) \tag{7.69}$$

where α is a coefficient for this self-adaptive strategy, with a value between 0 and 1 [60].

After determining the predicted equivalent isentropic index, the motoring pressure can be determined using Eq. (7.70).

$$p_{mot}(i) = p_{ref}\left(\frac{V_{ref}}{V(i)}\right)^{k_{e,p(i)}} \tag{7.70}$$

Using this method, the motoring pressure can be determined accurately with an error less than 0.02 bar [60].

Figure 7.31 illustrates the detection of SOC using self-adaptive correction of the polytropic index. The threshold value used for detection of low-temperature reactions is 0.02 bar. The method is able to determine SOC satisfactorily, and it can be used online. It can also be noted that this method does not use the pressure derivative term.

A real-time SOC detection method applying the first derivative of the equivalent isentropic index is proposed [57]. The polytropic index is calculated using Eq. (7.67) from the measured data. The first derivative of the polytropic index is calculated using Eq. (7.71), and its result is illustrated in Fig. 7.32.

$$dk_e(i) = 100 \times (k_e(i) - k_e(i-1)) \tag{7.71}$$

The original value of the derivative is magnified in order to be calculated in the real ECU [57].

Figure 7.32 depicts that first derivative of the equivalent isentropic index increases rapidly after the SOC which indicates that it can be utilized for the detection of the SOC. Figure 7.33 illustrates the determination of SOC using this method. To achieve the accurate results, the SOC detection begins only after fuel injection and two-stage threshold levels are used. The first point whose value is over the higher threshold level is obtained, and then the next point whose value is over the lower threshold level is obtained and treated as the detected SOC [57]. This method is capable of detecting multiple SOCs caused by multiple injections, and it can be used for engine operation in both in stationary and transient conditions.

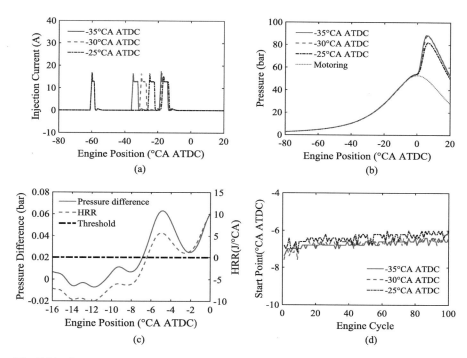

(a)

(b)

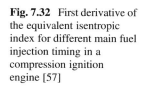

(c)

(d)

Fig. 7.31 Illustration of SOC detection method at 1200 rpm, IMEP 5.4 bar [60]. (**a**) Injection currents. (**b**) Cylinder pressure. (**c**) Pressure difference for second fuel injection timing at −25°CA ATDC. (**d**) Start of low temperature reactions

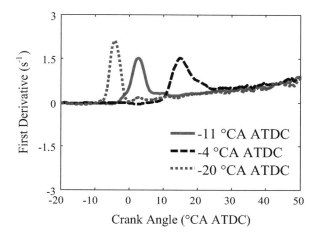

Fig. 7.32 First derivative of the equivalent isentropic index for different main fuel injection timing in a compression ignition engine [57]

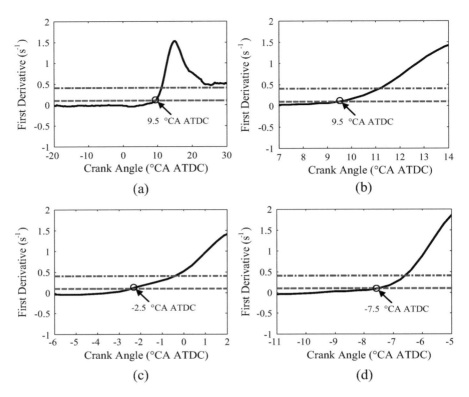

Fig. 7.33 Illustration of SOC detection method using the first derivative of isentropic index [57]. (**a**) −4°CA ATDC main fuel injection timing. (**b**) Partial view of (**a**). (**c**) −11°CA ATDC main fuel injection timing. (**d**) −20°CA ATDC main fuel injection timing

7.7.2 Ignition Delay

In a diesel engine, the ignition delay (ID) is typically defined as the duration between the SOI and the SOC. Ignition delay plays an important role in the operation and performance of diesel engines. Diesel engines tolerate a range of IDs with minimal variation in the performance. However, the ID longer than the designed value can cause significant impacts. Particularly, longer ignition delay leads to the very high-pressure rise rate which can be destructive to engine components [71, 72]. Figure 7.34 depicts the ignition delay and combustion process of typical diesel combustion. Diesel combustion can be divided into four phases (Fig. 7.34a) [73]. The fuel is injected between the start of injection (SOI) and end of injection (EOI) as diesel spray. The combustion phases can be distinguished from the heat release rate (HRR): initial jet development (physical delay and chemical delay) when the fuel starts to vaporize (SOV); triggering autoignition or start of combustion (SOC); premixed combustion; and diffusive combustion, also called mixing controlled, until end of

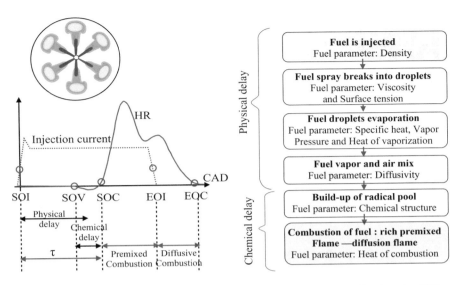

Fig. 7.34 Physical and chemical processes during ignition delay period in a diesel engine [71]

combustion (EOC). Diesel spray consists of three distinct zones: (1) spray evaporation, (2) mixing with surrounding hot air, and (3) combustion (Fig. 7.34b) [71]. Total time taken before the combustion and after the fuel injection can be divided into two major groups called physical delay and chemical delay. Physical delay comprises the time necessary for droplet breakup, air entrainment, and vaporization. Even though the fuel's physical properties such as density, specific heat, and volatility have an effect on the physical delay [74], in high-pressure injection jets, the fuel injection pressure has a deterministic impact on the spray breakup and vaporization [75]. Chemical delay is controlled by the fuel's autoignition kinetics which is dependent on fuel chemical composition, molecular structure, and associated reactivity. Thus, pressure, charge temperature and local equivalence ratio conditions affect chemical ignition delay [73, 74]. Physical delay and chemical delay usually overlap in time. For CI engines, the physical delay is much longer than chemical delay. The cylinder charge conditions at SOI affect the duration of the physical and chemical processes [73]. Higher temperature reduces both delays: the higher thermal energy entrainment for the same mass of air speeds up the vaporization [76]. In addition, the chemical kinetics are faster, and the chemical delay is shorter [77].

The ignition delay can be estimated by determination of SOI and SOC which can be determined by several methods (Sect. 7.7.2). Start of injection (SOI) can be determined from the injection current at a calibrated threshold where mass estimated by the difference between SOI and EOI, i.e., injection duration, agrees with measurements of injected mass in a spray rig. Start of vaporization (SOV) is when the absorption of energy for fuel vaporization starts to decrease the apparent HR. A good trade-off between detectability and accuracy was found when the derivative of the HR goes below -0.6 J/CAD^2 [73].

Fig. 7.35 Comparison of different ID criterion for the jet-A2 fuel data at three different SOI densities (15, 25, and 40 kg/m^3) versus temperature [72]

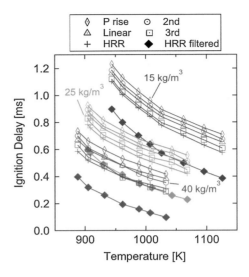

A study used six different criteria of SOC detection for the estimation of ignition delay [72]. The criterion uses are as follows: (1) the crank angle position of 50% of pressure rise (*P* rise) because of premixed burn combustion, (2) extrapolation of the peak slope of the pressure rise due to combustion to the zero-crossing point (denoted as linear in Fig. 7.35), (3) the crank angle position of the first peak of the second derivative of cylinder pressure (second), (4) the crank angle position of the first peak of the third derivative of cylinder pressure (third), (5) the crank angle position of 10% of the maximum HRR in the premixed burn (HRR), and (4) same as (5) except using the HRR computed from a low-pass filtered pressure data (HRR filtered).

Figure 7.35 illustrates the estimated ignition delay for three SOI densities using six different methods of SOC detection. The five methods based on the unfiltered pressure trace are grouped together, whereas, using the HRR determined from the filtered pressure gives ignition delay values that are 200–330 µs shorter than the other methods [72]. The figure shows that the criteria used for SOC detection strongly affect the ignition delay period. The use of filtering before ID estimation can lead to large shifts in the ID >200 µs and is not recommended.

7.7.3 Combustion Phasing

Combustion phasing in reciprocating engines typically means the location of combustion with respect to the piston/crankshaft position. The crank angle position corresponding to the 50% mass burned fraction is generally referred as combustion phasing. The combustion phasing typically affects the performance and emission characteristics of the engine. Too early combustion phasing leads to very high-pressure rise rate and combustion temperature, leading to high heat loss and also a loss in efficiency. Too retarded (delayed) combustion phasing may lead to incomplete combustion and higher unburned emissions. The delayed combustion phasing also leads to lower effective expansion ratio leading to a loss in engine efficiency. Therefore, engine operation at correct combustion phasing is required for optimal fuel economy. In premixed combustion where ignition is kinetically controlled, the closed-loop control of combustion phasing is essential.

Typically, combustion phasing is calculated by heat release analysis or mass burned fraction analysis. The crank angle position corresponding to 50% heat release can be computed using Eqs. (7.55) or (7.56). Sometimes other variables are also used as combustion phase like CA1 and CA5 (the crank angles corresponding 1% and 5%, burned mass fraction, respectively) which can be calculated using heat release analysis. The CA1 and CA5 are very sensitive parameters as they are easily affected by noise and other disturbances due to their small amounts of BMF. On the other hand, CA50 is more robust against noise or disturbances because the combustion has developed sufficiently at this point [78].

Combustion phasing estimation using heat release Eq. (7.55) takes a lot of computational time and, thus, not suitable for real-time applications. To reduce the computational time, the difference pressure apparent heat release equation is developed [61]. In this method, the total pressure is considered as summation of motoring pressure and difference pressure (P_{diff}) due to combustion, and heat release equation can be written as Eq. (7.63). Theoretically, there is no heat released during the motoring operation; therefore, the heat released contribution term by the motoring pressure term can be negligible in Eq. (7.63). The terms that remained after neglecting the motoring pressure terms are defined as the difference pressure apparent heat release equation, (Q_{diff}), which can be written as Eq. (7.72) [61].

$$Q_{\text{diff}} = \frac{1}{\gamma - 1} \int V \frac{dP_{\text{diff}}}{d\theta} d\theta + \frac{\gamma}{\gamma - 1} \int P_{\text{diff}} \frac{dV}{d\theta} d\theta \qquad (7.72)$$

The crank angle position of 50% of the difference pressure apparent heat release (MFB50$_{\text{diff}}$) can be calculated similar to conventional combustion phase detection algorithm using heat release analysis. Fig. 7.36 illustrates the calculation method of MFB50$_{\text{diff}}$ using different heat release. The MFB50$_{\text{diff}}$ is used to determine the combustion phase parameter using a linear fitting method [61].

Another study proposed a method of combustion phase parameter as a substitute for MFB50 for real-time application based on initial heat release (IHR) [79]. In this method, the Eq. (7.72) is further simplified by assuming volume change rate negligible near the TDC position. Thus, the heat release equation can be written as Eq. (7.73).

Fig. 7.36 Illustration of
MFB50$_{diff}$ calculation
method [61]

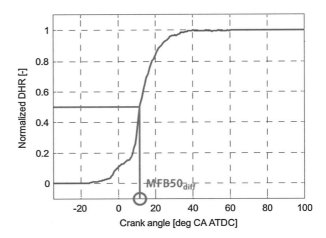

Fig. 7.36 Illustration of MFB50$_{diff}$ calculation method [61]

$$\frac{dQ}{d\theta} \approx \frac{1}{\gamma - 1} V \frac{dP_{\mathrm{diff}}}{d\theta} \tag{7.73}$$

The initial heat release (Fig. 7.30b) is divided into IHR before peak (IHR$_{bp}$) and IHR after peak (IHR$_{ap}$) and compared with the conventional heat release. It was found that the location of 20% of IHR$_{ap}$, (IHR$_{ap20}$), has a very good correlation with MFB50 [79]. The mass fraction burned (MFB) can be estimated using Eq. (7.74).

$$\mathrm{MFB50}_{\mathrm{estimated}} = a \times \mathrm{IHR}_{ap20}{}^3 + b \times \mathrm{IHR}_{ap20}{}^2 + c \times \mathrm{IHR}_{ap20} + d \tag{7.74}$$

Based on the analysis results, the IHR$_{ap20}$ found to be a viable real-time combustion phase indicator [79].

A method based on the normalized difference pressure (NDP) is also proposed for the estimation of combustion phasing in HCCI engine [78]. The difference pressure (DP) is calculated by subtracting the motoring pressure from the measured pressure. The normalized difference pressure can be calculated by Eq. (7.75).

$$\mathrm{NDP} = \frac{\mathrm{DP}}{\max(\mathrm{DP})} \tag{7.75}$$

Figure 7.37 shows the comparison between normalized heat release and NDP in an HCCI engine. The figure depicts that both curves follow closely during combustion. The combustion phasing parameter CA$_{NDP0.5}$ is defined as the crank angle position at which the NDP becomes 0.5. This parameter is a linear function of CA50, which is calculated by heat release. The combustion phase CA50 can be determined using Eq. (7.76) [78].

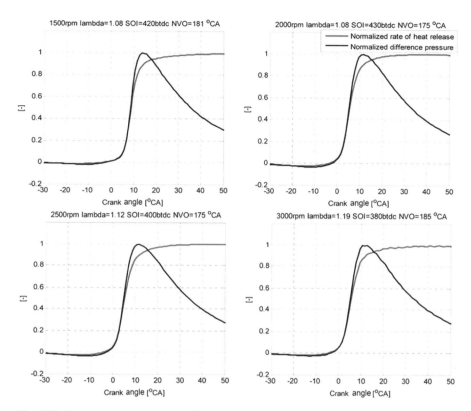

Fig. 7.37 Comparison between normalized heat release and normalized difference pressure [78]

$$CA_{50\%} = CA_{NDP0.5} - \Delta CA_{DD} \tag{7.76}$$

where CA_{DD} is a detection delay angle, which is a constant with a typical value of around 0.3 [78]. This method is proved simpler and faster than heat release analysis and demonstrated the very good accuracy of combustion phasing detection.

7.7.4   End of Combustion and Combustion Duration

Combustion duration is an important combustion parameter, which affects the engine performance. Combustion duration is dependent on the combustion rate in the cylinder. Combustion duration is typically defined as the time period between start of combustion (SOC) and end of combustion (EOC). Section 7.7.1 discussed the various methods for the detection of SOC. The EOC can also be detected using measured cylinder pressure. When the combustion process gradually leads to burn

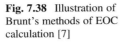
Fig. 7.38 Illustration of
Brunt's methods of EOC
calculation [7]

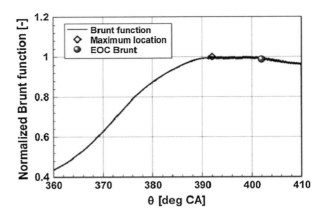

out phase, it is difficult to define suitable criteria that can adequately detect the
effective EOC position [7]. Typically, the crank angle positions corresponding to
99% or 95% or sometimes 90% mass burned fraction are qualitatively used to
represent EOC. A reasonably better approach is to define the ECO using $\log P - \log V$
curve (Fig. 7.1). In this curve, during expansion stroke after the end of combustion,
the curve should be straight line assuming the polytropic expansion process. The
inflection point (point e in Fig. 7.1) can be defined as the EOC position.

Another study [4] proposed the EOC detection by the crank angle position at
which $P \cdot V^{1.15}$ reached a maximum value, where P is the cylinder pressure and V is
the cylinder volume. The calculation of the EOC function was performed from $10°$
ATDC and continued to EVO-$10°$ CA and the average value for $10°$ being used to
minimize the noise effects [4]. The selected polytropic index (1.15) which is lower
than the typical expansion index (1.25–1.3) occurs. This is done to ensure that
reliable results are obtained even when large pressure errors are present and to
ensure a distinct maximum to the function. The addition of $10°$ is to partly compensate for the low value of the index and also to ensure that combustion is complete.
Figure 7.38 illustrates the EOC detection using this method. An improved method is
proposed based on the analysis of the HRR [7]. This method is based on the
threshold value, and a value 3% of HRR_{max} threshold is found suitable for the
detection of EOC. The EOC is defined as the crank angle at which the heat release
rate reaches to threshold value (3% of HRR_{max}) towards the completion of combustion process. The calculation of HRR involves the pressure derivative which is
sensitive to noise, and it can lead to oscillatory heat release toward the EOC. To
deal with this problem, moving average of heat release is proposed for evaluation of
EOC [7].

The SOC and EOC can be directly detected using $\log P - \log V$ (Fig. 7.1). The
polytropic index depends on engine operating conditions and differs for engine
models and from design to the other. The SOC and EOC can be determined from
the ratio of specific heat (γ) variations with crank angle position [80]. An accurate
$\gamma(T)$ model is a major parameter for an accurate heat release analysis because $\gamma(T)$

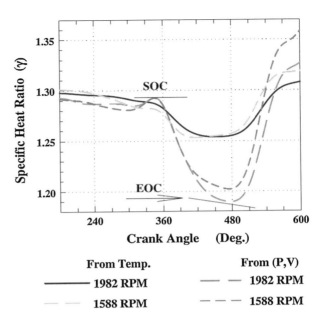

Fig. 7.39 Comparison between $\gamma(T)$ calculated from T_{Gas} and γ from (log P/ log V) for gasoline at different speeds [80]

From Temp.	From (P,V)
1982 RPM	1982 RPM
1588 RPM	1588 RPM

couples the system energy parameters to other thermodynamic quantities. Figure 7.39 depicts the $\gamma(T)$ calculated using two different methods (1) using temperature-dependent function and (2) using $\log(\Delta P)/\log(\Delta V)$ for two engine speeds. The $\gamma(T)$ curves have a similar trend but different numerical values. The inflection points in the $\gamma(T)$ curves are obtained based on $\log(\Delta P)/\log(\Delta V)$. The two inflection points can be considered as SOC and EOC from the Fig. 7.39.

Figure 7.40 illustrates the SOC and EOC combustion detection based on entropy determination method. The gas entropy can be calculated using a variation of gas temperature and pressure using Eq. (7.77).

$$dS = c_p \ln \frac{T}{T_r} - R \ln \frac{P}{P_r} \qquad (7.77)$$

During the compression process, when charge temperature is lower than the wall temperature, the heat transfers from wall to charge occur which leads to an increase in entropy. The entropy decreases the location where charge temperature is higher than the wall temperature. Similarly, during combustion, when the HRR is higher than the heat transfer rate to the wall, entropy begins to increase, and when the heat transfer rate exceeds the combustion rate, the entropy starts to decrease toward the EOC process [80]. Therefore, the minimum or maximum entropy values achieved at SOC and EOC position, which is illustrated in Fig. 7.40. Another study also showed that the point of maximum entropy of the cylinder gases closely matched the EOC predicted by the heat release curve, while SOC did not coincide with the minimum of entropy [81]. However, the SOC can be determined from the rate of change of entropy with crank angle, which showed a rapid change at SOC point.

Fig. 7.40 Entropy versus
gas temperature through
combustion duration for
gasoline at different speeds
[80]

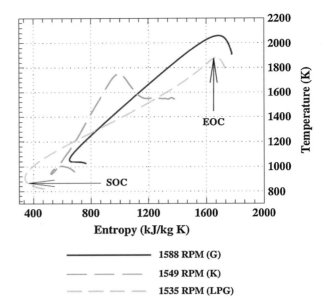

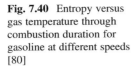

Figure 7.41 illustrates the variation of combustion duration detected using four
different methods: (1) burned mass fraction, (2) entropy change, (3) $\gamma(T)$, and
(4) $\log P - \log V$ scale. The figure shows that combustion duration increases with
engine speed irrespective of methods. The combustion duration estimated using
burned mass fraction shows the smallest value, but it depends on the selected
threshold value. The difference in the combustion duration predicted by other
three methods is small with respect to each other [80].

A method based on the change in polytropic volume is proposed for detection of
SOC and EOC [82]. The polytropic volume can be calculated using measured
cylinder pressure and polytropic process assumptions during compression and
expansion. The polytropic volume can be calculated by using Eq. (7.78).

$$V_{p,i(\phi)} = V_c\left(\phi_{\text{ref},i}\right)\left(\frac{p\left(\phi_{\text{ref},i}\right)}{p(\phi)}\right)^{1/n_i} \tag{7.78}$$

where V_p and V_c are the polytropic and real volume (calculated from geometry),
respectively. All crank angles (ϕ) where the polytropic cylinder volume $V_{p,i(\phi)}$ is not
equal to the real cylinder volume $V_{c(\phi)}$ do not belong to the compression process
(if i = compression) or expansion process (if i = expansion) of the engine cycle [82].

Figure 7.42 illustrates the variation of real volume and polytropic volume using
compression and expansion polytropic index. The crank angle location of the intake
valve closes (IVC), SOC, EOC, and exhaust valve opens (EVO) can be observed
clearly by the naked eye. The locations where the polytropic volume drifts away
from the real cylinder volume can be detected as IVC, SOC, EOC, and EVO.

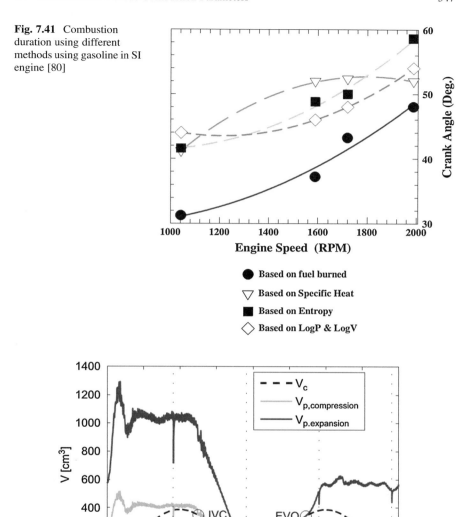

Fig. 7.41 Combustion duration using different methods using gasoline in SI engine [80]

Fig. 7.42 Variation of real cylinder volume (V_c) and polytropic cylinder volumes [82]

The detection of combustion events can be performed using the simplest possible fast wavelet transformation at the first level on the change in volume ($\Delta V = V_c - V_{p,i}$) signal. The change in volume can be calculated using Eq. (7.79) with i = compression or i = expansion.

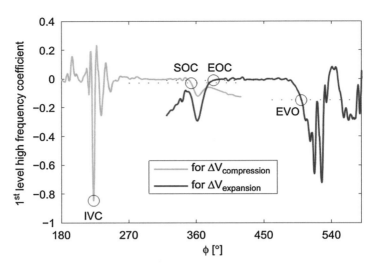

Fig. 7.43 High-frequency parts of the single-level discrete wavelet transformation for ΔV the signals [82]

$$\Delta V_i(\phi) = V_{c,\text{TDC}}\left[1 + \frac{\varepsilon - 1}{2}\left(\lambda + 1 - \cos\phi - \sqrt{R^2 - \sin^2\phi}\right)\right]$$
$$- V_c(\phi_{\text{ref},i})\left(\frac{p(\phi_{\text{ref},i})}{p(\phi)}\right)^{1/n_i} \tag{7.79}$$

Figure 7.43 illustrates the combustion event detection based on the high-frequency signals of compression and expansion using discrete wavelet transform. The combustion events can easily be detected using threshold values [82]. The detection of the first level frequency coefficient's minimum found to be a robust method for the data used because the signal is very noisy around the IVC event. The combustion events SOC and EOC can clearly be distinguished (Fig. 7.43). It is found that the SOC and EOC can be determined with less than 2% difference of the SOC or EOC based on the heat release method, and the method is robust with respect to pegging level error.

7.8 Thermal Stratification Analysis

Thermal stratification of the unburned charge before ignition plays a significant role in governing the heat release rates, particularly in HCCI engine. Understanding the conditions affecting the thermal stratification is essential for actively managing HCCI combustion rates and extending its operating range. A thermal stratification analysis (TSA) method is proposed for estimating the unburned temperature distribution prior to ignition [83–85]. The methodology divides the cylinder charge into a number of different regions or zones, where each region has different temperature. In this method, the regions are not assigned any spatial location like it is done in multi-zone modeling. The pressure and equivalence ratio of charge is assumed to be uniform in the entire combustion chamber.

For analyzing the temperature variations before ignition, it is important to estimate an upper and lower bound on the range of temperatures which can physically exist in the combustion chamber. To describe the hottest possible unburned gas temperature (corresponds to the adiabatic core of the charge) in the cylinder, the isentropic unburned temperature is computed using Eq. (7.80) [84, 85].

$$T_{isen, unburned} = T_{IVC} \cdot \left(\frac{P_{cyl}}{P_{IVC}} \right)^{(1-1/\gamma)} \tag{7.80}$$

where P_{cyl} is the measure cylinder pressure and γ is the ratio of specific heats of the charge, which is calculated at each time step based on the mixture properties and temperature at that time step.

In an HCCI combustion engine, as the first mass burns, it compresses the remaining unburned mass, increasing its temperature and accelerating the autoignition progression. The isentropic unburned temperature (Eq. 7.80) includes the compression effect from combustion elsewhere in the cylinder by using the measured cylinder pressure. Equation (7.80) includes compression effects from compression and combustion elsewhere in the cylinder, but it does not include any heat transfer effects between zones after combustion starts. An additional term is required to capture heat transfer between the burned and unburned gases [85]. To simulate the remaining (colder) regions of the cylinder, a linear combination of the wall temperature and the isentropic unburned temperature is used and represented by Eq. (7.81).

$$T_{zone} = (1 - NZT) \cdot T_{wall} + NZT \cdot T_{isen, unburned} \tag{7.81}$$

where NZT (normalized zone temperature profile) is a scaling variable that can vary from 0 to 1. When NZT equals 0, T_{zone} represents a region with the same temperature as the wall. When NZT is 1, T_{zone} represents the adiabatic core. By varying NZT from 0 to 1, all of the possible gas temperatures can be simulated.

A given NZT represents a self-similar temperature trajectory over the compression and expansion strokes. In the absolute temperature domain, ignition depends on both time and temperature. The ignition timing of each NZT is then determined using the autoignition integral proposed by Livengood and Wu [86]. The ignition timing as a function of normalized zone temperature (NZT) is shown in Fig. 7.44 (curve 1). The mass burned fraction can also be computed from measured cylinder pressure based on heat release analysis using Eq. (7.16). The heat release can be calculated using Eq. (7.55). The mass burned fraction as a function of crank angle is shown as curve 2 in Fig. 7.44. By setting the two variables equal to each other, the assumption is made that when a pocket of gas ignites, it burns instantaneously. The mass burned faction can be obtained as a function of NZT, which is presented as

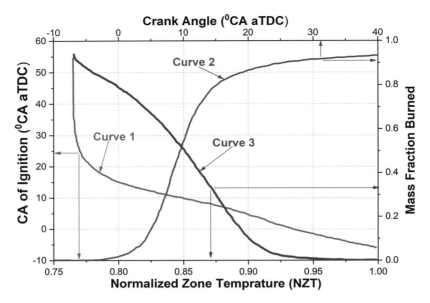

Fig. 7.44 Illustration of computation methodology of mass fraction burned as a function of NZT (adapted from [83–85])

curve 3 in Fig. 7.44. NZT relates to time and absolute temperature so that ignition timing can be plotted against one variable, instead of two. The curve 3 in Fig. 7.44 represents the distribution of mass and temperature in the cylinder, however, not in a way that is easy to visualize. Thus, the data of curve 3 in Fig. 7.44 can be rearranged in a more visually intuitive manner. The y-axis data in curve 3 of Fig. 7.44 is subtracted from 1, and the result is the mass cumulative distribution function (CDF) shown in Fig. 7.45.

The cumulative distribution function (CDF) is much easier to inspect the results visually. For example, the CDF in Fig. 7.45 depicts that about 90% of the mass is at a normalized temperature (NZT) of 0.90 or lower and about 20% of the mass is at NZT of 0.80 or lower. Thus, 70% of the mass is in the NZT range from 0.80 to 0.90.

The derivative of the CDF is defined as probability density function (PDF) in probability theory. The PDF of the mass burned function is calculated and presented in Fig. 7.45 (curve 2). The PDF reveals that the probability of finding mass at near-adiabatic conditions is relatively low. The highest probability of finding a mass in the cylinder occurs at a normalized temperature of about 0.88 (peak value) for this case (Fig. 7.45). As the NZT further decreases, the probability of finding mass at that temperature becomes less and less likely [85].

Chemical kinetics is ultimately sensitive to absolute temperatures. Thus, the same distributions can be plotted as a function of absolute temperature at TDC (as represented as upper x-axis in Fig. 7.45). Since a given NZT corresponds to an absolute temperature profile that varies with crank angle, to plot the absolute temperature distribution on a two-dimensional plot, an arbitrary crank angle needs

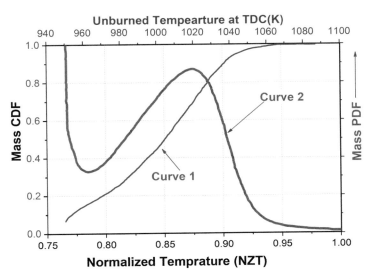

Fig. 7.45 Mass CDF and PDF with NZT and absolute temperature at TDC (adapted from [83, 85])

to be selected (TDC in this case). Then, wall temperature and isentropic unburned temperature are used to remove the normalization by Eq. (7.82).

$$NZT = \frac{T_{zone} - T_{wall}}{T_{isen, unburned} - T_{wall}} \tag{7.82}$$

The mass PDF as a function of temperature can be used to analyze the thermal stratification in the charge. A more complete detail of thermal stratification analysis can be found in original studies [83–85].

Figure 7.46 illustrates unburned temperature distribution (as mass PDF) at the different operating conditions in an HCCI engine. The mass PDFs show the density of mass over the range of temperatures that exist in the cylinder. Higher values of the mass PDF indicate more mass concentrated at that particular temperature. Figure 7.46 reveals that the higher intake temperature or higher glow plug voltage (early combustion phasing) has broader temperature distributions with a lower peak of the distribution. The near-wall regions of all test conditions overlap because the wall conditions are not changed between the cases (Fig. 7.46). The differences in the temperature distributions for the second latest CA_{50} pair (blue) and the mid-phase CA_{50} pair (red) are negligible. However, as the glow plug voltage increases, the differences become more significant [87]. The hottest leading edge of the temperature distribution is stretched toward higher temperatures because of the glow plug's ability to heat a fraction of the mass to a much higher temperature. These results illustrate that by actively controlling the glow plug voltage, the unburned temperature distribution can be controlled to some extent, which results in some level of control over HRR in the HCCI engine.

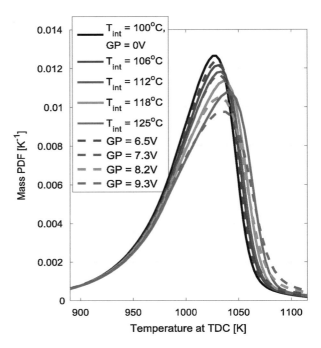

Fig. 7.46 Unburned temperature distributions for the glow plug voltage sweep and the intake temperature sweep [87]

Discussion/Investigation Questions

1. Discuss two methods for calculation of burned mass fraction (BMF) in reciprocating engines using measured in-cylinder pressure data. Describe a method that can be used for online calculation of mass burned fraction.

2. Describe the possible sources of error in the mass fraction burned calculation. Discuss why the error in mass burned fraction calculation is smaller than the error in heat release calculation.

3. Discuss the reasons why bass fraction burned calculation using Marvin's or Rassweiler and Withrow method is more suitable for conventional SI engine than diesel engines. Explain a method of mass fraction burnt calculation for a modern diesel engine.

4. Describe the experimental methods that can be used for measurement of gas temperature in the cylinder. Write the limitations of these methods and possible sources of errors in the measurement.

5. Discuss the operating conditions or combustion modes where calculated mean gas temperature using ideal gas equation is close to an actual gas temperature in the cylinder.

6. Describe a method for calculating the burned gas temperature in SI engine, and discuss the significance of the calculated burned gas temperature. Discuss its relevance over the average gas temperature calculated using single-zone model.

7. Discuss the possible reasons for spatial variations in heat flux in conventional SI and CI engines. Explain why heat flux measurement is normally conducted using the single sensor in motored condition and multiple sensors in fired operating conditions.

8. Describe various empirical models used for heat transfer coefficient in reciprocating engines. Describe why typically radiative heat transfer is not considered in SI engine but accounted in a diesel engine as it is significant. Investigate and find out the suitable heat transfer coefficient for SI, CI, and HCCI engines.

9. Discuss how fuel properties and air-fuel mixture quality affect the heat flux in the combustion chamber.

10. Investigate the effect of knocking on the heat transfer through the walls of the combustion chamber, and discuss the possible factors affecting the heat transfer.

11. Derive the heat release equation using a single-zone model, and also discretize the equation as measured pressure signal is typically digital. Discuss the assumptions made in the derivation of heat release calculation using a single-zone model. Differentiate between the net (apparent) and gross heat release rate.

12. Discuss the possible sources of error in heat release rate calculation using a single-zone heat release model. Explain the possible methods to minimize the errors in heat release calculation.

13. Figure P7.1 shows the heat release curve calculated by the single-zone model using measured cylinder pressure data from the SI engine. Heat release curves 2 and 3 have a shape different from 1 (correctly calculated) due to some error during calculation. Discuss the possible sources of error in heat release calculation in curves 2 and 3.

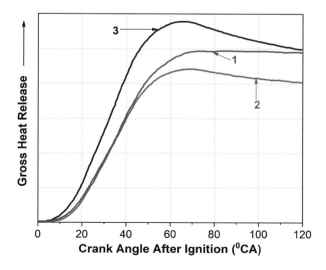

Fig. P7.1 Gross heat release as a function of crank position

14. Draw the heat release rate curves for SI, CI, and HCCI combustion engine. Discuss the difference between the heat release rate curve between conventional spark ignition and diesel combustion. How the heat release rate curve in low-temperature combustion modes such as HCCI and RCCI is different from the conventional combustion modes. Justify your answers with suitable reasoning.

15. Discuss the two methods to the tuning of gross heat release calculations. Describe the sources of error in each of the methods.

16. Discuss the combustion phase (ignition position, CA_{50}) detection algorithm for controlled autoignition (CAI) engines using in-cylinder pressure data. Why it is important to determine the combustion phasing in engine operating on LTC modes (CAI, HCCI, RCCI, etc.)?

17. Discuss the methods for detection of the start of combustion in reciprocating engines using pressure rise rate and heat release rate. Describe the merit and demerit of each of the methods.

18. Discuss the effect of SOC location on the engine combustion and performance. Explain the role of combustion duration on engine performance.

19. Explain why closed-loop control is required in modern combustion engines especially advance premixed combustion modes. Discuss the requirement that needs to be fulfilled by pressure variable that is used for closed-loop control.

20. Calculate the start of combustion, end of combustion, crank angle location for 50% burned mass fraction, and combustion duration for cylinder pressure shown in Fig. P7.2. The cylinder pressure measured at constant engine load for two different fuels (diesel and butanol blend) in a conventional four-stroke diesel engine. The bore/stroke and connecting rod length of the engine are 87.5/110 mm and 234 mm, respectively. Engine operation can be assumed at a compression ratio of 17.5. Clearly state your assumption.

Fig. P7.2 Cylinder pressure as a function of crank angle position in a diesel engine

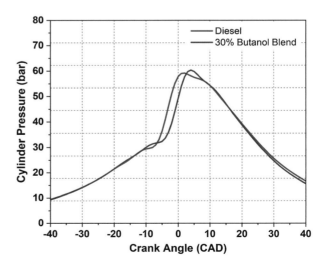

21. Discuss how the following parameters affect ignition delay in compression ignition engines: (1) decrease in compression ratio, (2) injection advance, (3) increase in air-fuel ratio, and (4) increase in coolant temperature.
22. Discuss the variation of ignition delay with brake mean effective pressure (BMEP) in a typical diesel engine. Draw the graph of ignition delay versus BMEP at two different cetane number (CN) fuel (e.g., CN = 60 and 40).

References

1. Maurya, R. K. (2018). *Characteristics and control of low temperature combustion engines: Employing gasoline, ethanol and methanol.* Cham: Springer.
2. Thor, M., Andersson, I., & McKelvey, T. (2009). *Parameterized diesel engine heat release modeling for combustion phasing analysis* (No. 2009-01-0368). SAE Technical Paper.
3. Brunt, M. F., Rai, H., & Emtage, A. L. (1998). *The calculation of heat release energy from engine cylinder pressure data* (No. 981052). SAE Technical Paper.
4. Brunt, M. F., & Emtage, A. L. (1997). *Evaluation of burn rate routines and analysis errors* (No. 970037). SAE Technical Paper.
5. d'Ambrosio, S., Ferrari, A., & Galleani, L. (2015). In-cylinder pressure-based direct techniques and time frequency analysis for combustion diagnostics in IC engines. *Energy Conversion and Management, 99*, 299–312.
6. Marvin Jr, C. F. (1927). *Combustion time in the engine cylinder and its effect on engine performance.* NACA Technical Report 276.
7. Baratta, M., & Misul, D. (2012). Development and assessment of a new methodology for end of combustion detection and its application to cycle resolved heat release analysis in IC engines. *Applied Energy, 98*, 174–189.
8. Rassweiler, G. M., & Withrow, L. (1938). Motion pictures of engine flames correlated with pressure cards. *SAE Transactions*, Paper no-380139, 185–204.
9. McCuiston, F. D., Jr., Lavoie, G., & Kauffman, C. W. (1977). Validation of a turbulent flame propagation model for a spark ignition engine. *SAE Transactions*, Paper No-770045, 200–223.
10. Arroyo, J., Moreno, F., Muñoz, M., & Monné, C. (2015). Experimental study of ignition timing and supercharging effects on a gasoline engine fueled with synthetic gases extracted from biogas. *Energy Conversion and Management, 97*, 196–211.
11. Matekunas, F. A. (1983). *Modes and measures of cyclic combustion variability.* SAE Technical Paper, 830337.
12. Asad, U., & Zheng, M. (2008). Fast heat release characterization of a diesel engine. *International Journal of Thermal Sciences, 47*(12), 1688–1700.
13. Vibe, I. I., & Meißner, F. (1970). *Brennverlauf und kreisprozess von verbrennungsmotoren.* Berlin: Verlag Technik.
14. Stone, C. R., & Green-Armytage, D. I. (1987). Comparison of methods for the calculation of mass fraction burnt from engine pressure—time diagrams. *Proceedings of the Institution of Mechanical Engineers, Part D: Transport Engineering, 201*(1), 61–67.
15. Eriksson, L., & Nielsen, L. (2014). *Modeling and control of engines and drivelines.* Chichester: Wiley.
16. Zhao, H., & Ladommatos, N. (2001). *Engine combustion instrumentation and diagnostics.* Warrendale, PA: Society of Automotive Engineers Warrendale.
17. Scaringe, R. J. (2009) *Extension of the high load limit in the homogenous charge compression ignition engine* (PhD thesis). Massachusetts Institute of Technology, Cambridge, MA.
18. Klein, M., & Eriksson, L. (2004). *A specific heat ratio model for single-zone heat release models* (No. 2004-01-1464). SAE Technical Paper.

19. Heider, G. (1996). *Rechenmodell zur Vorausrechnung der NO-Emission von Dieselmotoren* (Doctoral dissertation). Technische Universitaet Muenchen.
20. Heider, G., Woschni, G., & Zeilinger, K. (1998). 2-zonen Rechenmodell zur Vorausrechnung der NO-emission von Dieselmotoren. *MTZ-Motortechnische Zeitschrift, 59*(11), 770–775.
21. Kulzer, A., Lejsek, D., Kiefer, A., & Hettinger, A. (2009). *Pressure trace analysis methods to analyze combustion features and cyclic variability of different gasoline combustion concepts* (No. 2009-01-0501). SAE Technical Paper.
22. Carryer, J. E. (1992). *Estimating in-cylinder pre-combustion mixture temperatures using acoustic resonances* (PhD thesis). Stanford University, Stanford, CA.
23. Carryer, J. E., Roy, R. H., & Powell, J. D. (1996). Estimating in-cylinder precombustion mixture temperatures using acoustic resonances. *Journal of Dynamic Systems, Measurement, and Control, 118*(1), 106–112.
24. Broekaert, S., Demuynck, J., De Cuyper, T., De Paepe, M., & Verhelst, S. (2016). Heat transfer in premixed spark ignition engines part I: Identification of the factors influencing heat transfer. *Energy, 116*, 380–391.
25. Finol, C. A., & Robinson, K. (2006). Thermal modelling of modern engines: A review of empirical correlations to estimate the in-cylinder heat transfer coefficient. *Proceedings of the Institution of Mechanical Engineers, Part D: Journal of Automobile Engineering, 220*(12), 1765–1781.
26. Borman, G., & Nishiwaki, K. (1987). Internal-combustion engine heat transfer. *Progress in Energy and Combustion Science, 13*(1), 1–46.
27. Alkidas, A. C. (1980). Heat transfer characteristics of a spark-ignition engine. *Journal of Heat Transfer, 102*(2), 189–193.
28. Enomoto, Y., Furuhama, S., & Minakami, K. (1985). Heat loss to combustion chamber wall of 4-Stroke gasoline engine: 1st report, heat loss to piston and cylinder. *Bulletin of JSME, 28*(238), 647–655.
29. De Cuyper, T., Demuynck, J., Broekaert, S., De Paepe, M., & Verhelst, S. (2016). Heat transfer in premixed spark ignition engines part II: Systematic analysis of the heat transfer phenomena. *Energy, 116*, 851–860.
30. Eichelberg, G. (1939). Some new investigations on old combustion engine problems. *Engineering, 148*, 547–550.
31. Tsurushima, T., Kunishima, E., Asaumi, Y., Aoyagi, Y., & Enomoto, Y. (2002). *The effect of knock on heat loss in homogeneous charge compression ignition engines* (No. 2002-01-0108). SAE Technical Paper.
32. Lounici, M. S., Loubar, K., Balistrou, M., & Tazerout, M. (2011). Investigation on heat transfer evaluation for a more efficient two-zone combustion model in the case of natural gas SI engines. *Applied Thermal Engineering, 31*(2–3), 319–328.
33. Hohenberg, G. F. (1979). *Advanced approaches for heat transfer calculations* (No. 790825). SAE Technical Paper.
34. Woschni, G. (1967). *A universally applicable equation for the instantaneous heat transfer coefficient in the internal combustion engine* (No. 670931). SAE Technical Paper.
35. Broekaert, S., De Cuyper, T., De Paepe, M., & Verhelst, S. (2017). Evaluation of empirical heat transfer models for HCCI combustion in a CFR engine. *Applied Energy, 205*, 1141–1150.
36. Annand, W. J. D. (1963). Heat transfer in the cylinders of reciprocating internal combustion engines. *Proceedings of the Institution of Mechanical Engineers, 177*(1), 973–996.
37. Bargende M. (1990). *Ein Gleichungsansatz zur Berechnung der instationären Wandwärmeverluste im Hochdruckteil von Ottomotoren* (PhD thesis). Technische Universität Darmstadt.
38. Chang, J., Güralp, O., Filipi, Z., Assanis, D. N., Kuo, T. W., Najt, P., & Rask, R. (2004). *New heat transfer correlation for an HCCI engine derived from measurements of instantaneous surface heat flux* (No. 2004-01-2996). SAE Technical Paper.

39. Soyhan, H. S., Yasar, H., Walmsley, H., Head, B., Kalghatgi, G. T., & Sorusbay, C. (2009). Evaluation of heat transfer correlations for HCCI engine modeling. *Applied Thermal Engineering, 29*(2–3), 541–549.
40. Grandin, B., & Denbratt, I. (2002). *The effect of knock on heat transfer in SI engines* (No. 2002-01-0238). SAE Technical Paper.
41. Gingrich, E., Janecek, D., & Ghandhi, J. (2016). Experimental investigation of the impact of in-cylinder pressure oscillations on piston heat transfer. *SAE International Journal of Engines, 9*(3), 1958–1969.
42. Brunt, M. F., & Platts, K. C. (1999). *Calculation of heat release in direct injection diesel engines* (No. 1999-01-0187). SAE Technical Paper.
43. Gatowski, J. A., Balles, E. N., Chun, K. M., Nelson, F. E., Ekchian, J. A., & Heywood, J. B. (1984). *Heat release analysis of engine pressure data* (No. 841359). SAE Technical Paper.
44. Cheung, H. M., & Heywood, J. B. (1993). Evaluation of a one-zone burn-rate analysis procedure using production SI engine pressure data. *SAE Transactions*, Paper No-932749, 2292–2303.
45. Mattson, J. M., & Depcik, C. (2014). Emissions–calibrated equilibrium heat release model for direct injection compression ignition engines. *Fuel, 117*, 1096–1110.
46. Fathi, M., Saray, R. K., & Checkel, M. D. (2010). Detailed approach for apparent heat release analysis in HCCI engines. *Fuel, 89*(9), 2323–2330.
47. García, M. T., Aguilar, F. J. J. E., Lencero, T. S., & Villanueva, J. A. B. (2009). A new heat release rate (HRR) law for homogeneous charge compression ignition (HCCI) combustion mode. *Applied Thermal Engineering, 29*(17–18), 3654–3662.
48. Asad, U., & Zheng, M. (2008). *Real-time heat release analysis for model-based control of diesel combustion* (No. 2008-01-1000). SAE Technical Paper.
49. Srivastava, D. K., & Agarwal, A. K. (2014). Comparative experimental evaluation of performance, combustion and emissions of laser ignition with conventional spark plug in a compressed natural gas fuelled single cylinder engine. *Fuel, 123*, 113–122.
50. Maurya, R. K. (2012) *Performance, emissions and combustion characterization and close loop control of HCCI engine employing gasoline like fuels* (PhD thesis). Indian Institute of Technology, Kanpur, India.
51. Lee, J., Chu, S., Min, K., Kim, M., Jung, H., Kim, H., & Chi, Y. (2017). Classification of diesel and gasoline dual-fuel combustion modes by the analysis of heat release rate shapes in a compression ignition engine. *Fuel, 209*, 587–597.
52. Eriksson, L. (1998). *Requirements for and a systematic method for identifying heat-release model parameters* (No. 980626). SAE Technical Paper.
53. Maurya, R. K., & Agarwal, A. K. (2013). Investigations on the effect of measurement errors on estimated combustion and performance parameters in HCCI combustion engine. *Measurement, 46*(1), 80–88.
54. Tunestål, P. (2009). Self-tuning gross heat release computation for internal combustion engines. *Control Engineering Practice, 17*(4), 518–524.
55. Manente, V., Vressner, A., Tunestal, P., & Johansson, B. (2008). *Validation of a self tuning gross heat release algorithm* (No. 2008-01-1672). SAE Technical Paper.
56. Oh, S., Min, K., & Sunwoo, M. (2015). Real-time start of a combustion detection algorithm using initial heat release for direct injection diesel engines. *Applied Thermal Engineering, 89*, 332–345.
57. Fang, C., Ouyang, M., & Yang, F. (2017). Real-time start of combustion detection based on cylinder pressure signals for compression ignition engines. *Applied Thermal Engineering, 114*, 264–270.
58. Lee, K., Yoon, M., Son, M. H., & Sunwoo, M. (2006). Closed-loop control of start of combustion using difference pressure management. *Proceedings of the Institution of Mechanical Engineers, Part D: Journal of Automobile Engineering, 220*(11), 1615–1628.
59. Yang, F., Wang, J., Gao, G., & Ouyang, M. (2014). In-cycle diesel low temperature combustion control based on SOC detection. *Applied Energy, 136*, 77–88.

60. Fang, C., Ouyang, M., Yin, L., Tunestal, P., Yang, F., & Yang, X. (2018). Start of low temperature reactions detection based on motoring pressure prediction for partially premixed combustion. *Applied Thermal Engineering, 141,* 1101–1109.
61. Chung, J., Oh, S., Min, K., & Sunwoo, M. (2013). Real-time combustion parameter estimation algorithm for light-duty diesel engines using in-cylinder pressure measurement. *Applied Thermal Engineering, 60*(1–2), 33–43.
62. Assanis, D. N., Filipi, Z. S., Fiveland, S. B., & Syrimis, M. (2003). A predictive ignition delay correlation under steady-state and transient operation of a direct injection diesel engine. *Journal of Engineering for Gas Turbines and Power, 125*(2), 450–457.
63. Hariyanto, A., Bagiasna, K., Asharimurti, I., Wijaya, A. O., Reksowardoyo, I. K., & Arismunandar, W. (2007). *Application of wavelet analysis to determine the start of combustion of diesel engines* (No. 2007-01-3556). SAE Technical Paper.
64. Katrašnik, T., Trenc, F., & Oprešnik, S. R. (2006). A new criterion to determine the start of combustion in diesel engines. *Journal of Engineering for Gas Turbines and Power, 128*(4), 928–933.
65. Wilhelmsson, C., Tunestål, P., Widd, B. J. A., & Johansson, R. (2009). *A physical two-zone NO x model intended for embedded implementation* (No. 2009-01-1509). SAE Technical Paper.
66. Shahbakhti, M., & Koch, C. R. (2009). Dynamic modeling of HCCI combustion timing in transient fueling operation. *SAE International Journal of Engines, 2*(1), 1098–1113.
67. Christensen, M. (2002). *HCCI combustion-engine operation and emission characteristics* (PhD thesis). Lund University, Lund.
68. Solaka, H. (2014). *Impact of fuel properties on partially premixed combustion* (PhD thesis). Lund University, Lund.
69. Solaka, H., Tunér, M., & Johansson, B. (2012, May). Investigation on the impact of fuel properties on partially premixed combustion characteristics in a light duty diesel engine. In *ASME 2012 Internal Combustion Engine Division Spring Technical Conference* (pp. 335–345).
70. Truedsson, I., Tuner, M., Johansson, B., & Cannella, W. (2013). Pressure sensitivity of HCCI auto-ignition temperature for oxygenated reference fuels. *Journal of Engineering for Gas Turbines and Power, 135*(7), 072801.
71. McAllister, S., Chen, J. Y., & Fernandez-Pello, A. C. (2011). *Fundamentals of combustion.* New York: Springer.
72. Rothamer, D. A., & Murphy, L. (2013). Systematic study of ignition delay for jet fuels and diesel fuel in a heavy-duty diesel engine. *Proceedings of the Combustion Institute, 34*(2), 3021–3029.
73. Moreno, C. J., Stenlaas, O., & Tunestal, P. (2017). *Influence of small pilot on main injection in a heavy-duty diesel engine* (No. 2017-01-0708). SAE Technical Paper.
74. Groendyk, M. A., & Rothamer, D. (2015). Effects of fuel physical properties on auto-ignition characteristics in a heavy duty compression ignition engine. *SAE International Journal of Fuels and Lubricants, 8*(1), 200–213.
75. Siebers, D. L. (1998). *Liquid-phase fuel penetration in diesel sprays* (No. 980809). SAE Technical Paper.
76. Jung, D., & Assanis, D. N. (2001). *Multi-zone DI diesel spray combustion model for cycle simulation studies of engine performance and emissions* (No. 2001-01-1246). SAE Technical Paper.
77. Finesso, R., & Spessa, E. (2014). Ignition delay prediction of multiple injections in diesel engines. *Fuel, 119,* 170–190.
78. Lee, M., Oh, S., & Sunwoo, M. (2011). Combustion phase detection algorithm for four-cylinder controlled autoignition engines using in-cylinder pressure information. *International Journal of Automotive Technology, 12*(5), 645.
79. Oh, S., Chung, J., & Sunwoo, M. (2015). An alternative method to MFB50 for combustion phase detection and control in common rail diesel engines. *IEEE/ASME Transaction on Mechatronics, 20*(4), 1553–1560.

80. Shehata, M. S. (2010). Cylinder pressure, performance parameters, heat release, specific heats ratio and duration of combustion for spark ignition engine. *Energy, 35*(12), 4710–4725.
81. Tazerout, M., Le Corre, O., & Ramesh, A. (2000). A new method to determine the start and end of combustion in an internal combustion engine using entropy changes. *International Journal of Thermodynamics, 3*(2), 49–55.
82. Thurnheer, T., & Soltic, P. (2012). The polytropic volume method to detect engine events based on the measured cylinder pressure. *Control Engineering Practice, 20*(3), 293–299.
83. Lawler, B. J. (2013). *A methodology for assessing thermal stratification in an HCCI engine and understanding the impact of engine design and operating conditions* (Doctoral dissertation). University of Michigan, Ann Arbor.
84. Lawler, B., Hoffman, M., Filipi, Z., Güralp, O., & Najt, P. (2012). Development of a postprocessing methodology for studying thermal stratification in an HCCI engine. *Journal of Engineering for Gas Turbines and Power, 134*(10), 102801.
85. Lawler, B., Lacey, J., Dronniou, N., Dernotte, J., Dec, J. E., Guralp, O., . . . & Filipi, Z. (2014). *Refinement and validation of the thermal stratification analysis: A post-processing methodology for determining temperature distributions in an experimental HCCI engine* (No. 2014-01-1276). SAE Technical Paper.
86. Livengood, J. C., & Wu, P. C. (1955). Correlation of autoignition phenomena in internal combustion engines and rapid compression machines. In *Symposium (international) on combustion* (Vol. 5, No. 1, pp. 347–356). Amsterdam: Elsevier.
87. Lawler, B., Lacey, J., Güralp, O., Najt, P., & Filipi, Z. (2018). HCCI combustion with an actively controlled glow plug: The effects on heat release, thermal stratification, efficiency, and emissions. *Applied Energy, 211*, 809–819.

Chapter 8
Combustion Stability Analysis

Abbreviations and Symbols

ACF	Autocorrelation function
AMI	Average mutual information
ANN	Artificial neural networks
ASI	Advanced spark ignition
ATDC	After top dead center
BDC	Bottom dead center
BTDC	Before top dead center
CA	Crank angle
CAD	Crank angle degree
CCF	Cross-correlation function
CCV	Cycle-to-cycle variation
CEI	Controlled electronic ignition
CI	Compression ignition
CN	Cetane number
CO	Carbon monoxide
COI	Cone of influence
COV	Coefficient of variation
COV_{IMEP}	Coefficient of variation of indicated mean effective pressure
CR	Compression ratio
CWT	Continuous wavelet transform
Cyl_{Temp}	Cylinder surface temperature
DCO	Dual coil offset
DET	Determinism
DI	Direct injection
EEGR	External exhaust gas recirculation
EGR	Exhaust gas recirculation
EVO	Exhaust valve opening

© Springer Nature Switzerland AG 2019
R. K. Maurya, *Reciprocating Engine Combustion Diagnostics*, Mechanical
Engineering Series, https://doi.org/10.1007/978-3-030-11954-6_8

FNN	False nearest neighbors
GDI	Gasoline direct injection
GEV	Generalized extreme value
GIE	Gross indicated efficiency
GWS	Global wavelet spectrum
HC	Unburned hydrocarbon
HCCI	Homogeneous charge compression ignition
ICA	Independent component analysis
IMEP	Indicated mean effective pressure
IVC	Inlet valve closing
LAM	Laminarity
LES	Large eddy simulation
LNV	Lowest normalized value
LOI	Line of identity
LTC	Low-temperature combustion
MBT	Maximum break torque
MFB	Mass fraction burning
MFBR	Mass fraction burning rate
MHRR	Maximum heat release rate
MI	Mutual information
MSD	Mean square displacement
OBD	Onboard diagnostic
ON	Octane number
PDF	Probability density function
PFI	Port fuel injection
PPRR	Peak pressure rise rate
PRF	Primary reference fuel
RCCI	Reactivity controlled compression ignition
REGR	Rebreathed exhaust gas recirculation
RMS	Root mean square
rpm	Revolution per minute
RQA	Recurrence quantification analysis
RR	Recurrence rate
RSM	Response surface model
SD_{IMEP}	Standers deviation of indicated mean effective pressure
SI	Spark ignition
SOC	Start of combustion
SOI	Start of injection
STD	Standard deviation
STFT	Short time Fourier transform
TCI	Transistor coil ignition
TDC	Top dead center
THR	Total heat release
TP	Throttle position

TT	Trapping time
VCR	Variable compression ratio
WFT	Windowed Fourier transform
WOT	Wide open throttle
WPS	Wavelet power spectrum
$^{\circ}CA$	Degree crank angle
$_{max}$	Maximum
CA_{50}	Crank angle at which 50% of heat release occur
A_f	Flame front surface
K_c	Asymptotic growth rate
S_c	Turbulent combustion velocity
W_s	Global wavelet spectrum
R^n	Finite-dimension vector space
K	Kurtosis
P	Pressure (bar)
Q	Heat release (J)
S	Skewness
m	Embedding dimension
$f(\alpha)$	Singularity spectrum
ρ_u	Unburned density
$\Delta\alpha$	Broadness of a singularity spectrum
$\Delta\alpha_c$	Combustion duration
θ	Angle (degree)
λ	Relative air-fuel ratio
σ	Standard deviation
σ^2	Variance
σ_P	Standard deviation of pressure time series
τ	Lag, time delay
ϕ	Equivalence ratio
$\psi(t)$	Wavelet function
ω	Engine speed
$\Theta(x)$	Heaviside function

8.1   Combustion Stability in Reciprocating Engines

Emission legislation and automotive market demands have been a constant driving force for significant increases in vehicle fuel economy to reduce petroleum use and CO_2 emissions. Achieving this goal while also continuing to reduce emissions of traditional pollutants is a significant challenge. Increasing fuel conversion efficiency while maintaining emissions performance is a necessary component of any solution. An approach which has gained widespread adoption in the market is the combination of engine downsizing with gasoline direct injection (GDI) and turbocharging utilized to maintain high torque output [1]. Another potential approach for future powertrain is advanced low-temperature combustion (LTC) modes that meet the requirement of higher engine efficiency and lower emissions [2]. Stability and combustion control are potential roadblocks to the most efficient implementation of many advanced combustion concepts due to higher cyclic combustion variability.

The cylinder pressure measurement for consecutive engine cycles depicts significant variations on a cycle-to-cycle basis (Fig. 8.1). Figure 8.1a shows the variation in cylinder pressure with increasing amount of EGR (higher throttle opening angle) in a homogeneous charge compression ignition (HCCI) engine [3]. In HCCI engine, autoignition of well-mixed charge occurs in the combustion chamber. Increasing exhaust gas recirculation (EGR) changes (increases) the specific heat of charge, which lowers down the combustion temperature. Thus, autoignition reaction rate decreases in the cylinder, and combustion phasing is retarded. The combustion

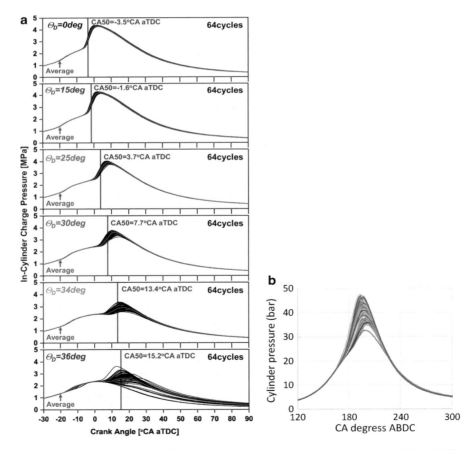

Fig. 8.1 (a) Variations in the cylinder pressure curve of 64 consecutive cycles for different EGR operation (varied by a throttle angle; Θ_D) in an HCCI engine [3] and (b) cylinder pressure for 100 consecutive engine cycle in an SI engine [4]

variability increases with retarded combustion phasing [2]. Increasing the EGR reaches to a point where cycle-to-cycle variations in the combustion process are very high, and engine operation becomes rough and unstable and unburned hydrocarbons increase rapidly. The engine operating condition at which this phenomenon occurs defines the stable operating limit of the engine. Figure 8.1a depicts that some of the engine cycles misfire or partially burn for highest EGR operating condition. Similarly, variations in the cylinder pressure can be observed in spark ignition (SI) engines (Fig. 8.1b). Significant variations in the cylinder processes (flow and ignition characteristics) lead to the significant variation in the combustion process because the pressure development is uniquely defined with the combustion process. The cycle-to-cycle variations in the combustion process are the result of cyclic variations of mixture motion in the combustion chamber at the time of spark, the variations in air-fuel ratio in each cycle (due to variations in the amounts of air or fuel or both), and cyclic variations in the mixing of fresh mixture and residual gases in the combustion chamber particularly in the vicinity of spark plug [5].

Analysis of cycle-to-cycle variations in the SI engine combustion is more important because engine operation is limited by extreme cyclic variations. Additionally, the optimum spark timing is set for the average cycle. Thus, combustion cycle faster than average cycle has effectively over-advanced spark timing, and cycles slower than average cycles have retarded spark timings. The advanced or retarded spark timings lead to a loss of power and efficiency [5]. Due to the spark timing set for the average cycle, the faster cycles are more likely to have knocking tendency, and the slower cycles are likely to be partial burn or misfire cycles. The slowest cycle actually defines the lean operating limit of the engine and also the maximum EGR tolerance of the engine.

Figure 8.2 illustrates the stable and unstable engine operating region and the edge of combustion stability. The stable combustion is characterized by acceptable cyclic variations in the combustion process. During engine operation in the lean and highly diluted mixture (with air or EGR) or at lower loads and engine speeds (such as under idle conditions), the cyclic variation increases and engine operation shifts to transition regime (partial burn), and in extreme conditions misfire can occur (Fig. 8.2). In the transition period, the cyclic variations in the combustion process (COV_{IMEP}) are very high, and the engine efficiency decreases rapidly. Unintended excursions to the unstable operating region may result in misfires and very strong "rebound" events which could damage the engine and/or catalyst system [7]. Thus, to avoid unintended excursions, practically engines need to operate well away from the edge of stability. Cyclic combustion variations are governed by stochastic (in-cylinder variations) and deterministic (cycle-to-cycle coupling) processes, and deterministic mechanisms act as a nonlinear amplifier to stochastic variations. Cyclic combustion variation can also amplify the cylinder-to-cylinder imbalances [7]. Thus, improved understanding of instability mechanism is required for better control of the engine.

Combustion stability is one of the main parameters on which researchers are keenly interested in improving the drivability, fuel economy, and emissions. Combustion stability (cyclic variations) can be quantified in terms of standard deviation

Fig. 8.2 Stable and
unstable engine operating
regions (adapted from [6, 7])

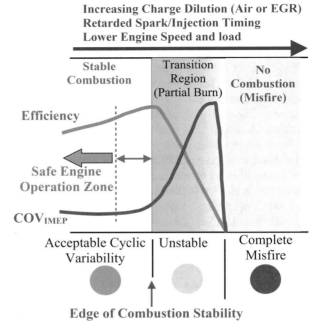

of indicated mean effective pressure (SD_{IMEP}), the coefficient of variation of IMEP (COV_{IMEP}), lowest normalized value (LNV), and standard deviation and coefficient of variation of speed [8]. Figure 8.3 illustrates the cycle-to-cycle combustion variations in HCCI engine during stable and unstable operations.

Figure 8.3a shows that the stable operation of HCCI engine has small cyclic combustion variation ($COV_{IMEP} = 3.56\%$) at average combustion phasing (CA_{50}) of 13.4 °CA after TDC. In the stable operating condition, cyclic variations of the mass and temperature of each gas content at intake valve closing (IVC) (m_{air}, m_{fuel}, m_{EEGR}, m_{REGR}, T_{air}, T_{fuel}, T_{EEGR}, T_{REGR}) and the T_{IVC} are also small. The EEGR denotes external exhaust gas recirculation, and REGR denotes rebreathed exhaust gas recirculation. Cyclic combustion variations for the stable operation are equally distributed to values higher and lower than the average (Fig. 8.3a). In contrast, a much larger spread around the average value in m_{air}, m_{EEGR}, m_{REGR}, T_{REGR}, and the resulting T_{IVC} is exhibited by several cycles during unstable engine operation (Fig. 8.3b). The unstable engine operation has larger variations in combustion as depicted by IMEP and CA_{50} in Fig. 8.3b. Some of the cycles partially burn in the unstable engine operating region as illustrated in Fig. 8.2. The partial burn cycles can be observed in unstable engine operation in Fig. 8.3b. The partial burn and misfire cycles are discussed in Sect. 8.2. The manifestation of combustion variability and sources of combustion variability are discussed in the next subsections.

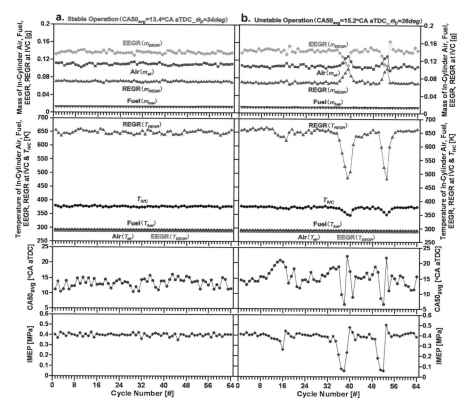

Fig. 8.3 Illustration of (**a**) stable and (**b**) unstable engine operation in HCCI combustion [3]

8.1.1   Manifestation of Combustion Variability

Cycle-to-cycle combustion variability phenomenon is significant during engine operation typically at low load and speed conditions (particularly idle conditions), high EGR operation, and highly diluted mixture. The two consecutive engine combustion cycles are not exactly same similar to human fingerprints. There exist cycle-to-cycle and cylinder-to-cylinder variations in developed torque along with fluctuations in the engine speed.

The signs of combustion variability in automotive engines are summarized as engine roughness, compromised torque/power, higher engine emissions, loss in engine efficiency, lower fuel economy, lower knock resistance, compromised dilution tolerance, compromised spark/injection timing (injection point diesel), and cyclic variations in torque and engine speed [9]. The significant cyclic variations in the developed engine torque directly affect the drivability of the vehicle. Therefore, minimization of cyclic combustion variations is essential for stable engine

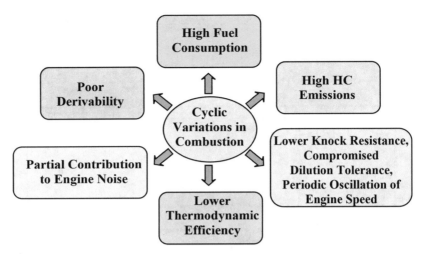

Fig. 8.4 Effect of cyclic variations in the combustion process on engine performance (adapted from [10])

operation along with optimal engine performance in terms of emissions and fuel economy.

Combustion variability influences the engine performance at all operating conditions. Idle instability typically is governed by variations in fuel flow and exhaust residuals. Part-load combustion variability is driven by fuel flow variations and EGR. Wide open throttle (WOT) combustion instability is typically dictated by variations in airflow [9]. The summarized general negative impacts of the cycle-to-cycle variations in the combustion process on engine performance are presented in Fig. 8.4. The cyclic variations in the combustion duration result in a condition where the combustion process in some of the cycles is faster while in others is slower than average cycle. The combustion variations are associated with losses in terms of power and thermodynamic efficiency and fluctuations in the amount of work done. The faster combustion cycles (than optimized) will have higher peak pressure and the tendency of knocking. Thus, these cycles impose the lower limit for the allowed fuel octane number and the upper limit for the engine compression ratio, which compromises the thermodynamic efficiency. In the slow combustion cycles (than optimized), the combustion may not be completed before exhaust valve opening (EVO) timing, which leads to a high unburned hydrocarbon (HC) emissions as well as high fuel consumption. This effect is significantly observed in with diluted mixtures (either with EGR, with lean mixtures, or under throttled conditions) [11]. The cyclic combustion variations also lead to the engine speed and torque fluctuations which affect the overall engine performance characteristics, such as the engine brake power and its specific fuel consumption. The speed and torque fluctuations also result in poor drivability of the vehicle for some kinds of transmissions, such as lockup torque converters and manual transmissions [12]. The cyclic variations also partially contribute to engine noise due to variations in cylinder pressure

[10–12]. Thus, reduction in cyclic variations in combustion may also help in mitigating the engine noise. Periodic oscillations in engine speed of a port fuel-injected (PFI) SI engine at idle conditions are also found to be affected by combustion perturbations [10, 11]. Considering the negative impacts of cyclic combustion variations, it is important to determine the sources of variability and quantify (characterize) the combustion variation. The origin of combustion variability in reciprocating engines is discussed in the next subsection.

8.1.2   Sources of Cyclic Combustion Variability

The cycle-to-cycle combustion variations are evident from the beginning of the combustion process (Fig. 8.1). Variations in burning rate are also apparent throughout the combustion process. Combustion variability may be caused by various factors including cyclic variations in gas motion in the cylinder during combustion; variations in the amount of fuel, air, and residual gases inducted in every engine cycle; gas charge motion and composition at the location and time of spark; mixing homogeneity; fuel preparation (targeting, droplet size, swirl, cone angle); disproportionate dilution (by EGR or air); long combustion duration because of poor combustion system hardware design; and low ignition energy or a small spark plug gap [9]. Origins of cyclic combustion variations can be divided into two groups: (1) prior-cycle effects (residual gas, etc., results of misfire and partial burn, wall temperature) and (2) same-cycle effects (in-cylinder flow, etc., results of random variations) [13]. Both prior-cycle and same-cycle effects are always present in reciprocating engines. However, one effect may dominate depending on engine operating conditions. The relative contribution of each process to overall cyclic combustion variability is not known and may be different for different engines depending on fuel injection system, engine geometry, and operating conditions. The specific reasons for cyclic combustion variations depend on the charge preparation process and combustion mode. The sources of cyclic variation in spark ignition and compression ignition engines are discussed in the next two subsections.

8.1.2.1   Causes of Cyclic Combustion Variations in SI Engines

Cyclic variations in the combustion process are generated due to the variations in mixture motion within the combustion chamber; variations in the amounts of air and fuel fed to the cylinder; variations in the mixing of fresh mixture and residual gases, particularly in the vicinity of the spark plug, which determine mixing, stratification, convection of spark kernel away from the electrodes, and heat loss from the kernel to the electrodes; etc. [13]. It is proposed to divide all the sources of cyclic combustion variation in SI engine into the four categories, namely, (1) mixture composition, (2) spark and spark plug, (3) in-cylinder mixture motion, and (4) engine operating factors/conditions [11, 12, 14]. Figure 8.5 presents the summary of all the major

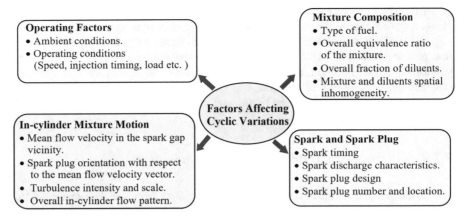

Fig. 8.5 Summary of factors influencing the cyclic combustion variation in spark ignition engine (adapted from [11, 12, 14])

factors influencing the cycle-to-cycle variations in spark ignition engine. These factors affect the different stages of combustion depending on their characteristics. In spark ignition engine, the combustion process can be divided into four main stages, namely, (1) sparking and flame initiation, (2) initial flame kernel development, (3) turbulent flame propagation (main combustion stage), and (4) flame termination [11]. Each of these stages of SI engine combustion may be influenced by a different set of dominating factors, while each preceding stage affects the subsequent ones. The last stage (flame termination) does not affect noticeably the cyclic combustion variations.

There are several known factor responsible for creating cyclic combustion variation in SI engine including turbulent nature of the in-cylinder flow, mixture spatial inhomogeneity, fluctuations in the mean flow vector at the spark gap, etc. (Fig. 8.5). Other factors, such as spark plug orientation, electrode shape, overall equivalence ratio, etc., though do not generate cyclic combustion variation but influence their intensity [10]. The complexity of the phenomenon of cyclic combustion variations is illustrated in Fig. 8.6. Solid lines in the figure represent the ways of the influence of the factors which contribute to cyclic variation generation, and dashed lines show the influence of the factors which only affect the extent of cyclic variations. Figure 8.6 shows how several factors affect the cyclic variations in the SI engine combustion through different mechanism and stages of the combustion development. For example, mixture spatial inhomogeneity causes cyclic variations in "first eddy burnt" composition and thereby affects the initial flame kernel development stage. Besides that, some of the factors are strongly interrelated [10]. The other observation is one group of factors such as overall equivalence ratio, spark plug location, or fuel type do not cause cyclic combustion variation themselves but affect the extent of cyclic variation generated by other factors. Thus, understanding of the contribution of a particular factor in cyclic variation and its mechanism of influence at a particular stage of combustion is important.

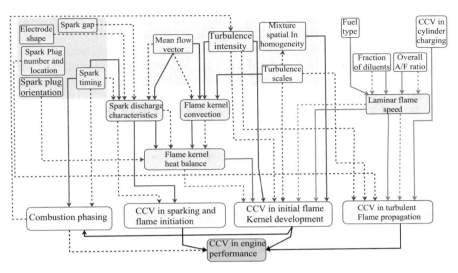

Fig. 8.6 Mechanisms of the influence of the various factors involved in cyclic combustion variations in SI engine performance (adapted from [10])

Mixture composition-related factors include the type of fuel, the overall equivalence ratio (Φ) of the mixture and its cyclic variations, the overall fraction of diluents and its cyclic variation, and the mixture equivalence ratio and diluent spatial inhomogeneity. The fuel type affects the cyclic combustion variation through laminar burning velocity, which significantly affects the initial flame kernel development [11]. Two major characteristics of the fuel type affecting the cyclic combustion variations are (1) magnitude of the maximum flame speed and (2) the equivalence ratio at which it occurs [12]. It was found that fuel with very high laminar burning velocity show closely repeated cycles (less cyclic variations) and the flame remained centered at the spark gap [15]. Fuels having higher laminar flame velocity leads to the higher burning rate, which results in relatively lower the cyclic combustion variations. One more parameter by which one fuel can differ from another is its heating value. This parameter can influence the burning rate through the expansion rate of the flame and adiabatic flame temperature [11]. The overall equivalence ratio of the mixture and its cyclic variations affect the combustion variations through the laminar burning speed, which is highest in stoichiometric or slightly enriched mixtures. Thus, any deviation from the engine operation at a stoichiometric ratio leads to a decrease in the laminar burning velocity with a consequent increase in the ignition delay time and the level of cyclic combustion variations [11, 12]. The overall fraction of diluents and its cyclic variations reduces charge flammability and the laminar burning rate. The amount of diluents can be varied by different methods including throttle opening variation, skipping ignition in cycles, variation of the extent of EGR, and the introduction of inert additives [12]. It is observed that higher dilution reduces the burning rate which leads to higher cyclic combustion variations. Mixture and diluent spatial inhomogeneity are created due to imperfect mixing of charge components which includes air, fuel, and residual gases (internal or external EGR). The spatial inhomogeneity in the charge leads to the cyclic variations in local equivalence ratio and the diluent quantity in the vicinity of the spark electrode gap which consequently leads to cyclic variations in the initial flame kernel development stage [11, 14].

Design of spark plug and the spark discharge characteristics are the dominant parameters in the initial flame kernel development. The evolution of cylinder pressure in the combustion chamber is strongly related with the initial flame kernel development stage, and thus, spark plug design and its characteristics strongly affect the engine performance [16, 17]. Spark and spark plug-related factors include spark timing, spark discharge characteristics, spark plug design (electrode gap and shape), and spark plug number and location. All the spark-related parameters affect the early stage of flame development. In the later stage of combustion process, both the location of the spark plug and their number play a significant role [11]. Minimum cyclic variations in combustion and pressure development generally take place at maximum brake torque (MBT) spark timing. The dependence of cyclic variations on the spark timing is affected by overall mixture strength and location of spark plug. The more prominent cyclic variations are observed for leaner mixtures or the large spark plug offset from the center of the combustion chamber. The spark discharge process also has an important influence on flame initiation. The initial flame development plays an important role in cyclic variations. The quicker (fast) the flame kernel reaches a certain critical size, the lower cyclic variations are observed. Enhanced ignition increases the early flame development rates and thus results in lower cyclic variations. The ignition system should be redesigned so that most of the electrical energy is dissipated in the breakdown phase. This would ensure that energy is transferred to the gas most efficiently and produce sharp temperature gradients and the right plasma geometry to enhance the initial flame speed [18]. Spark duration and spark energy are two characteristics that are typically used to characterize spark discharge. Generally, these two are strongly coupled with each other, and varying spark duration causes changes in the total spark energy. It is suggested that one of the methods of reducing cyclic variations is by increasing the energy of the conventional spark [17]. This implies that the spark energy has an effect on cyclic combustion variations. The design of spark plug can affect the cyclic combustion variations through several mechanisms. The number of spark plug electrode and their shape govern the flame contact area fraction, which controls the heat losses into the body. The electrode shape and number also influence the flow field characteristics within the spark gap [19]. More favorable kernel heat balance is expected from thin and/or sharply pointed electrodes. However, thin/sharply pointed electrodes are subjected to relatively stronger erosion and therefore have less durability. Additionally, the surface temperature of a thin or sharply pointed electrode will be higher. This creates a danger of preignition at high engine load operations. Spark gap affects the breakdown voltage and energy. Higher voltage is required for breakdown at larger spark gap. Additionally, the spark plug gap also affects the energy loss from the flame kernel to the electrodes by heat transfer. Increasing the spark gap leads to the larger ratio between flame kernel volume and "wetted" electrode surface area. Thus, the electrode gap of spark plug affects the flame initiation and the stage of flame kernel growth. Typically, the spark gap is required to be larger than the flame quenching distance.

Mixture motion and turbulence are the causes deemed most responsible for the cyclic variations in combustion and the subsequent pressure development. In-cylinder measurements of flow are difficult to obtain so that the role of turbulence and mixture motion in cyclic behavior of combustion has been largely inferred from the nature of the physical changes made to the engine. This evidence, although largely circumstantial, appears strong enough to single flow variations as being instrumental in determining the character of the combustion from cycle-to-cycle. Turbulence in the cylinder accelerates the combustion by increasing flame front area and improving heat and mass transport between the burned and unburned parts of the charge, which tend to decrease cyclic variations. On the other hand, turbulence leads to fluctuation in the magnitude and direction of the charge velocity due to random flow pattern in the spark gap vicinity which results in cyclic variations in early flame kernel development [11].

Factors related to the mixture motion are (1) mean flow velocity vector in the spark gap vicinity and its cyclic variations, (2) spark plug orientation with respect to the mean flow velocity vector, (3) turbulence intensity and scales, and (4) overall in-cylinder flow pattern. In the very early stage of sparking and flame initiation, the mean flow velocity near the spark plug lengthens the discharge channel and increases the electrical energy deposited into the flame kernel in the breakdown phase. Later, in the initial flame kernel development stage, mean flow velocity can convect flame kernel away from the electrodes which lower down the heat loss [20]. This convection toward the electrodes of a spark plug or combustion chamber walls is undesirable. The early flame kernel convection affects both ignition delay time and flame kernel radius at the particular crank position. The cyclic variation in the mean flow can lead to random convection which transports the kernel to different locations around the spark plug. The flame propagation in a particular cycle depends on the initial position of the early flame kernel center [11]. Gas dynamics conditions in the early flame kernel growth stage are influenced by local flow field around the electrodes which is governed by mean flow direction and location of the spark plug in the combustion chamber. Several types of flow pattern can be achieved in the combustion chamber of the reciprocating engine. The engine design parameters (such as the shape of the combustion chamber and induction ports, and the type and configuration of inlet valves) govern the actual flow in the cylinder. The instantaneous flow also depends on the crank angle position (piston position) and the engine operating conditions. Swirl, squish, and tumble are the three main macrostructures found in the in-cylinder flow. Early flame development was more affected than the main combustion period when swirl was present in the combustion chamber. It was found that cyclic variations reduced by 30% at part load with volumetric efficiency of 72% by swirl due to increase in burning rate [21]. Tumble flow generates turbulence in the combustion chamber more effectively than does swirl flow. Swirling flow reduces the cycle-by-cycle variability of the mean velocity in the combustion chamber, which tends to be generated by tumble motion [22].

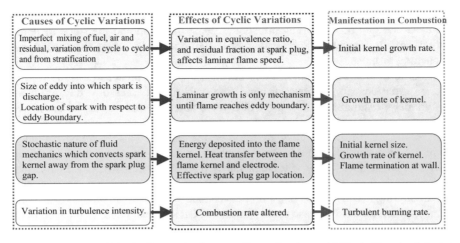

Fig. 8.7 Summary of physical factors leading to cyclic combustion variability in SI engine (adapted from [25, 26])

Ambient conditions (such as intake temperature, coolant temperature, cold start, etc.) and engine operating conditions (such as injection timings, engine speed, load etc.) also significantly affect the cyclic variations. Increasing the inlet air temperature improves the evaporation of fuel droplets and compensates for the reduction of charge temperature because of the latent heat effect of fuel. It is demonstrated that cyclic variations (COV_{IMEP}) have optimal values in the inlet temperature range of 60–70 °C [23]. Engine operating factors such as engine speed also affect the cyclic variations as it influences the turbulence intensity. Injection characteristics such as fuel spray patterns at the pressurized conditions affect the combustion stability [24].

The physical factors leading to cyclic combustion variability are summarized in Fig. 8.7. Cycle-to-cycle combustion variability in spark ignition engines limits the use of lean mixtures, the amount of recycled exhaust the engine will tolerate, and lower idle speeds because of increased emissions and poor engine stability.

Causes of cyclic combustion variations can be categorized into two groups: prior-cycle effects (residual gas, etc., results of misfire and partial burn, engine thermal state, EGR) and same-cycle effects (in-cylinder flow, etc., results of random variations) [13]. Prior-cycle effects are mainly due to residual gas in the cylinder and thermal state of the engine. The residual gases from the previous cycle alter the burning rates of burning and may even cause an engine misfire. The residual gases consist of exhaust gases remained in the clearance volume of cylinder (internal) and gases returned by backflow during the valve overlap period. Additionally, external exhaust gas recirculation (EGR) through intake manifold also leads to prior-cycle effect. Mass flow analysis of gases suggests that the major portion of residual gases are contributed by the gases that remained in the clearance volume of cylinder. The gases remained in the clearance volume are dependent on valve profile, flow coefficients, and engine speed. Additionally, the gases exchanged during valve overlap are governed by the gas dynamics in the manifold [27]. The pressure fluctuations during valve overlap can result in variations in the fresh air-fuel mixture and backflow of exhaust gases. This leads to variations in residual gases in the cylinder which affect the succeeding combustion event [28]. The amount of residual gases are influenced by several factors including the pressure difference across the valves, valve characteristic, pressure, engine speed, and the gas dynamics during valve overlap [29].

The same-cycle effects leading to cyclic variations are mainly governed by variation in the flow of the air-fuel mixture as well as the quality of mixture which is also partially based on the type of fuel injection system used in the engine. The port fuel injection (PFI) as well as direct injection (DI) is typically used in commercial vehicles using spark ignition engines. The fuel-air mixture changes by the way gases flow through inlet manifold during transportation to the cylinder, particularly in PFI engines. Therefore, the charge in the cylinder can vary in different cycles, which leads to combustion variability. Physical study of the mixture preparation in PFI engines showed that fuel puddles are located on the valves, downstream in the intake port near the valve, upstream near the injector, and on the cylinder wall [30]. These puddles of fuels lead to cyclic variations in the air-fuel mixture. Fuel properties affect the puddles of fuels due to the difference in evaporation characteristics [31]. Thus, different fuels will have a different level of variability in fuel-air mixture. In PFI engines, fuel transportation process significantly depends on the timing and duration of the fuel injector pulse. In some systems, the fuel spray is targeted on the back of the intake valve, and fuel is injected in fully closed valve conditions or partially open conditions. Thus, the backflow of hot residual gas leads to the vaporization of liquid fuel off the valve and wall. The fuel can be drawn into the cylinder as a liquid drop in some cases although the engine is running under fully warm conditions [5]. Fuel transportation mechanism in the intake manifold can be understood regarding impingement regimes of fuel. The impingement regimes are divided into different categories: (1) stick, (2) spread, (3) rebound, (4) rebound with a breakup, (5) boiling-induced with breakup, (6) breakup, and (7) splash [32]. These regimes are influenced by different fuel parameters (droplet velocity, size, temperature, the angle of wall hitting, and fuel properties) and manifold wall parameters (temperature, surface roughness, and remaining fuel droplets). These effects significantly influence the air-fuel ratio before the start of combustion.

Additional factors such as turbulence intensity and scale, mean flow velocity at spark plug, air-fuel ratio, variations in spark discharge characteristics (breakdown energy, timing, type of spark plug, and spark orientation), variations in mass of charge, leakage through valves, and crevice effects also influence the cyclic combustion variation in spark ignition engine [13].

8.1.2.2 Causes of Cyclic Combustion Variations in CI Engines

Conventionally in compression ignition (CI) engines, the diesel fuel ignites spontaneously following the injection event. The combustion and fuel injection often overlap with a very short ignition delay period in conventional CI diesel engines. Therefore, diesel engines offer superior combustion stability characterized by the low cyclic combustion variations [33]. Diesel engine usually exhibits lower significant cyclic combustion variations because non-premixed combustion dominates the overall combustion process, which is mainly governed by fuel-air mixing [34]. The fuel-air mixing is governed by fuel injection system, which is the highly repeatable system. Thus, there exists only a very small amount of time where combustion is uncoupled with the injection or uncontrolled combustion that results in significantly lower cyclic variations in comparison to homogeneous spark ignition engine. Modern high-speed diesel engines are equipped with multiple pilot injections potential with an objective of reducing the main injection ignition delay. The reduction in ignition delay leads to mainly diffusion-dominated combustion in a modern diesel engine. Cyclic variations in background turbulence (swirl, etc.) are minimal, and will not affect the diesel combustion rate significantly because the intensity of background turbulence is several times lower than the injection-generated turbulence [34].

However, cyclic combustion variability can also occur in CI engine mainly due to instabilities in the fuel injection system or prolonged ignition delay. Cyclic variations in a diesel engine are observed due to variations in injection timing between cycles because of rotary fuel pump used for fuel injection [35]. Recent studies also demonstrated the cyclic combustion variations in diesel engine due to the variations in the fuel path [36, 37]. Study showed that cyclic variations in the fuel injection pressure (at a command pressure of 1000 bar) are between 1 and 9% which results in variations in the needle lift that leads to variations in fuel delivery up to 23%. The variations in the fuel delivery are correlated with the corresponding cyclic variations in the IMEP and instantaneous angular velocity of the engine [36]. Cycle-to-cycle variations demonstrate very fast dynamics. Among all the controlled inputs available in diesel engines, the injector pulse or current profile, typically defined by start of injection (SOI), fuel ratio, dwell time, etc., has the most immediate dynamic effect on the combustion process [37].

In addition to instabilities of the fuel injection system, prolonged ignition delay contributes to increase in cycle-to-cycle combustion variations. It is demonstrated that the cyclic variations in the cylinder pressure cannot be fully explained by variations in the fuel injection [38]. Prolonged ignition delay conditions created by varying intake temperature, pressure (engine load), and fuel injection timings show higher cyclic combustion variation in diesel engine [34]. Ignition characteristics of fuel also affect the cyclic variations, and lower cetane number (CN) fuels exhibit larger cyclic combustion variations. Cold start studies in diesel engines show that colder in-cylinder conditions are resulting into a longer ignition delay period that leads to larger cyclic variations [39]. Diesel cold starting is influenced by many design and operating parameters, which affect the air temperature and pressure near the end of the compression stroke. Such parameters include the ambient temperature, cranking speed, injection parameters, and fuel properties particularly cetane number and volatility. Combustion instability during the cold start of diesel engines is influenced by several factors, most important of which are the ambient temperature, injection timing, and the instantaneous engine speed during the cycle [40].

Combustion instability was found to increase with the drop in temperature. Cycle-to-cycle analysis suggested that the number of misfiring cycles increases with the drop in ambient temperature. The cause of misfiring was found to be a mismatch between the injection parameters and the instantaneous engine speed at the time of misfiring. Combustion instability can be reduced if fast injection timing controls are developed. Ideally, the injection timing should be adjusted to suit the conditions in each cylinder, which requires fast response fuel injection controls. [40]. It is found that the cycle-to-cycle stability of IMEP is generally improved by using triple- or quad-pilots over single and possible twin pilots [41]. Additionally, increasing the glow plug temperature always improves stability, even when the best fuel injection strategy is used. The number of pilot injections is more important than the total pilot quantity. Total pilot quantity has been observed to have an influence which can produce a deterioration or improvement in stability, and the direction of change may depend on the number of pilot injections [41].

In well-mixed (homogeneous) charge compression ignition (HCCI) engine, the cyclic combustion variation is typically lower than conventional spark ignition (SI) engine, but it can be high in some of the operating conditions due to no direct control on ignition timings [2]. Ignition timing in HCCI engines is sensitive to inlet air temperature, equivalence ratio, residual gases/EGR, cylinder wall temperature, compression ratio, and chemical kinetics of fuel-air mixture. The HCCI combustion in the engine cylinder is achieved using different strategies for autoignition of charge. Typically, cyclic combustion variations and cylinder-to-cylinder variations in HCCI engines depend on combustion phasing and strategies employed to control the combustion [2, 42, 43]. Several factors affect the HCCI combustion, and the level of cyclic combustion variations is governed by variations in several factors including (1) equivalence ratio, (2) inlet air temperature and pressure, (3) mixture inhomogeneity (thermal and composition stratifications), (4) amount of EGR (external or internal), (5) coolant and lubricating oil temperatures, (6) intensity of intake charge motion and bulk turbulence, (7) fuel-air mixing and charge preparation strategies, and (8) combustion completeness in the previous engine cycle [43]. Several strategies (such as inlet air preheating, variable compression ratio (VCR), exhaust recompression, exhaust reinduction, employing multi-fuel and multiple injection strategies) can be used to control/minimize the cyclic variations in HCCI engine [44].

In HCCI engine, four major contributing sources of cyclic combustion variations are identified, namely, (1) mixture composition stratification, (2) thermal stratification, (3) cycle-to-cycle variations in fuel-air ratio, and (4) cycle-to-cycle variations in diluents (residual gases and/or EGR) [45]. Spatial mixture composition inhomogeneity occurs in the cylinder due to inadequate mixing of air, fuel, and residual/recirculated burned gas from the previous cycle. Homogeneity of charge is affected by fuel injection strategy used for fuel-air mixture preparation [2]. Cyclic variations in charge composition homogeneity (stratification) lead to the cyclic combustion variations. Temperature stratification is created by heat transfer from surfaces (piston, valves, and cylinder head) having different temperatures and from the insufficient mixing of air, fuel, and residual gas [46]. The inhomogeneous heat release from the low-temperature heat release (cool flame chemistry) can also contribute to the temperature stratification in the combustion chamber [47]. Delayed combustion phasing significantly enhanced the thermal stratification in the combustion chamber because of longer time available for developing the stratification [48]. Cyclic variations in the residual gas temperature can generate the cyclic thermal stratification in the cylinder that results in cyclic combustion variations in HCCI engine [2, 49]. Cyclic variations in the mean equivalence ratio of the charge can occur in PFI engine [50], which is typically used for fuel injection in HCCI engine. Cyclic variations in equivalence ratio can also be generated by variations in gas exchange process as well as incomplete evaporation of fuel droplets. The cyclic variations in mean equivalence ratio lead to cyclic combustion variations in HCCI engine.

The HCCI combustion phasing is typically controlled by internal (trapped residuals) as well as external cooled EGR. Due to the incomplete mixing of EGR, thermal stratification in the cylinder can be created. The level of dilution by EGR affects the HCCI combustion, and thus, cycle-to-cycle variations in the amount and composition of diluents result in cycle-to-cycle variations in combustion. Both chemical and thermal effects are responsible for cyclic combustion variation in HCCI engine when EGR is used for dilution [3].

Figure 8.8 illustrates the unstable HCCI engine operation using EGR. The number of cycles presented is zoomed version of Fig. 8.3b. It can be noticed from Fig. 8.8d that a partial burn cycle is often followed by another partial burn cycle due to a lower temperature at intake valve closing (T_{IVC}) which is the result of the lower temperature of rebreathed EGR. The lower temperature leads to delayed combustion phasing, which increases the possibility of a second partial burn cycle with further lower IMEP [3]. Figure 8.8d also shows the variations of IMEP do not well matched with the variations of CA_{50} (particularly for the partial burn cycles). Additionally, immediate cycle after an excessive partial burn cycle (the lowest IMEP cycle with number 38 and 52) improves the IMEP toward the average value at substantially lower T_{IVC} and highly retarded CA_{50} timings. Engine cycles number 39 and 53 (Fig. 8.8d) have maximum IMEP at the highly retarded CA_{50} instead of the lowest T_{IVC}. The recovery of IMEP just after misfire cycle cannot be described by thermal effects of the charge, and it suggests a chemical state of charge is responsible (chemical enhancement of autoignition) [3]. The low-temperature heat release is maximum for excessive partial burn cycle (Fig. 8.8c). It is suggested that trace species produced during low-temperature heat release period are recirculated in the next cycle, which is responsible for the sufficient enhancement of autoignition of charge. Therefore, cycle-to-cycle variations in the EGR can lead to cycle-to-cycle variations in combustion process through chemical effects also.

The low-temperature combustion (LTC) engines have higher sensitivity to cyclic variations input parameters than conventional diesel combustion. The LTC engines show higher cyclic variations than conventional diesel engines [2]. Figure 8.9 shows the normalized contribution to cyclic variations in output parameters (IMEP, CA_{50}, gross indicated efficiency (GIE), and peak pressure rise rate (PPRR)) by several input parameters in HCCI and reactivity controlled compression ignition (RCCI) engines. The considered input parameters are temperature and pressure at intake valve closing (T_{ivc}, P_{ivc}), PFI fuel mass $(PFI_{FuelMass})$, EGR, cylinder surface

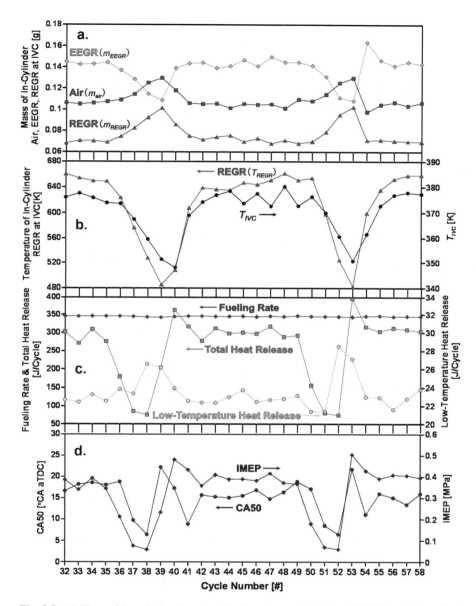

Fig. 8.8 (**a**) Mass of in-cylinder air, external exhaust gas recirculation (EEGR), and rebreathed exhaust gas recirculation (REGR) at IVC; (**b**) temperature of REGR and averaged charge temperature (T_{IVC}) at IVC; (**c**) fueling rate, total heat release, and low-temperature heat release; (**d**) CA$_{50}$ and IMEP as a function of consecutive cycles of #32–#58 for the unstable HCCI operation [3]

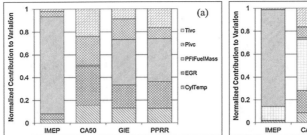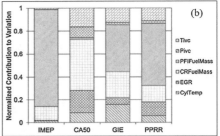

Fig. 8.9 Effect of several input parameters on the cyclic variations in output parameters (IMEP, CA_{50}, GIE, and PRRR) in (**a**) HCCI and (**b**) RCCI engines [51]

temperature (Cyl_{Temp}), and direct injected fuel mass ($CR_{FuelMass}$) in case of RCCI combustion. Individual contributions of every input parameter are calculated by the response surface model (RSM) by uncertainty propagation technique [51]. The major contribution in cyclic variations of the output parameters is from four sources: PFI mass, DI mass, trapped gas temperature, and EGR% (Fig. 8.9). The P_{ivc} has an only minor influence to the variation in RCCI combustion, but it is a significantly contributing factor in cycle-to-cycle variation HCCI engine [51]. The combustion phasing (CA_{50}) is mainly affected by T_{IVC}, P_{IVC}, EGR%, and cylinder surface temperature in the HCCI engine. The PFI fuel mass has a minimal effect on CA_{50} even though it has a very strong effect on IMEP and PPRR (Fig. 8.9a). The cyclic variations in CA_{50} are mainly influenced by T_{IVC}, direct injection fuel mass, and EGR% in RCCI combustion (Fig. 8.9b). The accurate and consistent fuel delivery systems are crucial to minimizing the cyclic variations in LTC engines. Additionally, management of the temperature and EGR sensitivity is required to take the full advantages of TLC engines [51].

8.2   Characterization of Cyclic Combustion Variations

The problem of cycle-to-cycle variations in engine operation is fundamentally a problem of variation in combustion from one cycle to the next. The cyclic variations can be characterized by the variations in the different combustion parameters. Increase in cyclic variations eventually leads to deterioration of combustion stability to unacceptable levels, which results in partial burning on some cycles, and in the extreme, misfires or weak burn cycles occur that produce no work output. In this section, indicators of cyclic variations and recognition of partial burn and misfire cycles are discussed.

8.2.1   *Indicators of Cyclic Combustion Variations*

Various parameters used as measures of combustion variability can be classified into four main categories [5, 52]: (1) parameters related to characteristics of engine cylinder pressure, (2) parameters related to characteristics of engine combustion, (3) parameters related to characteristics of flame front, and (4) parameters related to the concentration of engine exhaust gas.

The use of pressure-related parameters for cyclic variation analysis appears to be a natural choice because these parameters can easily be measured using a pressure transducer that can be usually integrated with the spark plug unit (SI engine) or glow plug unit (CI engine). The technique of using pressure-related characteristics provides a link between the flame propagation process and the thermodynamics of the SI engine cycle. Among the major pressure-related variables include (1) maximum pressure (P_{max}), (2) crank angle position corresponding to the maximum cylinder pressure $(\theta_{P_{max}})$, (3) maximum pressure rise rate $((dP/d\theta)_{max})$, (4) crank angle position corresponding to the maximum rate of pressure rise $(\theta(dP/d\theta)_{max})$, and (5) variations in the indicated mean effective pressure (IMEP).

The maximum cylinder pressure is the most widely characteristic variable for cyclic combustion variability because of the relative ease with which it is measured. The crank angle position of the maximum cylinder pressure $(\theta_{P_{max}})$ is also a useful index for determining cyclic variability. Cyclic variations in P_{max} respond to variations in the combustion phasing which is introduced by fluctuations in the initial part of combustion. It also responds to cyclic variations in peak burn rate and in fueling level or airflow (charging). This behavior suggests that care must be taken while relating variations in P_{max} to variation in the flame propagation process. The maximum pressure is typically higher with higher burn rates and vice versa. Using the location of peak pressure $(\theta_{P_{max}})$ as a measure of phasing of cyclic variation analysis is similar to using peak pressures. The relationship between the maximum in-cylinder pressure (P_{max}) and the crank angle at which it occurs (CAP_{max}) is shown in Fig. 8.10. The figure shows that the P_{max} reduces and the corresponding crank

Fig. 8.10 Variation in maximum cylinder pressure and its corresponding crank angle position at the various levels of hydrogen enrichments (α_{H_2}) at 1400 rpm in a gasoline engine [53]

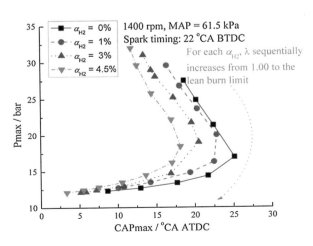

position for the $P_{\max}$ first retards and then advances with the increase of excess air ratio at a specified hydrogen enrichment (Fig. 8.10). This phenomenon illustrates that the effect of piston motion on the $P_{\max}$ is getting pronounced at extremely lean conditions due to the further prolonged combustion duration [53]. It is also found that the relationship between $P_{\max}$ and $\theta_{P_{\max}}$ is approximately linear for fast burn in the entire range, while for the slow burn case this linearity prevails only up to a certain retard. For further retarding, there is a "hook-back" region, which corresponds to a very late phasing of combustion, when the effect of expansion due to the piston movement becomes an important factor [5, 11].

The maximum pressure rise rate can be used for variation analysis as close to TDC position the piston movement is not significant and maximum pressure rise occurs close to constant volume condition. In this case, the maximum pressure rise rate would be expected to be related to burning rate and hence to flame speed. A combination of maximum pressure rise rate and its crank position in the cycle can provide some discrimination between burning rate variations and initiation period variations in much the same way that peak pressure and its location in the cycle correspond to a unique burn rate and phasing variations. The interpretation is more complex, however, and requires accurate estimation of the maximum pressure rise rate ($(dP/d\theta)_{\max}$) despite an inherent sensitivity to noise in the pressure data. Another measure of cyclic variability is the variation in the work output in each cycle or the variability in the IMEP. The coefficient of variation in IMEP (COV_{IMEP}) is a widely used parameter for determination of combustion stability. Figure 8.11 illustrates the variations in COV_{IMEP} with ignition timing at different excess air ratios and throttle positions (TP) in a spark ignition engine. The figure shows that COV_{IMEP} decreases firstly and then increases with the retard of ignition timing, and thus, the ignition timing should be set at the optimum timing with the lowest cyclic variations [54]. Figure 8.11 also demonstrates that COV_{IMEP} generally increases with the increase of excess air ratio due to the slower flame propagation speed caused by increased level of dilution mixture in-cylinder.

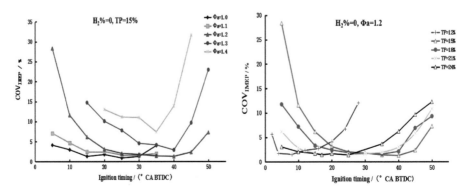

Fig. 8.11 The variations in COV_{IMEP} with ignition timing at different excess air ratios and throttle positions (TP) [54]

Practically, the combustion-related parameters are derived from the measured pressure data by using thermodynamic heat release model. The following are the methods used to quantify cycle-to-cycle variations based on the combustion characteristics: (a) maximum rate of heat release, $(dQ/d\theta)_{max}$, (b) total heat release rate, (c) combustion duration, (d) ignition delay, (e) time period in crank angle degrees from ignition to the point where a certain mass fraction is burned, and (f) maximum rate of mass burning or maximum rate of change of mass burnt fraction in the cylinder [5]. Combustion-related parameters, such as the ignition delay, the combustion duration, the total heat release, and the crank angle period from ignition to a certain heat release fraction (typically 10 and 50%), are frequently used. Ignition delay is a good parameter to indicate the extent of cyclic combustion variations in initial stages of combustion, and it is commonly accepted that both its mean value and its variance to a great extent determine the cyclic variations [11].

Typically, flame front-related parameters such as (1) flame front position, smoothed flame front area, or flame entrained volume at a specific crank angle, (2) crank angle lapse between flame front arrival and two pre-specified different locations in the cylinder, and (3) displacement of the flame kernel center from the spark gap at different crank angles are used for the analysis of cyclic variations. In exhaust gas-related parameters, generally the concentration of different components in the exhaust gases is used for the analysis [11].

8.2.2   *Recognition of Partial Burn and Misfire Cycles*

In reciprocating engines, terms such as misfire and partial burn are used to indicate an absence of combustion or weak combustion in the cylinder. Typically, engine operation with diluted mixture leads to the increase in the duration of all the stages of the combustion process, and in some of the engine cycles, there is no time to complete the combustion within the cylinder. Further increasing the mixture dilution may lead to a situation when the mixture never ignites and misfiring cycles occur [55]. A cylinder misfire occurs when the injected fuel mass (diesel engines) or the air-fuel mixture (port or direct injection gasoline engines) does not ignite, or it burns incompletely. Misfire events can occur due to several reasons such as a fault in the spark ignition or fuel injection control system, defective fuel injection and air intake systems, insufficient ignition energy, bad fuel quality, excessive EGR, low temperature, etc. [9, 56]. Besides the irregular engine running, further consequences of regular misfires are long-term increases mainly in hydrocarbon (HC) and carbon monoxide (CO) emissions. Sustained misfiring can also damage the catalytic converter in the exhaust system [57]. Misfire can also lead to a sudden engine speed decrease. Misfire cycles are undesirable since it can lead to speed and torque fluctuations [58]. The misfire detection is one of the most important aspects of onboard diagnostics (OBD). The diagnostic system must be able to detect a single misfire and also to determine the specific cylinder in which the misfire event has occurred. An alert must be generated whenever the rate of misfire exceeds a particular threshold (corresponding either to excessive catalyst temperature or increased HC emissions) [56]. Thus, partial burn and misfire can be detected by the cylinder pressure signal, and several methods are proposed for different combustion modes [59–62].

Typically, two types of constraints (misfire limit and cyclic variability limits) can be defined as combustion stability limits [2]. For a particular engine speed and fuel, the HCCI engine operation is limited by three boundaries: misfire, partial burn, and knock limit. Maximum achievable load in HCCI combustion is limited by misfire operating conditions. Figure 8.12 illustrates the normal burn, partial burn, and misfire cycles in unstable operating conditions of an HCCI engine. The figure shows that at cyclic combustion variations are very high at unstable operating conditions.

Figure 8.13 illustrates the three regions, normal, partial burn, and misfire, in the HCCI engine. The figure shows that the IMEP and heat release in misfiring cycles

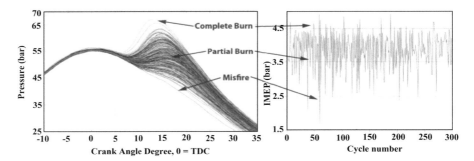

Fig. 8.12 Measured cylinder pressure and IMEP variations at an unstable operating condition in an HCCI engine for 300 consecutive cycles [63]

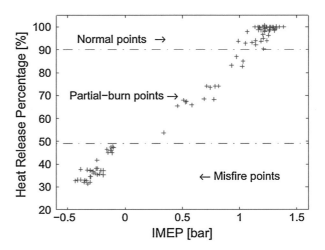

Fig. 8.13 Percentage of heat release as a function of IMEP for 120 consecutive cycles in HCCI engine [59]

are far below the well-burning cycle and heat release of the misfire cycle is less than 50% compared to a well-burning cycle [59]. Typically, when IMEP of a cycle is negative, then that particular operating cycle is considered a misfire cycle [60]. A cycle is defined as a partial burn cycle if its total heat release is reduced by 10% or more compared to a well-burning cycle [61]. A study defined partial burn and misfire operating condition in an HCCI engine based on the estimation of partial burn cycles [64]. An operating condition is defined as a partial burn operating condition if more than 20% of the cycles are partial burn cycles (heat release is reduced more than 10% from the previous cycle). Misfire operating condition is considered with more than 30% partial burn cycles at a particular engine operating condition. The COV$_{IMEP}$ is typically used to define the partial burn or misfire conditions. A method combined the COV$_{IMEP}$ and heat release method to determine the partial burn operating conditions. It is found that all the operating conditions with COV$_{IMEP}$ more than 6% and higher than 14% partial burn cycles can be defined as the partial burn operating conditions [62].

It was concluded that the operating conditions have a lower risk of misfire where the coefficient of variation (COV) of IMEP is below 3.5% in HCCI engine [2]. The value of IMEP is low at lower engine loads near idle operating conditions. In such low IMEP conditions, the stable operating limit is defined in terms of the standard deviation (STD) of IMEP instead of COV of IMEP, and the typical limit is 15 kPa of IMEP for stable combustion [65]. Another stable HCCI operating criteria is developed as the operating conditions where the standard deviation of IMEP is below 2% which correspond to the occurrence of less than one percent (<1%) partial burn cycles [61]. Figure 8.14 illustrates the stable operating criteria for stable HCCI combustion using isooctane and PRF80 fuel. The figure shows that the latest acceptable CA$_{50}$ is 373 °CA for isooctane HCCI operation. A high frequency of partial burn cycles occurs for CA$_{50}$ beyond 373 °CA. In contrast, the PRF80 fuel can tolerate more retarded combustion phasing (Fig. 8.14) while maintaining stable combustion due to low-temperature heat release (two-stage heat release fuel) [61].

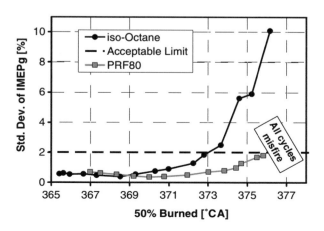

Fig. 8.14 Standard deviation of IMEP with combustion phasing in HCCI engine [61]

Similarly, in SI engines, the leaner engine operation (dilution by air, residual, or EGR) leads to an increase in flame development period and the duration of rapid combustion phase along with higher cyclic combustion variations. Leaning of the mixture can eventually lead to a condition where the engine becomes rough and unstable with the rapid increase in unburned hydrocarbon emission. These engine operating conditions are defined as a stable engine operating limit [5]. Leaner mixture operations slow down the combustion process, and it may happen that time is not sufficient for consuming the full charge by flame propagation before the exhaust valve opening. This condition is considered as partial burn. With the increase of EGR, slow burn, then partial burn, and then misfire cycles occur in SI engine. The combustion is able to complete in slow burn cycles (typically after 80° aTDC), and IMEP is low (between 85 and 46% of average value). The IMEP value is less than 46% of average IMEP in partial burn cycles [66]. Typically, misfire operating cycle has negative IMEP value. However, these values can be updated based on the current acceptable limits. A more recent study used the occurrence of partial burning cycles as burning cycle producing less than 70% of the mean IMEP output, and it starts to be significant once the COV_{IMEP} exceeds about 8% [67]. Further increase of COV_{IMEP} after 8% leads to an increase in partial burn frequency with a reasonably linear variation and with a weak dependence on engine speed and load. Misfire cycles are considered as cycles producing less than 5% of the mean IMEP output, and it starts to occur at COV_{IMEP} values of around 20%. Thus, characterization of partial burn and misfire is important for better engine performance. Several methods are used in published literature for misfire recognition including cylinder pressure sensor signal, crankshaft angular speed, ionization current or breakdown voltage, and temperature and oxygen concentration.

In modern engines, sometimes rare partial burn and misfire conditions occur at robust operating conditions also. Figure 8.15 illustrates the rare random partial burn and misfire cycle in a spray-guided spark-ignited direct injection SI engine with well-burned cycles comparable to those at spark timing (ST) at 31° bTDC. To investigate the causes of the partial burn and misfire, the spatial equivalence ratio and charge velocity near the spark plug are investigated along with spark energy and spark duration. Figure 8.16 shows the variations of spatial-average equivalence ratio

Fig. 8.15 Random partial burn and misfire in spray-guided direct injection SI engine [60]

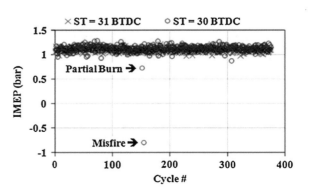

Fig. 8.16 (a) Spatial-
average equivalence ratio
and velocity and (**b**) range of
spark duration and spark
energy during well-burned,
misfire, and partial burn
cycles in spray-guided direct
injection SI engine [60]

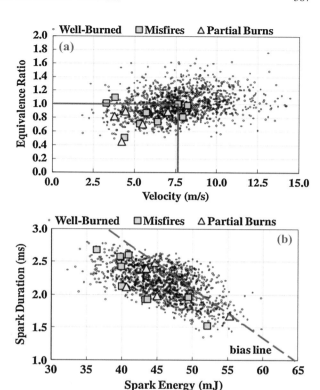

and velocity and range of spark duration and spark energy during well-burned, misfire, and partial burn cycles in spray-guided direct injection SI engine. The values for occurrence of partial burn and misfire are highlighted.

Figure 8.16a illustrates that there exists a very large cyclic variation of equivalence ratio and velocity near the spark plug at the beginning of the discharge that produces well-burning cycles. Additionally, the partial burn and misfires occur within the range of equivalence ratio and velocity for well-burned cycles but are heavily biased toward lean mixture and low velocities (less than 8 m/s). However, these metrics are not sufficient by themselves to recognize the causes of partial burn and misfire because these errant cycles occur in the same range of well-burning cycles [60]. Figure 8.16b shows that partial burn and misfires are somewhat biased toward the low-energy, short duration discharge (below the line).

Traditionally, partial burn and misfire detection are conducted by measuring the in-cylinder pressure signal analysis and calculating IMEP. The other methods based on cylinder pressure measurement are developed for determination of misfire and partial burns [55, 59, 68, 69]. The pressure angular ratio (ratio of pressure at two symmetric points before and after TDC) is used for the identification of partial burning and misfire events [55]. When the pressure angular ratio ($P(\theta)/P(-\theta)$) is greater than a pre-defined value, a partial burning event occurs and when less than unity for misfire event. This method is able to determine the misfire event more reliably than partial burn conditions. Skewness and kurtosis of in-cylinder pressure data are found to have a desirable correlation with maximum heat release rate and IMEP, and thus, it can be used for misfire detection in HCCI engine [69]. For misfire detection, the values of skewness and kurtosis are compared with the data of normal burn cycles.

Artificial neural networks (ANN) have been used for the detection of misfires in SI and HCCI engines [59, 70]. Misfiring is directly related to the maximum heat release rate in the cylinder [71]. The cylinder pressure values at different crank angle positions ($0°$, $5°$, $10°$, $15°$, $20°$ after TDC) well correlated with the maximum heat release rate (MHRR) as illustrated in Fig. 8.17. The pressure values at crank positions $0°$, $5°$, $10°$, $15°$, and $20°$ aTDC are designated as P0, P5, P10, P15, and P20.

Figure 8.17 shows that the regression line fits in the center of the data points for P10, P15, and P20 with a good correlation coefficient ($R^2 > 0.97$). For cylinder pressure at 5CAD after TDC (P5), the R^2 is less than the value of P10, P15, and P20, but for pressure at TDC position (P0), a scattering of the points is far from the regression line (Fig 8.17a). The in-cylinder pressure variation at the TDC position has significant variations (scatter in the data), and it is less sensitive to the variations of the maximum heat release rate [59]. The pressure values at these points are used as input of ANN model used for determination of misfire in HCCI engine. This model is able to well differentiate between the normal burn cycles and misfire cycles.

Understanding the dynamics of cyclic combustion variation particularly in partial burn and misfire regime helps in designing effective control of the engine. A comprehensive ignition timing metric is required for effective control of the engine, and that metric should be applicable over a wide range of engine operating conditions particularly in partial burn operating conditions [2]. An additional requirement of metric for engine control is that it needs to be computationally less expensive such as $(\theta_{P_{max}})$. Two combustion timing parameters $\theta_{P_{max}}$ and CA_{50} are compared for determination of accurate and robust HCCI combustion timing over a wide range of operating conditions (329 different engine operating conditions including both normal and partial burn) [72]. Figure 8.18 illustrates the cyclic variations of $\theta_{P_{max}}$ and CA_{50} in HCCI combustion in the particular operating condition. The figure depicts that the cyclic variation of $\theta_{P_{max}}$ is higher than CA_{50} that indicates a higher sensitivity of $\theta_{P_{max}}$. Particularly, the $\theta_{P_{max}}$ is able to register a cycle of early ignition timing with misfire occurs (cycle 44) which is not registered using CA_{50} (Fig. 8.18). This observation is further validated, and $\theta_{P_{max}}$ is found to be a good ignition timing criteria to differentiate between normal and misfire operation in HCCI engine [72, 73]. When the standard deviation of CA_{50} is greater than 2 CAD, the combustion phasing (CA_{50}) is found to be a poor measure of cyclic variations.

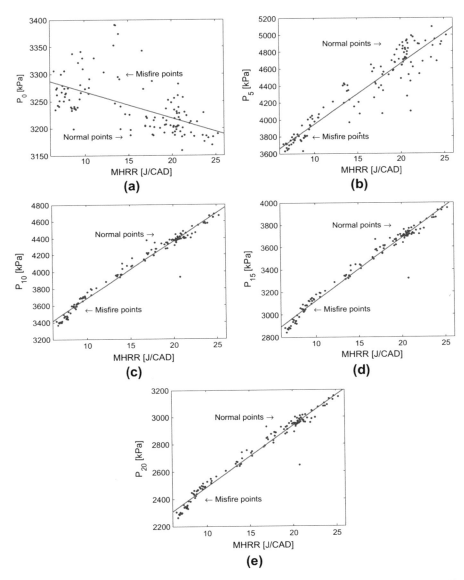

Fig. 8.17 Cylinder pressure at different crank angle positions, (**a**) 0°, (**b**) 5°, (**c**) 10°, (**d**) 15°, (**e**) 20° after TDC, as a function of the maximum heat release rate in HCCI engine [59]

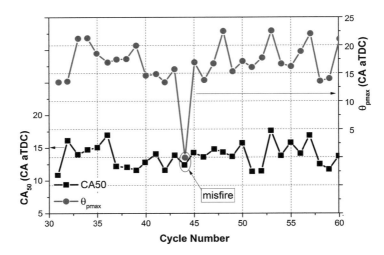

Fig. 8.18 The variation of $\theta_{P_{max}}$ and CA_{50} in a HCCI engine operating at PRF0, 1000 rpm, $T_i = 100\ °C$, $\phi = 0.57$, EGR = 0%, $T_{coolant} = 75\ °C$ (adapted from [72, 73] Courtesy of Ahmad Ghazimirsaied)

8.3 Combustion Stability Analysis by Statistical Methods

Several factors (chemical (gas composition), mixing, turbulence, and spark discharge-related) affect the cyclic combustion variation in engines (Sect. 8.1.2). Measured cylinder pressure data and subsequently calculated combustion parameters for a large number of consecutive engine cycles are used for analysis of combustion stability. Based on measured cylinder pressure data, several indicators are used for characterization of combustion stability such as peak pressure, IMEP, crank angle of peak pressure, and maximum heat release rate, different combustion phasing parameters, etc. Traditionally, statistical distributions of these indicators are used as measures of combustion stability or cyclic combustion variability. Typically, the statistical methods treat the indicators as independent random variables, which ignore the possible temporal correlations in the data. The most common methods to quantify cycle-to-cycle and cylinder-to-cylinder variability include the standard deviation (σ) of IMEP and the standard deviation of engine speed. Several statistical parameters generally used for quantification of cyclic and cylinder-to-cylinder variations are coefficient of variation (COV) of IMEP, lowest normalized value (LNV) of IMEP, root mean square (RMS) of the ΔIMEP, and IMEP imbalance [9]. These parameters can be defined using the following equations:

$$\text{Standard Deviation of IMEP}(\sigma_{\text{IMEP}}) = \sqrt{\frac{1}{N-1}\sum_{i=1}^{N}\left(\text{IMEP}_i - \overline{\text{IMEP}}\right)} \quad (8.1)$$

where i is the sample of interest and N is the number of the samples. The mean of IMEP is calculated by Eq. (8.2):

$$\overline{\text{IMEP}} = \sum_{i=1}^{N} \frac{\text{IMEP}_i}{N} \tag{8.2}$$

The COV of IMEP (COV$_{\text{IMEP}}$) is estimated using Eq. (8.3):

$$\text{COV of IMEP} = \frac{\sigma_{\text{IMEP}}}{\overline{\text{IMEP}}} \times 100 \tag{8.3}$$

For any combustion parameter, the COV can be calculated using Eq. (8.4):

$$\text{COV}(x) = \frac{\sigma}{\bar{x}} \times 100\% \tag{8.4}$$

$$\bar{x} = \sum_{i=1}^{n} \frac{x_i}{n}; \quad \text{and} \quad \sigma = \sqrt{\sum_{i=1}^{n} \frac{(x_i - \bar{x})^2}{(n-1)}} \tag{8.5}$$

The lowest normalized value (LNV) of IMEP is calculated by normalizing the lowest IMEP value in a data set by the mean IMEP value as shown in Eq. (8.6) [9]:

$$\text{LNV} = \frac{\text{IMEP}_{\min}}{\overline{\text{IMEP}}} \tag{8.6}$$

A measure of cylinder-to-cylinder variations is defined by IMEP imbalance, which is defined as Eq. (8.7). The IMEP imbalance is calculated by subtracting the average IMEP in the weakest cylinder from the average IMEP in the strongest cylinder and then normalizing by the mean IMEP [9]:

$$\text{IMEP}_{\text{imbalance}} = \frac{\overline{\text{IMEP}}_{\max} - \overline{\text{IMEP}}_{\min}}{\overline{\text{IMEP}}_{\text{engine}}} \tag{8.7}$$

The root mean square (RMS) of the ΔIMEP (i.e., the difference between highest and lowest IMEP reading in the cylinders) characterizes the difference in work performed in each cylinder event (in the firing order) and calculated by Eq. (8.8) [9]:

$$\text{RMS of } \Delta\text{IMEP} = \sqrt{\frac{\Delta\text{IMEP}^2}{n_c \cdot x - 1}} \tag{8.8}$$

where n_c is the number of cylinders and x is the number of cycles.

Commonly used statistical analysis of different combustion and performance parameters are discussed in the following subsections.

8.3.1 Time Series Analysis

The cyclic variation data series of different combustion parameters are calculated from measured cylinder pressure over a large number of consecutive engine cycles. The different types of pattern can be observed by simply plotting the data series with respect to cycle number. Figure 8.19 depicts the three different types of pattern in cycle-to-cycle variations of IMEP P_{max} and start of combustion (SOC) in an HCCI engine. The figure clearly illustrates that the cyclic combustion variations are not always an unstructured random event. Figure 8.19a illustrates the normal variation pattern, which does not follow a definite pattern and frequently occurs in the engine [45]. A periodic pattern fluctuating within two limits in combustion data series can appear (Fig. 8.19b), which can possibly due to fluctuation in equivalence ratio [45]. Figure 8.19c illustrates another type of pattern with several weak/misfired ignitions and some strong ignitions, and this engine operating condition has very large cyclic combustion variations. The weak/misfire cycles are sometimes followed

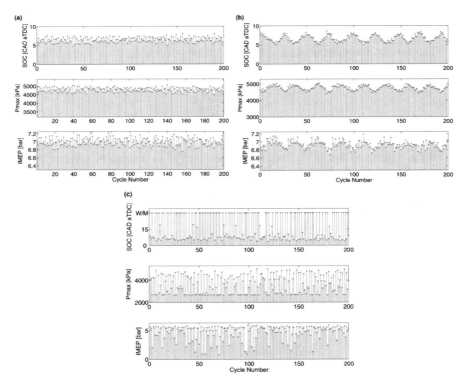

Fig. 8.19 Cycle-to-cycle variation patterns of IMEP, P_{max}, and SOC in HCCI combustion for (**a**) normal variation ($\varphi = 0.40$, $T_i = 121\,°C$, ON $= 20$, $P_i = 119$ kPa, 800 rpm); (**b**) periodic variation ($\varphi = 0.36$, $T_i = 133\,°C$, ON $= 20$, $P_i = 125$ kPa, 1003 rpm); (**c**) variations with weak (misfire) cycles ($\varphi = 0.42$, $T_i = 116\,°C$, ON $= 40$, $P_i = 124$ kPa, 907 rpm) [45]

by a strong cycle, and the values of peak pressure and IMEP are very low in weak/misfire cycles. The larger cyclic variations during these (weak/misfire pattern and periodic pattern) operating conditions need to be eliminated by engine control. However, for the normal cyclic variation pattern, the amplitude of the variation needs to be minimized to improve the engine stability.

Further analysis of time data series of different combustion parameter reveals important information regarding combustion characteristics. Variation in one parameter can be analyzed with respect to another parameter. Figure 8.20 depicts the variation of maximum pressure (P_{max}) with a crank angle corresponding to maximum pressure (CA(P_{max})), which is also sometimes referred as Matekunas diagrams [52]. The large eddy simulation (LES) and experimental results are compared for the stable and two unstable conditions (created by dilution by residual and dilution by air to produce lean mixture). The experimental results are colored by their probability of reaching their instantaneous values [74]. Three zones are identified by Matekunas in their study [52] which include (1) a linear zone where P_{max} and CA(P_{max}) vary linearly for fast burning cycles, (2) a hook-back area where P_{max} varies much more

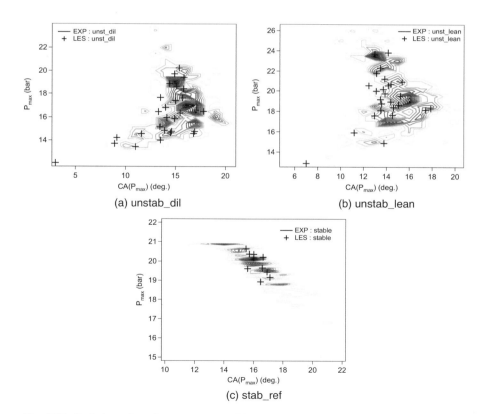

(a) unstab_dil

(b) unstab_lean

(c) stab_ref

Fig. 8.20 Variation of maximum pressure with crank angle position corresponding to maximum pressure in (**a**) unstable operating conditions by dilution with nitrogen, (**b**) unstable operating condition using lean mixture (reducing Φ), and (**c**) stable operating condition [74]

than CA(P_{max}), and (3) a return zone with only small variations of P_{max}. Figure 8.20 illustrates that the stable engine operation occurs in the linear zone in P_{max} versus CAP$_{max}$ plot. In the unstable engine operating conditions, the clear deviation from the linear region is observed for both the conditions (Fig. 8.20a, b).

The cyclic variations in stable and unstable operating condition can be explained in terms of the variations in flame initiation, flame development, and flame propagation stage of combustion. The variation in these stages of combustion leads to the cyclic variations in pressure- and corresponding combustion-related parameters.

Figure 8.21 illustrates the cyclic variations by depicting the flame surface variations at 35 crank angle degrees after ignition in six consecutive engine cycles for an unstable lean operating condition. The flame surface is visualized using an

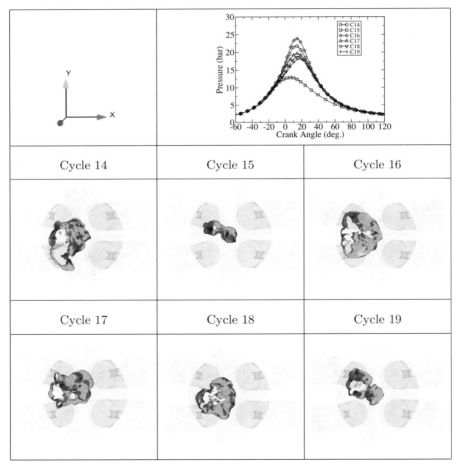

Fig. 8.21 Flame surface visualization using an iso-surface of the reaction rate (50 mol m^3 s^{-1}) colored by the velocity (blue, 0 m/s; red, 20 m/s) at 35 crank angle degrees after ignition for six consecutive cycles in an unstable lean operating condition [74]

iso-surface of the reaction rate (50 mol m^3 s^{-1}) colored by the velocity (blue, 0 m/s; red, 20 m/s). The cylinder pressure trace for respective cycles is also presented to differentiate between slow and fast burning cycles. Figure 8.21 shows typically faster propagation toward the exhaust side (negative x-direction) as a result of the tumbling flow. The cycle number 15 occupies a small volume of the cylinder and is very elongated toward the exhaust side. Another cycle (number 16) shows a more isotropic shape, which is also the fastest combustion cycle [74]. The trend variations in cylinder pressure follow the trend in the flame surface (area and velocity).

8.3.2 Frequency Distribution and Histograms

Data series of combustion parameters are typically characterized with both univariate statistics and linear temporal analysis. In the traditional statistical analysis, each measurement is assumed to be independent random events, and probability density function (PDF) of these events is characterized [75]. Typically, characteristics of PDF include skewness, kurtosis, LNV, and COV. However, these statistical parameters do not consider the temporal pattern in the data series, which might be present in the data series obtained from engine combustion.

Histogram is a graphical presentation of a time series showing the frequencies of different values in the data. Histograms typically reveal a qualitative assessment of the underlying distribution of the data series. Figure 8.22 shows the histograms of heat release data for different equivalence ratio in an SI engine. The distributions in the histogram have either a Gaussian or non-Gaussian shape. Figure 8.22 illustrates that the cyclic variations closely follow a Gaussian distribution for engine operation at the near-stoichiometric fuel-air ratio. Normally distributed data typically indicates a strong presence of independent random sources in the data [75]. It can be reasonably presumed that nearly normal distribution at stoichiometric conditions means the cyclic variations are stochastic in nature at these operating conditions. This would be possible according to the central limit theorem if the cyclic combustion variations are the result of a large number of different independent effects leading to a Gaussian distribution [76].

Relatively larger cyclic variations are observed as the fuel-air mixture becomes leaner than stoichiometric, and histogram significantly deviated from Gaussian distribution (Fig. 8.22). The non-Gaussian distribution is asymmetric with a peaked maximum and broad tail [76]. Further leaning of the charge ($\phi < 0.67$), the variations in heat release data tends to move back toward Gaussian distribution but with substantial asymmetry (Fig. 8.22g). Non-Gaussian distribution of the data indicates that there are not many dominant independent random sources in the data [75].

Three statistical parameters, namely, standard deviation (σ), skewness (S), and kurtosis (K), are used for quantitative estimation of the data distribution with respect to normal distribution [76, 77]. The standard deviation is a measure of the data spread about the mean of the distribution. Skewness quantifies the asymmetry of the data distribution about the mean and is defined as third moment about the mean of

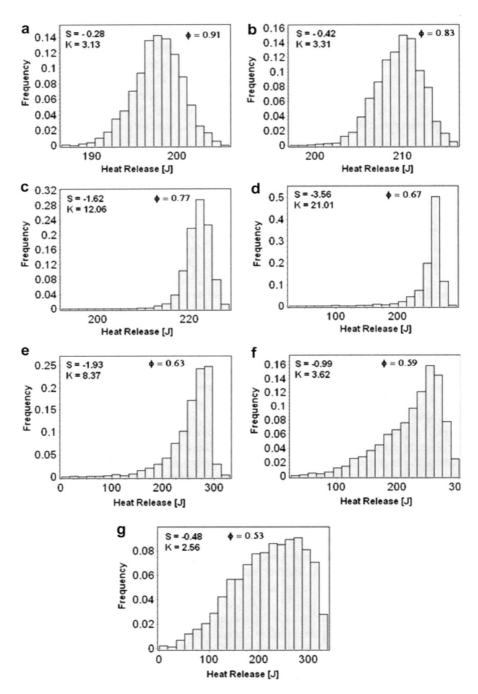

Fig. 8.22 Histograms of heat release data for different equivalence ratios (**a**) Φ=0.91, (**b**) Φ=0.83, (**c**) Φ=0.77, (**d**) Φ=0.67, (**e**) Φ=0.63, (**f**) Φ=0.59, (**g**) Φ=0.53 in an SI engine [76]

data distribution (Eq. 8.10). Variance (σ^2) is defined as the second moment about the mean (Eq. 8.9), and it also basically depicts the spread of distribution about the mean. Kurtosis characterizes the shape of distribution (flatness) around the mean with respect to normal distribution and defined by the fourth moment about the mean (Eq. 8.11).

$$\sigma^2 = \frac{\sum_{i=1}^{N} (x_i - \bar{x})^2}{N} \tag{8.9}$$

$$S = \frac{\sum_{i=1}^{N} (x_i - \bar{x})^3}{(N-1) \cdot \sigma^3} \tag{8.10}$$

$$K = \frac{\sum_{i=1}^{N} (x_i - \bar{x})^4}{(N-1) \cdot \sigma^4} \tag{8.11}$$

The values of skewness and kurtosis are usually zero for normally distributed data. Skewness values can be positive and negative. A positive value of skewness suggests the existence of right asymmetric tails longer than the left tail which means data distribution has the frequency bias above the mean value. The negative value of skewness shows the reverse trend. Kurtosis is a measure of whether the data is peaked (leptokurtic or super-Gaussian distribution) $(K > 3)$ or flat (platykurtic or sub-Gaussian distribution) $(K < 3)$ relative to a normal distribution [77]. Figure 8.22 shows the largest kurtosis and lowest skewness values at $\phi = 0.67$.

The probability distribution functions (PDF) can be further characterized by fitting the different probability distribution models in the experimental data [78, 79]. Figures 8.23 and 8.24 show the best fit probability distribution models for IMEP and combustion duration in RCCI and HCCI engines, respectively. The goodness of fit is analyzed using Kolmogorov-Smirnov test. Figure 8.23 shows that IMEP variations in RCCI engine are relatively close to a normal distribution (with certain deviation) at near TDC diesel injection timings, and deviation from normal distribution that increases at advanced diesel injection timings. Methanol/diesel RCCI combustion has a relatively more significant deviation from normal distribution than gasoline/diesel RCCI combustion. Among all distributions fitted to the experimental data, generalized extreme value (GEV) and Johnson SB distribution cover the entire range of distribution shapes observed in IMEP ensemble at different RCCI operating conditions.

The GEV distribution provides a continuous range of shapes by combining three simpler distributions (Gumbel, Frechet, and Weibull) into a single form. The probability density function for this model is based on three main parameters including a location parameter (mean), a scale parameter (standard deviation), and a shape parameter:

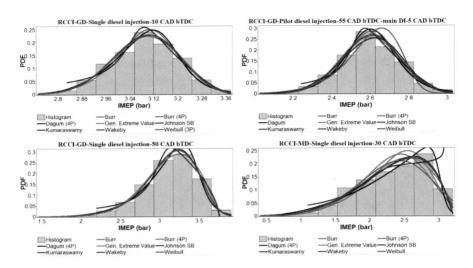

Fig. 8.23 The PDF plots for IMEP in gasoline/diesel RCCI and methanol/diesel RCCI operation at advanced diesel injection timings

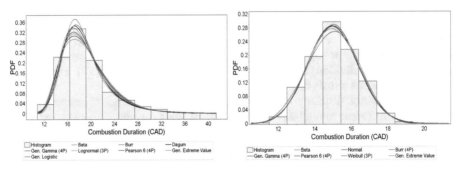

Fig. 8.24 The PDF plots for combustion duration in an HCCI engine for misfire and normal operating conditions [78]

$$
P(x) = \begin{cases} \dfrac{1}{\sigma} \exp\left[-\left[1 - k\dfrac{(x - \mu)}{\sigma}\right]^{\frac{1}{k}}\right]\left[\left[1 - k\dfrac{(x - \mu)}{\sigma}\right]^{\frac{1}{(k-1)}}\right], & k \neq 0 \\[2ex] \dfrac{1}{\sigma} \exp\left[-\exp\left(\dfrac{\mu - x}{\sigma}\right)\right]\exp\left(\dfrac{\mu - x}{\sigma}\right), & k = 0 \end{cases}
\tag{8.12}
$$

where σ is the scale parameter, μ is the location parameter, and k is the shape parameter. This distribution can be used to model the cyclic variation in RCCI engine.

Similarly, GEV distribution is also the best fitting distribution in the range of distribution shapes in HCCI combustion duration data series at different HCCI operating conditions (Fig. 8.24) [78].

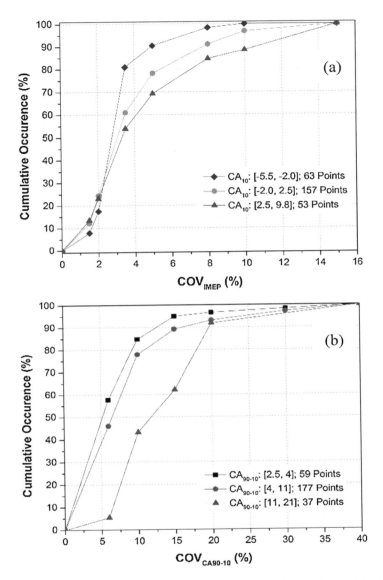

Fig. 8.25 Cumulative occurrence for (**a**) COV$_{IMEP}$ for different CA$_{10}$ positions and (**b**) COV of combustion duration for different CA$_{90-10}$ [78]

In case of large data set, plotting of cumulative percentage occurrence of the coefficient of variations with a certain parameter can reveal some useful global information regarding the cyclic variability. Figure 8.25 illustrates the cumulative percentage of occurrence for COV$_{IMEP}$ for a different range of CA$_{10}$ (crank angle position corresponding to 10% heat release) and COV of combustion duration for different combustion duration for 273 engine operating conditions in an HCCI engine by varying fuel, engine speed, inlet temperature, and air-fuel ratio. The occurrence percentage is calculated by counting the number of engine operating conditions out of all the test conditions with a particular interval of CA_{10}, or combustion duration has the same range of cyclic variations.

Figure 8.25a depicts that advanced CA_{10} operating conditions have lower cyclic variations in IMEP in comparison to the retarded CA_{10} positions. This observation is confirmed by the fact that 80% operating conditions having CA_{10} in the range of -5.5 and -2 CAD show the COV_{IMEP} less than 3.5% and 60% operating conditions with CA_{10} in the range of -2 and 2.5 CAD show COV_{IMEP} lower than 3.5% (Fig. 8.25a). The COV_{IMEP} less than 3.5% is considered as an acceptable range. Similarly, Fig. 8.25b shows the cumulative occurrence for COV of combustion duration, and shorter combustion duration seems to have lower cyclic variations in combustion duration. The longer combustion duration is the lowest curve showing lower cumulative occurrences at a particular variation (Fig. 8.25b).

8.3.3   Normal Distribution Analysis

To describe the patterns of cycle-to-cycle variations in reciprocating engines, the data series combustion parameters (IMEP, combustion phasing, combustion duration, etc.) using a large number of consecutive engine cycles are calculated and further analyzed. This ensemble of combustion parameters shows different types of distribution shapes depending on engine operating conditions. Knowledge of the distributions provides valuable information to be able to find high cyclic variability regions of engines. The time series of any combustion parameter at a particular engine operating condition can be used to form a probability distribution. For statistical analysis, the normal distribution is the most commonly used probability distribution function. The normal probability plot is a good graphical tool to test whether or not a data series follows a normal distribution. Experimental data points are plotted against a theoretical normal distribution. The data series have normal distribution that closely follow the theoretical normal distribution line, and the level of departures from normality is judged by how far the points vary from the straight line [2, 80]. Figure 8.26 shows the normal probability plots of IMEP for gasoline/ diesel RCCI and methanol/diesel RCCI operation at very advanced diesel injection timings. A large deviation from the normal distribution is found for the very advanced diesel injection timings for both gasoline and methanol RCCI (Fig. 8.26). However, relatively close to a normal distribution is observed for close to TDC diesel injection timings [79].

Similarly, Fig. 8.27 depicts the normal probability graphs for combustion duration time series at normal stable, misfire, and knocking conditions in an HCCI engine. A large deviation from the normal distribution is observed in combustion duration distribution during knocking (RI $=$ 12.9 MW/m^2) and misfire range ($COV_{IMEP} =$ 12.55%) operating conditions in HCCI engine (Fig. 8.27). During

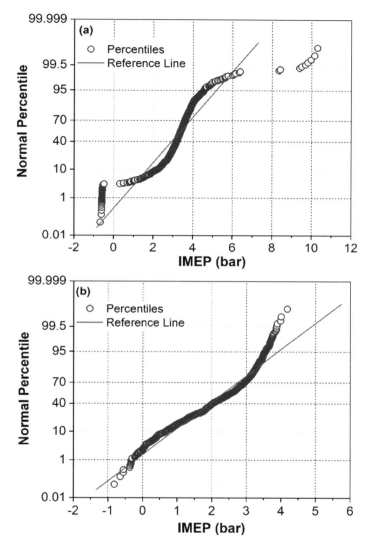

Fig. 8.26 Normal probability plots of IMEP for gasoline/diesel RCCI and methanol/diesel RCCI operation at very advanced diesel injection timings. (**a**) RCCI-GD-single diesel injection—60 CAD bTDC. (**b**) RCCI-MD-single diesel injection—40 CAD bTDC

knocking operating condition, a stretched C-shape distribution and in misfire operating condition a V-shape distribution of combustion duration are observed. However, in the stable HCCI engine operating condition, close to a normal distribution is found (Fig. 8.27c). As discussed in Sect. 8.3.2, the normal distribution conditions may have a strong presence of independent random sources in the data. Thus, cyclic variations are stochastic in nature. The deviation from the normal distribution depicts

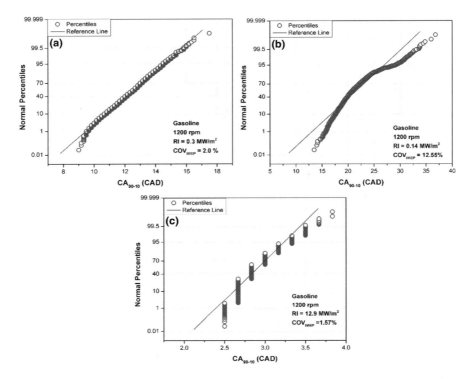

Fig. 8.27 Normal probability plots using time series of burn duration (**a**) normal stable, (**b**) misfire, and (**c**) knocking engine operating conditions in the HCCI combustion [78]

dependency of sources in the data, or some deterministic pattern is present in the data. Thus, HCCI operation in misfire or knocking condition or RCCI combustion with advanced diesel injection timings (Figs. 8.26 and 8.27) has some dependent sources of variation, and not many dominant independent random sources in the data.

8.3.4   Coefficient of Variability and Standard Deviation

Standard deviation and coefficient of variability (COV) are statistical parameters used for analysis of combustion variability, and their values are calculated using Eqs. (8.4) and (8.5). Typically, the standard deviation and COV of IMEP are used as a measure of the statistical instability of combustion. It was demonstrated that the COV_{IMEP} values greater than 10% in SI engine leads to drivability issues [5] and more stringent values are suggested for idle operating conditions [81]. However, the acceptable COV_{IMEP} for advanced premixed combustion mode is typically 3.5%

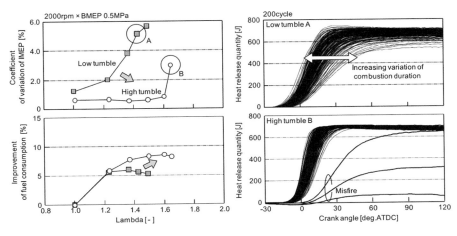

Fig. 8.28 The effect of tumble flow on heat release, combustion stability, and fuel economy [82]

[2]. The COV and standard deviation of IMEP are measures of the roughness or unsteadiness of engine combustion [81]. The fast burn engine cycles typically have acceptable levels of COV and standard deviation of IMEP. Burning rate depends on the several engine operating parameters such as equivalence ratio, turbulences, overall flow pattern, etc.

Figure 8.28 demonstrates the effect of flow field (high and low tumble) on the COV_{IMEP}, dilution limit, heat release, and fuel economy of SI engine. The improvement of fuel consumption is constrained at excessive dilution by reducing the laminar burning speed, which finally reaches to a dilution limit where flame propagation fails. Cylinder flow structures (such as tumble) are used to intensify the turbulence intensity and enhance the flame propagation. Figure 8.28 shows that the high tumble has stable combustion (lower COV_{IMEP}) and improves the fuel economy. High tumble flow increases the dilution limit in comparison to low tumble by stabilizing the combustion. However, further increase of dilution may lead to misfire in some of the cycles (Fig. 8.28).

Typically, the COV of maximum pressure (P_{max}) and IMEP is used to characterize the cyclic variations in the combustion. Figure 8.29 shows the $COV_{P_{max}}$ and COV_{IMEP} with excess air ratio at different hydrogen enrichment fractions. The figure shows that the variations of $COV_{P_{max}}$ and COV_{IMEP} with excess air ratio have different trends.

The $COV_{P_{max}}$ first increases with excess air ratio and reaches to a peak and then starts decreasing at particular hydrogen enrichment condition. However, the COV_{IMEP} always increases with excess air ratio (Fig. 8.29). This interesting trend can be explained by the fact that the COV_{IMEP} is affected by the entire combustion process and $COV_{P_{max}}$ is only symbolized by the maximum cylinder pressure. The combustion rate decreases with an increase in excess air ratio, and prolonged combustion duration increases the cyclic combustion variations. Thus, COV_{IMEP} distinctly increases with leaner mixtures. Comparatively, the maximum cylinder

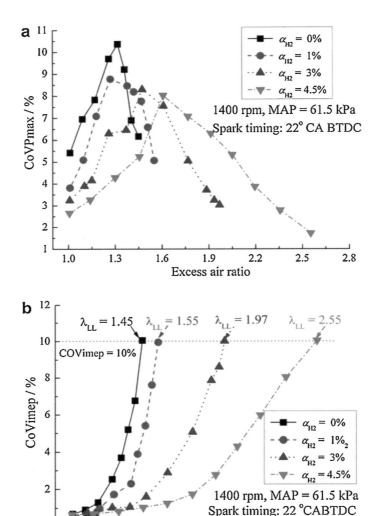

Fig. 8.29 The variation of (**a**) $COV_{P_{max}}$ and (**b**) COV_{IMEP} with excess air ratio for different hydrogen enrichment (α_{H_2}) at 1400 rpm and a MAP of 61.5 kPa in SI engine [53]

pressure is influenced by both combustion and piston motion [53]. Due to different trends in variations of $COV_{P_{max}}$ and COV_{IMEP}, the combustion stability limits should be decided carefully looking at the operating conditions.

The values of COV for the stable operating condition can also depend on a parameter selected for the analysis. Figure 8.30 illustrates the variations of COV

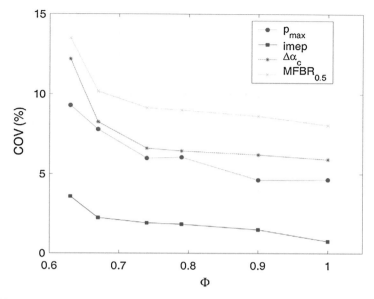

Fig. 8.30 The variations of COV with equivalence ratios for different combustion parameters in SI engine [83]

for four different combustion parameters (P_{max}, IMEP, mass fraction burning rate (MFBR), and combustion duration ($\Delta\alpha_c$)) in SI engine at 1500 rpm. The figure clearly illustrates that the COV values of all the parameters increase with leaner mixture. The trend of $COV_{P_{max}}$ is different from Fig. 8.29 due to the selected range of equivalence ratio of engine operation for the study. The COV of the mass fraction burning rate (MFBR) is calculated for 50% mass fraction burned, and it has the highest cyclic variation relative to other three parameters. The mass fraction burning rate is related to the combustion velocity, and it is calculated by Eq. (8.13) [83]:

$$\text{MFBR}(\alpha) = \frac{d(\text{MFB}(\alpha))}{d\alpha} = \frac{\dot{m}_b}{m \cdot \omega} = \frac{\rho_u(\alpha)A_f(\alpha)}{m \cdot \omega}S_c(\alpha) \qquad (8.13)$$

where ρ_u is the unburned density, A_f is the flame front surface, S_c is the turbulent combustion velocity, m is the total mass, and ω is the engine speed.

Using Eq. (8.13), the turbulent fluctuation of the combustion speed (S_c) can be determined from the variations in MFBR. Figure 8.31 illustrates the variations in the standard deviation in MFBR at different engine speed and equivalence ratio. The figure illustrates that the equivalence ratio has almost no effect (very weak dependence) on the variation of MFBR, but significant differences with variation in engine speed for a particular mass fraction burned value. The variation in MFBR is a function of mass fraction burned (MBF), and it reaches to maxima in the range of 30–60% of MFB of the charge (Fig. 8.31).

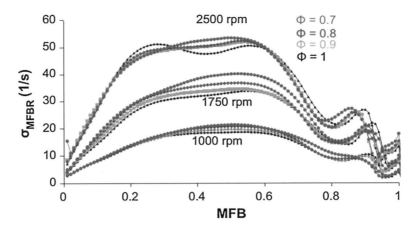

Fig. 8.31 Variations in the standard deviation of MFBR with mass fraction burned for different engine speeds and equivalence ratio [83]

The standard deviation of MFB is also dependent on the ignition system. Figure 8.32 shows the variations in combustion phasing (5% MFB position) with transistor coil ignition (TCI), advanced spark ignition (ASI), and corona ignition with combustion duration (the difference between 5 and 50% MBF). The main difference between conventional TCI and ASI system is that ASI has longer applicable spark duration and it can provide sufficient current level for the longer duration [84]. The ASI system has relatively higher dilution tolerance than a conventional system. The TCI and ASI systems become unstable for the longer main duration at 12 °CA and 14 °CA, respectively. However, with corona ignition system, stable region lasts for even more longer duration, and almost no scattering is observed in combustion phasing (Fig. 8.32). Thus, the ignition system also has the significant effect on the combustion of spark ignition, and it can be effectively observed by statistical parameters.

The cyclic variations in IMEP also depend on the combustion phasing (CA_{50}) and combustion duration (CA_{90-10}). Important trend can be extracted by observing simultaneously. Figure 8.33 illustrates the variations of COV_{IMEP} as a function of combustion phasing and combustion duration in a HCCI engine employing gasoline and ethanol. Both fuels have higher cyclic variations of IMEP for longer combustion duration and late combustion phasing (Fig. 8.33). The contour lines represent the constant variation lines. Contour lines of COV_{IMEP} are almost horizontally inclined for ethanol, which suggest that cyclic variation in IMEP has more dependency on combustion duration than combustion phasing (Fig. 8.33). However, in case of gasoline HCCI combustion, the variation depends on both combustion phasing and duration.

In modern direct injection spark ignition, the injection and ignition timings, engine load and speed, compression ratio, and injector configuration significantly affect cyclic combustion variations. Since the COV_{IMEP} characterizes the engine

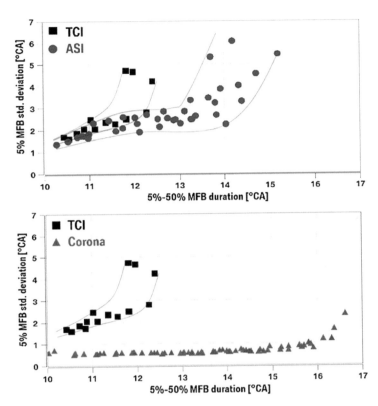

Fig. 8.32 Variations in combustion phasing with transistor coil ignition (TCI), advanced spark ignition (ASI), and corona ignition with combustion duration [84]

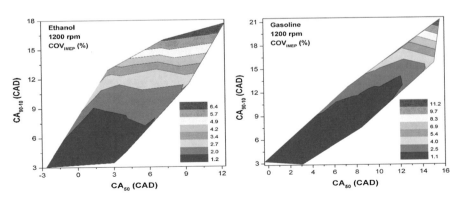

Fig. 8.33 The COV$_{IMEP}$ as a function of combustion phasing and combustion duration in an HCCI engine using gasoline and ethanol [78]

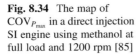

Fig. 8.34 The map of $COV_{P_{max}}$ in a direct injection SI engine using methanol at full load and 1200 rpm [85]

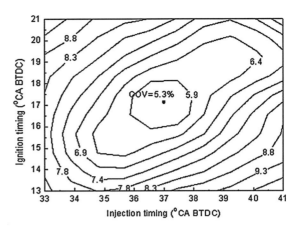

roughness, it can be used to optimize the injection and ignition timings. Figure 8.34 shows the variation of COV with injection timing and spark timing in a direct injection SI engine using methanol at 1200 rpm. The figure depicts that the COV value is minimum in the middle of the map (at an injection timing of 37° bTDC and an ignition timing of 17° bTDC) and the COV increases with deviation in the ignition or injecting timing corresponding to the minimum value. This is possibly due to the fact that the mixture distribution in the cylinder is ideal and the flame propagation is the fastest at an injection timing of 37° bTDC and an ignition timing of 17° bTDC [85].

Combustion stability can be quantified in terms of variation in engine speed. Weak combustion events are considered to be one of the primary causes for poor combustion stability. Severe weak combustion events can lead to misfire, in which there will be no work done by the piston. This will act as an impulse load to the structure of the vehicle which starts to vibrate on its natural frequency giving an unpleasant feeling to passengers and the driver [8]. Figure 8.35 illustrates the variation of idle engine speed with ignition timing and air-fuel ratio. The ignition timing is varied from 0° (TDC) up to an advance of 16° bTDC. Higher idle speed shows higher engine stability with variation in both ignition timing and an air-fuel ratio (Fig. 8.35). Fluctuations in speed are higher at lower idle engine speed. Ignition timing close to TDC and air-fuel ratio close to stoichiometric are found to be more stable. Advancing the ignition timing leads to misfire and, thus, increases in COV of idle speed [8].

8.3.5   Lowest Normalized Value

The lowest normalized value (LNV) is defined as the lowest single IMEP in the total cycles divided by the mean IMEP (Eq. 8.6). LNV can predict the misfire or partial burning tendency of the engine. The concept of LNV in judging the combustion

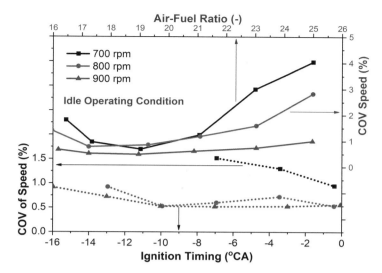

Fig. 8.35 Variation of COV of engine speed with ignition timing and air-fuel ratio for different idle speed (adapted from [8])

stability found that the minimum permissible value of LNV is around 75% [81]. To achieve acceptable levels of LNV, typically "fast burn" engines are required. The LNV is of great importance for transient engine operation because it reveals better the driver feeling during acceleration. In transient engine operation, increased tolerance of the mixture in the combustion chamber can be witnessed because the gas exchange flow varies from stationary conditions. Temperature and pressure relevant to the gas exchange are away from their equilibrium state. This can lead to EGR rate overshoots; nevertheless a continuous evolution of the torque is expected [86]. Typically the LNV value should drop below 80%. Figure 8.36 shows the typical variations of LNV and COV with an increase in EGR at different ignition energy and ignition system. The TCI (transistor coil ignition) is still dominating ignition system of a passenger car in spite of the tremendous variety of ignition concepts and their competition. Figure 8.36 compares the TCI system with DCO (dual coil offset) and CEI (controlled electronic ignition) systems at a different spark energy. The CEI and DCO systems facilitate shifting the combustion stability limit by around $\Delta EGR = 5\%$ applying a COV limit of 5%, but benefit vanishes compared to the 80 mJ case if the full engine COV limit of 3% is selected (Fig. 8.36).

8.3.6   *Autocorrelation and Cross-Correlation*

Statistical parameters such as skewness, kurtosis, LNV, and COV do not consider the temporal variation in the data. The autocorrelation and power spectrum functions

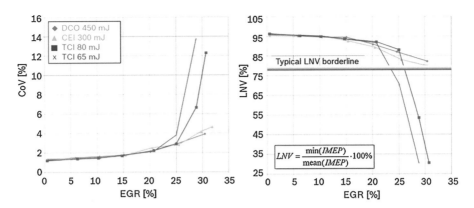

Fig. 8.36 Variation of COV and LNV with external EGR in a SI engine at 1500 rpm [86]

can reveal temporal structure effectively in linear systems, but the complete description is not revealed when underlying dynamics are inherently nonlinear [75]. The autocorrelation is defined as the amount of linear correlation a time series has with itself. Serial coupling between elements of a data series is quantified by the autocorrelation function (ACF), which can be calculated by Eq. (8.14):

$$\text{ACF}(\tau) = \left(\sum_{i=1}^{N-\tau} (x_i - \bar{x})(x_{i+\tau} - \bar{x}) \right) \cdot \left(\sum_{i=1}^{N} (x_i - \bar{x})^2 \right)^{-1} \tag{8.14}$$

where τ is lag and N is the number of elements in the data series.

Non-zero values of autocorrelation suggest the degree of serial coupling in time. Some important features of a time series (e.g., power spectrum, stationarity, or decorrelation time) can be extracted by the autocorrelation function. The autocorrelation function is analogous to the power spectrum, which can be employed to observe oscillations in a time series [75]. According to the Wiener-Khinchin theorem, the power spectrum of a data series can be obtained by applying the Fourier transform to the autocorrelation function [87]. When autocorrelation of a data series exponentially decays toward zero as the lag approaches infinity, the data series is most likely to be stationary. A system or process is said to be stationary if its statistical and dynamical properties remain constant over time. A nonstationary or periodic time series has a non-zero autocorrelation value at very large lag times [75, 88]. The decorrelation time is defined as the first zero crossing (or minimum in the case of a nonstationary time series) of the autocorrelation function [75, 89]. Especially this feature of the autocorrelation function is helpful in highlighting the "memory" effect between cycles of combustion data.

Figure 8.37 shows the autocorrelation of heat release data series while transitioning from stable combustion to increased levels of internal EGR. An element in a time series will correlate perfectly with itself, and thus, the correlation

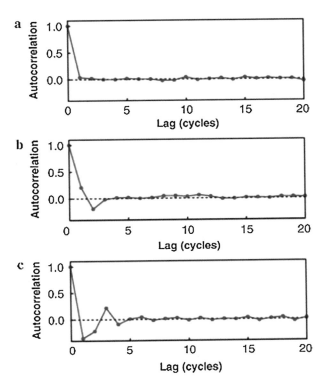

Fig. 8.37 Autocorrelation of heat release data series while transitioning from (**a**) stable combustion to (**b**) and (**c**) increased levels of internal EGR [90]

coefficient at lag zero is one (Fig. 8.37). Figure 8.37 shows the persistent anticorrelated oscillations start developing between successive cycles when EGR is increased from the point of stable combustion. This oscillating trend depicts a clearly deterministic "memory" between cycles [90]. The anticorrelation occurs due to the small variation in the degree of dilution of the inlet charge by residual gases from prior cycles. This dilution by residual can lead to either better or worse SI in the following cycles. Additionally, other factors such as pressure fluctuations in the intake and exhaust manifolds in the injection system can also contribute to the correlation of heat release.

Figure 8.38 shows the autocorrelation in the IMEP data series at different engine speeds in a diesel engine. The figure shows that the autocorrelation reduces at a different rate depending on the engine speed. The IMEP data series has a relatively low and high degree of autocorrelation between adjacent and near-adjacent data at 1200 rpm and 2000 rpm, respectively. The autocorrelation undergoes fast modulation at the intermediate speed of 1600 rpm (Fig. 8.38). The slow decay of the autocorrelation curve suggests the presence of nonstationarity in the IMEP data [91].

The cross-correlation function reveals information very similar to the autocorrelation function but between two data series. Cross-correlation is used to quantify the temporal coupling between two variables instead of one which is used in autocorrelation. The cross-correlation can be calculated using Eq. (8.15) [75]:

Fig. 8.38 Autocorrelations in the IMEP for different engine speeds in a diesel engine [91]

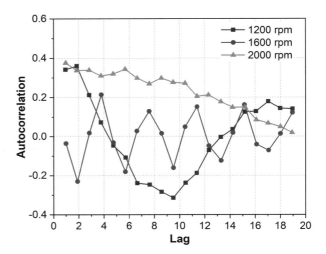

$$\mathrm{CCF}(\tau) = \left(\sum_{i=1}^{N-\tau} (x_i - \bar{x})(y_{i+\tau} - \bar{y}) \right) \cdot \left(\sum_{i=1}^{N} (x_i - \bar{x})(y_i - \bar{y}) \right)^{-1} \quad (8.15)$$

where CCF (τ) is the cross-correlation function at lag τ, $\bar{x}$ and $\bar{y}$ is the average of the first and second data series, and N is the number of samples in the data series.

8.3.7 Principal Component Analysis

To further analyze the combustion variations and possible sources, statistical methods such as principal component analysis, factor analysis, and independent component analysis are typically used [37, 92]. The principal component analysis uses an orthogonal transformation to convert a set of observations of possibly correlated variables into a set of values of linearly uncorrelated variables called principal components. Assuming X is a dataset with the row is the variable, the column is the observations. Its transpose X^{T} is an $n \times m$ matrix. The singular value decomposition of X defined in Eq. (8.16) is a principal component analysis of these variables:

$$X = W\Sigma V^{\mathrm{T}} \quad (8.16)$$

where W is an $n \times m$ matrix of eigenvectors of the covariance matrix XX^{T}; the nonnegative real entries on the diagonal of matrix Σ are the eigenvalues; and V is an $n \times n$ matrix of eigenvectors of $X^{\mathrm{T}}X$. Each eigenvector represents a variation pattern in X. The ratio of its eigenvalue over the sum of all eigenvalues gives the important information of this variation pattern [37].

Independent component analysis (ICA) is applied to cycle-resolved images of luminosity in SI engine combustion [92]. The independent components related to the underlying phenomena of the combustion process are identified using ICA. The components (and corresponding coefficients) are used to characterize the morphological evolution of the luminous combustion during a particular cycle and over a number of cycles. The three components identified from the images are representative of ignition and radial-like flame propagation (first component) and erratic luminous combustion (second and third component) occurring subsequently [92].

8.4 Combustion Stability Analysis Using Wavelets

Cyclic combustion variations are typically quantified by using statistical and chaotic methods [78, 93–96]. These methods are used for determination of variation patterns and their possible correlation as well as deterministic contents. Conventional statistical approaches normally use the coefficient of variation (COV) to measure the cycle-to-cycle variations in combustion parameters such as IMEP, peak cylinder pressure (P_{max}), heat release, etc. The key limitation of traditional statistical approaches is that they only provide the temporal variations present in the data series. Traditional statistical methods are unable to consider the spectral characteristics (frequency domain) of the data. Typically, the frequency content of the data is analyzed using Fourier transform. Constituting frequencies of the signal is revealed (and calculated) by the Fourier transform. Inverse Fourier transform is typically applied to back convert the signal from frequency domain to the time domain. Using Fourier transform, measured signal can be represented in terms of sine and cosine functions. Information regarding the frequency content of a signal is generated by Fourier transform. However, Fourier transform is unable to provide the information on variation of frequency with time if the frequency content of the signal is changing with time. Fourier transform of a transient or rapidly varying signal can provide the information regarding the frequencies present in the signal, but it cannot reveal any facts about the time of appearance of these frequencies as well as the duration of occurrences of frequencies [97]. Fourier analysis is sufficient if the signal to be analyzed is stationary and if the time period is accurately known. However, Fourier analysis may not be appropriate if the signal has nonstationary characteristics such as drifts and frequency trends [98]. To overcome the disadvantages of Fourier transform, short time Fourier transform (STFT) or windowed Fourier transform (WFT) is proposed to get the information on the frequencies present in the signal at different time locations. The STFT or WFT divides the original signal into smaller segment signal of equal time length, which subsequently apply the Fourier transform on each smaller segments of original signal. For each shorter segment, Fourier spectrum is calculated and presented as a function of time [99]. The sampling of signals may produce leakages, which are also denoted as aliases. Sampling (for creating segments for STFT) may lead to aliasing, which makes different signals to become indistinguishable. The STFT uses a fixed window size leading to fixed frequency and the time resolution, which results in poor temporal or spectral resolution. Aliasing of low- and high-frequency constituents, which does not occur in the frequency range of window, may lead to inaccuracies [100]. The fixed size of the window is the main drawback of STFT.

The wavelet transform is introduced to overcome the difficulties of Fourier transform. Fourier transform does not represent the abrupt changes and functions are not localized in space and time. This limits their applications for signals with slowly changing and transient fluctuating trends. Wavelet analysis eliminates the difficulties related to STFT by using adaptive usage of long windows for retrieving low-frequency information and short windows for high-frequency information. The ability to perform the flexible localized analysis is one of the main features of the wavelet transform. Wavelet analysis is successfully used in characterizing the cyclic combustion variations in reciprocating engines [79, 101–103]. Wavelets are used to determine the amplitude as well as periodicities of cycle-to-cycle variations in combustion engines because wavelet transform offers good spectral and temporal resolution.

A wavelet function is defined as rapidly decaying small oscillation or wave. A function with zero mean and finite energy can be characterized as a wavelet, and the admissibility condition of a function $\psi(t)$ to be considered as the wavelet is shown in Eq. (8.17):

$$\int_{-\infty}^{\infty} \psi(t)dt = 0; \quad \text{and} \quad \int_{-\infty}^{\infty} |\psi(t)|^2 dt < \infty \qquad (8.17)$$

Wavelet transform is fundamentally an integral transform. The basis functions which are localized in both frequency as well as time domains are used in wavelet transform. It decomposes the original signal into frequency bands (or scales) at various resolutions by scaling the basis functions. The original signal is projected on a set of basis functions called mother wavelets [104]. The scale and translation parameters make the difference between various wavelet functions. Originally, Morlet thought the wavelets as a family of functions generated from translations and dilations of a single function (known as mother wavelet) [105]. Daughter wavelet in terms of mother wavelet is presented by Eq. (8.18) [104]:

$$\psi_{a,b}(t) = \frac{1}{\sqrt{|a|}} \psi\left(\frac{t-b}{a}\right), \quad a,b \in R, \quad a \neq 0 \qquad (8.18)$$

where $\psi(t)$ is the mother wavelet with unit energy, $\psi_{a,b}(t)$ is daughter wavelet created by mother wavelet ($\psi(t)$) by scaling and translating, and parameters "a" and "b" are the scaling (or dilation) and translating factors, respectively. Compressed or compacted version of the mother wavelet is created when $|a| > 1$, while enlarged or widened version of the mother wavelet is created when scaling factor is less than one ($|a| < 1$). Stretched wavelet or small scales are used for slowly changing signal, and compressed wavelet is used for abrupt changes in signal. The translating parameter "*b*" governs the time location of the wavelet, and wavelet is shifted depending upon the sign of translating parameter. The shifting of the function on the real axis is termed as translating the function. The factor $\frac{1}{\sqrt{|a|}}$ is used, so that the function $\psi_{a,b}(t)$ has the same energy for all scale "*a*." Wavelets can vary their time width by varying the scale "*a*," which means wavelets are adaptive of their frequencies. Scaling parameter allows the wavelets to fine-tune the width at higher frequencies in the signal and increase the width while focusing on smaller frequencies, similar to a zoom lens [99]. Wavelet transform can have combinations of time-frequency representations with different resolutions of the same signal. Thus, wavelet transform is preferably used over Fourier transform.

The continuous wavelet transform (CWT) with respect to a wavelet $\psi(t)$ is given by Eq. (8.19) [99]:

$$CWT(a,b) = \frac{1}{\sqrt{|a|}} \int_{-\infty}^{\infty} x(t)\psi^*\left(\frac{t-b}{a}\right), \quad a,b \in R, \quad a \neq 0 \tag{8.19}$$

where $x(t)$ is continuous signal, $\psi(t)$ is a mother wavelet with unit energy, and ψ^* indicates its conjugate. Equation (8.19) can be rewritten in terms of a daughter wavelet as in Eq. (8.20):

$$CWT(a,b) = \int_{-\infty}^{\infty} x(t)\psi_{a,b}^*(t) \tag{8.20}$$

Practical applications such as cyclic variations in engines involve a discrete time series signal. Continuous wavelet transform (CWT) [106] on a discrete time series x_n is represented in Eq. (8.21):

$$CWT_n(s) = \left(\frac{\delta t}{a}\right)^{\frac{1}{2}} \sum_{n'=0}^{N-1} x_{n'}\psi^*\left[\frac{(n'-n)\delta t}{a}\right] \tag{8.21}$$

where $x_{n'}$ is the discrete time series, "*a*" is the scaling parameter, $\psi(t)$ is the wavelet function, ψ^* is its conjugate, and "*n*" is the localized time index. From Eq. (8.21) "*N*" different signals are achieved that are combined using convolution to get a single continuous wavelet transform [100]. To approximate the continuous wavelet transform, the convolution (8.21) should be done "*N*" times for each scale, where "*N*" is the number of points in the time series. "*N*" simultaneous convolutions performed using discrete Fourier transform in Fourier space [100]. The discrete Fourier transform is used on the time series $x_{n'}$ which is shown in Eq. (8.22):

$$\hat{x}_k = \frac{1}{N} \sum_{n=0}^{N-1} x_n e^{-2\pi ikn/N} \tag{8.22}$$

where k ranges from 0, 1, 2, ..., $N-1$. Using Eq. (8.22), Eq. (8.21) can be represented as Eq. (8.23):

$$\text{CWT}_n(a) = \left(\frac{\delta t}{a}\right)^{\frac{1}{2}} \sum_{k=0}^{N-1} \hat{x}_k \hat{\psi}^*(s\omega_k) e^{i\omega_k n\delta t} \qquad (8.23)$$

where the angular frequency ω_k is given by Eq. (8.24) [100]:

$$\omega_k = \begin{cases} \dfrac{2\pi k}{N\delta t} & k \le \dfrac{N}{2} \\[2mm] -\dfrac{2\pi k}{N\delta t} & k > \dfrac{N}{2} \end{cases} \qquad (8.24)$$

The CWT at all the n time indices can be calculated using Eqs. (8.22)–(8.24).

Mexican hat wavelet and the Morlet wavelet more often used continuous wavelets. The mathematical representation of the Morlet wavelet [100, 102] is represented in Eq. (8.25):

$$\psi(\eta) = \pi^{-1/4} e^{i\omega_0 \eta} e^{-\eta^2/2} \qquad (8.25)$$

where ω_0 is chosen as 6 to satisfy the admissibility condition and $\pi^{-1/4}$ is a normalizing factor. The chosen value of ω_0 allows to obtain good time and frequency localization, and in this case Fourier period and the scale are equal. Similar value of ω_0 was also chosen for yielding good results in reference [102].

The wavelet power spectrum (WPS) reveals the information about the fluctuations of variances at different scales or frequencies. The magnitude of signal energy at a particular scale "a" and certain position "n" is computed by the squared modulus of CWT. This is denoted as WPS and is also presented as scalogram. The WPS is normalized by dividing with σ^2 such that the power relative to white noise is achieved. The WPS is calculated by Eq. (8.26):

$$\text{WPS} = |\text{CWT}_n(a)|^2 \qquad (8.26)$$

The normalized WPS is depicted in Eq. (8.27):

$$\text{WPS}_n = \frac{|\text{CWT}_n(a)|^2}{\sigma^2} \qquad (8.27)$$

where σ is the standard deviation. The continuous wavelet transform is typically a complex function (real and imaginary part), and thus, modulus would actually mean the amplitude of continuous wavelet transform. The WPS depends on the time and scale (frequency) represented by a surface. The contours of the surface can be plotted on a plane to obtain a time scale representation of WPS. Through WPS important information can be obtained such as events with higher variances and the frequency of their occurrences and the duration of time for which they persist. This information can be used to modify and control the system. The wavelet power spectrum is the distribution of energy within the data, so by observing the WPS regions of large power can be identified, which will provide a better understanding as to the features that are important in the signal. As wavelet has a changing window size in comparison to that of a fast Fourier transform, the variations in frequency of occurrence can be visualized in a WPS. The WPS is contour plot, which has the cycles number (data series) on the x-axis and the Fourier period on the y-axis, and the intensity of the variations in the data series represented in the contour plot. The stronger color in the contour plot indicates a higher variation of the parameter (Fig. 8.39). The period

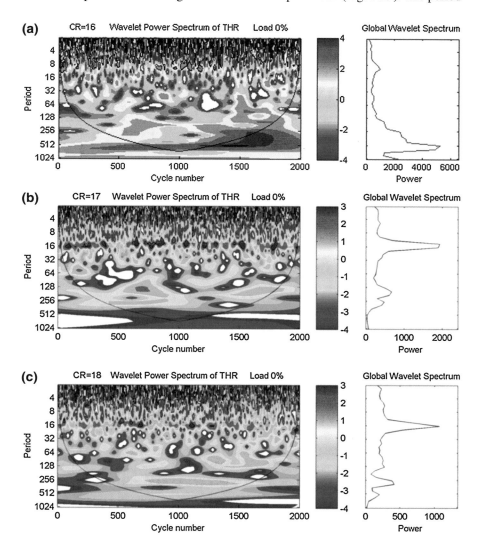

Fig. 8.39 WPS and GWS of THR at compression ratios (**a**) 16, (**b**) 17, and (**c**) 18, respectively, at no load conditions

(on y-axis) at which the higher variation occurs, indicates the frequency of stronger variations and the x-axis shows the occurrence of these stronger variations. The color bar in the figures is the logarithm (base 2) of the WPS. For example, a value of -4 on the color scale bar indicates that the WPS is $2^\wedge(-4) = 1/16$. Similarly, if the value is $+4$, it indicates the WPS of 16.

Typically, measured data is discrete and finite length time series. In this particular case, time series of combustion parameters such as IMEP, P_{max}, $\theta_{P_{max}}$, etc. of finite number of cycles are typically used for wavelet analysis. The finite length time series can lead to errors at the start and end of the wavelet power spectrum due to the assumption of cyclic data by Fourier transform. To overcome this issue, the original signal is padded with zeroes at the end of the data series, and these are eliminated after applying wavelet transform. The data series is padded with zeroes to create the total length of data series equal to next nth power of 2, which speed up the Fourier transform and limit the edge effects [100]. Zero padding introduces some discontinuities, and it leads to a decrease in amplitude at the edges. The edge effects become important in cone of influence (COI) region. The COI is characterized as the e-folding time for autocorrelation of wavelet power spectrum at every scale [100]. The e-folding time is considered in such that the magnitude of wavelet power decreases by a factor e^{-2} at the edges where discontinuities are present. Therefore, the region inside the COI is considered for wavelet analysis and the outside region is not used for extracting information about time series [100].

Global wavelet spectrum (GWS) is the time average of the WPS and calculated by Eq. (8.28):

$$\text{GWS} = W_s = \frac{1}{N} \sum_{n=1}^{N} |\text{CWT}_n(a)|^2 \qquad (8.28)$$

The global wavelet spectrum is represented by W_s. The peak locations in the global wavelet spectrum give an indication about the dominant periodicities in the time data series.

Figure 8.39 shows the time scale representation of the wavelet power spectrum of THR at no load conditions for three different compression ratios 16, 17, and 18. The thick contour lines represent the 5% significance level below which denotes the cone of influence (COI). The COI is the region where edge effects become important and the region inside the COI is considered for analysis and the region outside COI is ignored from analysis.

Figure 8.39a shows that the strongest intensity of variance occurs in the periodic band 221–625 cycles along the COI during 690–1665 cycles. Other weaker bands are 128–315 and 39–157 spanning in the ranges of 545–920 and 1440–1800 cycles, respectively. The presence of strong periodic band over a large number of cycles indicates a high cyclic variability in the no load condition at compression ratio (CR) 16. With the increase in the compression ratio (from 16 to 17), the maximum GWS power decreases from 5250 to 1930, which indicates a decrease in cyclic variations. The period at which maximum GWS power is obtained also shifts from

512 to 16 with an increase in compression ratio (Fig. 8.39), which suggests that frequency of variation also increases with compression ratio. Figure 8.39b shows the WPS and GWS of CR 17, and the strongest intensity bands are found in the period 14–20 cycles in cycles ranging from 50 to 410, 570 to 880, 925 to 955, 1020 to 1070, 1180 to 1550, and 1655 to 1880. As the CR increased to 18, the cycle-to-cycle variations shift to higher period indicating the cyclic variations are occurring with lower frequencies. The WPS of THR at CR 18 is depicted in Fig. 8.39c. The strongest intensity band is observed in the 10–16 periodic band intermittently in 161–347, 453–731, 909–987, 1057–1122, 1230–1319, and 1391–1631 cycles. The GWS indicates a power of 1025 which is lower than the previous cases and symbolizes a decrease in the cycle-to-cycle variations with an increase in compression ratio at no load condition. With the increase in compression ratios, combustion temperature increases, which results in better combustion stability (lower cyclic variations). With the increase in compression ratio, combustion temperature increases along with advanced combustion phasing. The sensitivity of combustion variation decreases at advanced combustion phasing (near TDC) due to high temperature and slow piston speed. Higher combustion temperature in particular combustion cycle leads to higher residual temperature and wall temperature, which affects the next consecutive cycles.

Figure 8.40 shows the time scale representation of the wavelet power spectrum of P_{max} at no load conditions for different compression ratios. For CR 16, the strongest intensity periodic band 128–625 period occurs in the cycle range of 433–1834 adjacent to the COI, and other weaker periods are observed in the period 32–64 for the cycles ranging in between 467–692, 815–1035, and 1801–1886, respectively (Fig. 8.40a). Maximum GWS power of 4.89 at the period around 350 is shown by GWS in Fig. 8.40a. For an increase in CR to 17, the bands with the highest power are scattered over the entire WPS (Fig. 8.40b). Few of the strong intensity bands are observed in the periods 55–78, 55–96, 96–156, and 156–315 cycles ranging from 1316 to 1420, 826 to 952, 212 to 419, and 1085 to 1631 cycles, respectively. A GWS power of 1 (at period 256) is observed in this case (CR 17), which indicates lower cycle-to-cycle variations, in comparison to 16 compression ratio. Figure 8.40c indicates that the maximum GWS power further decreases to 0.75 (at period 128) for CR 18, which implies that cycle-to-cycle variations have been further diminished. The period at which maximum GWS occurs also decreases with an increase in compression ratio, and a similar trend is observed for THR. Very few strong intensity bands are observed and occur in the period of 46–64, 55–110, and 256–312 cycles and stretch in between 85–215, 1305–1720, and 1327–1536 cycles, respectively, at compression ratio 18.

The WPS and GWS of IMEP for various diesel start of injection (SOI) timings for gasoline/diesel RCCI operation is illustrated in Fig. 8.41. In the WPS, the horizontal axis depicts the number of cycles (time scale), and the vertical axis shows the periodicity (frequency scale) of the time series. In the GWS, the peaks of the power depict the prevailing periodicities in the time series. The areas above this COI are only significant and considered in the analysis. Figure 8.41 reveals that variations in IMEP occur at multiple time scales in RCCI engines. Figure 8.41a

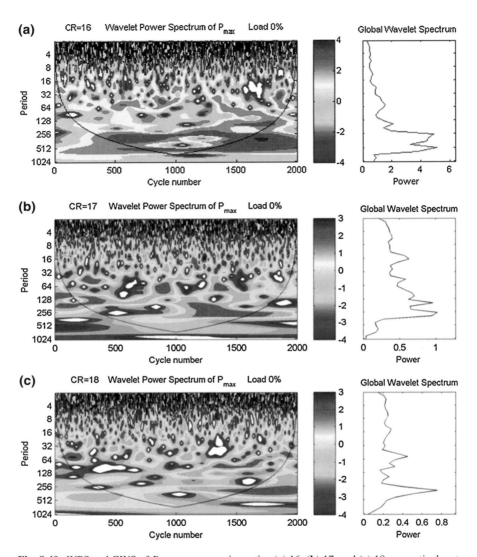

Fig. 8.40 WPS and GWS of P_{max} at compression ratios (**a**) 16, (**b**) 17, and (**c**) 18, respectively, at no load conditions

illustrates the WPS and GWS of IMEP for diesel SOI timing at 10 CAD bTDC for gasoline/diesel RCCI operation. The figure depicted that periodic band of the 4–8 period having higher variations in the cycle ranging from 48 to 68 and 398 to 422 (red color patches represent the higher cyclic variations). Similarly, a periodic band of the 8–16 period has higher variations between the cycles ranging from 70 to 84, 121 to 142, 271 to 284, and 732 to 747. A periodic band of 128–256 period shows higher variations during the cycles 538–660. The periodic band with strong intensity (dark red color) reveals higher variations.

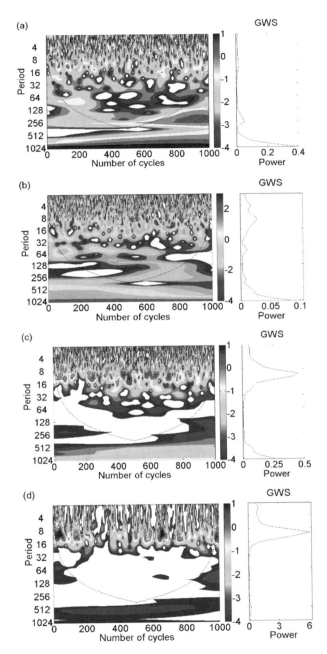

Fig. 8.41 WPS and GWS of IMEP for various diesel SOI timings for gasoline/diesel RCCI operation. (**a**) WPS of IMEP; Diesel SOI—10 CAD bTDC. (**b**) WPS of IMEP; Diesel SOI—30 CAD bTDC. (**c**) WPS of IMEP; Diesel SOI—50 CAD bTDC. (**d**) WPS of IMEP; Diesel SOI—60 CAD bTDC

Figure 8.41b shows the WPS and GWS of IMEP for diesel SOI timing at 30 CAD bTDC. The figure shows that a periodic band of the 4–16 period with moderate intensity is observed intermittently throughout the cycles. Peak power in the GWS with advanced injection timing, i.e., 30 CAD bTDC (Fig. 8.41b), in comparison to diesel SOI timing at 10 CAD bTDC (Fig. 8.41a) is slightly higher, which indicates higher cyclic variations. Higher cyclic variations are mainly due to lower mean combustion temperature and retarded combustion phasing with advanced diesel injection timing. Further advancing in diesel injection timing from 30 CAD bTDC to 50 CAD bTDC, the peak of the GWS increases which depicts higher cyclic variations. Additionally, the band of period 4–16 with strong intensity of variance observed in the cycle ranging from 24 to 57, 213 to 263, 418 to 486, 563 to 662, 695 to 738, and 829 to 863. Periodicity with strong intensity throughout the combustion cycles reveals to higher cyclic variations. Similarly, Fig. 8.41d shows the periodic band with strong intensity of period 8–16 occurs in the cycles between 13–70, 122–162, 630–686, and 743–765. The peak power in the GWS is also increased for diesel SOI timing at 60 CAD bTDC. It is interesting to note that in the condition of 50 CAD bTDC and 60 CAD bTDC, a significant portion of the spectrogram is empty (white), which means these frequencies are not present. This may be due to misfires (very high cyclic variations or poor combustion efficiency) occurring intermittently throughout the combustion cycles [79]. It is very interesting to note that (Fig. 8.41) the period at which the peak GWS power obtained for conventional dual fuel combustion (retarded combustion timing condition) is shifting from 256 period to 8–16 periodic band for gasoline/diesel RCCI operation. This indicates that the frequency of variations increases with advanced diesel injection timing. In conventional dual fuel operation, combustion initiates with diesel pilot injection, which means the variations in the start of combustion is possible due to the variation in injection parameters. In modern CRDI system, the variations are very minimal at constant demand/setting. Hence, in conventional dual combustion operation, the variations are mainly due to long-term effect (such as wall temperature), which leads to higher period band (lower-frequency) variations. However, in premixed RCCI combustion (advanced DI timings), even combustion initiation is controlled by local equivalence ratio, reactivity, and temperature, where cyclic variations are very likely, which leads to higher-frequency variations. In premixed case, the variations in local equivalence ratio, reactivity, and temperature depends on various conditions such as flow conditions, evaporation of diesel, mixing of droplets and distribution of droplets, etc. even at constant DI timings.

WPS and GWS of IMEP for multiple diesel injections in gasoline/diesel RCCI operation are presented in Fig. 8.42. By keeping pilot diesel injection timing constant at 55 CAD bTDC, main diesel injection timing is swept from 10 to 50 CAD bTDC. Figure 8.42a shows the WPS and GWS of IMEP for diesel main injection timing 10 CAD bTDC. WPS depicts that the strong intensity periodic band of 16–32 and 32–64 period occurs in the cycles ranging from 665 to 708, 722 to 747, and 692 to 734, respectively. The WPS and GWS of IMEP for diesel main injection timing 30 CAD bTDC are illustrated in Fig. 8.42b. The WPS indicates that periodic band with strong intensity of 16–32 and 32–64 period occurring in the

Fig. 8.42 WPS and GWS of IMEP for various main diesel SOI timings for gasoline/diesel RCCI operation. (**a**) WPS of IMEP; Main DI—10 CAD bTDC. (**b**) WPS of IMEP; Main DI—30 CAD bTDC. (**c**) WPS of IMEP; Main DI—50 CAD bTDC

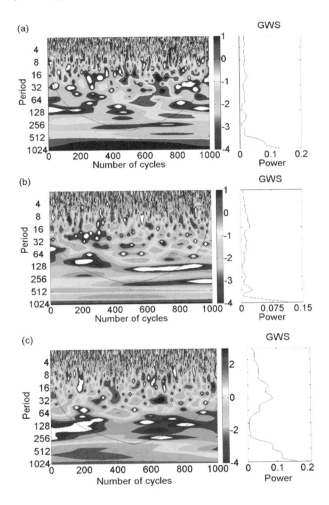

cycles ranging from 450 to 575 and 754 to 786, respectively. It is interesting to note that for the same diesel injection timing (i.e., after 30 CAD bTDC) of single and double injection strategy, double injection strategy has a lower peak for GWS. This depicts that double diesel injection has lower cyclic variations in comparison to single injection strategy, which is also confirmed by statistical technique [79]. Similarly, a strong intensity periodic band of 32–64 period occurs in the cycle ranging from 600 to 698 (Fig. 8.42c). For same diesel injection timing of 50 CAD bTDC, double diesel injections have a lower peak of GWS in comparison to single injection strategy. Figures 8.41 and 8.42 also show that in the case of a single diesel injection strategy, the cyclic variations are mainly concentrated in the periodic band of 8–16 (for advanced DI timing), while in the case of double injection strategy, the cyclic variations are distributed in the different periodic bands. This indicates that frequency of variation is higher for single diesel injection strategy in comparison to double injection strategy. In double injection strategy, the fuel injected by pilot injection at very early timings mixes with gasoline mixture and creates partial reactivity stratification, but combustion initiation is mainly triggered by the main injection. Thus, periods of variations will be higher, but periods decrease with advanced main injection timings (Fig. 8.42) because of mainly premixed combustion.

8.5 Nonlinear and Chaotic Analysis of Combustion Stability

The nonlinear dynamics in the combustion variations can be revealed through nonlinear analysis. Traditional statistical methods are not able to provide the information regarding the temporal variations. The autocorrelation and Fourier transform can provide the linear temporal correlations. However, nonlinear factors can govern the combustion variation dynamics in the reciprocating engines. Thus, nonlinear time series analysis methods such as return maps, Poincare sectioning, mutual information, modified Shannon entropy, data symbolization, etc. can provide additional information/features which are not revealed by traditional measures. Table 8.1 compares the linear and nonlinear signal processing methods [107].

Table 8.1 Comparative analysis of linear and nonlinear signal processing [107]

Linear signal processing	Nonlinear signal processing
Finding the signal—**signal separation** Separate broadband noise from the narrowband signal using spectral characteristics. System known: make a matched filter in the frequency domain	*Finding the signal*—**signal separation** Separation broadband signal from broadband "noise" using deterministic nature of signal. System known: Use manifold decomposition. Separate two signals using statistics on attractor
Finding the space—**Fourier transform** Use Fourier space method to turn differential equations or recursion relations into algebraic forms $X(n)$ is observed $X(f) = \sum x(n)\exp[i2\pi nf]$ is used	*Finding the space*—**phase space reconstruction** Time lagged variables form coordinates for a phase space in d_E dimensions: $Y(n) = [x(n), x(n+T), \ldots, x(n+(d_E-1)T)]$ d_E and time lag T are determined using mutual information and false nearest neighbors
Classify the signal Sharp spectral peaks Resonant frequency of the system *Quantity independent of initial conditions*	*Classify the signal* Invariants of orbits. Lyapunov exponents; various fractal dimensions; linking number of unstable periodic orbits; *Quantities independent of initial conditions*
Make models, predict $X(n+1) = \sum c_j x(n-j)$ Find parameters c_j consistent which invariant classification—Location of spectral peaks	*Make models, predict* $y(n) \rightarrow y(n+1)$ At time evolution $y(n+1) = Fy(n), a_1, a_2, \ldots, a_p]$ Find parameter a_j consistent with invariant classifiers—Lyapunov exponents and fractal dimension. Models are in local dynamical dimensions d_L; form local false nearest neighbors. Local or global models

Nonlinear analysis techniques are used to detect the information linked to temporal correlations, and it is used to discern prior-cycle effects. Nonlinear analysis methods for combustion stability analysis in reciprocating engines are discussed in the following subsections.

8.5.1   Phase Space Reconstruction

The phase space (or state space) of a dynamical system is a mathematical space with orthogonal coordinate directions representing each of the variables needed to specify the instantaneous state of the system [108]. Several approaches in nonlinear data analysis fundamentally start with the construction of a phase space portrait of the considered system. Phase space of a dynamical system is defined as a finite-dimension vector space R^n, and a state is specified by a vector $x \in R^n$ [109]. A point in phase space diagram represents a completely defined state of the system [87]. Thus, the time series of the system occurs as an orbit or trajectory in the phase space representing the time evolution, the dynamics, of the system. The shape of the trajectory of points in the phase space provides guidance regarding the characteristics of the dynamical system such as periodic or chaotic systems. Chaos has a structure in phase space [107].

In systems like reciprocating engines, a time series of measured in-cylinder pressure or calculated combustion parameters on crank angle basis or cycle-to-cycle basis is available for analysis. The observed time series needs to be converted into state vectors for phase space reconstruction. The conversion into state vectors is typically done by time delay embedding, derivative coordinates, or principal component analysis [109]. A d-dimensional system is possible to be reconstructed in a m-dimensional phase space by using time delays and $m \geq 2d + 1$ for an adequate reconstruction [110]. Time delay embedding comprises of creating a state space trajectory matrix $\mathbf{X}$ from the measured time series $\mathbf{x}$ by the time delay (τ) coordinates as shown in Eq. (8.29):

$$
\begin{cases}
X_1 = x(t), \\
X_2 = x(t + \tau), \\
\quad \vdots \\
X_m = x(t + (m - 1)\tau).
\end{cases}
\tag{8.29}
$$

Alternatively, derivative coordinates can be used for phase space reconstruction (Eq. 8.30). The derivative coordinates have the advantage of their clear physical meaning [109]:

$$
\begin{cases}
X_1 = x(t), \\
X_2 = \dfrac{dx(t)}{dt}, \\
\quad \vdots \\
X_m = \dfrac{d^{m-1}x(t)}{dt^{m-1}}.
\end{cases}
\tag{8.30}
$$

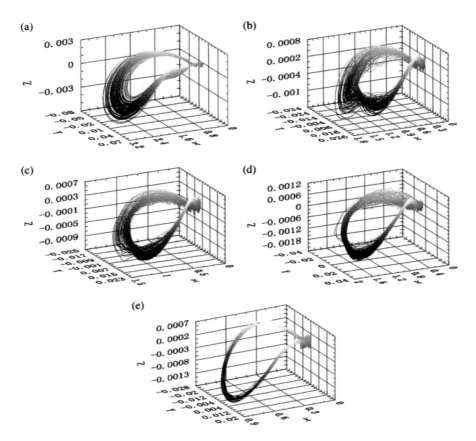

Fig. 8.43 Phase spaces reconstructed using measured pressure time series at different equivalence ratio [109]. (**a**) Case A1. (**b**) Case A2. (**c**) Case A3. (**d**) Case A4. (**e**) Case A5

Figure 8.43 shows the phase spaces generated using measured cylinder pressure data of 100 cycles for different equivalence ratios ($\Phi = 0.781, 0.677, 0.595, 0.588$, and 0, respectively) in a spark ignition engine. Phase spaces are created using derivative coordinates as shown in Eq. (8.31):

$$
\begin{cases}
X = p(\varphi), \\
Y = \dfrac{dp(\varphi)}{d\varphi}, \\
Z = \dfrac{d^2 p(\varphi)}{d\varphi^2}.
\end{cases}
\tag{8.31}
$$

Figure 8.43a depicts that the part of the phase space diagram shows very small variations of different trajectory cycles and close to the peak pressure position, with

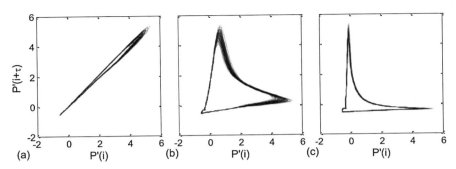

Fig. 8.44 Phase space portrait of normalized cylinder pressure for time lag of (**a**) $\tau = 1$, (**b**) $\tau = 48$, and (**c**) $\tau = 92$ in SI engine operated at stoichiometric mixture [111]

an obvious divergence between each cycle. Figure 8.43a, b shows well-developed combustion at the point of peak pressure because the trajectories disperse adequately at that point. However, the well dispersion part in projection plans is pushed to the area after the point of peak pressure (Fig. 8.43c, d). With the decrease in equivalence ratio, both the beginning and the end of combustion phase are delayed. Figure 8.43a–d shows that, although the curves in phase space diagram show obvious cyclic variations, these curves are rather well confined in the phase space and well organized. Therefore, cylinder pressure evolution could be possibly governed by a dynamics with deterministic components [109]. Poincaré section can be used to reduce the dimension of the phase space and reveal a structure, which can further identify the patterns of the dynamical behavior of time series.

Two of the most important parameters are time lag (τ) and embedding dimension (m) in the process of phase space reconstruction. Figure 8.44 shows the effect of time delay on the phase space portrait of cylinder pressure of a spark ignition engine. The normalized in-cylinder pressure time series is used to avoid the influence of pressure units. In case of too short time delay, the coordinates $P(i)$ and $P(i + \tau)$ will not be independent enough, which means not enough time will have evolved for the dynamical system to have explored enough of its state space to generate the new information about that state space. In case of too large time delay, any connection between the measurements $P(i)$ and $P(i + \tau)$ is numerically equivalent to being random with respect to each other because chaotic systems are intrinsically unstable [107].

The simple way to determine the time delays is the autocorrelation function method, and it is commonly used although a linear method [87, 111]. This method is relatively simple for calculation (Eq. 8.32), and it can be used as the phase space reconstruction is not so sensitive to time delay [111]:

$$C(\tau) = \frac{\sum [P'(i + \tau)P'(i)]}{N} \qquad (8.32)$$

The optimal time lag (τ) value is chosen when the value of the autocorrelation function ($C(\tau)$) decreases to $C(0)/e$. It was found that the value of autocorrelation function decreases to $C(0)/e$ and zero at 48 and 92, respectively (Fig. 8.44). The figure shows that the attractor of the combustion process is fully unfolded when time delay $\tau = 48$, and thus, the optimal τ value can be determined at which $C(\tau) = C(0)/e$ [111].

An alternative method of time lag determination is the average mutual information (AMI) [112]. However, AMI needs a larger scale of calculations. The time lag at which the first minimum of mutual information occurs can be considered as time lag for phase space construction [107]. Mutual information is an analysis tool based on information theory that measures univariate temporal coupling (i.e., the predictability in a signal). Mutual information-based method for detecting temporal relationships is more powerful than autocorrelation because it is equally sensitive to linear and nonlinear structure [75].

The mutual information between two time series at time delay τ in bits can be calculated by Eq. (8.33):

$$\mathrm{MI}(\tau) = \sum_{x_i=1}^{N} \sum_{x_j=1}^{N} p(x_i, x_j) \log_2 p(x_i, x_j) - \sum_{x_i=1}^{N} p(x_i) \log_2 p(x_i) - \sum_{x_j=1}^{N} p(x_j) \log_2 p(x_j)$$

$$(8.33)$$

where x_i is the time series value at time t, x_j is the time series value at time $t + \tau$, $p(x_i)$ is the individual probability density for x_i, $p(x_j)$ is the individual probability density for x_j, and $p(x_i, x_j)$ is the joint probability density for x_i and x_j. The probability functions can be calculated by binning the data and constructing histograms [75].

One of the most common methods to determine the value of the embedding dimension (m) is the false nearest neighbors (FNN) [113]. The m value is selected as the embedding dimension at which the percentage of FNN decreases to approximately 0%. The FNN can be calculated as shown in Eq. (8.34):

$$f_m(i) = \left[\frac{R_{m+1}^2(i) - R_m^2(i)}{R_m^2(i)} \right]^{1/2} = \frac{\left| p'(i + mn) - p'^{\mathrm{MM}}(i + mn) \right|}{R_m(i)} \tag{8.34}$$

where $p'(i)$ is the normalized time series of pressure data, n is equal to τ/t, t is the sampling time interval, $p'^{\mathrm{MM}}(i)$ is the nearest neighbor of $p'(i)$, and $R_m(i)$ represents the distance between $p'(i)$ and $p'^{\mathrm{MM}}(i)$ when the embedding dimension is m [111].

8.5.2   Poincaré Section

To determine the structure in the attractor, the Poincare section is used to reduce the dimension of the phase space. This method reduces one dimension of the phase space portrait. An invertible map can be constructed on the section by following the trajectory of the flow. The iterates of the map are given by the points where the trajectory intersects the section in a specified direction [109]. It is easier to analyze the distribution of these points because the Poincaré section has a lower dimension. Figure 8.45 shows the Poincare section of phase spaces shown in Fig. 8.43. Poincaré section reduces the dimension of the earlier phase space (Fig. 8.43) to a two-dimensional representation, which can be used to identify patterns of the dynamical behavior. Poincaré sections defined by Eq. (8.35) are shown in Fig. 8.45:

$$\sum\nolimits_{XZ} = \left\{(X, Y) \in \mathfrak{R}^2 \middle| Y = 0, Z \leq 0\right\}. \qquad (8.35)$$

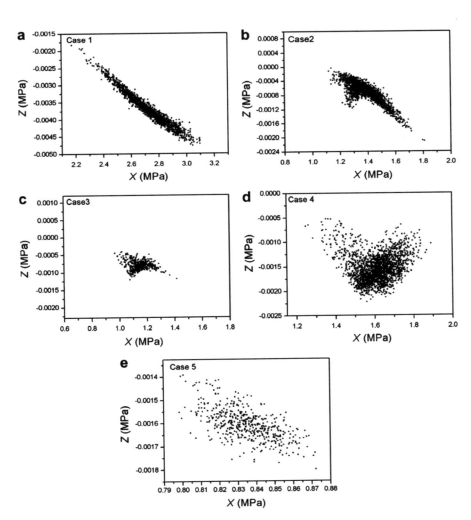

Fig. 8.45 Poincare sections in corresponding phase spaces in Fig. 8.43 for case1-case5 [109]

Figure 8.45a shows that the combustion strongly dominates the cylinder pressure development and a straight band of points with inverse ratio indicating linear deterministic characteristics in the Poincaré section. The Poincaré section reveals a bifurcation structure (Fig. 8.45b), in which a short branch appears at the upside of the main band. This structure suggests an unstable trend of combustion [109]. A more obvious bifurcation occurs with a decrease in equivalence ratio, which suggests that the domination of deterministic components to the cylinder pressure development is weakened further (Fig. 8.45c). Further reduction in equivalence ratio, the Poincaré section displays a more erratic structure with high levels of dynamic noise indicating stronger stochastic characteristics (Fig. 8.45d) [109].

8.5.3   Return Maps

Most commonly used phase spaces are return maps, which have embedding dimension of two and delay time selected to one. Return map provides a simple way to study the interactions between consecutive events. In reciprocating engines, the return maps can be used to determine the inherent deterministic interaction between combustion cycles qualitatively [2, 114]. Consecutive engine cycles are not interrelated in random time series of combustion events, and the return map shows an unstructured cloud of data points collected around a fixed point. The return map reveals more structures such as dispersed data points about a diagonal line when the deterministic coupling between combustion cycle exists [90]. In return maps, pairs of consecutive data series values of combustion parameters (cycle i versus cycle $i + 1$) are plots with each other (Fig. 8.46). Data point of each cycle relates to the next successive cycle through the general statistical picture of the whole cycles interrelation using return map plots [71].

Figures 8.46 and 8.47 show the return maps $\theta_{P_{max}}$ and heat release for HCCI combustion using primary reference fuel (PRF) where octane number (ON) varies from 3 to 7. Return maps for octane number three (ON = 3) show an unstructured cluster of circular data collected around a fixed point (Figs. 8.46a and 8.47a), indicating stochastic variations in the combustion cycle. The combustion is relatively stable with octane number three as relatively lower dispersion. Further increasing the octane number, the fixed concentrated points (as with ON = 3) start to destabilize in certain directions of the return map, and the highest levels of destabilization occur for ON 6 and 7 (Figs. 8.46 and 8.47). The data points scattered over diagonal line. The structured patterns of return map can be attributed to the deterministic coupling between consecutive cycles [71]. The functional form (Eq. 8.36) at cycle (i) using previous cycles can be used to characterize the dynamics of a combustion parameter (e.g., $\theta_{P_{max}}$):

$$\theta_{P_{max}}(i) = f(\theta_{P_{max}}(i - 1), \theta_{P_{max}}(i - 2), \ldots, \theta_{P_{max}}(i - (L - 1))) \qquad (8.36)$$

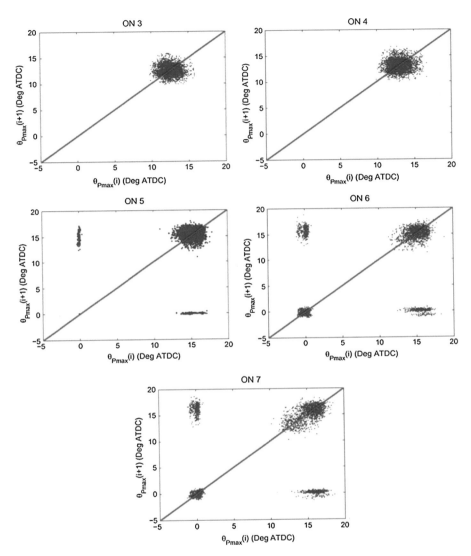

Fig. 8.46 Return map of $\theta_{P_{max}}$ for different octane numbers (ON) in HCCI combustion for 3000 consecutive engine cycles [71]

Chaotic methods like return maps and symbol sequence techniques (see Sect. 8.5.4) are used to estimate the approximate function f and value of L. A random time data series having an unstructured cluster of data points on return map tends to produce a high-dimensional function f [90]. Thus, the return maps of ON 5–7 (Fig. 8.46) suggest a relatively low value of L. Additionally, Fig. 8.46 indicates that the function f is a nonlinear function [71]. Pattern in the return map is also

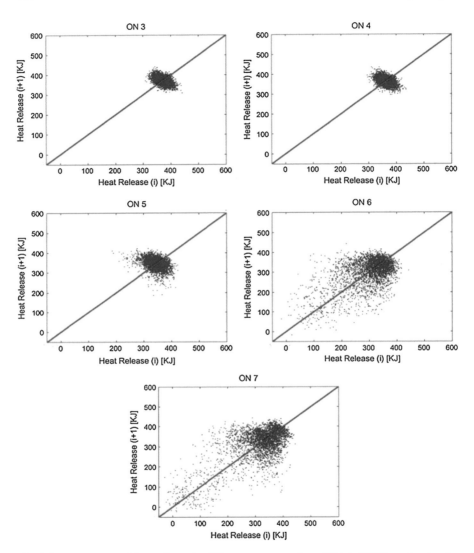

Fig. 8.47 Return map of heat release for different octane numbers (ON) in HCCI combustion for 3000 consecutive engine cycles [71]

dependent on the combustion parameters (Figs. 8.46 and 8.47) in the unstable combustion zone. Combustion parameters such as P_{max}, IMEP, combustion duration, combustion phasing, etc. can also be used for generating a return map for combustion stability analysis.

Typically, low-temperature combustion (LTC) engines have intrinsically high sensitivity for small fluctuations in the engine running conditions. Thus, the control system is required which should be able to satisfactorily respond to such

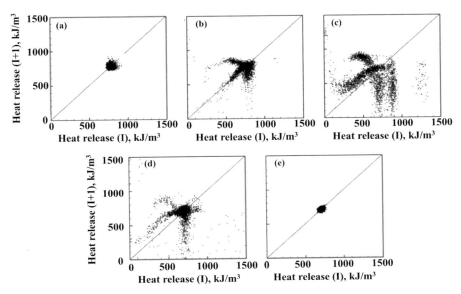

Fig. 8.48 Heat release return maps during the transition from (**a**) stable SI combustion to (**e**) HCCI operation using increased levels of internal EGR [90]

disturbances on cycle-to-cycle basis [2]. Figure 8.48 shows the complex dynamics of heat release using return maps during SI to HCCI combustion mode transition. The differences in combustion modes can be observed from the heat release return maps (Fig. 8.48). Data points are collected in a small unstructured cluster of around a fixed point in SI combustion (Fig. 8.48a) which shows the nominal flame propagation heat release [90].

In typical port fuel-injected SI combustion, data is slightly dispersed on the return map indicating the stochastic or high-dimensional component (Fig. 8.48a). With the increase in internal EGR during the transition, the cyclic variations increases, and data points on return map scatter in particular directions (Fig. 8.48b), which indicates unstable manifolds in low-dimensional phase space [90, 107]. The level of destabilization reaches to a maximum with a further increase in EGR (Fig. 8.48c), and combustion again starts to become more stable (Fig. 8.48d). Finally, in HCCI combustion mode, combustion is stabilized, and scatters in return map is eliminated with concentrating the data points around a fixed point (Fig. 8.48e).

Figure 8.49 shows the return map of combustion phasing (CA_{50}) at different relative air-fuel ratio (λ) in an HCCI engine. For lean combustion ($\lambda = 2.8$), return map of combustion phasing shows a circular cloud indicating stochastic component. For rich combustion ($\lambda = 2.3$), the return map of the combustion phasing distribution shows flat distribution along the diagonal line. In this case, it is difficult to find correlation as skewed distributions can also produce dispersed data leading to complication in the return map analysis [2]. Generation of return map using normalized value (or quantile values) can solve this issue of visualization on the return

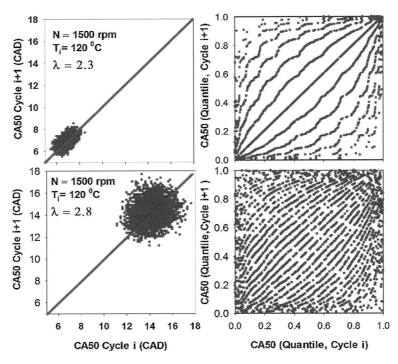

Fig. 8.49 Return maps for combustion phasing with a lag of one cycle for different λ in HCCI engine

map. The dataset of N values can be converted into another data set of N quantile values by replacing the lowest data value to $1/N$, next lowest to $2/N$, such that the maximum data value is assigned a value of 1 ($N/N = 1$). However, it is essential to ensure that the sequence of data is not disturbed. During this process, all the data points are converted into values between 0 and 1. In this process, highly skewed distribution is converted to a uniform distribution of data points evenly distributed between 0 and 1 [115]. Thus, an uncorrelated data will show a uniform density of data points on the return map using quantile values, and correlated data have nonuniform (higher and lower) densities of data points on the map [115]. Return maps using quantile values of combustion phasing for the lean and rich condition is shown in the right of Fig. 8.49. The figure shows that the leaner mixture has a relatively uniform distribution of data points and richer mixture has uneven distribution in the density. The nonuniform densities of data points indicate a deterministic dependency on the previous combustion cycles [2].

8.5.4 Symbol Sequence Statistics

Symbol sequence statistics method is very useful for time series data analysis, which can provide important insight into the behavior of different combustion parameters of reciprocating engines. Deterministic and stochastic behaviors can be identified with this approach. The presence of determinism in combustion parameters indicates that intelligent control of the system can extend the limits of engine operation over a wide range [2]. The deterministic information can be effectively used by controllers, and small variations in control input parameters/actuators can shift back the engine to stable operating conditions [116]. Control of cyclic variations in HCCI engine is demonstrated using symbol-statistics predictive approach [71]. Symbol generation process and histogram analysis are discussed in the following subsections.

8.5.4.1 Symbolization

Symbolization of combustion data of the engine can be very advantageous and effective to analyze the pattern when data contains measurement noise and/or has measurement errors. For correctly selected partition number, symbolization can also correctly estimate the deterministic effect of the previous cycle or inherent structure in the time series. A dynamic noise appears if the number of partition is higher. Symbolization of data can also act as a data compression methodology, which leads to relatively faster data processing during data acquisition. These characteristics of symbolization make it an effective tool for real-time control and onboard diagnostics of the engine [117, 118].

Symbolization of data converts it into a series of data with symbols. Figure 8.50 illustrates the process of symbolization of data using the binary partition. The times series data first divides the data into two equiprobable partitions in such a way that both the partitions contain the same number of data points. The data above the partition line is assigned a symbol "1," and below is assigned symbol "0" (Fig. 8.50). This data conversion results in a series of binary symbols. After data conversion into symbols, the frequency of occurrence of particular sequence length (3 in Fig. 8.50) is computed. The total number of possible sequences "N_{seq}" is dependent on the number of partition (n_{part}) or number of symbols and selected symbol sequence length "L." The total number of possible sequences can be calculated by Eq. (8.37):

$$N_{seq} = \left(n_{part}\right)^{L} \tag{8.37}$$

For three sequence length in the binary partition (Fig. 8.50), six (2^3) sequences are possible that can be presented in a binary or decimal format (as shown in Fig. 8.50). To generate a histogram, the number of occurrences of every sequence is counted and presented as bar as illustrated in Fig. 8.50b.

Figure 8.51 further illustrates the symbolization of 1000 combustion phasing (CA_{50}) data using the binary partition. Binary partition is the simplest partition,

Fig. 8.50 Illustration of symbol generation process and presenting a symbol sequence histogram [117]. (**a**) Data series. (**b**) Symbol-sequence histogram

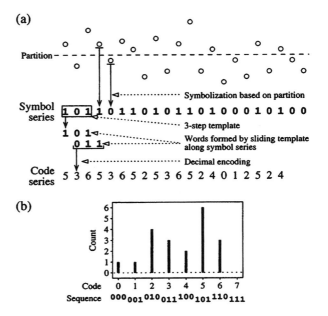

where data is divided into two equiprobable partitions using the median of the data. The combustion phasing values below the median are assigned symbol "0" and above the median "1." More than two partitions can also be possible leading to a different number/symbol system (e.g., 0, 1, 2, 3, 4, 5, 6, and 7 for $n_{part} = 8$). The sequence of symbols depicts important information regarding the combustion dynamics. The relative frequency corresponding to each possible sequence number is shown in Fig. 8.51 for sequence length of six. Sixty-four possible numbers (2^6) are represented in the decimal format on the x-axis of Fig. 8.51b. In this method, the relative frequency of truly random data is equal due to an equal number of values in each partition. All histogram bins will be equally probable within the uncertainty due to the finite data set. Therefore, significant deviation from equiprobability (for truly random data) indicates the deterministic structure or time correlation in the data [118]. The baseline frequency F_b for purely random data can be calculated using Eq. (8.38) [116]:

$$F_b = \left(\frac{1}{n_{part}}\right)^L \tag{8.38}$$

This baseline frequency for the combustion phasing data in Fig. 8.51b is represented by the thick red line. Therefore, the sequences appearing as peaks rising above the red line (F_b) corresponds to repeating deterministic events on the histogram plot (Fig. 8.51b). However, selection of appropriate sequence length and partition is essential for accurate analysis using symbolization method.

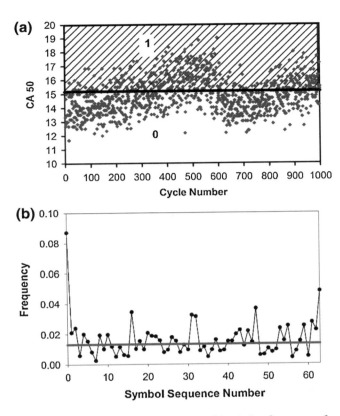

Fig. 8.51 (**a**) Symbolization of combustion phasing, (**b**) relative frequency of symbols using sequence length of six

8.5.4.2   Modified Shannon Entropy

Shannon entropy is a statistic derived from information theory that measures the degree of predictability in a time series, and it is useful for detecting dynamic patterns [75]. The optimal sequence length for creating a symbol sequence histogram can be computed using a modified form of Shannon entropy. A modified form of Shannon entropy is used to quantify the deviation of symbol statistics from randomness [116, 119]. Modified Shannon entropy value "1" indicates a purely random data series and for values less than "1" suggests a correlation between sequential points. The modified Shannon entropy (H_s) can be calculated by Eq. (8.39):

$$H_s = \frac{1}{\log n_{seq}} \sum_k p_k \log p_k \qquad (8.39)$$

where p_k is the probability with which sequence "k" occurs and n_{seq} is a total number of sequence with non-zero probability.

Fig. 8.52 The variation of modified Shannon entropy with symbol sequence length for different octane numbers in HCCI engine [71]

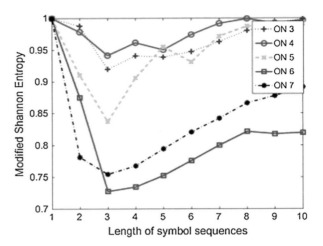

Figure 8.52 shows the variation of modified Shannon entropy with symbol sequence length for different octane numbers in HCCI engine (conditions shown in Fig. 8.46). In this calculation, the octal partition is used. The value of modified Shannon entropy varies as the sequence length, and it is found minimum at a sequence length of 3. Thus, it can be assumed that it is an optimal sequence length for this partition and combustion parameter. Another method based on joint probability distribution to predict the next cycle occurrence using previous cycle information is used to determine the optimal sequence length [71]. The joint probability histograms provide the maximum likelihood probability of the next cycle based on the occurrence of previous cycles. Then, based on a comparison of the one-cycle ahead predictions for different values of L, an optimal value of L can be determined.

The optimal combination of a number of partition and sequence length is required in symbol sequence method for effective control strategy. To determine the optimal combination of number of partition and sequence length, a matrix of values for both ranging 2–10 is calculated as shown in Fig. 8.53. The figure shows that for binary partition the optimal sequence length (minimum Shannon entropy) seems to be 8 or 9 for all the test condition. For 6–8 number of partitions, the optimal sequence length seems to be 3 where Shannon entropy is minimum. Figure 8.53 shows that the optimal combination of a number of partition and sequence length depends on the engine operating conditions [120].

8.5.4.3   Symbol Sequence Histograms

The symbol sequence histogram is used to find the sequence with a higher frequency above the baseline frequency for purely random data. The highest frequency provides the most repeated pattern in the data series. Figure 8.54 shows the symbol sequence histogram for combustion duration and IMEP data series at different

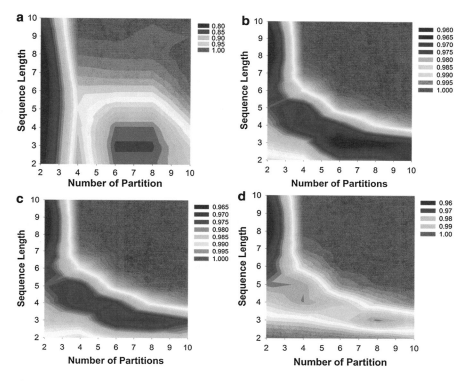

Fig. 8.53 Variation of modified Shannon entropy with sequence length and number of partition at (**a**) $\lambda = 2.0$, (**b**) $\lambda = 2.2$, (**c**) $\lambda = 2.4$, and (**d**) $\lambda = 2.6$ in a HCCI engine [120]

relative air-fuel ratio using octal partition and sequence length 3. Sequence having higher relative frequency than baseline frequency shows the determinism in the data series. Figure 8.85a shows that frequency of sequence number increases more as the engine is operated at the richer mixture. The relative frequency at richer mixture is highest for the richest mixture ($\lambda = 2.0$) in case of combustion duration data series. However, with IMEP data series, the number of sequence numbers having a higher frequency than baseline are higher for leanest mixture operation ($\lambda = 2.6$), indicating a higher number of deterministic patterns. Thus, two combustion parameters combustion duration and IMEP behave differently for the determination of deterministic patterns. Therefore, it is important to select a right number of partitions and sequence length for a particular combustion parameter [120].

Figure 8.55 shows the symbol sequence histograms of combustion phasing for different λ in HCCI engine at 1800 rpm using octal partition and sequence length 3. The figure shows that the richest mixture has a higher number of sequences above baseline frequency indicating more deterministic patterns. The sequence codes that occur more frequently with higher frequency are 0, 8, 16, 64, 128, 276, and 511. These numbers when converted from decimal to octal number corresponds to sequence 0–0–0, 0–1–0, 0–2–0, 1–0–0, 2–0–0, 4–2–4, and 7–7–7. The two

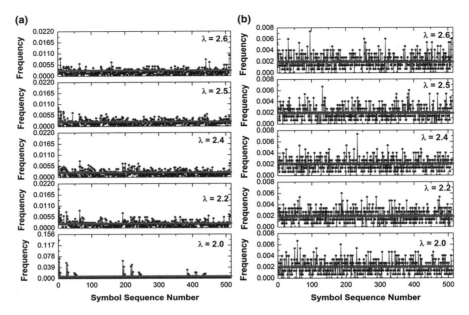

Fig. 8.54 Symbol sequence histograms of (**a**) combustion duration (**b**) IMEP for different λ using octal partition and sequence length 3 [120]

sequences 0–0–0 and 7–7–7 represent the steady behavior of combustion timing. Other patterns indicate that advanced-to-retard and retard-to-advanced timing combustion events are dominant.

8.5.4.4   Time Irreversibility

Time irreversibility is defined such that a qualitative or quantitative description of a time series is indistinguishable from a time-reversed version of itself [75]. Qualitatively time irreversibility can be depicted from the return map, which shows symmetry about diagonal for reversible data and significant bias for irreversible data. Nonstationarity in the data increases the possibility of observing time irreversibility.

Time irreversibility can be quantified using symbol sequence histograms because the relative frequencies will shift when the data series is used in backward time. There should be no significant variation in the histogram for the backward time series if the data measures are time symmetric. To compare forward- and reverse-time histograms, a quantified statistic can be defined by Eq. (8.40) [75, 121]:

$$T_{\mathrm{irr}} = \sqrt{\sum_{i} (F - R)^2} \qquad (8.40)$$

Fig. 8.55 Symbol sequence histograms of combustion phasing for different λ at 1800 rpm using octal partition and sequence length 3

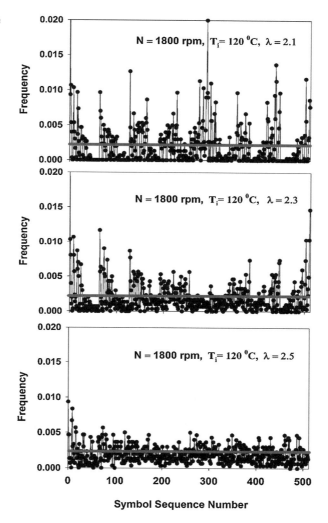

where "i" is indexed over all possible sequence codes. F and R are the symbol sequence histogram frequencies for the forward- and reverse-time analyses, respectively. The magnitude of T_{irr} quantifies the level of time irreversibility.

8.5.5 Recurrence Plot and Its Quantification

Recurrence plot (RP) method is first introduced by Eckmann et al. [122] in 1987 to visualize the time-dependent behavior of the dynamics of systems, which were used later for the identification of nonlinear systems with various possible behaviors [123–126]. A recurrence plot is a useful tool to visualize recurrences of phase space trajectories, and it provides a qualitative description of the dynamics of a time series. The recurrence plot efficiently visualizes recurrences mathematically expressed by matrix given by Eq. (8.41) [123, 127]:

$$R_{i,j}^{m,\epsilon} = \Theta\left(\epsilon - \|X_i - X_j\|\right), \quad i,j = 1, \ldots, N \tag{8.41}$$

where N is the number of measured points X_i, ϵ is a threshold distance, Θ ($\cdot$) the Heaviside function (i.e., Θ $(x) = 0$, if $x < 0$, and Θ $(x) = 1$ otherwise), and $\|\cdot\|$ is a norm. For the analysis of combustion cycles of reciprocating engines, X_i can be heat release, combustion phasing, or IMEP based time series data. The phase space vectors for one-dimensional cycle-based time series observations can be reconstructed by using time delay method, $X_i = (x_i, x_{i+\tau}, \ldots, x_{i + (m-1)\tau})$, where τ is the time delay and m is the embedding dimension [110]. The dimension m can be determined using the method of false nearest neighbors [125]. Since $R_{i,i} = 1$ ($i = 1, \ldots, N$) by definition, the recurrence plot has a black main diagonal line, the line of identity (LOI), with angle of $\pi/4$ [127]. It is possible to classify the dynamics of the system by its characteristic patterns showing diagonal, vertical, or horizontal structure of lines using recurrence plot method [126, 128]. This method (applied to time series) is capable of distinguishing chaotic and stochastic behavior [126]. A pattern for a stochastic system is based on uniform distribution of points in the recurrence plot, while a chaotic system possesses structure of lines with finite lengths. On the other hand, in a case of the intermittent motion [129], a vertical stripe structure is expected [121–131]. In recurrence plot, the abrupt changes in dynamics and extreme events are characterized by white areas or bands, and oscillating systems have diagonally oriented or periodic recurrent structures (i.e., diagonal lines or checkerboard patterns) [91].

Figure 8.56 shows the recurrence plot for IMEP at different engine speeds of 1000, 1200, 1400, 1600, 1800, and 2000 rpm, respectively, in a diesel engine. Figure 8.56a, c–e shows several vertical lines identifying the presence of intermittency in the IMEP time series. A series of lines with unit slope parallel to the main diagonal line is observed at engine speed of 1200 rpm (Fig. 8.56b), which indicate a more regular oscillatory behavior [91]. Interestingly, the recurrence plot has a checkerboard structure indicating a regular oscillatory behavior also at the maximum speed of 2000 rpm (Fig. 8.56f) similar to Fig. 8.56b.

Recurrence quantification analysis (RQA) provides measures of complexity quantifying structures in a recurrence plot [123]. The first measure of RQA is the recurrence rate (RR) or percent of recurrences given by Eq. (8.42), which counts the black dots in the recurrence plot [127]. It is a measure of the density of recurrence points:

$$\mathrm{RR}(\epsilon) = \frac{1}{N^2} \sum_{i,j=1}^{N} R_{i,j}^{m,\epsilon} \tag{8.42}$$

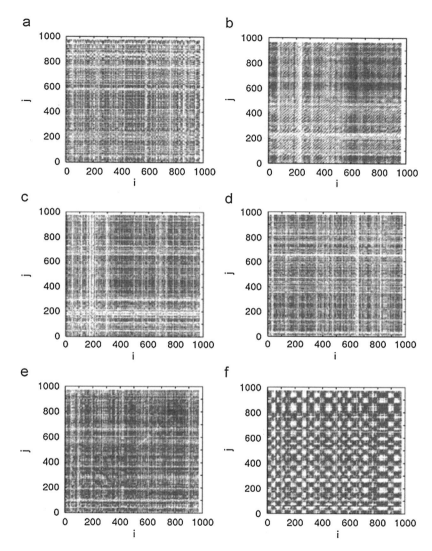

Fig. 8.56 Recurrence plots of the IMEP for different engine speeds ($n =$ (**a**) 1000, (**b**) 1200, (**c**) 1400, (**d**) 1600, (**e**) 1800, and (**f**) 2000 rpm) in a diesel engine [91]

RQA can be used to identify vertical or diagonal lines through the maximal lengths $L_{\max}$ and $V_{\max}$ for diagonal and vertical lines, respectively, given by Eqs. (8.43) and (8.44) [123]:

$$L_{\max} = \max(\{l_i; \quad i = 1, \ldots, N_l\}) \qquad (8.43)$$

$$V_{\max} = \max(\{v_i; \quad i = 1, \ldots, N_v\}) \qquad (8.44)$$

RQA also enables to perform probability $p(l)$ or $p(v)$ distribution analysis of lines according to their length l or v (for diagonal and vertical lines). Practically they are calculated by Eq. (8.45) [126, 127]:

$$p(x) = \frac{P^\epsilon(x)}{\sum_{x=x_{\min}}^{N} P^\epsilon(x)} \tag{8.45}$$

where $x = l$ or v depending on diagonal or vertical structures in the specific recurrence diagram and $P^\epsilon(x)$ is probability for a given threshold value ϵ.

The measures of entropy refer to the Shannon entropy of the frequency distribution along the diagonal and vertical line length given by Eqs. (8.46) and (8.47) and reflect the complexity of the deterministic structure in the system [123, 126, 127]. The larger value indicates more complex deterministic structure.

$$L_{\text{ENTR}} = -\sum_{l=l_{\min}}^{N} p(l)\ln p(l) \tag{8.46}$$

$$V_{\text{ENTR}} = -\sum_{v=v_{\min}}^{N} p(v)\ln p(v) \tag{8.47}$$

Other measures of RQA like determinism (DET) are given by Eq. (8.48), calculated using probabilities. Determinism (DET) is the measure of the predictability of the examined time series that gives the ratio of recurrent points formed in diagonals to all recurrent points. It can be noted that in a periodic system, all points would be included in the lines [123].

$$\text{DET} = \frac{\sum_{l=l_{\min}}^{N} lP^\epsilon(l)}{\sum_{i,j=1}^{N} R_{i,j}^{m,\epsilon}} \tag{8.48}$$

Laminarity (LAM) is a similar measure which corresponds to points formed in vertical lines, given by Eq. (8.49). This measure is telling about dynamics behind sampling point changes. For small point-to-point changes, the consecutive points form a vertical line [123]. It indicates the extent of laminar phases or intermittency in the time series [91].

$$\text{LAM} = \frac{\sum_{v=v_{\min}}^{N} vP^\epsilon(v)}{\sum_{v=1}^{N} vP^\epsilon(v)} \tag{8.49}$$

Trapping time (TT) is given by Eq. (8.50) calculated using probabilities:

$$\text{TT} = \frac{\sum_{v=v_{\min}}^{N} vP^\epsilon(v)}{\sum_{v=v_{\min}}^{N} P^\epsilon(v)} \tag{8.50}$$

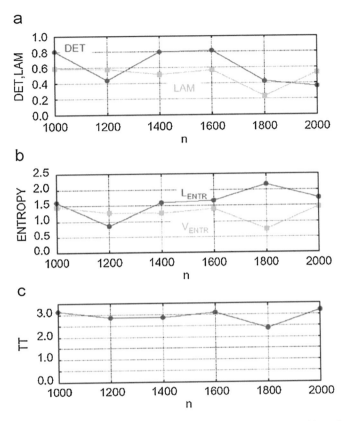

Fig. 8.57 Variations of (**a**) DET and LAM, (**b**) L_{ENTR} and V_{ENTR}, and (**c**) TT with engine speed, for the same value of RR = 0.2 [91]

Trapping time (TT) refers the average length of vertical lines measuring the time scale (in terms of sampling intervals) of these small changes in the examined time history [123]. This parameter describes how long the system remains in a specific laminar phase.

Figure 8.57 shows the different RQA parameters with engine speed for a recurrence plot shown in Fig. 8.56. The figure shows that L_{ENTR} is largest at 1800 rpm engine speed indicating higher structural complexity [91]. Additionally, at this speed (1800 rpm), the parameters LAM and TT have their minimum values, while DET shows its minimum value at 2000 rpm. The small value of DET indicates low predictability, whereas the small values of LAM and TT, respectively, indicate a dominance of large fluctuations and a short duration of time spent in a laminar phase in the intermittent dynamics [91].

8.5.6   0–1 Test

The 0–1 test is proposed for investigation of the dynamical system and tests whether the system is regular or chaotic [132–134]. In comparison to conventional methods, the advantages of 0–1 test are as follows: (1) it can be applied to any deterministic dynamical system directly and universally, (2) this method does not require the phase space reconstruction, (3) this method requires relatively less computational effort, and (4) the output is intuitive with only one value, approximately equal to 0 or 1, based on whether the system is regular or chaotic [135]. The errors due to the selection of embedding dimensions and time lag are eliminated in this method because it does not require the phase space construction.

Combustion instability is typically analyzed using measured cylinder pressure. The measured cylinder pressure time series $P(j)$ is converted into new coordinates ($u(n)$, $v(n)$) by Eq. (8.51) [135]:

$$u(n) = \sum_{j=1}^{n} P(j) \cos(jc) \quad \text{and} \quad v(n) = \sum_{j=1}^{n} P(j) \sin(jc) \qquad (8.51)$$

where c in range $(0, \pi)$ is a random frequency and $n = 1, 2, 3. \ldots$

The form of coordinates is changed to reduce the effect of potential resonance caused by the choice of c frequency. Suitable choice of c is crucial in the application of 0–1 test. Normalized pressure values are used to remove the effect of pressure units. The final coordinates used for combustion analysis is computed using Eq. (8.52) [135]:

$$u(n) = \sum_{j=1}^{n} P'(j) \cos(\psi(j)) \quad \text{and} \quad v(n) = \sum_{j=1}^{n} P'(j) \sin(\psi(j)) \qquad (8.52)$$

$$\psi(j) = jc + \sum_{h=1}^{j} P(h) \quad \text{and} \quad P'(j) = \frac{P(j) - P_{\text{mean}}}{\sigma_P} \qquad (8.53)$$

where P_{mean} is the mean value of the pressure time series and σ_P is the standard deviation of pressure time series.

Quantitative characterization of function in Eq. (8.52) is performed by the mean square displacement (MSD), which is calculated by Eq. (8.54) in the new (u, v) plane:

$$\text{MSD}(n) = \lim_{N \to \infty} \frac{1}{N - n} \sum_{j=1}^{N} \left\{ [u(j + n) - u(j)]^2 + [v(j + n) - v(j)]^2 \right\} \qquad (8.54)$$

Distribution characteristics of MSD characterized the regular and chaotic time series. The MSD is bounded in time for regular time series, while it grows linearly with time for chaotic time series. The dynamics of the time series can be further characterized by computing the asymptotic growth rate K_c, which can be calculated by Eq. (8.55):

$$K_c = \frac{\text{cov}(n, \text{MSD}(n))}{\sqrt{\text{Var}(n)\text{Var}(\text{MSD}(n))}} \tag{8.55}$$

where $\text{cov}(n, \text{MSD}(n))$ is the covariance of two different time series and $\text{var}(n)$ is the $\text{cov}(n, n)$. Averaging the K_c values on c in $(0, \pi)$, the final 0–1 test result K is calculated. The K values provide the quantitative information regarding the dynamics of the process. The K value is close to 0 for regular time series, while it approaches to 1 for chaotic time series. Typically, $K \geq 0.7$–0.8 is considered as the chaotic regions [133, 135–137].

Figure 8.58 shows the cylinder pressure time series (u, v) planes at $c = 2.5$ for motoring and different fuel injection timings in natural gas SI engine. Figures 8.59 and 8.60 show the corresponding MSD versus n (at various c values) and K_c (asymptotic growth rate) versus c. In the motoring engine operating conditions (Fig. 8.58a), the pressure time series shows the bounded behavior indicating a regular time series. However, the cylinder pressure time series at different injection timings (Fig. 8.58b–h) show unbounded Brownian (random) behavior in u–v plane. Similarly, all the MSDs for different injection timings (Fig. 8.59b–h) show an approximately linear increase (unbounded) with n for various c values. The MSD values show bounded behavior for motoring condition (Fig. 8.59a), indicating regular behavior. The K_c values for different injection timings (Fig. 8.60b–h) show close to 1 for all the cases with a certain level of fluctuations. The observations from Figs. 8.58 to 8.60 clearly indicate that the combustion process of this engine is a

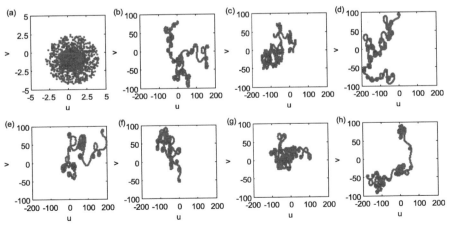

Fig. 8.58 The (u, v) planes of cylinder pressure time series for motoring condition (**a**) and different injection timings (**b**) 0 °CA, (**c**) 15 °CA, (**d**) 30 °CA, (**e**) 45 °CA, (**f**) 60 °CA, (**g**) 75 °CA, and (**h**) 90 °CA, respectively, after intake TDC in SI engine [135]

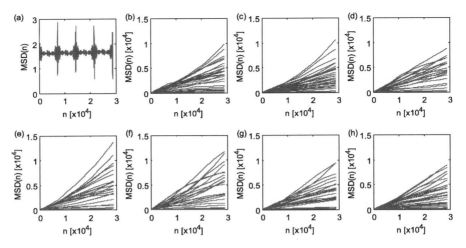

Fig. 8.59 The variation of MSD (*n*) with *n* for different injection timings in SI engine. The cases (**a**)–(**h**) are defined in Fig. 8.58 [135]

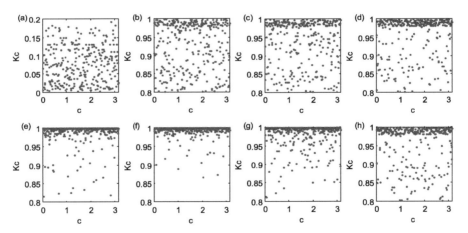

Fig. 8.60 The value of K_c for different values of *c* for different injection timings in an SI engine. The cases (**a**)–(**h**) are defined in Fig. 8.58 [135]

chaotic dynamical system [135]. In motoring engine operation, the *K* value is 0.077, which clearly indicates that this process is regular. It can be further observed that *K* value first increase and then decreases with increasing injection timings and has larger *K* values for 45–60 °CA injection timing after intake TDC. The larger *K* values indicate the stronger chaotic behavior at those particular fuel injection timings.

8.5.7 Multifractal Analysis

The multifractal analysis provides a method to analyze the signals having nonlinear power-law behavior that depends on higher-order moments and scale. Multifractals are used to explain the dynamics of complex processes exhibiting multiple time scales in addition to a range of oscillation amplitudes [76]. A monofractal process is self-similar in such a way that its dynamics can be fully explained by one power-law time-scaling exponent (such as Hurst exponent) and its complexity can be described by single fractal dimension. The multifractal process needs a spectrum of scaling exponents to fully describe the complex dynamics, and it can be considered as locally self-similar. One method to explain such multifractal spectrum is to use Hölder exponent [76, 138]. The singularity spectrum ($f(\alpha)$) is created by calculating the average relative contribution of the different possible Hölder exponents over the observation window (i.e., the overall time series). The broadness of the singularity spectrum is one measure of the complexity of the process [76]. Engine combustion process is analyzed using multifractals based on cylinder pressure measurement [76, 139–141].

Figure 8.61 shows the multifractal spectrum at different engine operating conditions in SI engine. Figure 8.61a illustrates the method to determine the coefficient of the multifractal spectrum. Calculated exponential distribution $f(\alpha)$ (red points) is found using multifractal analysis, and then it is interpolated with a polynomial trend line. Two main characteristics of multifractal spectrum are typically used for analysis. First is the value of the Hölder exponent, $\alpha = \alpha_0$, at the peak of $f(\alpha)$, and second is the broadness, $\Delta\alpha$, that is defined as the distance between the (extrapolated) points of intersection of the spectral curve with the α-axis [76, 141]. The parameter, α_0, signifies the most dominant fractal exponent in the data series, and it shows the correlation or degree of persistence in the data series. Persistent or positively correlated process has $\alpha_0 > 0.5$, and anti-persistent or negatively correlated process has $\alpha_0 < 0.5$. The two cases, $\alpha_0 = 0.5$ and $\alpha_0 = 0$, represent a Brownian random walk

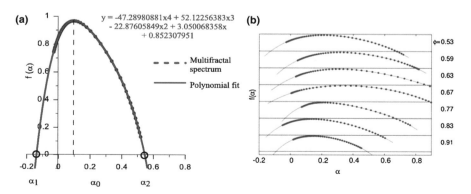

Fig. 8.61 (a) Illustration of method for determination of coefficients of multifractal spectrum using IMEP time series [141] and (b) multifractal spectrum at different equivalence ratio $\phi = 0.91, 0.83, 0.77, 0.67, 0.63, 0.59,$ and 0.53 using heat release time series [76]

and Gaussian random process, respectively. The Gaussian process is considered as
the series of random numbers, while the Brownian walk is characterized by random
steps. The broadness of a singularity spectrum ($\Delta\alpha$) reflects the range of possible
fractal exponents. Therefore, the complexity of the process is characterized by the
distribution of α, and $\Delta\alpha$ provides a measure of multifractality or complexity in the
data [76, 141]. A large value of $\Delta\alpha$ shows a richer multifractal structure, whereas a
small value approaches a monofractal limit. Figure 8.61b depicts that as the fuel-air
mixture becomes leaner (decrease in ϕ), the broadness ($\Delta\alpha$) first increases to reach a
maximum at $\phi = 0.67$, and further leaning of charge decreases the broadness. This
observation suggests the highest degree of multifractal complexity in heat release
time series when the engine is operated at $\phi = 0.67$.

Figure 8.62 shows the variation of α_0 and $\Delta\alpha$ (obtained by multifractal analysis)
at different engine operating conditions with hydrogen enrichment. The engine
operating conditions at which experiments are conducted are shown in the right of
Fig. 8.62. The optimal conditions are indicated by a full line on the left-hand side of
Fig. 8.62. Engine operating points 8 and 10–12 have higher $\Delta\alpha$ (broadness) with
higher peaks, which suggests the IMEP time series is characterized by higher cycle-
to-cycle interactions [141]. It means the IMEP oscillations depend on the IMEP in
the previous cycle when the engine is fueled with hydrogen at an engine load of
85 kPa (Fig. 8.62). The highest value of α_0 indicates the lowest negative correlations
in the anti-persistent walk. The most expected value of α_0 is below 0.2 and a range of
critical $\Delta\alpha$ below 0.7 for the IMEP time series. Figure 8.62 clearly indicates that
there is no change in the combustion process with hydrogen addition at low engine
load (operating points 1–4). Thus, in summary, the multifractal analysis provides the
quantitative information of the combustion process using persistence and complexity
measures. This information can be used in the design of effective engine control
strategy [141].

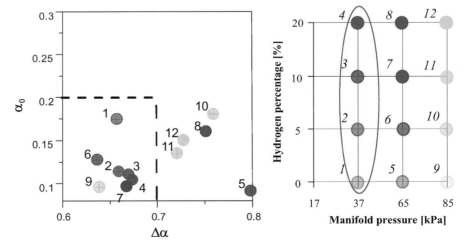

Fig. 8.62 Result of multifractal analysis of IMEP data series with the most expected exponent
α_0 and a range of critical exponents $\Delta\alpha$ (right) with measurement points (left) [141]

8.6   Steps to Improve Combustion Stability

Combustion stability in reciprocating engines can be analyzed using statistical methods or chaotic methods using measured cylinder pressure data. The quantitative measures are defined to determine the levels of the cyclic combustion variability. After characterizing the cyclic combustion variations, the methods can be suggested to improve the combustion stability and control strategy of engine to minimize the cyclic variations. Generally, the combustion stability in reciprocating engines improve with (1) higher engine load and speed, (2) higher compression ratio, (3) lower valve overlap, (4) higher energy of ignition system, (5) higher temperatures, and (6) lower humidity [9]. Typically, combustion stability improves with parameters, which tends to increase the combustion rate of the engines, and faster combustion has typically lower cyclic variations. Therefore, the cylinder pressure-based combustion stability methods can help to see the effect of design changes on the combustion stability. The engine operating parameters and combustion mode can be tuned for lower cyclic variability by combustion stability analysis using cylinder pressure measurement.

Discussion/Investigation Questions

1. Why measurement and characterization of cycle-to-cycle variations are important in internal combustion engines? Discuss the reason for cycle-to-cycle variations in SI, CI, and LTC (HCCI, RCCI, etc.) modes of engine operation.
2. Define the combustion stability in a reciprocating internal combustion engine. Discuss how cyclic combustion variability manifests in the engine performance.
3. Discuss the mechanism on how mixture composition (a type of fuel, overall equivalence ratio, and an overall fraction of diluents) affects the cycle-to-cycle variations (CCV) in SI engine. Describe the stage of combustion (sparking and flame initiation, initial flame kernel development, and turbulent flame propagation) most affected by each parameter. Explain how each parameter can be adjusted to minimize the CCV in SI engine.
4. Explain why cycle-to-cycle variations in conventional diesel engines are relatively lower than the conventional spark ignition engines.
5. Discuss the mechanism on how spark characteristics (spark plug gap vicinity, electrode shape, spark plugs number, and location) affects the cycle-to-cycle variations (CCV) in SI engine. Describe the stage of combustion (sparking and flame initiation, initial flame kernel development, and turbulent flame propagation) most affected by each parameter. Explain how each parameter can be adjusted to minimize the CCV in SI engine. Discuss the advantage in terms of cyclic variations gained by advanced ignition systems such as laser ignition.
6. Discuss the mechanism on how in-cylinder mixture motion (mean flow vector at the spark plug gap vicinity, overall in-cylinder flow pattern, turbulence intensity,

and scale) affects the cycle-to-cycle variations (CCV) in SI engine. Describe the stage of combustion most affected by each parameter and also explain how each parameter can be adjusted to minimize the CCV in SI engine. Suggest the engine design changes for minimization of cyclic combustion variability.

7. Discuss the reasons why low-temperature combustion modes (HCCI, RCCI, etc.) typically have relatively higher cyclic combustion variations than conventional diesel engines.

8. Discuss the effect of cyclic variations in equivalence ratio (ϕ), diluents, mean flow vector, spark discharge characteristics, and in-cylinder charging on cycle-to-cycle combustion variations in SI engine. Comment on the combustion stage which is most affected by the variations in each parameter.

9. Discuss the prior-cycle effects and same-cycle effects contributing to the cyclic combustion variations in homogeneous spark ignition engines and compression ignition engines.

10. Describe the sources of cyclic combustion variability in homogeneous and heterogeneous charge compression ignition engines.

11. Describe why fast burn operating conditions typically have lower cycle-to-cycle variations in reciprocating engines.

12. Explain the different indices based on in-cylinder pressure and heat release used for characterization of the combustion stability in reciprocating internal combustion engines. Discuss the significance of each parameter with respect to information revealed by each parameter.

13. Write the causes of partial burn and misfire cycles in reciprocating internal combustion engines.

14. How can you determine partial burn operation or misfire in SI and HCCI engine using in-cylinder pressure-based measurement methods?

15. Discuss the difference between statistical and nonlinear dynamics approach of combustion stability analysis. Write the advantages of both the approaches of cyclic variation analysis.

16. Define skewness and kurtosis of a data series, and discuss the significance of these statistical parameters with respect to the kind of information revealed. Discuss the different methods that can be used to test whether an engine data series has Gaussian or non-Gaussian distribution.

17. Define the coefficient of variation and lowest normalized value of IMEP data for a large number of engine cycles. Discuss the information revealed by COV_{IMEP} and LNV parameter at particular engine operating conditions.

18. Discuss the advantages of wavelet transform over Fourier transform. Write the typical applications of continuous wavelet transform and discrete wavelet transform. Explain how wavelets can be used for characterization of cycle-to-cycle variations under different engine operating conditions.

19. Discuss the differences between linear and nonlinear signal processing methods. Describe the additional features of engine cyclic combustion variations which can be extracted from the nonlinear analysis that are not available in the traditional statistical analysis.

20. Describe the methods to find the optimal embedding dimension and the time lag during phase space reconstruction. Describe the information depicted by phase space reconstruction, Poincare section, and return map.
21. Discuss the patterns by which return maps can reveal deterministic or stochastic component in the variation. Discuss the significance of quantile return map with normalized value.
22. Explain the methods that can be used to determine whether cyclic variations are random or deterministic in nature. Why this information is important for the engine designer and how these methods can be used to improve engine performance?
23. Explain the advantages of data symbolization. Discuss the methods to find the optimal sequence length and number of partition in symbol sequence analysis. Write the major issues with an inappropriate number of partition in the data.
24. Discuss the steps that can be taken during engine design and calibration to improve the combustion stability in SI, CI, and HCCI engines. Describe the effects of fuel on the combustion stability of engine and suggest fuel characteristics/properties for improving the combustion stability.

References

1. Kaul, B. C., Finney, C. E., Wagner, R. M., & Edwards, M. L. (2014). Effects of external EGR loop on cycle-to-cycle dynamics of dilute SI combustion. *SAE International Journal of Engines, 7*(2), 606–614.
2. Maurya, R. K. (2018). *Characteristics and control of low temperature combustion engines: Employing gasoline, ethanol and methanol.* Cham: Springer.
3. Jung, D., & Iida, N. (2017). Thermal and chemical effects of the in-cylinder charge at IVC on cycle-to-cycle variations of DME HCCI combustion with combustion-phasing retard by external and rebreathed EGR. *Applied Thermal Engineering, 113*, 132–149.
4. Kosmadakis, G. M., Rakopoulos, D. C., Arroyo, J., Moreno, F., Muñoz, M., & Rakopoulos, C. D. (2018). CFD-based method with an improved ignition model for estimating cyclic variability in a spark-ignition engine fueled with methane. *Energy Conversion and Management, 174*, 769–778.
5. Heywood, J. B. (1988). *Internal combustion engine fundamentals.* New York: McGraw-Hill.
6. Kaul, B., Jatana, G., & Wagner, R. (2016). High-dilution stoichiometric gasoline direct-injection (SGDI) combustion control development. In *DOE Hydrogen Program and Vehicle Technologies Annual Merit Review Presentation June 7, 2016.*
7. Wagner, R. (2011). Enabling high efficiency combustion through an improved understanding of cyclic dispersion. In *ERC Symposium on Future Engines and Their Fuels, University of Wisconsin, June 8, 2011.*
8. Thomas, S., Kannan, R. N., Saroop, A., & Sharma, S. (2013). *A study on the idle combustion stability of a CNG powered naturally aspirated engine* (No. 2013-26-0003). SAE Technical Paper.
9. Atkins, R. D. (2009). *An introduction to engine testing and development.* Warrendale: SAE International.
10. Ozdor, N., Dulger, M., & Sher, E. (1996). *An experimental study of the cyclic variability in spark ignition engines* (No. 960611). SAE Technical Paper.
11. Ozdor, N., Dulger, M., & Sher, E. (1994). Cyclic variability in spark ignition engines a literature survey. *SAE Transactions, 940987*, 1514–1552.

12. Young, M. B. (1981). Cyclic dispersion in the homogeneous-charge spark-ignition engine—A literature survey. *SAE Transactions, 810020*, 49–73.
13. Dai, W., Trigui, N., & Lu, Y. (2000). Modeling of cyclic variations in spark-ignition engines. *SAE Transactions,* (2000-01-2036), 1851–1861.
14. Mostafa, M. I. A. (2004). *Characterisation and modelling of cycle-to-cycle variations in spark-ignition engines* (PhD thesis). University of Waterloo, Ontario, Canada.
15. Heywood, J. B., & Vilchis, F. R. (1984). Comparison of flame development in a spark-ignition engine fueled with propane and hydrogen. *Combustion Science and Technology, 38*(5–6), 313–324.
16. Kalghatgi, G. T. (1985). Early flame development in a spark-ignition engine. *Combustion and Flame, 60*(3), 299–308.
17. Kalghatgi, G. T. (1987). Spark ignition, early flame development and cyclic variation in IC engines. *SAE Transactions*, 149–161.
18. Maly, R., & Vogel, M. (1979, January). Initiation and propagation of flame fronts in lean CH4-air mixtures by the three modes of the ignition spark. In *Symposium (International) on Combustion* (Vol. 17(1), pp. 821–831). Elsevier.
19. Ho, C. M., & Santavicca, D. A. (1987). Turbulence effects on early flame kernal growth. *SAE Transactions, 872100*, 505–512.
20. Pischinger, S., & Heywood, J. B. (1990). *How heat losses to the spark plug affect flame kernel development in an SI-engine*. SAE Technical Paper 900021.
21. Whitelaw, J. H., & Xu, H. M. (1995). Cyclic variations in a lean-burn spark ignition engine without and with swirl. *SAE Transactions*, 1202–1220.
22. Urushihara, T., Murayama, T., Takagi, Y., & Lee, K. H. (1995). Turbulence and cycle-by-cycle variation of mean velocity generated by swirl and tumble flow and effects on combustion. *SAE Transactions, 950813*, 1382–1389.
23. Chiu, C. P., & Horng, R. F. (1994). Effects of intake air temperature and residual gas concentration on cycle-to-cycle combustion variation in a two-stroke cycle SI engine equipped with an air-assisted fuel injection system. *JSME International Journal Series B Fluids and Thermal Engineering, 37*(4), 957–965.
24. Nogi, T., Shiraishi, T., Nakayama, Y., Ohsuga, M., & Kurihara, N. (1998). *Stability improvement of direct fuel injection engine under lean combustion operation* (No. 982703). SAE Technical Paper.
25. Brehob, D. D., & Newman, C. E. (1992). *Monte Carlo simulation of cycle by cycle variability* (No. 922165). SAE Technical Paper.
26. Shen, H., Hinze, P. C., & Heywood, J. B.. (1996). *A study of cycle-to-cycle variations in SI engines using a modified quasi-dimensional model* (No. 961187). SAE Technical Paper.
27. Schwarz, F., & Spicher, U. (2003). *Determination of residual gas fraction in IC engines* (No. 2003-01-3148). SAE Technical Paper.
28. Suyabodha, A. (2012). *Simulation of cyclic variability in gasoline engine under cold start conditions* (PhD thesis). University of Bath, UK.
29. Winterbone, D. E., & Pearson, R. J. (2000). *Theory of engine manifold design: Wave action methods for IC engines*. London: Professional Engineering Publication.
30. Batteh, J. J., & Curtis, E. W. (2003). *Modeling transient fuel effects with variable cam timing* (No. 2003-01-3126). SAE Technical Paper.
31. Batteh, J. J., & Curtis, E. W. (2005). *Modeling transient fuel effects with alternative fuels* (No. 2005-01-1127). SAE Technical Paper.
32. Bai, C., & Gosman, A. D. (1995). Development of methodology for spray impingement simulation. *SAE Transactions*, (950283), 550–568.
33. Divekar, P., Han, X., Yu, S., Chen, X., & Zheng, M. (2014). *The impact of intake dilution and combustion phasing on the combustion stability of a diesel engine* (No. 2014-01-1294). SAE Technical Paper.
34. Kyrtatos, P., Brückner, C., & Boulouchos, K. (2016). Cycle-to-cycle variations in diesel engines. *Applied Energy, 171*, 120–132.

35. Wing, R. D. (1975). The rotary fuel-injection pump as a source of cyclic variation in diesel engines, and its effect on nitric oxide emissions. *Proceedings of the Institution of Mechanical Engineers, 189*(1), 497–505.

36. Zhong, L., Singh, I. P., Han, J. S., Lai, M. C., Henein, N. A., & Bryzik, W. (2003). *Effect of cycle-to-cycle variation in the injection pressure in a common rail diesel injection system on engine performance* (No. 2003-01-0699). SAE Technical Paper.

37. Yang, Z., Steffen, T., & Stobart, R. (2013). *Disturbance sources in the diesel engine combustion process* (No. 2013-01-0318). SAE Technical Paper.

38. Schmillen, K., & Wolschendorf, J. (1989). *Cycle-to-cycle variations of combustion noise in diesel* (No. 890129). SAE Technical Paper.

39. Rakopoulos, C. D., & Giakoumis, E. G. (2009). *Diesel engine transient operation: Principles of operation and simulation analysis.* London: Springer.

40. Han, Z., Henein, N., Nitu, B., & Bryzik, W. (2001). *Diesel engine cold start combustion instability and control strategy* (No. 2001-01-1237). SAE Technical Paper.

41. McGhee, M., Shayler, P. J., LaRocca, A., Murphy, M., & Pegg, I. (2012). The influence of injection strategy and glow plug temperature on cycle by cycle stability under cold idling conditions for a low compression ratio, HPCR diesel engine. *SAE International Journal of Engines, 5*(3), 923–937.

42. Yun, H., Kang, J. M., Chang, M. F., & Najt, P. (2010). *Improvement on cylinder-to-cylinder variation using a cylinder balancing control strategy in gasoline HCCI engines* (No. 2010-01-0848). SAE Technical Paper.

43. Li, H., Neill, W. S., & Chippior, W. L. (2012). An experimental investigation of HCCI combustion stability using n-heptane. *Journal of Energy Resources Technology, 134*(2), 022204.

44. Jungkunz, A. F. (2013). *Actuation strategies for cycle-to-cycle control of homogeneous charge compression ignition combustion engines* (Doctoral dissertation, Stanford University).

45. Shahbakhti, M., & Koch, C. R. (2008). Characterizing the cyclic variability of ignition timing in a homogeneous charge compression ignition engine fuelled with n-heptane/iso-octane blend fuels. *International Journal of Engine Research, 9*(5), 361–397.

46. Pan, J., Sheppard, C. G. W., Tindall, A., Berzins, M., Pennington, S. V., & Ware, J. M. (1998). *End gas inhomogeneity, autoignition and knock* (No. 982616). SAE Technical Paper.

47. Yu, R., Bai, X. S., Lehtiniemi, H., Ahmed, S. S., Mauss, F., Richter, M., ... & Hultqvist, A. (2006). *Effect of turbulence and initial temperature inhomogeneity on homogeneous charge compression ignition combustion* (No. 2006-01-3318). SAE Technical Paper.

48. Dronniou, N., & Dec, J. (2012). Investigating the development of thermal stratification from the near-wall regions to the bulk-gas in an HCCI engine with planar imaging thermometry. *SAE International Journal of Engines, 5*(3), 1046–1074.

49. Chiang, C. J., & Stefanopoulou, A. G. (2007). Stability analysis in homogeneous charge compression ignition (HCCI) engines with high dilution. *IEEE Transactions on Control Systems Technology, 15*(2), 209–219.

50. Grünefeld, G., Beushausen, V., Andresen, P., & Hentschel, W. (1994). A major origin of cyclic energy conversion variations in SI engines: Cycle-by-cycle variations of the equivalence ratio and residual gas of the initial charge. *SAE Transactions*, 882–893.

51. Klos, D., & Kokjohn, S. L. (2015). Investigation of the sources of combustion instability in low-temperature combustion engines using response surface models. *International Journal of Engine Research, 16*(3), 419–440.

52. Matekunas, F. A. (1983). Modes and measures of cyclic combustion variability. *SAE Transactions*, 1139–1156.

53. Wang, S., & Ji, C. (2012). Cyclic variation in a hydrogen-enriched spark-ignition gasoline engine under various operating conditions. *International Journal of Hydrogen Energy, 37*(1), 1112–1119.

54. Yu, X., Wu, H., Du, Y., Tang, Y., Liu, L., & Niu, R. (2016). Research on cycle-by-cycle variations of an SI engine with hydrogen direct injection under lean burn conditions. *Applied Thermal Engineering, 109*, 569–581.

55. Cesario, N., Tagliatatela, F., & Lavorgna, M. (2006). Methodology for misfire and partial burning diagnosis in SI engines. *IFAC Proceedings Volumes, 39*(16), 1024–1028.
56. Cavina, N., Luca, P., & Sartoni, G. (2011). Misfire and partial burn detection based on ion current measurement. *SAE International Journal of Engines, 4*(2), 2451–2460.
57. Moro, D., Azzoni, P., & Minelli, G. (1998). *Misfire pattern recognition in high performance SI 12-cylinder engine*. SAE Technical Paper 980521.
58. Ponti, F. (2005, January). Development of a torsional behavior powertrain model for multiple misfire detection. In *ASME 2005 Internal Combustion Engine Division Spring Technical Conference* (pp. 237–251). American Society of Mechanical Engineers.
59. Bahri, B., Aziz, A. A., Shahbakhti, M., & Said, M. F. M. (2013). Understanding and detecting misfire in an HCCI engine fuelled with ethanol. *Applied Energy, 108*, 24–33.
60. Peterson, B., Reuss, D. L., & Sick, V. (2011). High-speed imaging analysis of misfires in a spray-guided direct injection engine. *Proceedings of the Combustion Institute, 33*(2), 3089–3096.
61. Sjöberg, M., & Dec, J. E. (2007). Comparing late-cycle autoignition stability for single-and two-stage ignition fuels in HCCI engines. *Proceedings of the Combustion Institute, 31*(2), 2895–2902.
62. Ghazimirsaied, A., Shahbakhti, M., & Koch, C. R. (2010, May). Recognizing partial burn operation in an HCCI engine. In *2010 Combustion Institute-Canadian Section (CICS) Spring Technical Conference* (pp. 9–12).
63. Saxena, S., & Bedoya, I. D. (2013). Fundamental phenomena affecting low temperature combustion and HCCI engines, high load limits and strategies for extending these limits. *Progress in Energy and Combustion Science, 39*(5), 457–488.
64. Ghazimirsaied, A., Shahbakhti, M., & Koch, C. R. (2009, May). Partial-burn crank angle limit criteria comparison on an experimental HCCI engine. In *Proceeding of Combustion Institute-Canadian Section Spring Technical Meeting, University of Montreal, Quebec* (pp. 11–13).
65. Johansson, T., Johansson, B., Tunestål, P., & Aulin, H. (2010). *Turbocharging to extend HCCI operating range in a multi cylinder engine-benefits and limitations*. FISITA2010, (F2010A037).
66. Kuroda, H., Nakajima, Y., Sugihara, K., Takagi, Y., & Muranaka, S. (1978). *The fast burn with heavy EGR, new approach for low NOx and improved fuel economy* (No. 780006). SAE Technical Paper.
67. Shayler, P. J., Winborn, L. D., Hill, M. J., & Eade, D. (2000). *The influence of gas/fuel ratio on combustion stability and misfire limits of spark ignition engines* (No. 2000-01-1208). SAE Technical Paper.
68. Komachiya, M., Kurihara, N., Kodama, A., Sakaguchi, T., Fumino, T., & Watanabe, S. (1998). *A method of misfire detection by superposing outputs of combustion pressure sensors* (No. 982588). SAE Technical Paper.
69. Bahri, B., Aziz, A. A., Shahbakhti, M., & Said, M. M. (2012). Misfire detection based on statistical analysis for an ethanol fuelled HCCI engine. *International Review of Mechanical Engineering (IREME), 6*(6), 1276–1282.
70. Wu, Z. J., & Lee, A. (1998). *Misfire detection using a dynamic neural network with output feedback* (No. 980515). SAE Technical Paper.
71. Ghazimirsaied, A., & Koch, C. R. (2012). Controlling cyclic combustion timing variations using a symbol-statistics predictive approach in an HCCI engine. *Applied Energy, 92*, 133–146.
72. Ghazimirsaied, A., Shahbakhti, M., & Koch, C. R. (2011, May). Ignition timing criteria for partial burn operation in an HCCI engine. In *Proc. of CI/CS Conference*.
73. Ghazimirsaied, A. (2012). *Extending HCCI low load operation using Chaos prediction and feedback control* (Doctoral dissertation, University of Alberta).
74. Truffin, K., Angelberger, C., Richard, S., & Pera, C. (2015). Using large-eddy simulation and multivariate analysis to understand the sources of combustion cyclic variability in a spark-ignition engine. *Combustion and Flame, 162*(12), 4371–4390.

75. Green, J. B. (2000). *Application of deterministic chaos theory to cyclic variability in spark-ignition engines* (PhD thesis). Georgia Institute of Technology, USA.
76. Sen, A. K., Litak, G., Finney, C. E., Daw, C. S., & Wagner, R. M. (2010). Analysis of heat release dynamics in an internal combustion engine using multifractals and wavelets. *Applied Energy, 87*(5), 1736–1743.
77. Medina, A., Curto-Risso, P. L., Hernández, A. C., Guzmán-Vargas, L., Angulo-Brown, F., & Sen, A. K. (2014). *Quasi-dimensional simulation of spark ignition engines.* Berlin: Springer.
78. Maurya, R. K., & Agarwal, A. K. (2013). Experimental investigation of cyclic variations in HCCI combustion parameters for gasoline like fuels using statistical methods. *Applied Energy, 111*, 310–323.
79. Saxena, M. R., & Maurya, R. K. (2018). *Effect of diesel injection timing on peak pressure rise rate and combustion stability in RCCI engine* (No. 2018-01-1731). SAE Technical Paper.
80. NIST/SEMATECH. (2012, April). *e-Handbook of statistical methods.* Retrieved from http://www.itl.nist.gov/div898/handbook.
81. Hoard, J., & Rehagen, L. (1997). *Relating subjective idle quality to engine combustion* (No. 970035). SAE Technical Paper.
82. Suzuki, K., Uehara, K., Murase, E., & Nogawa, S. (2016, November). Study of ignitability in strong flow field. In *International Conference on Ignition Systems for Gasoline Engines* (pp. 69–84). Cham: Springer.
83. Reyes, M., Tinaut, F. V., Giménez, B., & Pérez, A. (2015). Characterization of cycle-to-cycle variations in a natural gas spark ignition engine. *Fuel, 140*, 752–761.
84. Schenk, M., Schauer, F. X., Sauer, C., Weber, G., Hahn, J., & Schwarz, C. (2016, November). Challenges to the ignition system of future gasoline engines—An application oriented systems comparison. In *International Conference on Ignition Systems for Gasoline Engines* (pp. 3–25). Cham: Springer.
85. Gong, C., Huang, K., Chen, Y., Jia, J., Su, Y., & Liu, X. (2011). Cycle-by-cycle combustion variation in a DISI engine fueled with methanol. *Fuel, 90*(8), 2817–2819.
86. Brandt, M., Hettinger, A., Schneider, A., Senftleben, H., & Skowronek, T. (2016, November). Extension of operating window for modern combustion systems by high performance ignition. In *International Conference on Ignition Systems for Gasoline Engines* (pp. 26–51). Cham: Springer.
87. Kantz, H., & Schreiber, T. (1997). *Nonlinear time series analysis.* Cambridge: Cambridge University Press.
88. Chatfield, C. (2016). *The analysis of time series: An introduction.* Boca Raton, FL: CRC Press.
89. Finney, C. E. A. (1995). *Identification and characterization of determinism in spark-ignition internal combustion engines* (Master's thesis). University of Tennessee, Knoxville.
90. Daw, C. S., Wagner, R. M., Edwards, K. D., & Green, J. B., Jr. (2007). Understanding the transition between conventional spark-ignited combustion and HCCI in a gasoline engine. *Proceedings of the Combustion Institute, 31*(2), 2887–2894.
91. Sen, A. K., Longwic, R., Litak, G., & Gorski, K. (2008). Analysis of cycle-to-cycle pressure oscillations in a diesel engine. *Mechanical Systems and Signal Processing, 22*(2), 362–373.
92. Bizon, K., Continillo, G., Lombardi, S., Sementa, P., & Vaglieco, B. M. (2016). Application of independent component analysis for the study of flame dynamics and cyclic variation in spark ignition engines. *Combustion Science and Technology, 188*(4–5), 637–650.
93. Maurya, R. K., & Agarwal, A. K. (2012). Statistical analysis of the cyclic variations of heat release parameters in HCCI combustion of methanol and gasoline. *Applied Energy, 89*(1), 228–236.
94. Foakes, A. P., & Pollard, D. G. (1993). Investigation of a chaotic mechanism for cycle-to-cycle variations. *Combustion Science and Technology, 90*(1–4), 281–287.
95. Daily, J. W. (1988). Cycle-to-cycle variations: A chaotic process? *Combustion Science and Technology, 57*(4–6), 149–162.
96. Daw, C. S., Kennel, M. B., Finney, C. E. A., & Connolly, F. T. (1998). Observing and modeling nonlinear dynamics in an internal combustion engine. *Physical Review E, 57*(3), 2811.

97. Maurya, R. K., & Nekkanti, A. (2017). Combustion instability analysis using wavelets in conventional diesel engine. In *Mathematical Concepts and Applications in Mechanical Engineering and Mechatronics* (pp. 390–413). IGI Global.
98. Rajagopalan, V. (2007). *Symbolic dynamic filtering of complex systems* (PhD thesis). The Pennsylvania State University.
99. Kaiser, G. (2010). *A friendly guide to wavelets*. London: Springer.
100. Torrence, C., & Compo, G. P. (1998). A practical guide to wavelet analysis. *Bulletin of the American Meteorological Society, 79*, 61–78.
101. Sen, A. K., Litak, G., Edwards, K. D., Finney, C. E., Daw, C. S., & Wagner, R. M. (2011). Characteristics of cyclic heat release variability in the transition from spark ignition to HCCI in a gasoline engine. *Applied Energy, 88*, 1649–1655.
102. Sen, A. K., Litak, G., Taccani, R., & Radu, R. (2008). Wavelet analysis of cycle-to-cycle pressure variations in an internal combustion engine. *Chaos, Solitons & Fractals, 38*(3), 886–893.
103. Maurya, R. K., & Akhil, N. (2018). *Experimental investigation on effect of compression ratio, injection pressure and engine load on cyclic variations in diesel engine using wavelets* (No. 2018-01-5007). SAE Technical Paper.
104. Barford, L. A., Fazzio, R. S., & Smith, D. R. (1992). *An introduction to wavelets*. Hewlett-Packard Laboratories, Technical Publications Department.
105. Sifuzzaman, M., Islam, M. R., & Ali, M. Z. (2009). Application of wavelet transform and its advantages compared to Fourier transform. *Journal of Physical Sciences, 13*, 121–134.
106. Kumar, P., & Foufoula-Georgiou, E. (1997). Wavelet analysis for geophysical applications. *Reviews of Geophysics, 35*(4), 385–412.
107. Abarbanel, H. (2012). *Analysis of observed chaotic data*. New York: Springer.
108. Baker, G. L., Baker, G. L., & Gollub, J. P. (1996). *Chaotic dynamics: An introduction*. Cambridge: Cambridge University Press.
109. Li, G. X., & Yao, B. F. (2008). Nonlinear dynamics of cycle-to-cycle combustion variations in a lean-burn natural gas engine. *Applied Thermal Engineering, 28*(5–6), 611–620.
110. Takens, F. (1981). Detecting strange attractors in turbulence. In *Dynamical systems and turbulence, Warwick 1980* (pp. 366–381). Berlin: Springer.
111. Ding, S. L., Song, E. Z., Yang, L. P., Litak, G., Yao, C., & Ma, X. Z. (2016). Investigation on nonlinear dynamic characteristics of combustion instability in the lean-burn premixed natural gas engine. *Chaos, Solitons & Fractals, 93*, 99–110.
112. Fraser, A. M., & Swinney, H. L. (1986). Independent coordinates for strange attractors from mutual information. *Physical Review A, 33*(2), 1134.
113. Kennel, M. B., Brown, R., & Abarbanel, H. D. (1992). Determining embedding dimension for phase-space reconstruction using a geometrical construction. *Physical Review A, 45*(6), 3403.
114. Daw, C. S., Finney, C. E. A., Green, J. B., Kennel, M. B., Thomas, J. F., & Connolly, F. T. (1996). *A simple model for cyclic variations in a spark-ignition engine* (No. 962086). SAE Technical Paper.
115. Scholl, D., & Russ, S. (1999). *Air-fuel ratio dependence of random and deterministic cyclic variability in a spark-ignited engine* (No. 1999-01-3513). SAE Technical Paper.
116. Kaul, B. C., Vance, J. B., Drallmeier, J. A., & Sarangapani, J. (2009). A method for predicting performance improvements with effective cycle-to-cycle control of highly dilute spark ignition engine combustion. *Proceedings of the Institution of Mechanical Engineers, Part D: Journal of Automobile Engineering, 223*(3), 423–438.
117. Daw, C. S., Finney, C. E. A., & Tracy, E. R. (2003). A review of symbolic analysis of experimental data. *Review of Scientific Instruments, 74*(2), 915–930.
118. Finney, C. E. A., Green, J. B., & Daw, C. S. (1998). *Symbolic time-series analysis of engine combustion measurements* (No. 980624). SAE Technical Paper.
119. Kaul, B., Wagner, R., & Green, J. (2013). Analysis of cyclic variability of heat release for high-EGR GDI engine operation with observations on implications for effective control. *SAE International Journal of Engines, 6*(1), 132–141.

120. Maurya, R. K. (2018). Experimental investigation of deterministic and random cyclic patterns in HCCI engine using symbol sequence approach. *Iranian Journal of Science and Technology, Transactions of Mechanical Engineering*, 1–12.

121. Green, J. B., Daw, C. S., Armfield, J. S., Finney, C. E. A., Wagner, R. M., Drallmeier, J. A., . . . Durbetaki, P. (1999). *Time irreversibility and comparison of cyclic-variability models* (No. 1999-01-0221). SAE Technical Paper.

122. Eckmann, J. P., Kamphorst, S. O., & Ruelle, D. (1987). Recurrence plots of dynamical systems. *EPL (Europhysics Letters), 4*(9), 973.

123. Marwan, N., Romano, M. C., Thiel, M., & Kurths, J. (2007). Recurrence plots for the analysis of complex systems. *Physics Reports, 438*(5–6), 237–329.

124. Casdagli, M. C. (1997). Recurrence plots revisited. *Physica D: Nonlinear Phenomena, 108* (1–2), 12–44.

125. Marwan, N., Wessel, N., Meyerfeldt, U., Schirdewan, A., & Kurths, J. (2002). Recurrence-plot-based measures of complexity and their application to heart-rate-variability data. *Physical Review E, 66*(2), 026702.

126. Litak, G., Kamiński, T., Czarnigowski, J., Żukowski, D., & Wendeker, M. (2007). Cycle-to-cycle oscillations of heat release in a spark ignition engine. *Meccanica, 42*(5), 423–433.

127. Marwan, N. (2003). *Encounters with neighbours: Current developments of concepts based on recurrence plots and their applications* (PhD thesis). Universität Potsdam, Potsdam.

128. Webber, C. L., Jr., & Zbilut, J. P. (1994). Dynamical assessment of physiological systems and states using recurrence plot strategies. *Journal of Applied Physiology, 76*(2), 965–973.

129. Chatterjee, S., & Mallik, A. K. (1996). Three kinds of intermittency in a nonlinear mechanical system. *Physical Review E, 53*(5), 4362.

130. Marwan, N., & Meinke, A. (2004). Extended recurrence plot analysis and its application to ERP data. *International Journal of Bifurcation and Chaos, 14*(02), 761–771.

131. Wendeker, M., Litak, G., Czarnigowski, J., & Szabelski, K. (2004). Nonperiodic oscillations of pressure in a spark ignition combustion engine. *International Journal of Bifurcation and Chaos, 14*(05), 1801–1806.

132. Gottwald, G. A., & Melbourne, I. (2004, February). A new test for chaos in deterministic systems. In *Proceedings of the Royal Society of London A: Mathematical, Physical and Engineering Sciences* (Vol. 460(2042), pp. 603–611). The Royal Society.

133. Gottwald, G. A., & Melbourne, I. (2009). On the validity of the 0–1 test for chaos. *Nonlinearity, 22*(6), 1367.

134. Gottwald, G. A., & Melbourne, I. (2009). On the implementation of the 0–1 test for chaos. *SIAM Journal on Applied Dynamical Systems, 8*(1), 129–145.

135. Ding, S. L., Song, E. Z., Yang, L. P., Litak, G., Wang, Y. Y., Yao, C., & Ma, X. Z. (2017). Analysis of chaos in the combustion process of premixed natural gas engine. *Applied Thermal Engineering, 121*, 768–778.

136. Litak, G., Syta, A., Budhraja, M., & Saha, L. M. (2009). Detection of the chaotic behaviour of a bouncing ball by the 0–1 test. *Chaos, Solitons & Fractals, 42*(3), 1511–1517.

137. Krese, B., & Govekar, E. (2012). Nonlinear analysis of laser droplet generation by means of 0–1 test for chaos. *Nonlinear Dynamics, 67*(3), 2101–2109.

138. Peng, C. K., Havlin, S., Stanley, H. E., & Goldberger, A. L. (1995). Quantification of scaling exponents and crossover phenomena in nonstationary heartbeat time series. *Chaos: An Interdisciplinary Journal of Nonlinear Science, 5*(1), 82–87.

139. Sen, A. K., Litak, G., Kaminski, T., & Wendeker, M. (2008). Multifractal and statistical analyses of heat release fluctuations in a spark ignition engine. *Chaos: An Interdisciplinary Journal of Nonlinear Science, 18*(3), 033115.

140. Curto-Risso, P. L., Medina, A., Hernández, A. C., Guzman-Vargas, L., & Angulo-Brown, F. (2010). Monofractal and multifractal analysis of simulated heat release fluctuations in a spark ignition heat engine. *Physica A: Statistical Mechanics and its Applications, 389*(24), 5662–5670.

141. Gęca, M., & Litak, G. (2017). Mean effective pressure oscillations in an IC-SI engine after the addition of hydrogen-rich gas. *Measurement, 108*, 18–25.

Chapter 9
Knocking and Combustion Noise Analysis

Abbreviations and Symbols

AEFD	Average energy in frequency domain
AEHR	Average energy of the heat release
AEHRO	Average energy of heat release oscillations
ATDC	After top dead center
BDC	Bottom dead center
BTDC	Before top dead center
CA	Crank angle
CAD	Crank angle degree
CDC	Convention diesel combustion
CDF	Cumulative distribution function
CI	Compression ignition
CNL	Combustion noise level
COV	Coefficient of variation
CPL	Cylinder pressure level
dB	Decibel
DI	Direct injection
DISI	Direct injection spark ignition
DKI	Dimensionless knock indicator
DMP	Derivative at maximum pressure position
EGR	Exhaust gas recirculation
EMD	Empirical mode decomposition
EOC	End of combustion
FFT	Fast Fourier transform
GCI	Gasoline compression ignition
GIMEP	Gross indicated mean effective pressure
HCCI	Homogeneous charge compression ignition
HRR	Heat release rate

© Springer Nature Switzerland AG 2019
R. K. Maurya, *Reciprocating Engine Combustion Diagnostics*, Mechanical
Engineering Series, https://doi.org/10.1007/978-3-030-11954-6_9

ID   Ignition delay
IMEP   Indicated mean effective pressure
IMPG   Integral of modulus of pressure gradient
IMPO   Integral of the modulus of pressure oscillations
IVC   Intake valve closing
KDI   Knocking damage index
KI   Knock intensity
KI1   Knocking index
KLSA   Knock-limited spark advance
KO   Knock onset
LSPI   Low-speed preignition
MAHRO   Maximum amplitude of heat release oscillations
MAPO   Maximum amplitude of pressure oscillations
MIAA   Main injection advanced angle
MNL   Mechanical noise level
MON   Motor octane number
MPRR   Maximum pressure rise rate
NCS   Noise-canceling spike
NVH   Noise vibration and harshness
OH   Hydroxyl radical
OI   Octane index
ON   Octane number
ON   Overall noise
PCCI   Premixed charge compression ignition
PDF   Probability density function
PFI   Port fuel injection
PIC   Pilot injection combustion
PPC   Partially premixed combustion
PPRR   Peak pressure rise rate
PRF   Primary reference fuel
PRR   Pressure rise rate
PTP   Peak to peak
RCCI   Reactivity controlled compression ignition
RI   Ringing intensity
RMS   Root mean square
$ROHR_u$ Rate of heat release in unburnt charge
RON   Research octane number
RSD   Relative standard deviation
S   Sensitivity
SA   Structure attenuation
SACI   Spark-assisted compression ignition
SEHRO   Signal energy of heat release oscillations
SEPO   Signal energy of pressure oscillations
SER   Signal energy ratio

SI	Spark ignition
SNR	Signal-to-noise ratio
SPL	Sound pressure level
ST	Spark timing
STFT	Short-time Fourier transform
TDC	Top dead center
TN	Toluene number
TRF	Toluene reference fuels
TVE	Threshold value exceeded
WD	Wigner distribution
A	Area of the reaction front
a	Speed of sound
$B_{i,j}$	Bessel constant
D	Cylinder diameter
E_{res}	Signal energy of the resonance pressure oscillations
$f_{i,j}$	Resonance frequency
f_{Nyq}	Nyquist frequency
f_s	Sampling frequency
I_1	Combustion indicator
I_2	Resonance indicator
n_{idle}	Idle engine speed
P	Pressure
p_{bp}	Band-pass filtered pressure
P_{in}	Intake pressure
P_{max}	Maximum pressure
P_{RMS}	Root mean square (RMS) value of the filtered pressure
p_{ub}	Unburned gas pressure
R^2	Correlation coefficient
T	Temperature
T_{comp15}	Compression pressure of 15 bar
t_{IVC}	Time of intake valve closing
t_{KNOCK}	Time of knock onset
T_{max}	Maximum temperature
T_{ub}	Unburned temperature
u_a	Velocity relative to the unburned gas
V	Volume
τ	Autoignition delay time
θ	Crank angle position
ε	Dimensionless reactivity parameter of hot spot
ξ	Dimensionless resonance parameter
μ	Mean
γ	Specific heat ratio
σ	Standard deviation

9.1 Introduction

Knock is one of the key phenomena which requires close observation during the engine development and calibration process. During the engine development process, combustion chamber design needs to ensure some resilience to knocking combustion particularly for engine operation over a wide range of fuels. During engine production stage, the tuning of the engine control system needs to ensure the non-knocking combustion over a wide range of operating conditions. In spark ignition (SI) engines, knock represents one of the major constraints on performance and efficiency because it limits the maximum value of the engine compression ratio. Additional reasons for investigating engine knock are that when it's very severe, it can also quickly lead to engine damage [1–3]. Furthermore, being a source of noise, engine knock is typically considered a drivability issue. The occurrence of engine knock over a long period of time leads to adverse effects such as breakage of piston rings, piston crown and top land erosion, engine cylinder head erosion, melting of piston, increasing emissions, significant increase in fuel consumption, piston ring sticking, cylinder head gasket leakage, cylinder bore scuffing, reduction in engine efficiency, limiting the vehicle acceleration performance, and possibility of structural harms to engine [1, 3].

Combustion in SI engine can proceed as a normal or abnormal combustion process depending on engine operating conditions and engine design. Figure 9.1 illustrates the normal and abnormal combustion phenomena occurring in the SI engines. In normal combustion, the combustion of the fuel-air mixture begins at the correct time, and solely via the ignition system, that produces a controlled and

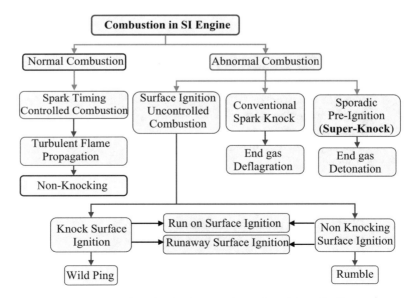

Fig. 9.1 Normal and abnormal combustion in SI engines (adapted from [2, 4])

predictable burning (energy release) of the in-cylinder charge. In the normal combustion of SI engine, the entire charge is consumed by turbulent flame propagation, and combustion is controlled by the spark timing, and via charge state (turbulence).

Abnormal combustion in SI engine can refer to a variety of conditions where normal combustion does not take place in the combustion chamber. The abnormal combustion includes modes wherein the turbulent flame fails to consume all of the charge—known as partial burns or misfires—as well as conditions such that the flame front is initiated before or after the timed spark ignition by other means (i.e., pre- or post-ignition) or wherein some or all of the charge is combusted spontaneously (Fig. 9.1) [2, 5]. The autoignition of the fuel-air mixture ahead of the propagating flame front leads to uncontrolled burning and high instantaneous energy release which results into local over-pressurization in the gas. This produces damaging, high-amplitude pressure waves that impart the acoustic resonance, which is typically referred as knock. Knocking is generally referred to the noise transmitted to the engine structure due to the autoignition of unburned end gas charge [2].

There are different types of the engine knock distinguished by the source of the ignition. Preignition is autoignition of the fuel mixture well before the spark ignition. Surface ignition is occurred when the engine surface is hot enough to ignite the fuel mixture. Preignition can occur due to several reasons including too hot spark plug, carbon deposits that remain incandescent, sharp edges in the combustion chamber, overheating, and valves operating at higher than normal temperature because of excessive guide clearance or improper seal with valve seats. Surface ignition is typically initiation of the flame front by a hot surface other than the spark, and it may occur before the spark (preignition) or after the spark (post ignition) [2]. Knocking due to surface ignition is initiated by preignition caused by glowing combustion chamber deposits or other sources, and the severity of knock depends on the timing of preignition. Run-on surface ignition occurs when engine continues to fire after the spark is switched off. Run-away surface ignition condition appears when surface ignition occurs earlier and earlier in the cycle which leads to overheating and damage to the engine. Wild ping is defined as irregular sharp combustion knock caused by early surface ignition from deposits. Non-knocking surface ignition is typically associated with the surface ignition that occurs late in the combustion cycle. Rumble is a relatively stable low profile noise (600–1200 Hz) caused by deposit surface ignition generating high rates of pressure rise in the cylinder [2]. The surface ignition problems can be addressed by adequate attention to engine design as well as fuel and lubricant quality.

Spark knock is recurrent and repeatable (occurring more than occasionally) in terms of audibility, and it is controlled by spark timing. The knock varies significantly cycle to cycle and cylinder to cylinder (in multicylinder engine). The knocking occurs due to end-gas autoignition of unburned charge in the front of the turbulent propagating flame. Figure 9.2 illustrates the knocking engine cycle and depicts the different combustion parameters. In the knocking engine cycle, the combustion process occurs in two stages, i.e., turbulent flame propagation initiated by a spark and end-gas autoignition causing pressure oscillations in the combustion

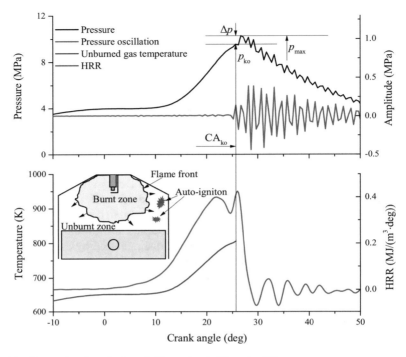

Fig. 9.2   Conventional engine knock illustration in SI engine [1]

chamber. The flame propagation stage is considered between spark timing to crank angle position of the onset of pressure oscillation (CA_{ko}) (Fig. 9.2). The heat release rate (HRR) increases during flame propagation stage with a possible short-term drop due to the movement of the piston toward BDC and heat transfer [1]. The unburned gas temperature stably increases due to the compression heating effect of the burned gas and propagating flame, as well as the compression induced by piston motion. Knocking is the abnormal combustion occurring in SI engine by the autoignition of the unburned fuel and air mixture before the flame front consumes it. This occurs after spark plug ignition and causes the high-frequency pressure oscillation that can be audibly detected. In the autoignition stage of the combustion process, the pressure signal in the cylinder at first increases dramatically reaching a peak value, and then it oscillates with decaying amplitude. The amplitude of this oscillation after a high-pass filter is used as an indicator to characterize the severity of the knock event. Thus, engine knocking is usually defined as the "clanging" or "pinging" sound that vibrations of the engine's block (and head) produce when these structures are excited by pressure waves generated inside the cylinder. Knock limitation is severe in case of low engine speeds due to lower turbulence levels in the cylinder and hence low flame speeds. This gives the end gas longer residence time at higher temperature to overcome the chemical delay. It can be noted that knocking occurs at the end of combustion in SI engine.

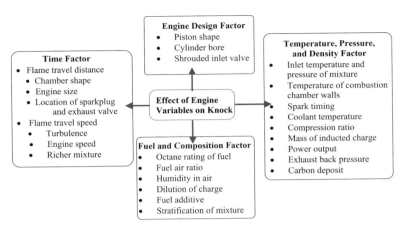

Fig. 9.3 Summary of factors affecting SI engine knock

There are certain engine operating conditions under which knock is most likely: generally high-load conditions, particularly at low engine speed. There are several variables which affect the engine knocking in SI engine, are summarized in Fig. 9.3. The parameters are categorized in four factors, namely, (1) engine design factor; (2) temperature, pressure, and density factor; (3) time factor; and (4) fuel and its composition factor.

In general, any action which tends to lower the temperature of unburned charge will tend to prevent the knocking by reducing the possibility of unburned charge to reach to the autoignition temperature. Factors reducing the density of charge tend to lower the possibility of knocking by providing lower energy release (lower rates of pre-combustion reactions in the end charge). Factors listed in temperature, pressure and density factor (Fig. 9.3) affects the knocking. Time spent by turbulent flame to consume entire charge affects the tendency of knocking, and the factors involved are presented as a time factor. Flame travel distance and flame travel speed are responsible for time required for consuming the charge by turbulent flame. The factors increase the time for entire charge consumption by turbulent flame, increases the tendency of knocking combustion. In shorter combustion duration, less time is available for which the unburned charge is exposed to high temperature and pressure. The factors such as small engine bore, compact combustion chamber, and central spark plug location decrease the flame travel distance. A high flame speed provides faster combustion leading to shorter flame propagation time. The flame speed increases with higher turbulence, which increases with the engine speed. Additionally, the engine design factors such as piston shape or shrouded valves can also be used to increase the turbulence in the combustion chamber which affects the engine knocking.

Fuel chemical composition and air-fuel ratio are the two main factors which affect the chemical characteristics of charge. Dilution of charge (by EGR or residual gases) affects the charge composition and knocking behavior. Isoparaffins, aromatics, and olefinic hydrocarbons have higher octane number (ON), while longer chain n-paraffins have poor knock resistance [2]. The octane number is typically used to define autoignition or knock-resistant quality of fuel. Alcohols such as methanol or ethanol are high-octane fuels and thus, they have lower tendency of knock in SI engine. Dilution by residual gases lowers the tendency of knocking as residuals are inert gases and acts as sink absorbing some energy released which reduces the chemical reaction rate and retards the combustion. Humidity affects the engine knock similar to charge dilution. In full-load operation, cooled EGR is shown to reduce knock tendency. The main mechanism for this improvement is by an increase of the specific heat capacity of the charge air and reduced combustion temperatures, thereby enabling a more favorable combustion phasing [6]. The benefit of EGR is that stoichiometric operation is maintained and a three-way catalyst can be used while diluting the mixture.

Substantial cycle-to-cycle variability also exists in the knock phenomenon. Cyclic variability is largely caused by cycle-to-cycle variations in the burn rate. The burn rate is affected by several engine operating parameters, which are susceptible to cycle-to-cycle variations. Flame geometry also varies cycle-to-cycle basis, and thus, end-gas location and shape will vary. With four-valve cylinder heads with a central spark plug location, the circumferential location of the end-gas region is unclear and may vary significantly cycle by cycle. Additionally, end-gas composition and temperature nonuniformities can be significant which results in variability of the autoignition process and subsequent rate of chemical energy release. All of these factors result in the cycle-to-cycle variability in the knocking phenomenon being substantially larger than the cycle-to-cycle combustion variability [7].

9.2   Knock Fundamentals

9.2.1   Knock Onset

Knocking combustion in SI engine is caused by autoignition of the end gas ahead of the advancing flame front, and it depends on the pressure and temperature history of the end gas. Knock onset is determined by chemical kinetics as the pressure and temperature in the end gas increase with crank angle (as combustion proceeds). The autoignition of end-gas charge during knocking operation is validated using optical experimental measurements. Figure 9.4 depicts the series of images (corresponding to light emission from the high-temperature burned gases) of combustion during knocking and non-knocking operating conditions. The figure shows the normal flame front propagation in non-knocking operating conditions. During knocking operating conditions, image A shows the normal flame front propagation, and the position of the combustion flame with the dark crescent-shaped end-gas region ahead of it, before any autoignition. In image B, the hot spots occur at the upper left of the frame, which creates the autoignition region in the end gas, the autoignition region moves upward, and it is brighter and hotter. The autoignition

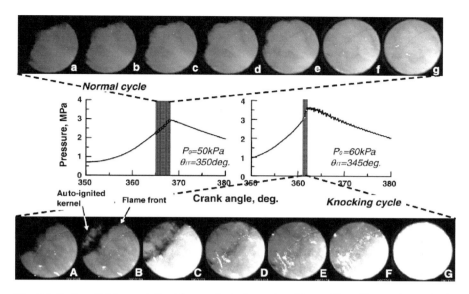

Fig. 9.4 Series of high-speed direct images for both non-knocking and knocking engine cycles in SI engine [8]

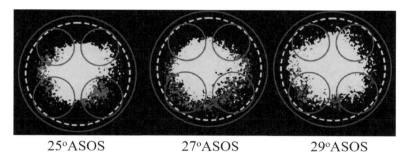

Fig. 9.5 HCO (red) and OH (cyan) spatial distributions during the knocking combustion cycle [9]

region has shifted to the right with propagation of the autoignition region, and, finally, the end gas is burned completely [3, 8]. Knocking combustion is further investigated in optical engines using spectroscopic measurements by visualizing flame structures, intermediate species, and pressure evolution during combustion in SI engine [9]. Figure 9.5 shows the spatial distributions of radicals HCO and OH during the knocking combustion cycle. Green dashed line (Fig. 9.5) indicates optical window, and blue solid line indicates the cylinder liner. Radical species such as OH and HCO are correlated to the onset and duration of knock and the occurrence of hot spots in the end-gas region [9]. The occurrence of the HCO radical in the end gas indicates the start of autoignition (knocking) combustion, and OH radical characterizes the burned zone during combustion (Fig. 9.5).

Engine knock analysis can be divided into two parts: (1) the phenomena leading to knock or knock onset and (2) what happens after the knock onset or modes of knocking (Sect. 9.2.2). Investigation for knock analysis after the knock onset (how it progress) includes the visualization of the knocking phenomenon and the numerical simulation of knock in a particular fuel-air mixture. Knock onset analysis include (1) studying ignition delay (ID) through measurements in shock tubes and rapid compression machines, (2) modeling by basic chemical kinetics, (3) developing empirical correlations using engine data, and (4) assessing the fuel effects [10].

The chemical kinetics of autoignition leading up to autoignition and knock onset has been reviewed [11]. With the progress of combustion process after spark, the pressure and temperature of the end gas (charge) increase which generates the radicals from stable fuel species by initiation reactions. Knock onset occurs once chain reactions in the charge lead an exponential growth in temperature resulting into the eventual consumption of the fuel to release its chemical energy as sensible energy [11, 12]. Different reactions become significant at different pressures and temperatures in the combustion chamber. Chemical kinetic models can be used to analyze the chemistry of autoignition at different thermodynamic conditions. These models are useful in characterizing autoignition in terms of an autoignition delay time (τ). The autoignition delay time is the induction time required for a fuel and air mixture to react and autoignite.

Phenomenological models typically use a one-step reaction instead of detailed chemical kinetics, and the autoignition delays are expressed as the Arrhenius equation as presented by Eq. (9.1) [13]:

$$\tau = C_1 \cdot p^{-C_2} \cdot e^{\frac{C_3}{T}} \tag{9.1}$$

where p and T are the pressure and temperature of the end gas and C_1 to C_3 are the fitting coefficients. The phenomenological models are extensively used in simulations because of their simplicity and low computing cost. A number of correlations have been developed to predict autoignition of end gas. The typical correlation for autoignition delay time can be presented by Eq. (9.2) [14]:

$$\tau = C_1 \cdot \left(\frac{ON}{100}\right)^{C_2} \cdot p^{-C_3} \cdot e^{\frac{C_4}{T}} \tag{9.2}$$

where ON is fuel octane number and the model constants C_1, C_2, C_3, and C_4 are 17.69, 3.402, 1.7, and 3800, respectively. This model does not include the variables of EGR and excess air ratio (λ), and thus, the prediction of knock onset at engine operating with EGR or fuel enrichment contains error. Thus, the parameters including these variables are added to improve Eq. (9.2) in reference [13, 15].

The Livengood and Wu integral [16] is the simplest method for predicting the onset of knock in an engine. The knock integral can be presented as Eq. (9.3), which relies on an expression for the ignition delay for the current (and changing) thermodynamic state of the end-gas mixture:

$$I = \int_{t_{IVC}}^{t_{knock}} \left(\frac{1}{\tau}\right)_{P,T} dt \tag{9.3}$$

where t_{IVC} and t_{knock} refer to the time of intake valve closing (IVC) and knock onset, respectively. Knock occurs at the time t_{knock} when the integral I equals to 1.

A more comprehensive method of calculating the onset of knock in the end gas is to directly integrate a kinetic mechanism under the end-gas thermodynamic conditions. This method will allow for the evolving composition of the end gas to affect the ignition calculation, whereas the knock integral method ignores this effect [17].

Fuel's tendency to autoignite and cause knock is quantified in terms of an octane index (OI). The idea behind the OI development is that fuel's complex autoignition chemistry can be compared to a binary fuel blend which would have a predictable autoignition behavior. The binary blend used in these comparisons is a primary reference fuel (PRF), a blend of isooctane and n-heptane. As the percentage of isooctane increases in the blend, the fuel-air mixture is less prone to autoignition. The OI of a fuel is defined as the volumetric percentage of isooctane in the PRF that knocks at the same intensity at the same conditions [2, 18]. The OI depends on the engine operating conditions, and thus, it is measured at two standardized test conditions for fuel rating purposes. These two conditions are the Research and Motor Octane Number (RON and MON) tests. The octane scale is based on two paraffins, n-heptane and isooctane. The two main differences between the RON and MON tests are the intake air temperature and the engine operating speed. Modern engines have intake air temperatures that are well below the RON and MON tests [18].

The difference between the RON and MON is termed the sensitivity (S). The sensitivity of a PRF is 0. However, most commercial fuels have a sensitivity of about 8 [2].

$$S = RON - MON \tag{9.4}$$

The autoignition chemistry of non-paraffinic components in gasoline is different from that of PRF, and RON or MON describes the anti-knock behavior of the gasoline only at the RON or MON test condition [19]. Aromatics and olefins, however, tend to have S values that far surpass those of paraffins [20]. Figure 9.6 illustrates the variations of sensitivity with a different paraffinic fraction of different fuels.

Since the RON and MON are the OI of fuel at two set conditions, the OI can be interpolated from these values using a weighing factor (K). The true anti-knock quality of gasoline is best described by an octane index (OI) defined by Eq. (9.5) [21–24]:

$$OI = (1 - K) \cdot RON + K \cdot MON = RON - K \cdot S \tag{9.5}$$

The value of K is assumed to be independent of the fuel, and depending only on the engine's operating condition [21]. The value of K is zero for the RON test, and K is one for MON test. The value of K tends to become negative for the most knock-

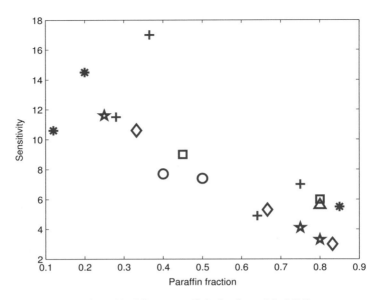

Fig. 9.6 Sensitivity variation with different paraffinic fractions of fuel [20]

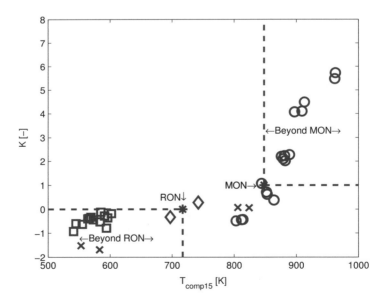

Fig. 9.7 Variation of K vs. T_{comp15} for different studies [20]

limited regions in modern SI engines [1]. There is a good correlation for K with the unburnt gas temperature (e.g., at a compression pressure of 15 bar (T_{comp15})). Figure 9.7 illustrates the variation of K with T_{comp15} from various studies [21–24]. At 15 bar, the K values are found to decrease from around 6 at 1000 K to 0 at

700–800 K, down to at 500 K [25]. For modern engines, much of the autoignition chemistry occurs in the transition region between high- and low-temperature combustion regimes, corresponding to end-gas temperatures between 775 and 900 K. The MON test captures the chemical kinetics of fuel autoignition in the high-temperature regime (above 900 K). Modern engines operate with end-gas temperatures well below the high-temperature regime, and hence the MON test temperatures should be lowered to better capture fuel autoignition chemistry in modern engines [18].

A study proposed that the octane scale based on primary reference fuels, mixtures of isooctane and n-heptane, can be replaced by a scale based on toluene/n-heptane mixtures (TRF, toluene reference fuels) for ranking practical fuels [19]. A calibration curve relating the toluene number (TN), the volume percent of toluene in TRF, to the head position at knock in the RON test is established. Another study developed a simple model which can be used to find the composition of a TPRF surrogate (a mixture of toluene, isooctane, and n-heptane) to match a given RON and sensitivity [26]. The RON and MON of the fuel describe knock behavior only at RON and MON test conditions. Fuels of different chemistry are ranked differently depending on temperature and pressure development in the end gas. In real engines, anti-knock quality of practical fuels depends both on fuel chemistry and on engine design and operating conditions.

Knock onset (KO) detection using in-cylinder pressure measurement is only possible in the time domain because the Fourier transform (frequency domain) does not preserve a signal's timing information [27]. There are a variety of methods for the determination of KO mostly based on the threshold value exceeded (TVE) method. The TVE method for knock onset detection uses the first crank angle where the high- or band-pass filtered pressure signal or the first or third derivatives exceeds a predetermined threshold as knock onset [13, 27]. This method is intuitive and simple; however, it typically detects the KO late by up to a few hundred microseconds (up to 3° CA for typical speeds) for weak to intermediate knock cycles because the threshold value needs to be set high enough to avoid false detection [27]. Additionally, these methods are either sensitive to the sampling rate or predict the crank angle late for the knock onset dependent on the predetermined threshold value [13]. For strong knock cycles, the TVE method detects knocking combustion within ±0.5° CA.

Knock onset detection methods have been proposed based on the first and third derivatives of the pressure signal [28, 29]. The first occurrence at which the first derivative of the pressure signal exceeds a threshold is denoted as KO, and the third derivative of low-pass filtered pressure signal is scanned for the highest negative value, and the corresponding crank angle is defined as KO. Another method is proposed to compute the knock onset based on the signal energy ratio (SER), which is presented as Eq. (9.6) [30]:

$$\text{SER} \equiv \frac{\text{SEPO}_{\text{fwd}}^2}{\text{SEPO}_{\text{bwd}}^{1/2}} = \frac{\left(\int_{\theta_o}^{\theta_o+\Delta\theta} P_{\text{filt}}^2 d\theta\right)^2}{\left(\int_{\theta_o-\Delta\theta}^{\theta_o} P_{\text{filt}}^2 d\theta\right)^{1/2}} \tag{9.6}$$

This method is based on the signal energy of pressure oscillations (SEPO) and embodies a modified signal-to-noise ratio by looking both forward and backward in time. The $\Delta\theta$ was optimized to be 5°CA, SEPO$_{fwd}$ is the SEPO for the next 5°CA, and SEPO$_{bwd}$ is the SEPO for the previous 5°CA. The SER is calculated for a wide range of crank angles around TDC, and the knock onset is defined as the crank angle at which the SER function is a maximum. This method is inadequate for real-time knock detection, but for a posteriori diagnostic purposes, it is considered more accurate [27, 30].

9.2.2   Modes of Knock

Knock occurs when the autoignition of the unburned mixture takes place in the combustion chamber. There are a number of qualitatively different ways in which a nonuniform fuel-air mixture can autoignite [31, 32]. For fully homogeneous air-fuel mixture, the chemical reactions occur at the same rates in the entire mixture. The pressure rises rapidly and uniformly throughout the whole mixture when the reactions and the associated heat release rate become very fast. There are no propagating pressure pulses, and the combustion occurs essentially at constant volume. This mode of autoignition with no spatial gradients comprises a thermal explosion [31]. In real engine operating conditions, perfect homogeneity (of temperaure and fuel concentration spatially) is not possible to achieve. In real engine conditions, the gradients of temperature, mixture strength, and active radicals exist, which produce gradient in ignition delay of the charge. Large temperature gradients can occur through charge stratification, or near the walls. Temperature stratification may be required to ensure that the autoignition (even in the thermal explosion mode) does not happen instantaneously but over a few degrees of crank angle. Additionally, local hot spots or regions of enhanced chemical activity can develop because of the locally high-temperature or active species concentrations. Reaction develops more rapidly at a hot spot, and an autoignition front propagates outward from it. Localized ignitions occur at different instants at different positions [31]. Three modes (deflagration, thermal explosion, and developing detonation) of post-knock combustion have been identified due to the temperature and composition nonuniformity in the end gas (unburned charge) [33].

The local pressure buildup mechanism and modes of propagation are discussed in reference [10] using the first of law of thermodynamics in autoignition region. The local pressure that builds up in the autoignition region is the result of the competition between the heat release rate and the pressure relief due to the volumetric expansion of the burned mixture. Figure 9.8 illustrates the pressure buildup in the autoignition region of radius (R). Assuming burned gas as ideal gas, the pressure rise rate can be written by Eq. (9.7) by rearranging Eq. (7.56):

$$\frac{dP}{dt} = (\gamma - 1)\dot{q} - \gamma P \frac{\dot{V}}{V} \tag{9.7}$$

Fig. 9.8 Illustration of pressure buildup in autoignition region

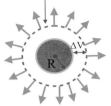

Acoustic Expansion at Local Speed of Sound

Autoignition Region

where Q is heat release rate per unit volume and P and V are pressure and volume of autoignition region, respectively.

The local pressure will build up $(dP/dt > 0)$ if the heat release term is large compared to the volumetric expansion term as shown in Eq. (9.8) [10]:

$$\dot{q} \gg \frac{\gamma}{\gamma - 1} P \frac{\dot{V}}{V} \tag{9.8}$$

For acoustic expansion with speed of sound "a," the volumetric change can also be written as Eq. (9.9):

$$\frac{\Delta V}{\Delta t} = 4\pi R^2 a \tag{9.9}$$

For a spherical autoignition region (Fig. 9.8), the criteria in Eq. (9.8) can be written as Eq. (9.10):

$$\dot{q} \gg \frac{\gamma}{\gamma - 1} P \frac{3a}{R} \tag{9.10}$$

It can be noted that the criterion on the size R of the autoignition region in relation to the volumetric heat release rate and the local speed of sound (a) is dependent on temperature.

For very small autoignition radius (R), very little pressure builds up, and thus, there will be no or very weak acoustic wave developed. The small R is associated with steep gradients (temperature and/or composition) in the nonuniform end gas, which appears as the regions of ignition are small islands. The radicals and high local temperature of the ignited region initiate a flame (a deflagration or subsonic propagation of the heat release front) in the unburned mixture. Figure 9.9 illustrates the combustion process during a super-knock initiated by preignition. The figure depicts the three stages of the combustion process including deflagration, detonation, and the resulting pressure oscillations. In the deflagration region, the chemical heat released leads to thermal expansion of the burned zone, which compresses the

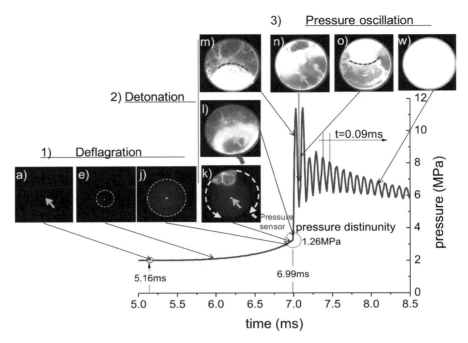

Fig. 9.9 Pressure trace with synchronous images showing three stages of the super-knock process [34]

surrounding unburned mixture to high pressure and high temperature [1]. The flame velocity is subsonic during deflagration, and transport processes involving simultaneous conduction of heat and diffusion of radicals govern the speed of combustion wave or flame.

For moderately large autoignition region (R) which satisfies the criteria in Eq. (9.10), significant local pressure builds up, and pressure waves (acoustic waves) or even weak shock waves are excited. This mode of combustion is termed "thermal explosion" in reference [33] due to the fast heat release from a sizable region. The pressure wave generated in this mode can be noisy or quite intense that can lead to damage of engine components by repeated pounding on the chamber surfaces by the local high pressure and high temperature [10]. The major manifestation of the pressure wave is the excitation of the engine structural vibration; hence, the phenomenon is term acoustic knock. However, these pressure waves are not strong enough to initiate Chapman-Jouguet type of denotation [35]. Depending on the temperature/composition nonuniformity in the end gas, there could be sequential autoignition of isolated regions or successive ignition of connected regions [31, 33]. Figure 9.10 illustrates the non-autoignition and sequential autoignition combustion modes in the end-gas region. In sequential autoignition mode, a flame is initiated at the outer boundary of the autoignited region. The flame speed is much

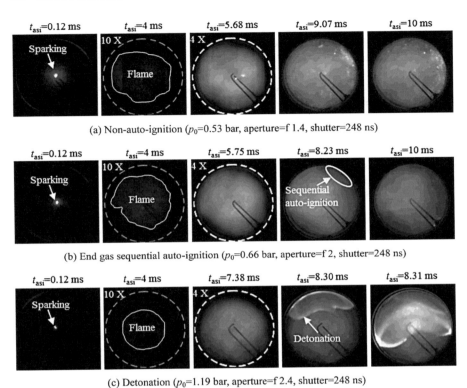

t_{asi}=0.12 ms t_{asi}=4 ms t_{asi}=5.68 ms t_{asi}=9.07 ms t_{asi}=10 ms

Sparking

10 X   Flame   4 X

(a) Non-auto-ignition (p_0=0.53 bar, aperture=f 1.4, shutter=248 ns)

t_{asi}=0.12 ms t_{asi}=4 ms t_{asi}=5.75 ms t_{asi}=8.23 ms t_{asi}=10 ms

Sparking

10 X   Flame   4 X   Sequential auto-ignition

(b) End gas sequential auto-ignition (p_0=0.66 bar, aperture=f 2, shutter=248 ns)

t_{asi}=0.12 ms t_{asi}=4 ms t_{asi}=7.38 ms t_{asi}=8.30 ms t_{asi}=8.31 ms

Sparking

10 X   Flame   4 X   Detonation

(c) Detonation (p_0=1.19 bar, aperture=f 2.4, shutter=248 ns)

Fig. 9.10 Images of different end-gas combustion modes [36]

slower than that of the pressure wave, and there is no interaction between the heat release and the pressure wave.

For sufficiently large autoignition region (R), a significant pressure ratio between the local pressure and the end gas pressure is developed, and resulting shock wave could induce fast heat-releasing chemical reactions in the end gas at the wave front [10]. Alternatively, the successive ignition along a gradient may be rapid enough to create a combustion wave with speed comparable to the local pressure wave propagation (sonic) speed. Then there is a significant interaction between the heat release reaction and the pressure wave. The latter phenomenon has been termed developing detonation [31, 33]. In both cases, the local post-combustion pressure is higher than the isochoric value at the end-gas condition because of compression by the pressure wave [10].

The name "developing" detonation is little confusing with detonation. When combustion wave undergoes a transition from subsonic to supersonic speed, then it is known as detonation. The speed of detonation is not controlled by heat conduction and radical diffusion like deflagration. Rather, the shock wave causes the temperature and pressure to increase to such extent that it can lead to an explosion and a large amount of energy is released during this process. McKenzie and Cheng [10] suggested that if definition of detonation is broaden from a combustion wave which is induced by the pressure wave to that which has significant interaction with the pressure wave, then both the normal and the developing detonation can be described by the term "detonation" since, in both cases, the combustion wave and the pressure wave travel together, and there is significant interaction between them. Detonation of end gas in super-knock condition is illustrated in Figs. 9.9 and 9.10.

The mode of combustion is affected by thermodynamic conditions (pressure and temperature) of end gas which is affected by several engine and operating parameters such as intake pressure, temperature, compression ratio, etc. The charge density can be used to represent the coupling effect of the ambient pressure and temperature ($\rho \propto p/T$), and it affects the end-gas combustion mode. Figure 9.11 illustrates the

Fig. 9.11 Relationship between energy density at the end of compression and combustion mode [36]

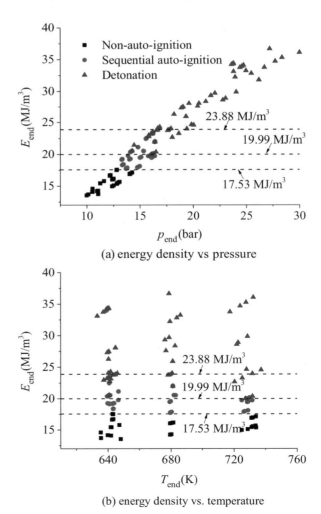

(a) energy density vs pressure

(b) energy density vs. temperature

relationship between energy density at the end of compression and combustion mode. The mixture density is shown as energy density (E_{end}) to consider the lower heating value of the mixture. The boundary between non-autoignition and autoignition modes is very clear and shown by the dashed line of 17.53 MJ/m^3 (Fig. 9.11). The figure shows that with an increase in energy density, the end-gas combustion mode gradually transits from sequential autoignition to detonation if the energy density exceeds 17.53 MJ/m^3. Only the sequential autoignition can be induced if the energy density is below 19.99 MJ/m^3, while the detonation always occurs if the energy density is higher than 23.88 MJ/m^3 [36].

A number of strategies may be adopted to avoid entering the into detonation mode of knocking combustion. A study proposed two dimensionless parameters, ε and ξ, for investigating the hot spot-induced combustion mode using the $\varepsilon - \xi$ diagram [31]. The ε is the ratio of the residence time of the acoustic wave in the hot spot to the short excitation time in which most of the chemical energy is released, and ξ is the ratio of the acoustic speed to the localized autoignitive velocity. Super-knock in boosted SI engines is investigated by the relative differences between the spatial distribution of ignition delay and the ignition wave propagation speed [37, 38]. In a particular situation, when an autoignition front is propagating from a hot spot and the autoignition delay time (τ_i), increases with the distance from the hot spot (r), the autoignition front propagates at a velocity relative to the unburned gas (u_a) that is characterized by Eq. (9.11), which can also be related to a temperature gradient:

$$u_a = \left(\frac{\partial \tau_i}{\partial r}\right)^{-1} = \left(\frac{\partial \tau_i}{\partial T}\frac{\partial T}{\partial r}\right)^{-1}. \tag{9.11}$$

When the temperature gradient attains a critical value, $(\partial T/\partial r)_c$, such that the autoignition front moves into the unburned mixture at approximately the acoustic speed (a), the front of the pressure wave generated by the rate of heat release can couple with the autoignition front. The fronts are mutually reinforced to create a damaging pressure spike propagating at high velocity in a developing detonation [31]. The critical temperature gradient for such a chemical resonance between the chemical and acoustic waves, with $u_a = a$, can be presented by Eq. (9.12):

$$\left(\frac{\partial T}{\partial r}\right)_c = \frac{1}{a(\partial \tau_i/\partial T)} \tag{9.12}$$

The actual temperature gradient is normalized by this critical value, and parameter ξ is defined by Eq. (9.13):

$$\xi = \left(\frac{\partial T}{\partial r}\right) \bigg/ \left(\frac{\partial T}{\partial r}\right)_c = \frac{a}{u_a} \tag{9.13}$$

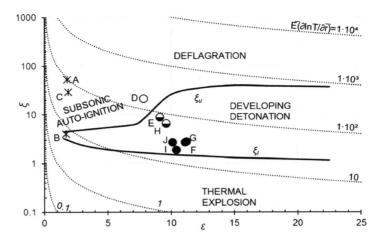

Fig. 9.12 ξ/ε regime diagram, for hotspot autoignition (increasing black fill of the symbols indicates the increasing severity of knock from no knock through mild knock to super-knock) [41]

In the case of $\xi \geq 1$, the pressure pulses run ahead of an autoignitive deflagration ($u_a \leq a$). Values of ξ are only readily known at the initial value, based on the initial boundary condition temperature gradient. In practice, heat conduction, species diffusion, and some reaction modify the initial boundary before autoignition occurs. It is demonstrated numerically that there are lower and upper limits, ξ_l and ξ_u, based on the initial boundary condition, between which a developing detonation can occur (Fig. 9.12) [31].

In severe knocking combustion, large mass fraction of the original charge is burned instantaneously to create a rapid increase of the pressure and rapid rate of change of the volume (dV/dt). The instantaneous sound pressure $p(t)$, at the distance d from the source, can be provided by Eq. (9.14) using acoustic theory [39]. The volumetric expansion rate of the hot spot with the area of the reaction front (A) propagating to the unburned gas at a velocity (u_a) is $A*u_a$. The burned and unburned gas density difference should be accounted to calculate the net rate of the volume expansion as given in Eq. (9.15), where σ is the ratio of unburned to burned gas densities. Assuming a small spherical hot spot of radius r and differentiating Eq. (9.15) with respect to t give the volume expansion rate as stated in Eq. (9.16). The nondimensional form of pressure oscillation is given by Eq. (9.17) where $\xi = a/u_a$ [37, 38, 40]:

$$p(t) = \frac{\rho}{4\pi d} \left| \frac{d}{dt}\left(\frac{dV}{dt}\right) \right|_{t-t_a} \tag{9.14}$$

$$\frac{dV}{dt} = Au_a(\sigma - 1) = 4\pi r^2 u_a(\sigma - 1) \tag{9.15}$$

$$\left|\frac{d}{dt}\left(\frac{dV}{dt}\right)\right|_{t-t_a} = 4\pi r(\sigma - 1)\left(2\sigma u_a^2 + r\frac{du_a}{dt}\right) \tag{9.16}$$

$$\frac{\Delta p(t)}{p} = \left[\frac{r\gamma}{d}(\sigma - 1)\left(2\sigma\xi^{-2} + \frac{r}{a}\frac{d\xi^{-1}}{dt}\right)\right]_{t-t_a} \tag{9.17}$$

An additional parameter affecting the probability of a detonation is likely to be one associated with the rate at which chemical energy is unloaded into the developing acoustic wave. During detonation condition, there are large changes in density at the pressure front, and chemical energy can be fed into the developing strong pressure wave at a hot spot. Under these conditions, the principal energy release occurs during an excitation time (τ_e), which is orders of magnitude less than τ_i [38]. By using an excitation time for energy release energy (τ_e), another dimensionless parameter $\varepsilon = (r_0/a)/\tau_e$ is defined, which is a measure of the hot spot reactivity. Figure 9.12 illustrates the ξ/ε regime diagram, for hot spot autoignition which characterizes the modes of combustion. Conditions of developing detonation are marked with the upper and lower limit. No developing detonation is observed in two regions of the parameter space (Fig. 9.12). The first is when the mixture is close to homogeneity, and ξ is low. A low value of ξ also is aided by a small dependence of ignition delay on temperature $\partial\tau_i/\partial T$, which is a characteristic of paraffinic hydrocarbons. The second region of low detonability occurs at high values of ξ or spatial temperature gradient, but only in less reactive mixtures, with $\varepsilon < 10$ [31]. Reactive hot spots leading to autoignition/ knock can arise by a number of factors such as partial mixing with hot gas or burned products, heat transfer from hot surfaces, and turbulent energy dissipation in flowing reactants. The size of hot spots may be of the order of millimeters [41].

The detonation diagram identifies the knock-limited conditions. Figure 9.13 shows the rate of heat release in unburned charge (RoHR-u) and the detonation peninsula along with the calculated $\xi - \varepsilon$ data for different spark timings at constant initial NO content (287 ppm) in the mixture. The figure shows that advancing of spark timing increases the propensity for autoignition. The value of the first peak in the RoHR-u trace is not significantly affected by spark timings, and only position where the first peak occurs changes with spark timing (ST). The spark timing (ST = −11.5 CAD ATDC) seems to be engine knock limit, and further advancing of ST by 0.5 CAD leads to a transition from the regime of subsonic autoignition to the regime of developing detonation (Fig. 9.13) [42]. Further advancement of spark timing moves the $\xi - \varepsilon$ solution toward the lower limit of the resonance parameter (ξ_l) and slightly higher values of the reactivity parameter (ε). Figure 9.13 also depicts that knock-limited operating conditions typically shows the occurrence of two peaks in the RoHR-u. Hence, the RoHR-u trace can also be used for very first and rough estimation of the end-gas conditions with respect to knocking occurrence [42].

Typically, external EGR is used for the mitigation of knock in the engines. The EGR can also add the initial NO in the air-fuel mixture. The presence of NO in the charge strongly affects the knock onset [42]. Figure 9.14 illustrates the effect of

Fig. 9.13 (a) Rate of heat release in unburned charge (RoHR-u) and (b) the detonation peninsula [31] along with the calculated $\xi - \varepsilon$ data for different spark timings at constant initial NO content (287 ppm) in the mixture [42]

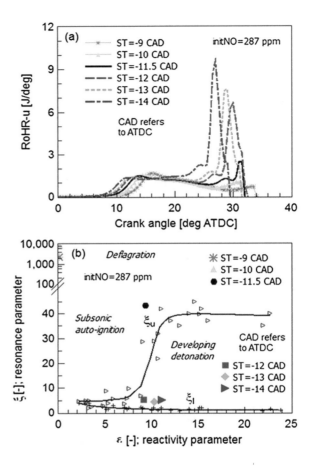

initial NO content in the mixture on knocking combustion mode. Figure 9.14 depicts that the RoHR-u is very sensitive to the changes of NO in the end gas. Generally, the reactivity of the charge is improved by increasing NO which is depicted by significantly higher values of the second peak of RoHR-u curve. Thus, the knock margin can be increased by lowering NO content in the charge.

9.2.3 Super-Knock and Preignition

To increase the fuel conversion efficiency, the SI engine design utilizes the downsized and turbocharged concepts. In turbocharged conditions, extremely high-intensity knock events informally described as "megaknock" or "super-knock" are found to occur occasionally even when spark timings have been expressly chosen to avoid knock. Super-knock is a type of autoignition phenomenon

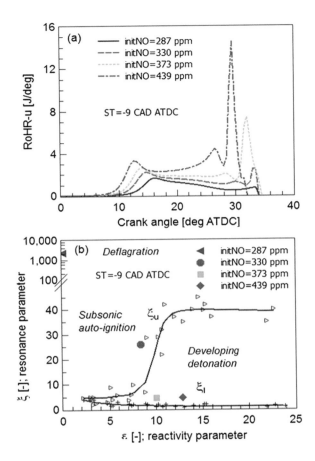

Fig. 9.14 (a) Rate of heat release in unburned charge (RoHR-u) and (b) the detonation peninsula [31] along with the calculated $\xi - \varepsilon$ data for different initial NO content in the mixture at constant spark timing [42]

that can occur in highly boosted SI engines under the conditions of low speed and high intake pressure. Additionally, super-knock events occur randomly with little direct relation with engine combustion control parameters, such as ignition timing, intake temperature, equivalence ratio, etc. Thus, typical knock mitigation techniques, such as retarding spark timing, cooling the intake charge, and enhancing heat transfer, are not very effective at eliminating super-knock [1]. Super-knock combustion can result into very high peak pressure (~30 MPa) and pressure oscillation (~10 MPa), which can significantly damage the cylinder or piston. Therefore, the super-knock occurrence is a major challenge in improving the fuel conversion efficiency of SI engine [43].

The super-knock combustion process is significantly different from traditional knocking process. Super-knock combustion can originate from the preignition in which a stable flame kernel is set off by a hot spot prior to the spark timing. The hot spot can possibly be induced by one of the following sources including fuel, particles, lubricant oil droplets, and surface ignition [44]. Additionally, the super-knock combustion mechanism is constituted by hot spot-induced deflagration to detonation followed by high-pressure oscillation [34]. Super-knock with a high-pressure fluctuation is typically produced by the occasional preignition that might depend on the critical conditions of autoignition within the combustion chamber [38].

Figure 9.15 illustrates the super-knock, heavy knock, and normal combustion in an SI engine with the same spark timing of 8° bTDC. The figure shows that under the same operating conditions, normal combustion, conventional knock, and super-knock (spark ignition-induced super-knock and preignition-induced super-knock) appeared in different combustion cycles [43]. The spark timing retarded because of knock limit and the peak cylinder pressure is observed to be 2.98 MPa during normal combustion. In the super-knock caused by preignition (shown by red color), the peak pressure reaches up to 25 MPa along with a strong pressure oscillation with an oscillation amplitude of 13.1 MPa (Fig. 9.15) due to pressure wave propagation in the cylinder. The extremely high knock intensities can be explained in terms of developing detonation (Sect. 9.2.2). Super-knock induced by spark ignition is indicated by the black line in Fig. 9.15. In this mode, combustion follows the pressure profile of normal combustion for several crank angles after the spark timing, and compression of flame propagation leads to high pressure and temperature in the end gas (charge). The high-temperature and high-pressure in the end gas results in autoignition at 10° CA aTDC (Fig. 9.15), and pressure curve deviates from normal combustion. A severe pressure oscillation with an amplitude of 16.8 MPa occurs,

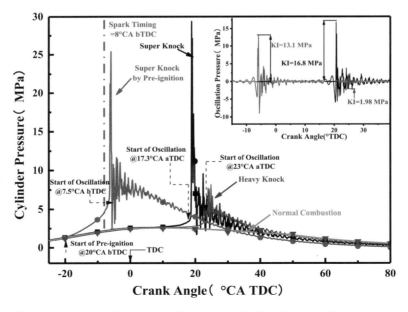

Fig. 9.15 Pressure traces and pressure oscillations for typical knock combustion at spark timing of 8° CA bTDC [43]

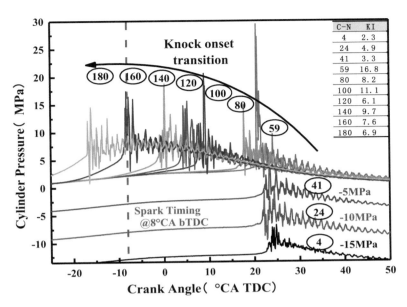

Fig. 9.16 Pressure traces for different cycles at a particular engine operating condition [43]

and knock intensity-based criteria meet the condition of super-knock [43]. Thus, the super-knock can be induced by spark ignition not only by preignition, and the difference between the two is the onset of super-knock.

The various combustion modes are sensitive to the different engine operating conditions, like cooling water temperature, wall temperature, cycle-to-cycle variations, etc. Figure 9.16 depicts the pressure trace and knock intensity (KI) of numerous typical super-knock combustion cycles between 200 consecutive engine cycles. To eliminate the cylinder pressure trace overlap, the pressure traces of super-knock cycle numbers of 4, 24, and 41 were offset by −15, −10, and − 5 MPa, respectively [43]. The figures illustrate that the onset of super-knock progressively advances with an increase in a number of cycles. This occurs possibly due to the enhancement of thermodynamic state of the end gas (charge) leading to shorter ignition delay. The onset of pressure oscillation advances with the number of cycles (80–180). The knock intensity values in different cycles are 8.2, 11.1, 6.1, 9.7, 7.6, and 6.9 MPa for cycle numbers 80, 100, 120, 140, 160, and 180, respectively (Fig. 9.16), which indicates the nonlinear relationship between knock onset and knock intensity. The knock intensity is governed by heat release rate of the unburned charge, and there is no direct relation between knock onset and intensity. The super-knock or knock combustion is dominantly induced by the spark ignition and is later induced by the preignition after several cycles. The results show that the occurrence of super-knock can be caused by spark ignition at particular thermodynamic conditions, not only by preignition [43].

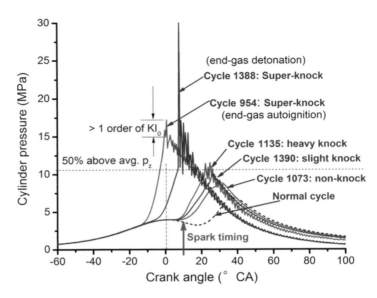

Fig. 9.17 Cylinder pressure profiles of typical preignition cycles and normal combustion cycle [4]

Although preignition can lead to the occurrence of super-knock combustion, it is not always necessary. Preignition can also lead to non-knocking, slight knock, or heavy knock combustion. Figure 9.17 illustrates the normal combustion cycle along with five different preignition lead combustion cycles. The figure shows that no engine knock is found in cycle 1073, which is a preignition cycle. However, it is typically expected that early preignition will lead to more severe engine knock combustion, which is not always true. The earliest preignition timing cycle 954 has lower knock intensity than cycle 1388 which has relatively late preignition timing (almost at the TDC). Furthermore, another study showed that the stationary autoignition events with a very early combustion phasing could take place without knocking [45], which is illustrated in Fig. 9.18. A reliable detection of this type of autoignitions by the knock control system is not guaranteed because the events occurred outside the knock detection window or the maximum permissible knock adjustment had already been used up [45].

For preignition, the temperature in a local area has to be increased sufficiently so that runaway chemical reactions start and an incipient flame is formed. The start of preignition was always associated with hot surfaces such as spark plug electrodes or hot spots on exhaust valves in older engines. However, in modern DISI engines, the flame initiation appears to be mostly away from internal surfaces. In any condition, the initial hot spot cannot be produced from the autoignition of the fuel-air mixture because the ignition delays are too large during the compression stroke when preignition is seen to be initiated [46]. Thus, fuels RON or MON is not expected to influence the onset of preignition. The most reasonable description is that, in modern engines, droplets of mixtures of oil and fuel fly into the combustion chamber

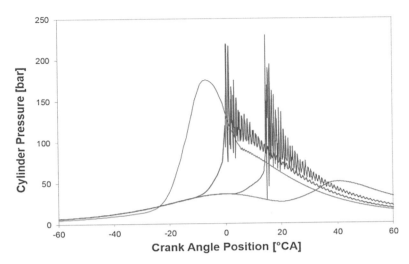

Fig. 9.18 Pressure profiles containing autoignition of different intensity/combustion phasing [45]

and mix with oxygen and autoignite, probably aided by some catalytic reactions, to form the initial ignition center. Lubricant, like diesel fuel, can be expected to be very prone to autoignition since it is made up large, straight chained hydrocarbons. There is also plausible experimental evidence that metallic lubricant additives, particularly calcium additives, promote preignition [46].

The abnormal combustion due to preignition at low engine speed is denoted as low-speed preignition (LSPI) in modern engines. The occurring pattern of LSPI is different from traditional preignition which is classified as a surface ignition. The causes of traditional one are unsuitable spark plug, heat spot of combustion chamber edge, or carbon deposit. In spite of them, LSPI occurs suddenly and continues for a while together with normal combustion. Then, it disappears and recovers to normal combustion [48].

Figure 9.19 summarizes mechanistic pathways for LSPI. The primary steps in LSPI are influenced by the extent of fuel adhesion on the liner wall and the properties of the lubricant oil that it interacts with on the liner wall. Although several pathways to LSPI are demonstrated, common to all is the involvement of lubricant oil-derived contents as preignition sources [48]. Figure 9.20 illustrates the well-known mechanism which assumes lubricant oil-containing droplets from piston crevice as a candidate of preignition source. Droplet of lubricant oil and fuel mixture, caused by adhesion of fuel spray on liner wall, will fly and preignite before spark ignition.

Studies have shown fuel/lubricant droplets to be a much more probable primary ignition source, and the effects of fuel and lubricant properties, as well as the interaction of fuel sprays and lubricating oil in the top crevice region, are investigated. Piston ring motion and turbulence variations have also been proposed as possible transport-related causes of the apparently stochastic nature of LSPI [49]. Preignition, which is also governed by stochastic processes, establishes a

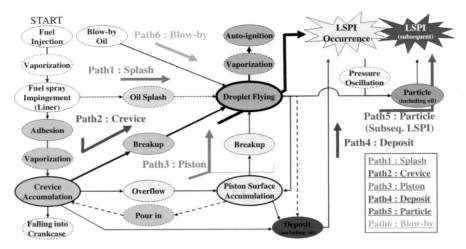

Fig. 9.19 Mechanistic pathways for low-speed preignition (LSPI) occurrence [47]

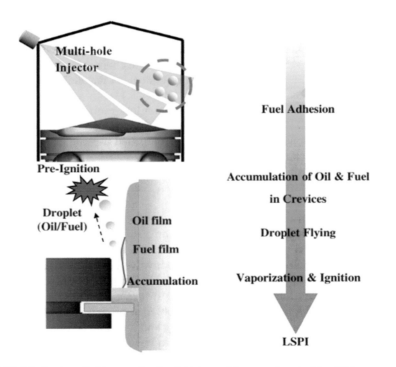

Fig. 9.20 Mechanism of piston crevice droplet triggered low-speed preignition [47]

flame before the spark fires and can cause knock onset to occur at high pressures and increase the chances of super-knock [46]. The preignition cycle occurs randomly maybe after thousands of cycles. Each individual LSPI cycle are typically characterized using three parameters: LSPI event start location, LSPI dwell duration, and LSPI intensity which is illustrated in Fig. 9.21a. LSPI dwell duration is denoted as the time required to achieve maximum pressure after preignition has started, and LSPI intensity quantifies the increase in maximum cylinder pressure relative to the maximum pressure of the median trace. Additionally, the total LSPI event number count and the cluster count of each fuel are also compared [49]. The LSPI dwell duration affects the intensity of super-knock. Figure 9.21b illustrates that longer LSPI cycles with a long dwell time exhibit negligible end-gas knock.

Figure 9.22 illustrates the distribution of preignition origins evaluated using high-speed camera. The figure shows the view through the endoscope into the combustion chamber, and each dot represents the origin of one optically detected preignition. The figure illustrates that the preignition origins are spread over a wide range in the plane of the cylinder head gasket. Detailed analyses presented no significant correlation between the engine operating conditions and the distributions of preignition origin [1]. Figure 9.22 also shows that there is no accumulation of preignition positions at any component of the engine cylinder. Preignition sometimes occurs in the form of intermittent sequences consisting of multiple cycles with preignition alternating with regular burning cycles [50].

Figure 9.23 presents the examples of preignition sites in different combustion cycles. The figure shows that in some cycles, several preignition locations occur in different positions approximately at the same time. Other cycles have movable preignition kernels, while preignitions outside of the viewing area of the combustion chamber can also be observed. It is found that the locations of autoignition can occur with enormous variability. In preignition condition, the flame propagation starts normally at the wrong location and additionally too early. These two effects of preignition flame provoke the fulminating knocking combustion [50]. Thus, the preignition should be avoided to improve the engine performance.

9.2.4 Characteristic Knock Frequencies

To analyze engine combustion during knocking, the determination of the characteristic frequency of pressure oscillation and the mode of oscillation is essential. During knocking combustion, intensive pressure waves and even shock waves are propagating and reflecting within the combustion chamber. The pressure waves in the cylinder may lead to resonance based on its natural frequency. Thus, cavity resonances in the combustion chamber further help describe the mechanism of knock or super-knock [43]. The engine knock frequencies are determined by the acoustic vibration modes specific to the combustion chamber. The high rate of heat is released during autoignition of charge in the combustion chamber. The energy release will in turn excite specific vibration modes. These modes cause pressure waves to traverse

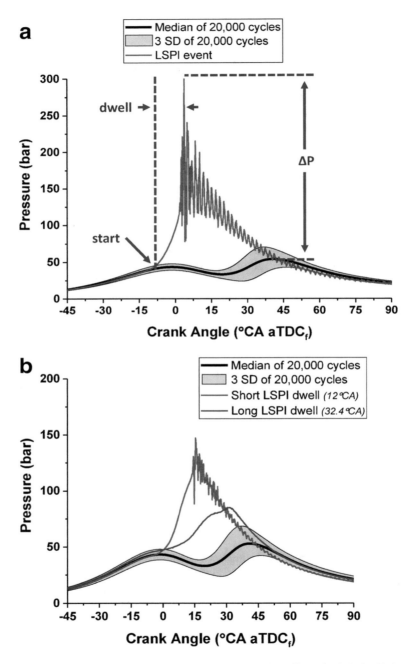

Fig. 9.21 (**a**) LSPI event characterization parameters and (**b**) effect of LSPI dwell times on knocking behavior [49]

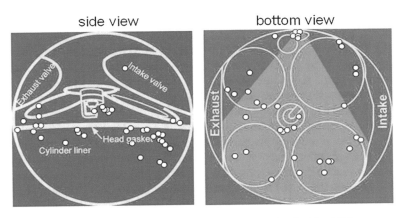

Fig. 9.22 Individual combustion cycles with several locations of preignition [50]

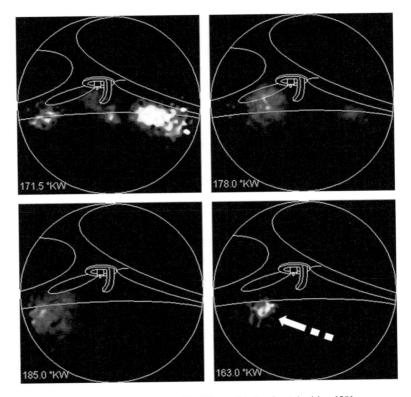

Fig. 9.23 Individual combustion cycles with different kinds of autoignition [50]

the combustion chamber at the local speed of sound, i.e., the acoustic velocity. The modes can be of both circumferential and radial nature, and the axial mode is typically negligible because the cylinder height at TDC is small compared with the engine bore diameter. Thus, the cavity resonance in the combustion chamber is dependent on shape size of the chamber and the local speed of sound [51].

Pressure resonance consists of high-frequency (few kHz) oscillations produced by pressure waves in the engine combustion chamber. The analytical solution of the wave equation can be used to determine the frequencies of different resonant modes. To estimate the resonance frequencies, Draper approach [52] can be used by a cylindrical combustion chamber assumption. Draper solved the wave equation with Bessel functions for a cylindrical geometry, and the resonance frequency is determined using Eq. (9.18). This equation relates the resonance frequency ($f_{i,j}$) with the cylinder diameter (D), the speed of sound (a), and a Bessel constant ($B_{i,j}$) related with radial modes. The axial modes (g) can be neglected because the frequencies associated with h (height of the combustion chamber) are too high.

$$f_{i,j} = a\sqrt{\frac{B_{i,j}^2}{(\pi D)^2} + \frac{g^2}{(2h)^2}} = \frac{aB_{i,j}}{\pi D} \tag{9.18}$$

The speed of sound can be determined by measuring the trapped mass m and the in-cylinder pressure P and estimating the instantaneous volume of the chamber V.

$$a = \sqrt{\gamma PV/m} = \sqrt{\gamma RT} \tag{9.19}$$

The specific heat ratio γ is a temperature-dependent variable, and it decreases as the temperature increases. Thus, local speed of sound depends on the crank angle position.

Figure 9.24 illustrates the variation of frequency content with a crank angle position in a HCCI engine. The evolution of the amplitude of the harmonics has been computed through a short-time Fourier transform (STFT) in the middle plot and through a Wigner distribution (WD) in the bottom plot. The frequency evolution derived from the cylindrical theory, by Eq. (9.18), is shown with a continuous black line, and three windows are highlighted with vertical lines [53].

A study used image-intensified high-speed video to determine the acoustic modes or pressure oscillations, in the HCCI combustion chamber by using an image analysis technique that extracts the different acoustic modes [54]. Figure 9.25 illustrates the amplitude of different frequencies depicting pressure oscillations or acoustic modes in the combustion chamber. Acoustic modes estimated from image analysis show good agreement with modes determined from measured cylinder pressure. In Fig. 9.25a, the nodes and antinodes of the four acoustic modes are apparent and illustrated schematically in Fig. 9.25b. This particular cycle is selected as an example as it is one of the few cycles where all four modes occurred [54]. Typically, the first circumferential mode (A1) is the dominant mode in HCCI combustion.

Fig. 9.24 Pressure oscillations using a band-pass filter between 3 and 10 kHz (top), STFT (center), and WD (bottom) of a cycle in HCCI combustion engine at 2000 rpm and low load [53]

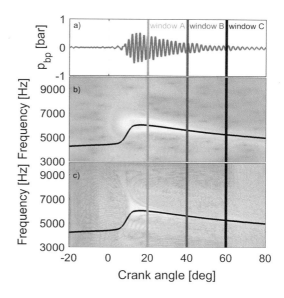

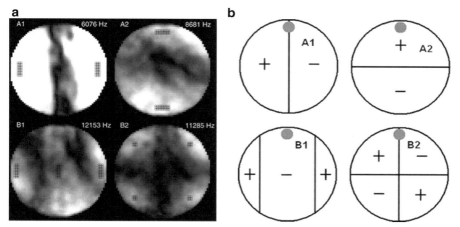

Fig. 9.25 (**a**) Spatially resolved amplitudes (a brighter color indicates higher amplitudes) of four different frequencies extracted from the high-speed video sequence showing the acoustic modes [54]. (**b**) Schematic of the four acoustic modes (lines represent the nodes) [54]

Another study on the HCCI engine concluded that the pressure waves manifested largest intensities for the first vibration mode, a mode suggesting radial propagation of the pressure waves in the combustion chamber [51]. The frequency for this mode was 4 kHz based on the pancake-shaped combustion chamber geometry. Altering the engine's combustion chamber geometry affects the properties of the oscillations. For the bowl-shaped combustion chamber geometry, the frequency for the first vibration mode was about 5 kHz. The directions of the pressure waves seem to be

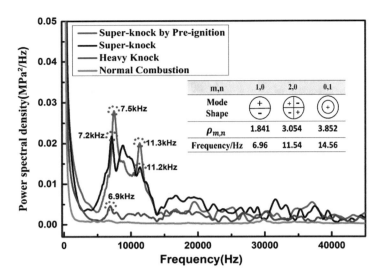

Fig. 9.26 Fast Fourier transform of in-cylinder pressure at different knock combustion modes in SI engine [43]

random. A consequence of this is that cyclic variations in the frequency domain are very prominent [51]. It was also found that the first two circumferential resonant frequencies that occur in the ranges of 2–4 kHz and 4–6 kHz accounted for more than 80% of the integrated power in the acoustic vibrational range for all HCCI, RCCI, and CDC cases. This suggests that in general, the audible ringing noise generated by compression ignition (CI) strategies on this engine platform primarily originates from circumferential pressure waves [55].

Figure 9.26 illustrates the fast Fourier transform (FFT) of cylinder pressure data for two typical super-knocks, the normal combustion and heavy-knock. The calculation of different resonance modes using characteristic frequencies reveals a correlation with experimentally measured frequency by using cylinder pressure data (Fig. 9.26). The characteristic frequencies of the first radial mode (1, 0) and the second radial mode (2, 0) have calculated resonant frequencies of 6.96 kHz and 11.55 kHz, respectively, which is consistent with the experimentally determined frequencies. Thus, the pressure wave generated during knocking resonated with the natural frequency of the engine combustion chamber to enhance the pressure wave energy, which leads to strong pressure oscillation [43]. However, the high energy in the high-frequency part during the super-knock conditions cannot be explained by acoustic theory. Therefore, it can be speculated that the pressure wave caused by super-knock may be a shock wave or detonation wave [43].

The combustion mode of the CI diesel engine is mostly the diffusion combustion. However, premixed combustion dominates the reactions in the preliminary stage of diesel combustion, and thus, the pressure oscillation is still an important part of the combustion process in diesel engine [56]. Modern diesel engines are typically using the multiple injections in one cycle. The pressure oscillations (frequency content) are typically dependent on engine load, engine speed, and injection timings. In the diesel engine with pilot-main injection strategy, the pilot injection plays a significant role in every combustion stages. The amplitude of pressure oscillation in pilot injection combustion (PIC) is the maximum, and the proportion of oscillation energy in PIC is significantly high. Figure 9.27 shows the time-frequency distribution and heat release rate at different engine loads. The figure shows that there are significant differences between the time-frequency distribution of the pressure oscillation at full-load and half-load operating conditions. At full load, the oscillation energy is mainly distributed in 13–16 kHz, while that under half load is distributed in a lower frequency band of 4–10 kHz. The pressure oscillations start at the beginning of pilot injection combustion at both the engine loads.

Figure 9.28 shows the time-frequency distribution of the pressure oscillation with a main injection advanced angle (MIAA) of $-10°$ CA and $-5°$ CA, respectively. The oscillation components below 10 kHz attenuate with injection timing of $-10°$ CA to $-5°$ CA, while those at 12–18 kHz are enhanced. Additionally, the oscillation energy around 12 kHz in the post-combustion stage almost disappears. The main pressure oscillation energy distributing between 12 and 18 kHz from the early pilot injection combustion to main injection combustion also moves to a slightly higher frequency band with the advance of MIAA [56]. Study summarizes that the advance of the MIAA not only increases the amplitude of pressure oscillation in the time domain but also amplifies the signals at main oscillation frequency band (12–16 kHz), which is evident from cylinder pressure signal [56].

9.3 Cylinder Pressure-Based Knock Analysis

The detection of the abnormal combustion (knock) onset and the determination of knock intensity are crucial issues in engine and fuel development. Various methods have been proposed and used for the detection of knock onset and intensity such as methods based on (1) cylinder pressure analysis, (2) block vibration analysis, (3) gas ionization analysis, and (4) heat transfer analysis [57]. The cylinder pressure-based method allows direct measurements of primary knock effects and detailed investigations on the abnormal combustion process. This is the most widely used method in laboratory tests for engine and fuel development. However, the necessity of one sensor for each cylinder and the high cost of the pressure transducers limit the use of this detection method to laboratory research, while most of the techniques that are currently employed on mass-produced engines for knock detection and control are based on engine block vibration analysis. The high-frequency pressure oscillations inside the cylinder, caused by abrupt end-gas autoignition, are transmitted through the engine structure, thus causing vibrations that can be detected by means of an accelerometer installed on the engine block. The knock detection methods involve different signal processing methods and algorithms depending on the sensor used for analysis.

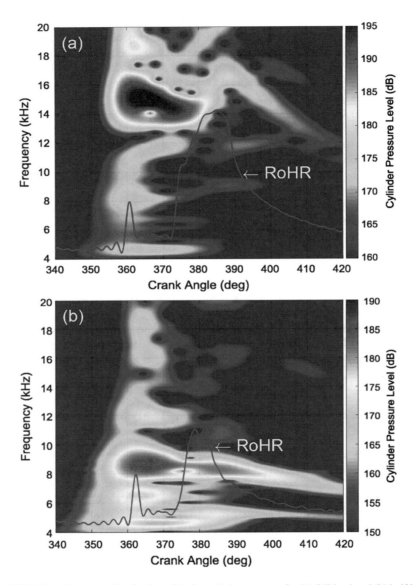

Fig. 9.27 Time-frequency distribution of the in-cylinder pressure for (**a**) full load and (**b**) half load at 1800 rpm [56]

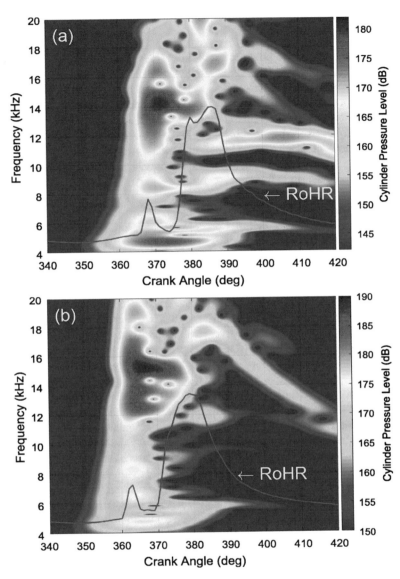

Fig. 9.28 Time-frequency distribution of the in-cylinder pressure for main injection advanced angle (**a**) −10° CA and (**b**) −5° CA at full load and 1400 rpm [56]

9.3.1 Signal Processing

Before calculating knock pressures and intensities from the measured cylinder pressure data, it is essential to ensure that the signals are free from unwanted effects such as transducer and it's installation cavity resonance, signal noise, and signal aliasing. Particularly, it is important to check when a relatively low sampling frequency is being used (mainly at low engine speed with crank angle based measurements). Non-flush mounting may lead to the problems with either cavity resonance or low sensitivity to the knock vibrations. Additionally, it is important to ensure that the transducer is actually producing signals when the engine is knocking and that these signals are at frequencies corresponding to the main acoustic vibration modes [58].

Sampling (or acquisition) frequency is an important parameter that needs to be considered while measuring cylinder pressure data for knocking combustion analysis. The acquisition frequency is required to ensure that the raw curve is sampled at a high enough rate that none of the high-frequency components are lost due to aliasing or under-sampling of the signal [5]. Generally, higher frequency that can be used is better, but the absolute minimum is set by the Nyquist criterion which suggests at least twice the highest frequency of significance needs to be used. Although digital filters and FFT algorithms may need even higher frequencies to function correctly. Since knock frequencies for the different modes are typically up to 20 kHz, which means an absolute minimum of 40 kHz sampling frequency requirement assuming that higher resonances and vibrations are not present. Typically, for crank angle-based cylinder pressure measurement, the resolution of the angle encoder is the most important factor. The calculated knock intensity is reduced when coarser crank angle resolution is employed, especially at low engine speeds, and a resolution of $0.2°$ or better resolution is suggested for routine knock measurement [58]. The disadvantage of high sampling rates is that a higher specification data acquisition system is required and a greater volume of data is acquired for processing and perhaps storage.

The effect of sample size also affects the knock intensity calculation and analysis. The variability of peak knock pressure data is very large, and a sample size of approximately 1000 cycles is recommended for the calculation of knock intensity [58]. Another important parameter which has a significant effect on the volume of data to be processed is the knock window. In addition to the filtering of the measured raw pressure data, the dataset is often windowed, which means that it is acquired only within a specific crank angle range on the pressure curve where knock typically occurs. The cylinder pressure data outside the window range will then be ignored, or acquired at a lower crank degree resolution, in order to optimize the available memory on the acquisition hardware [5]. A knock window crank angle range of TDC to $40°$ ATDC is recommended for general applications although most peak knock pressures should occur between $10°$ and $30°$A TDC [58].

Pressure transducers should have a high natural frequency to prevent knock-induced transducer resonance, and typically a natural frequency above 100 kHz is recommended. The pressure transducer should ideally be flush mounted (Chap. 2) to avoid resonance in the connecting passage; long connecting passages must be

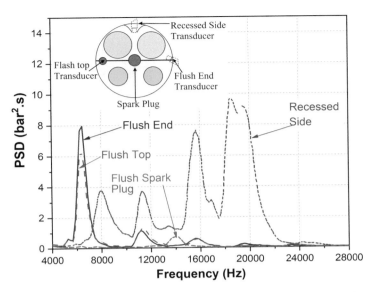

Fig. 9.29 Comparison of the power spectral densities from the four different transducers at the same engine operating conditions (adapted from [58])

avoided. Figure 9.29 illustrates the effect of flush and recessed mounting as well as the effect of sensor location on frequency measurement. The pressure transducer mounting position has a significant effect on the magnitude and characteristics of the knock signals obtained and, thus, the calculated knock pressures and intensities. Location of the transducer at the center of the combustion chamber has been shown to give low sensitivity to the main knock vibration modes [58]. To measure the maximum pressure oscillations, the position of the pressure transducer should be far from the center of the combustion chamber, however, not close to the quench zone.

9.3.2   Knock Indices

The knock-detection methods based on in-cylinder pressure analysis can be classified as methods based on (1) the evaluation of a single pressure value, (2) pressure derivatives, (3) frequency domain manipulations, and (4) heat release analysis [57]. Two main important parameters related to engine knock analysis is knock onset and its intensity, which can be evaluated using different methods and sensors. Figure 9.30 illustrates the different knock intensity (KI) metrics used for analysis in different engine combustion modes. The knock intensity calculation can be performed in the time domain or in the frequency domain. Further, the KI can be based on the pressure signal and its derivatives or on the heat release rate and its derivatives. Finally, knock intensity metrics can be based on a single value, (for example, the maximum value of a quantity) or an average or integrated value

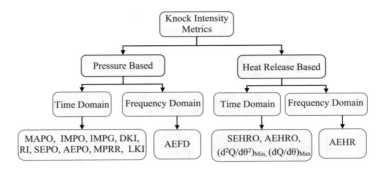

Fig. 9.30 Various knock intensity parameters for reciprocating engines

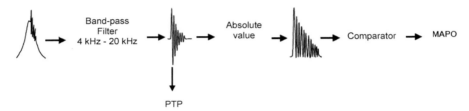

Fig. 9.31 Illustration of MAPO calculation method using in-cylinder pressure signal [60]

[30]. All time-domain knock metrics are typically based on a filtered pressure trace; either high- or band-pass filtering is performed to eliminate the low-frequency content of the signal associated with compression and the standard flame propagation heat release. Thus, filter characteristics such as the filter type and order, its roll-off characteristics, etc. (Chap. 5) affect the results of knock intensity calculations.

9.3.2.1   Pressure-Based Methods

During knocking combustion, the sharp pressure peak usually occurs, and thus, the simplest way to detect and characterize knock can be by the peak pressure itself [57, 59]. However, only high knock intensities can be reliably detected using this method, and the physical quantity measured is significantly affected by the engine parameters and operating conditions. A more effective knock index can be estimated by using the increase of pressure from the normal combustion conditions. The determination of the knock-related pressure peak from the normal pressure curve corresponds to the band-pass filtering of the pressure signal over the knock-characteristic frequency range. The most common time-domain pressure-based method to characterize KI is often known as the maximum amplitude of pressure oscillations (MAPO). The peak-to-peak (PTP) value of pressure oscillations is also used to quantify the KI. Figure 9.31 illustrates the calculation methods of PTP and MAPO.

The cylinder pressure signal is measured in a time window around TDC (from θi to $\theta i + W$) that is filtered by a band-pass filter 4–20 kHz to obtain the knocking pressure oscillations [60]. In another study, the high-frequency component was determined by first applying an optimized moving average process to smooth out most of the higher-frequency component and then subtracting the smoothed pressure curve from the original data to generate the high-frequency knocking pressure wave [58]. Two significant advantages of using the moving average smoothing routine instead of a high-pass digital filter are that it should be more tolerant to the coarser crank angle resolution (i.e., lower sampling rate) and that it captures all knock mode signals. The smoothing algorithm is also numerically very efficient.

The MAPO calculation can be mathematically represented by Eq. (9.20):

$$\text{MAPO} = \max\{P_{\text{filt}}\}|_{\theta i, \theta i + w} \tag{9.20}$$

Single value can be affected by noise, and it does not signify the duration of knock or further oscillation. Integral-based methods are used to represent the average value of pressure oscillation. This method involves summing up of the individual amplitudes of pressure oscillations to quantify the magnitude of knock and is typically referred as integral of the modulus of pressure oscillations (IMPO) [3]:

$$\text{IMPO} = \frac{1}{N} \sum_{1}^{N} \int_{\theta_i}^{\theta_{i+w}} |P_{\text{filt}}| \cdot d\theta \tag{9.21}$$

where N is the number of computed cycles.

Another pressure-based method of knock intensity is defined from the signal energy of pressure oscillations (SEPO), which is determined by integrating the square of P_{filt} over that crank angle window from knock onset, θ_0, to $\theta_0 + \Delta\theta$ [30]:

$$\text{SEPO} = \int_{\theta_0}^{\theta_0 + \Delta\theta} \{P_{\text{filt}}\}^2 \cdot d\theta \tag{9.22}$$

where $\Delta\theta$ is typically 5–20 CAD.

A similar method is to average the signal's energy over $\Delta\theta$; this method is called average energy of pressure oscillations (AEPO) [30, 40]:

$$\text{AEPO} = \frac{1}{\Delta\theta} \int_{\theta_0}^{\theta_0 + \Delta\theta} \{P_{\text{filt}}\}^2 \cdot d\theta = \frac{\text{SEPO}}{\Delta\theta} \tag{9.23}$$

A dimensionless knock indicator (DKI) is developed from existing knock indices (IMPO and MAPO) [61] which is calculated by Eq. (9.24):

$$\text{DKI} = \frac{\text{IMPO}}{\text{MAPO} \cdot W} \tag{9.24}$$

where W is the width of the computational window. The DKI is the ratio between two values that can be interpreted as two surfaces. The DKI value will decrease with the increase of KI independent of the engine geometrical characteristics and settings.

The pressure signal can be decomposed into a series of harmonic waves using a Fourier transform. The frequency range spans $0 \leq f \leq f_{Nyq}$, where f is frequency and f_{Nyq} is the Nyquist frequency given by $f_s/2$ with f_s as the sampling frequency of data acquisition. Since Fourier transform of a signal is a complex-valued function, the real-valued power spectrum needs to be used instead, which has units of kPa^2/Hz for pressure reported in kPa [30]. Naturally, the result comes as energy, and thus, the average energy of pressure computed in the frequency domain, over the range of frequencies defined by $f_1 < f < f_2$, is given by Eq. (9.25):

$$\text{AEFD}_{f_1-f_2} = \int_{f_1}^{f_2} \wp(f) \cdot df \tag{9.25}$$

where $\wp$ is the pressure power spectrum.

9.3.2.2   Pressure Derivative-Based Methods

The pressure rise rate (PRR) can be used for both knock onset detection and knock intensity measurements as it is greatly increased by end-gas autoignition. Even though the change in the pressure first derivative is well related to the amount of the cylinder charge that undergoes autoignition, the several knock-independent factors, which can affect the PRR, represent the greatest disadvantage of this method [57]. Maximum pressure rise rate (MPRR) is used to limit high load boundary or knock boundary in HCCI engines [62]:

$$\text{MPRR} = \left(\frac{dP}{d\theta}\right)_{max} \tag{9.26}$$

Alternative knock indicators based on further pressure derivative have been proposed. For example, due to the abrupt curvature change which accompanies the sharp knock pressure peak, the minimum value position of the third derivative of cylinder pressure can be used as a knock indicator [57, 63].

Based on the understanding of the energy associated with pressure oscillation in HCCI engine, Eng [64] proposed another metric called ringing intensity (RI) by using maximum pressure (P_{max}), maximum temperature (T_{max}), and MPRR in the cylinder. The RI is given by Eq. (9.27):

$$\text{RI} = \frac{\sqrt{\gamma R T_{max}}}{2\gamma P_{max}} \left[\beta\left(\frac{dP}{dt}\right)_{max}\right]^2 \tag{9.27}$$

This metric is commonly used for knock intensity evaluation in low-temperature combustion mode engines [62].

Integral of the modulus of pressure gradient (IMPG) is also defined similar to IMPO for KI calculation and represented by Eq. (9.28):

$$\text{IMPG} = \frac{1}{N} \sum_{1}^{N} \int_{\theta_i}^{\theta_{i+w}} \left| \frac{P_{\text{filt}}}{d\theta} \right| \cdot d\theta \tag{9.28}$$

9.3.2.3   Heat Release Analysis-Based Methods

The intense local heat release (due to end-gas autoignition) causes a substantial shift from the normal combustion heat release profile. Thus, heat release rate can also be used as knock indicators. The detailed calculation method of heat release is presented in Chap. 6. Analogous to the pressure-based metrics, knock can be defined as the maximum amplitude of heat release oscillations (MAHRO), signal energy of heat release oscillations (SEHRO), and average energy of heat release oscillations (AEHRO), which can be defined by Eqs. (9.29) to (9.31) [30, 40]:

$$\text{MAHRO} = \max \left\{ \left(\frac{dQ}{d\theta} \right)_{\text{filt}} \right\} \tag{9.29}$$

$$\text{SEHRO} = \int_{\theta_0}^{\theta_0 + \Delta\theta} \left(\frac{dQ}{d\theta} \right)_{\text{filt}}^{2} \cdot d\theta \tag{9.30}$$

$$\text{AEHRO} = \frac{1}{\Delta\theta} \int_{\theta_0}^{\theta_0 + \Delta\theta} \left(\frac{dQ}{d\theta} \right)_{\text{filt}}^{2} \cdot d\theta = \frac{\text{SEHRO}}{\Delta\theta} \tag{9.31}$$

Because of linearity, the filtered heat release can be acquired either by using the measured pressure and filtering the heat release or by filtering the pressure before calculating the heat release [30].

In the frequency domain, the average energy of the heat release (AEHR_{f1-f2}) can be defined by Eq. (9.32) similar to pressure-based metric:

$$\text{AEHR}_{f_1 - f_2} = \int_{f_1}^{f_2} dQ(f) \cdot df \tag{9.32}$$

A study concluded that methods based on a calculated heat release are redundant with other metrics [30]. It is also demonstrated that the superposition of multiple resonant waves can lead to interference and beating phenomena, which render the magnitude of the resulting time-domain signal not representative of the total energy of the base signals. As such, single-value metrics, e.g., MAPO, can be biased. Integrated energy-based quantities provide the most bias-free estimates of the knock intensity. The integration can be performed either in the time or frequency domain [30].

9.4   Methods of Knock Detection and Characterization

Knocking combustion in the reciprocating engine cannot be precisely predicted due to its random nature. The random nature of engine knocking is created by the cyclic pressure fluctuations as well as some unobservable effects, such as residual mass variations or temperature hot spots [65]. Thus, an accurate prediction of engine knocking is almost unaffordable for real applications. Presently employed knock control algorithms are based on the combination of open-loop control actions with some stochastic rules, which allows adapting the engine operation on the basis of the knock detection events in the previous combustion cycles. The engine knock detection is essential for conventional controller performance because failure in knock detection can result in engine damage and a false knock detection of particular cycles can lead to a reduction in the engine efficiency. For commercial applications, ion sensors or accelerometers are used. However, knock detection methods based on cylinder pressure are the most precise. Most of the methods based on measured cylinder pressure quantify the pressure oscillations in the time domain or frequency domain for engine knocking analysis.

Engine knock prediction methods typically focus on the two important attributes, namely, (1) exact detection of the knock onset and (2) quantification of knock intensity. The identification of the knock location in the combustion is also an important attribute, which needs further analysis and possible by installing multiple pressure sensors [66] or using optical methods. Typically band-pass filtered pressure signal is used for knock intensity and knock onset detection. Knock onset is defined as the crank angle position at which a threshold value is exceeded or maximum amplitude of oscillation occurs. Knock intensity is defined as maximum amplitude (MAPO) or integral over a range or energy density over a range (Sect. 9.3.2). First or third derivative-based methods are also used for knock onset and intensity determination.

In the threshold exceeding based index, the value threshold needs to be sufficiently high to avoid pressure oscillations produced by combustion process [67] but still must be as small as possible to detect weak knocking cycles [68]. The detection of weak knocking cycles can significantly improve knock control algorithms [69]. On the other hand, normal SI combustion can be confused with end-gas autoignition if a low threshold is selected. Spark ignition combustion and knock both can excite in-cylinder pressure resonance (oscillations). Though, the normal spark ignition combustion is controlled by spark timing and has significant but bounded cyclic variations, while knocking has a random nature creating larger variations of resonance excitation [65]. Figure 9.32 illustrates the pressure oscillations in normal and knocking combustion cycles in a SI engine. In Fig. 9.32 the heat

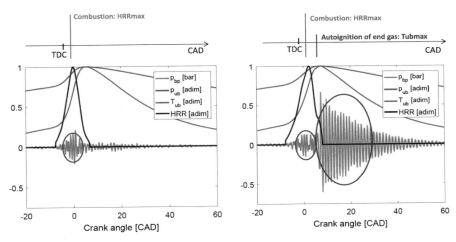

Fig. 9.32 In-cylinder pressure oscillations for normal cycle (left) and knocking cycle (right) at 1500 rpm in a SI engine [65]

release rate (HRR), the unburned temperature (T_{ub}), and the unburned gas pressure (p_{ub}) are presented as normalized value, while the band-pass filtered pressure (p_{bp}) data is presented in bar.

Figure 9.32 reveals that the excitation of the cylinder resonance due to combustion is similar in both normal and knocking combustion cycles. However, the oscillations due to autoignition of the end gas (charge) are positioned in the vicinity of the end of combustion (EOC). It is also found that oscillations due to autoignition differ significantly from cyclic basis [65]. The time domain does not provide the information regarding the resonant frequencies, which can be differentiated using short-time Fourier transform (STFT) or the wavelet transform. Figure 9.33 shows the short-time Fourier transform (STFT) spectrum of a normal cycle and a knocking cycle shown in Fig. 9.32. The figure depicts that the frequencies are only excited near the maximum HRR for the non-knocking cycle, while in the knocking cycle important resonance excitations generated near the end of combustion (around 10 CAD after TDC). After the frequencies are excited, only the resonant modes vibrate in the chamber, particularly the first radial mode (between 6 and 8 kHz) [65].

A knock event definition is proposed based on the comparison of the resonance oscillation at the maximum HRR and at the maximum unburned temperature (near EOC). It is assumed that the higher resonance at the EOC is result of knocking process (autoignition of end gas). Figure 9.34 illustrates the knock detection method based on the resonance at two locations. Two windows at different locations (one centered at maximum HRR and other centered at maximum unburned gas temperature) are used to separate combustion by autoignition. The resonance amplitude is calculated using fast Fourier transform (FFT). Two indices (I_C and I_A) are calculated by integrating the signal over the resonance frequencies. Knock event is detected only when the resonant content of the I_c is greater than I_A [65].

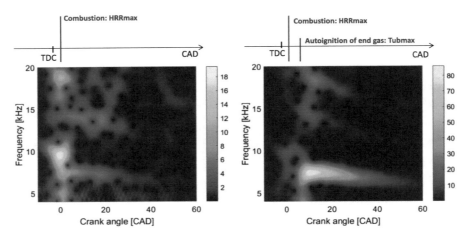

Fig. 9.33 Short-time Fourier transform (STFT) spectrum of a normal cycle (left) and a knocking cycle (right) [65]

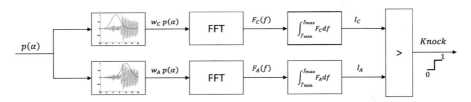

Fig. 9.34 Schematic diagram of knock determination procedure [65]

A knock indicator known as derivative at maximum pressure (DMP) is proposed for knocking analysis in dual-fuel engine [60]. The value of the pressure derivative is almost zero at peak pressure location for a non-knocking cycle. In the knocking condition, the value of this derivative increases due to pressure oscillations. Gradually, as the knocking intensifies, pressure oscillations become more important, and the value of this derivative becomes higher [60]. The DMP is calculated by Eq. (9.33).

$$\text{DMP} = \left(\frac{dp}{d\theta}\right)_{\theta_{p\text{max}}} \tag{9.33}$$

where $\theta_{p\text{max}}$ is the crank angle corresponding to the maximum pressure of engine cycle. This parameter correlates very well with the commonly used MAPO knock indicator. Advantages of this method are an easy and fast calculation.

An engine combustion cycle can be characterized by a numeric value correlated to the knock event using MAPO. However, cycle-to-cycle variations exist in SI engine, and thus, statistical methods are used to characterize the knock intensity (KI) of an engine operating point [67]. A critical value of MAPO (MAPO$_{th}$) is defined for

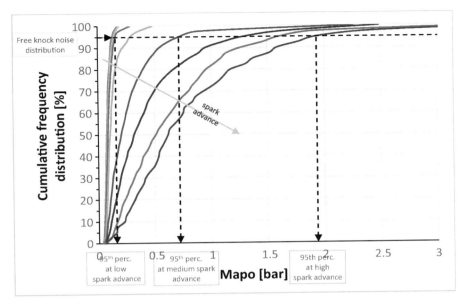

Fig. 9.35 Cumulative frequency distributions of MAPO values during knocking combustion [70]

the evaluation of knock intensity for a particular engine operating condition. For individual engine cycle, the knock intensity is usually calculated by Eq. (9.34) [70]:

$$I = \max\{(\text{MAPO} - \text{MAPO}_{\text{th}}), 0\} \tag{9.34}$$

where MAPO_{th} differentiates between knocking and non-knocking combustion cycles.

The MAPO distribution is very uneven because of cyclic variations at particular engine operating conditions. Figure 9.35 illustrates the cumulative frequency distribution plot of MAPO at different spark timings. The 95th percentile of MAPO distribution has been used to characterize the KI of each engine operating point. Figure 9.35 provides good discrimination between the different knock levels with varying spark timing. The 95th percentile can be used to characterize every engine operating point by means of a single numerical value [67]. Figure 9.36 illustrates the variation of the 95th percentile of MAPO distribution. The figure shows that for low spark advance timing (at particular test condition), the 95th percentile increases very slowly with the spark advance with almost constant value. This region is typically a non-knocking condition of engine operation. At higher spark advance timing, the 95th percentile increases rapidly with the spark advance timing indicating more and more heavy knock phenomena [70].

In the non-knocking region, the 95th percentile of MAPO distribution shows a constant value, which can be considered as the threshold to differentiate between

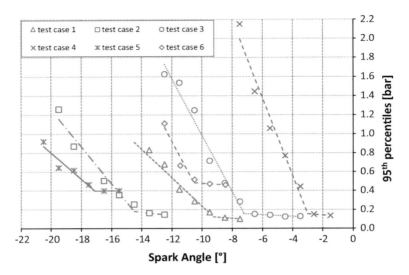

Fig. 9.36 The 95th percentile of MAPO distribution behavior [70]

non-knocking and knocking combustion [67, 70]. This threshold value depends on the engine operating conditions (Fig. 9.36) such as engine speed and load.

The KI of a particular engine cycle (j) can be calculated by Eq. (9.35), while the KI of an engine operating point can be calculated by Eq. (9.36) with N_{cyc} number of acquired cycles for knocking analysis [67, 70]:

$$I1_j = \frac{\max\{(\text{MAPO} - \text{MAPO}_{th}), 0\}}{\text{MAPO}_{th}} \tag{9.35}$$

$$KI1 = \frac{\sum_{j=1}^{N_{cyc}} I1_j}{N_{cyc}} \tag{9.36}$$

To overcome the dependency of the threshold value on engine operating conditions, a dynamic method has been proposed to determine the KI of a particular engine cycle [67]. For every cycle, a specific threshold is calculated by Eq. (9.37):

$$\text{MAPO}_{th,j} = k \cdot \max\left\{ p_{knock,j}(\theta) \big|_{SA}^{\theta_{60\%}} \right\} \tag{9.37}$$

where SA is the spark angle, $\theta_{60\%}$ is the angle corresponding to the 60% of the mass burned fraction, and k is a constant value.

The KI of a particular engine cycle (j) can be calculated by Eq. (9.38), while the KI of an engine operating point can be calculated by Eq. (9.39) with N_{cyc} number of acquired cycles [70]:

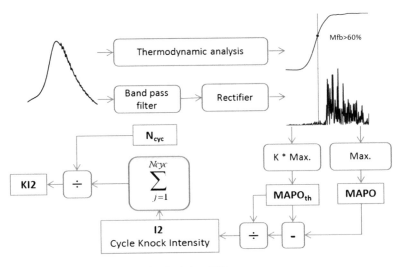

Fig. 9.37 Schematic diagram for KI2 calculation [70]

$$I2_j = \frac{\max\left\{\left(p_{\text{knock},j}(\theta)\big|_{\theta_{60\%}}^{\theta_{99\%}} - \text{MAPO}_{\text{th}}\right),0\right\}}{\text{MAPO}_{\text{th}}} \tag{9.38}$$

$$KI2 = \frac{\sum_{j=1}^{N_{\text{cyc}}} I2_j}{N_{\text{cyc}}} \tag{9.39}$$

where $\theta_{99\%}$ is the angle corresponding to the end of the combustion.

Figure 9.37 schematically demonstrates the calculation of dynamic knock intensity in a spark ignition engine using cylinder pressure measurement. It can be noted that these knock indices (KI1 and KI2) are dimensionless numbers which allow comparing engine operating conditions characterized by a different value of speed or load.

A critical first step in the analysis of the knock intensity data is to investigate the extent to which the data behave as a cyclically independent random process. Knock intensities are fundamentally uncontrollable if it is random, and deterministic control theory no longer applies. By adopting a more stochastic approach, however, it is still possible to regulate some statistical property of the data. Most knock controllers, for example, regulate the knock event rate or empirical knock probability rather than the (random) knock intensity signal itself [71]. An initial analysis has provided strong evidence, both from autocorrelation functions and from the Pearson chi-squared test statistic, that knock intensities closely approximate a cyclically uncorrelated process. The process is then completely described by its marginal probability density function (PDF) whose mean and variance were observed to increase as the spark timing is advanced [71].

The knock intensity is strongly influenced by cycle-to-cycle variations for a fixed operating condition. It is demonstrated that a strong relationship exists between knock intensity index and the charge turbulent motion near the spark plug or the in-chamber temperature distribution. Physical knock models derived from the studies can be efficiently calibrated to determine the knock onset and the knock-limited spark advance (KLSA) or, more recently, to predict the trend of a MAPO percentile with respect to spark timing [72].

The observation that knock intensities closely approximate a cyclically independent random process means that the process is completely characterized by its probability density function (PDF) or equivalently by its cumulative distribution function (CDF). A study demonstrated that the log-normal distribution can be used to characterize the knock intensity distributions in SI engines for both the pressure and accelerometer and determine quantitative metrics of mean, standard deviation, knock percentiles, skewness, and peakedness [73].

Figure 9.38 shows the probability distributions of gasoline compression ignition (GCI) and SI combustion modes from non-knocking to knocking conditions using 200 consecutive engine combustion cycles. In SI engines, at retarded spark timing conditions, the distribution of MAPO is concentrated at relatively low knock intensities. For SI combustion, knock and non-knock operating conditions are obviously categorized by a particular spark timing (ignition timing of 22° CA bTDC), which has knock borderline value of 0.1 MPa (Fig. 9.38b). Additionally, the probability distribution of MAPO for the two combustion modes is quite different. In GCI mode, advancing injection timing from 20 to 30° CA bTDC only results in slightly higher MAPO mean. However, further advanced injection timing from 30 to 34° CA bTDC, significant increase in MAPO mean can be observed (Fig. 9.38a). With advanced injection timing, the distribution shifts toward higher intensities, broadening and becoming increasingly skewed. The upper tail of the distribution therefore increasingly protrudes into the very high-intensity region, causing the knock event rate to rise rapidly.

Figure 9.39 further illustrates the knocking characteristics of GCI and SI mode by presenting the cyclic variations of MAPO for knock-free, critical knock, and knock conditions. Three statistics terms, mean (μ), standard deviation (σ), and relative standard deviation (RSD) which is also coefficient of variation (CoV), are used to describe the MAPO distributions. Figure 9.39 a1 to a3 shows that advancing the injection timings leads to an increase in mean and standard deviation of MAPO. However, the RSD (coefficient of variation) remains almost unaffected from non-knocking to knocking operating conditions, which suggests that the increase of KI in GCI combustion mode doesn't change the MAPO distribution pattern fundamentally. Moreover, the results show that the in-cylinder pressure oscillations have negligible effects on combustion mode, and only burning rate, HRR, and the amplitude of pressure oscillations are changed as a result of rapid combustion [74]. Figure 9.39 b1 to b3 shows the MAPO variation for SI combustion mode. In this combustion mode, higher variations of MAPO distribution are observed, and significant change in RSD value is found from non-knocking to knocking operating

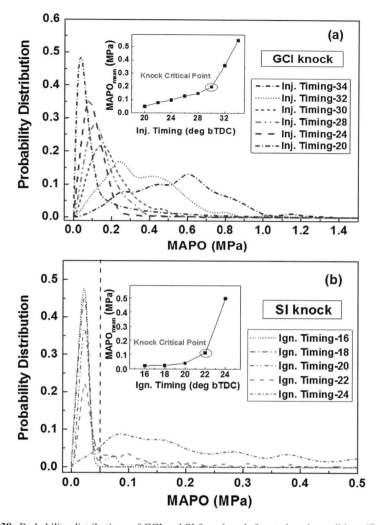

Fig. 9.38 Probability distributions of GCI and SI from knock-free to knock conditions [74]

conditions. The extremely high-pressure oscillation can be observed occasionally (Fig. 9.39b3). The reason for the distribution difference between the two combustion modes is supposed to be the nature of random end-gas autoignition under SI knocking conditions. To further illustrate the knocking differences in SI and low-temperature combustion modes, Fig. 9.40 shows the MAPO distribution for homogeneous charge compression ignition (HCCI), spark-assisted compression ignition (SACI), and SI combustion. The figure shows that the KI distributions of HCCI and SI are quite different. The average KI of SI knock is less than the HCCI knock, but randomly very high KI can be observed (Fig. 9.40). The RSD value of SI

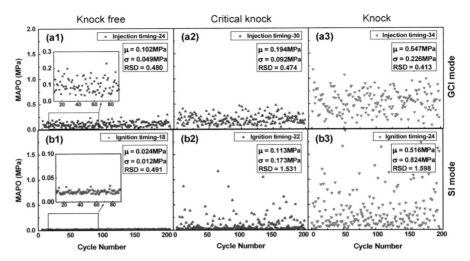

Fig. 9.39 Detailed MAPO variation of GCI (a1 to a3) and SI (b1 to b3) combustion from knock-free to knock conditions [74]

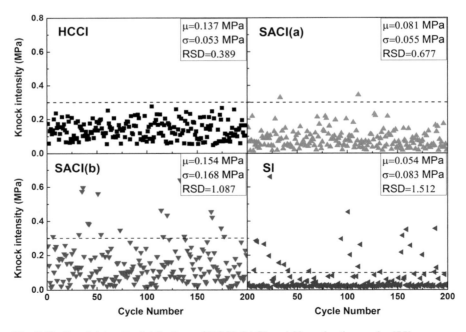

Fig. 9.40 Knock intensity distributions of HCCI, SACI, and SI combustion modes [75]

knock condition is 1.512, which is almost four times the value of HCCI (0.389) due to the differences in the process leading to knock phenomenon in HCCI and SI engine. The HCCI knock is generated by intensive multipoint autoignition and too fast burning rate, while SI knock is due to the random autoignition of end gas, which greatly relies on flame propagation [75]. In the two SACI conditions, KI distribution combines the characteristics of both HCCI and SI knock conditions (Fig. 9.40).

The engine knock event is not uniquely defined by the time history of the end gas's thermochemical history in SI engine. In a study, it is demonstrated that three cycles are identical up to the point of knock onset combustion event, but the resulting KI differed by a factor of 8 [76]. The KI is not uniquely described by the unburned mass fraction at knock onset. However, the upper limit of KI, as defined as the 95th percentile of the cumulative distribution in a narrow window of unburned mass fraction, did correlate with KI. A methodology to predict the maximum possible KI based on the volumetric expansion rate of the end gas was developed, and the expansion rate was modeled based on blast wave theory, which is proportional to the energy of the initial explosion. The pressure rise was predicted to be linear with the energy available at knock onset [76]. Another study showed that the ratio of KI and gross indicated mean effective pressure (GIMEP) follows a log-normal distribution with mean μ and standard deviation σ, which is estimated using experimental data [10]. The value of μ decreases linearly with the ignition delay at the knock point, and σ is found to be a constant. The log-normal distribution can be used to determine the statistical distribution of KI in an engine simulation.

9.5 Knock Detection by Alternative Sensors

Presently, there are several methods for knock detection such as autoignition detection, ion current detection, in-cylinder combustion pressure detection, engine block vibration detection, and combustion noise detection. Figure 9.41 illustrates the different knock detection approaches based on the knocking process and its effects.

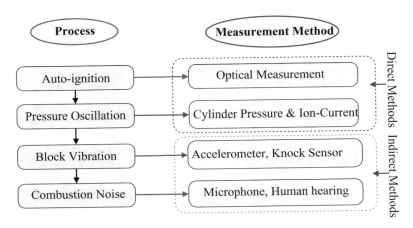

Fig. 9.41 Knock detection methods in reciprocating engines

In-cylinder combustion pressure detection is the most widely used approach in the laboratory for precise knock detection in each cylinder, because of its high signal-noise ratio, high sensitivity, but high cost.

Methods for knock detection can be classified into two broad categories, namely, direct and indirect methods [77]. Direct methods typically include the knock detection by directly measuring the knocking in the combustion chamber such as optical methods, ion current measurement, and cylinder pressure measurement. The in-cylinder pressure measurement can detect the typical high-frequency knock signature, related to the combustion chamber resonant frequencies. The typical drawback of cylinder pressure-based indices is that they are sensitive to local information provided by the pressure transducer. As knock occurs, huge pressure non-homogeneities take place inside the combustion chamber [59]. Thus, the measured pressure wave amplitude has only a local meaning, and cannot be considered as the maximum value of the whole combustion chamber pressure wave. Another intrusive (direct) knock sensor is based on ion current measurement. With years of development, ion current detection is more and more widely used due to its high detection precision, high signal-noise ratio, low cost, and rich combustion information content from each cylinder [78].

The indirect methods of knock detection include block vibration detection, engine noise detection, and exhaust temperature detection. Engine block vibration detection method (knock sensor) is broadly equipped in production vehicles. One engine normally mounts only one knock sensor. Thus, it is not so effective to distinguish each knocking cylinder or other vibrations caused by such as high-speed valves hitting valve seats, due to its fixed mounting point. The combustion noise detection is not frequently used, due to its low signal-noise ratio and low detecting precision. The exhaust gas temperature measurement can be utilized to identify knocking combustions because it reduces in knocking conditions as a consequence of increased wall heat fluxes inside the cylinder. This technique, moreover, is not affected by mechanical noise and can be applied for all types of engines [79]. Main drawbacks are related to the need of comparing the temperature level provided by the thermocouple to the one stored in a map, representative of a knock-free operation in the same engine speed, load, and air-fuel ratio operating conditions [77]. Table 9.1 shows the comparison of the different knock sensors on the basis of engine modification, cost, durability, and signal quality.

It can be summarized that all sensors respond to the resonant characteristics of knocking combustion but differ in robustness, cost, bandwidth, monitoring capability, and signal-to-noise ratio [80]. On the one hand, sensors like accelerometers are relatively cheap and robust sensor capable of detecting knock event but limited by the poor signal-to-noise ratio (SNR), particularly at high engine speeds and loads. On the other hand, cylinder pressure sensors provide good-quality signals adequate for detailed combustion analysis, but at the expense of cost and robustness. Ionization probes and production cylinder pressure sensors are now becoming viable as a compromise between these two extremes [80].

Table 9.1 Knock sensor comparison (Courtesy of Spelina, J. M, adapted from [80])

Sensor	Engine modification	Durability	Signal-to-noise ratio (SNR)	Cost	Remarks
Accelerometer	Nonintrusive	Good	Poor at high speed	Fairly cheap	Easy to mount and less accurate and not a fundamental measurement
Cylinder pressure sensor	Intrusive	Fair to poor	Good	Very expansive	Accurate and fundamental measurement. Difficult to mount
Ionization probe	Intrusive	Good	Poor at high speed	Fairly cheap	
Optical sensor	Intrusive	Poor	Good		Fundamental studies
Strain gauged cylinder head bolts	Nonintrusive	Fairly good	Vibrational interference	Fairly expansive	

Typically, the major factors considered for knock detection includes the cost of measurement, repeatability, and accuracy, time of measurement and analysis, and the required engine modifications for sensor installation. A brief description of these alternative sensors is provided in following subsections.

9.5.1 Optical Methods

Optical methods of knock detection are typically conducted on optical engines. Photographic observations of knocking combustion have been performed based on the progress of high-speed imaging techniques [8]. Photographic images of combustion chamber have been an important source of insight into the fundamentals of the knocking process because of the combustion process leading to the occurrence of knocking proceeds extremely fast. The optical measurement by the combustion light is a method for knocking measurement without being affected by mechanical noises. Widely used in the optical measurement is such a method that measures hydroxyl (OH) radical emission intensity by diffracting the combustion light and converting OH radical emission intensity to the voltage using a photomultiplier tube [81]. Emissions from chemical luminescence have been used to analyze the chemical reactions caused by autoignition of the end gas [8, 82]. A study used high-speed Schlieren photography and multi-optical fiber techniques to investigate the flame propagation and shockwaves on cycle-to-cycle resolved knocking combustion [82].

The methods for measurement of combustion light or image in the engine cylinder include such that a part of the cylinder liner or the piston is made of transparent glass, or an optical probe is installed in the spark plug or the combustion chamber. In either way, the measurable operating conditions are often limited from the mechanical strength of the light transmission route. There is a need of means for

Fig. 9.42 Optical pressure
wave measurement sensor
and optical fiber
configuration [85]

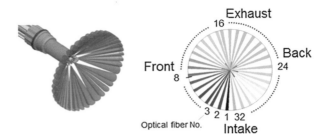

optical measurements under a wide variety of engine operating conditions including
high-load conditions of commercial engines taking advantages of the outcomes from
studies of optical measurement using experimental engines in order to develop more
efficient engines. The high-durability optical probe has been developed and demonstrated as an instrument for optical measurement of combustion [83]. The optical
probe has enabled measurement of knocking under high-speed and high-load conditions while minimizing restrictions to engine operating conditions. The optical
probe is demonstrated for the detection of knocking, and based on the measurement,
knocking damage index (KDI) is developed [81].

The localization of knock events in the combustion chamber of a SI engine is a
key diagnostic activity to support its full-load development. A plug built-in-type
optical sensor (AVL visio knock) is also used for knock detection [84]. Figure 9.42
illustrates the optical sensor-based knock measurement system. The sensor consists
of 32 channels of optical fibers located in the radial direction of the spark plug and
piston. Through these optical fibers, the sensor takes sample of the optical signals in
the radial direction in the combustion chamber [85]. The signal intensity of optical
sensor and the cylinder pressure have a linear relationship, which is used to get the
local changes in the cylinder pressure during the knocking process. The knock onset
direction is determined at this time by detecting the origin of the high-frequency
vibration. Typically, a reference light source is used to calibrate each optical fiber
channel to the same output characteristics.

9.5.2 Ion Current Sensor

The prevailing conditions in the engine cylinder during combustion lead to gas
ionization. A current can be produced by applying a voltage (of the order of a
hundred volts) between two electrodes because the ionized combustion gases are
conductive. Generally, the voltage inside the cylinder is applied using the spark plug
tips as electrodes. The principle of the ion current sensor is discussed in Sect. 2.3.1.
As the ion density increases as the cylinder pressure increases, the ion current
provides another measure of cylinder pressure signal. The problem with ion current
measurement is that it represents the combustion intensity in a very small volume
around the spark plug/ion probe and not the entire combustion chamber, and it also

Fig. 9.43 Ion current signal (blue) along with pressure oscillation (green) in a heavy knock condition [87]

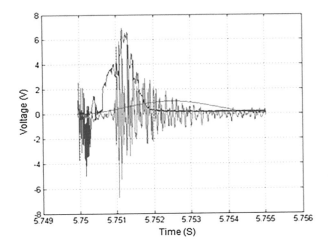

requires significant additional circuitry [80]. Thus, knock detection depends on the location of spark plug or ion current sensor.

Several studies used ion current sensor for the knock detection and characterization [78, 86–88]. Figure 9.43 illustrates the ion current signal during heavy knock conditions. The figures show that there is a high-frequency oscillation with larger amplitudes in the ion current signal during heavy knock conditions. Thus, it can be used for the knock detection and control. Ion current signal shows a good correlation with in-cylinder combustion process and the cylinder pressure signal. Ion current peak value can distinguish knock cycles from normal cycles fast and easily [78]. Both ion current band filter signal and integral value can distinguish different knock cycles from normal combustion cycles and show high linear correlation with the knock intensities during light knock and heavy knock. Ion current signal proves to be an effective method to discriminate different knock intensities [78, 86].

9.5.3 Engine Block Vibration

The engine knock can be sensed through accelerometers (piezo-ceramic sensors), mounted on the engine block, which evaluate the vibration transmitted from the knocking cylinder. Combustion noise is primarily transferred through the piston and connecting rod. This is dominant only below 1000 Hz [89]. A secondary combustion noise path which proves more useful for knock detection is through the cylinder block and cylinder head, which is utilized by the accelerometer. These knock detection methods can suffer from disturbance caused by other noise sources (valve closing, piston slap, etc.). Poor noise transmission from the knocking cylinder to the sensor can happen because of the sensor's location [86]. Additionally, the disturbances increase with engine speed. Due to the location issue, four-cylinder engines need one or two sensors, but six and more cylinder engines need at least two sensors [90]. The main disadvantages of this method are (1) the dramatic deterioration in the signal-to-noise ratio (SNR) of the sensor response at higher engine speed due to the increase of other vibrations related to the operation of the valve-train system and of other rotating/moving parts and (2) the impossibility to recognize the knocking cylinder, which clearly excludes the individual control of spark timing [86].

Engine knock detection using vibration signals is normally more accurate than using noise signals. However, the vibration signal can have signals from other vibration sources or electrical noise. Thus, the vibration signal cannot be directly used for knock detection, and an effective signal processing method needs to be used [91]. A recent study proposed the knock detection method based on the wavelet denoising and empirical mode decomposition (EMD) methods using vibration signal [92]. The wavelet denoising is used to eliminate the high-frequency noise which is not produced by the engine knock. After the wavelet denoising, EMD can effectively identify the knock characteristics (including the light knock) from a vibration signal. This method can significantly improve computational time while maintaining reliable analysis.

9.5.4 Microphone

The combustion noise detection method using a microphone can be used to determine the engine knock by analyzing the noise frequency spectrum of the engine combustion. The knock intensity can be estimated by evaluating sound pressure level in a certain frequency band. This method of knock detection is low cost and does not require changing the engine structure. However, the combustion noise signal can easily interfered with other noises, and the detection accuracy is difficult to guarantee, and therefore this method is rarely used [92]. A study used microphones for combustion timing detection for feedback control of HCCI engines with reliable results at medium and higher engine loads [93].

9.6 Knock Mitigation Methods

In conventional SI engine, the knock occurs due to competition between the flame propagation speed and ignition delay of end gas. Thus, the basic principle for mitigation of knock is to ensure that the flame propagation time is less than ignition delay. Figure 9.44 summarizes the measures and technologies that can be utilized to reduce the knocking tendency of gasoline engine.

Apart from the influence of oil and fuel quality, the majority of measures leads to a lower working gas temperature during compression and combustion process

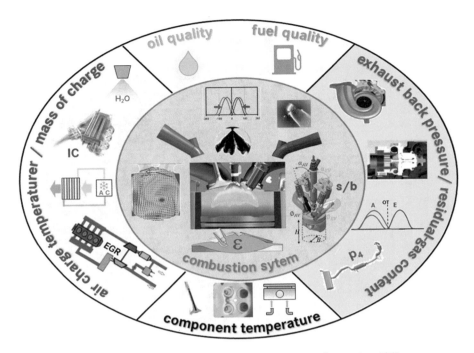

Fig. 9.44 Technologies for avoiding knocking combustion in gasoline engines [94]

[94]. Lowering the final compression temperature is considered as an effective measure for avoiding gasoline engine knock.

In conventional SI engine, the knock can be avoided by retarding spark timing; improving octane number and enriching fuel-air mixture, exhaust gas recirculation (EGR), and dual-fuel spark ignition with direct injection (DI) and port fuel injection (PFI); increasing turbulence; increasing cooling of the components and water injection; and lowering effective compression ratio [1].

Super-knock in a modern gasoline engine can be avoided by two methods: (1) by eliminating preignition and (2) by decoupling the shock wave and the heat release [1]. Figure 9.45 summarizes the measures that can be employed for avoiding the preignition in modern engines.

Water injection and cooled EGR are proposed practical solution for avoiding knocking combustion in a gasoline engine. Knocking is significantly influenced by the unburned zone temperature in the cylinder, and reduction in temperature leads to a reduction in knocking tendency. Table 9.2 depicts three possible technology approaches which are capable of decreasing combustion temperature and thus avoiding knocking combustion. Reduced knocking tendency allows the earlier combustion phasing operation, which improves the combustion efficiency and reduces the exhaust gas temperature. Additionally, lower combustion temperature results in reduced heat loss, lower NOx emission, and increased polytropic coefficient leading to a higher thermal efficiency [95]. In contrast to the Miller cycle, cooled EGR and water injection benefit from an increased mass within the

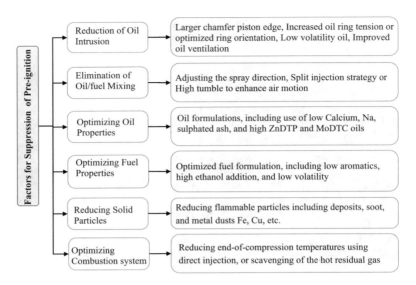

Fig. 9.45 Measures for avoiding preignition in modern gasoline engines (adapted from [1])

Table 9.2 Technology approaches for knock mitigation [95]

Water injection	Miller cycle	Cooled EGR
Increasing efficiency		
MFB$_{50\%}$ ↓ Gas property (κ ↑↓) Water heat losses ↓	MFB$_{50\%}$ ↓ Gas property (κ ↑) Water heat losses ↓	MFB$_{50\%}$ ↓ Gas property (κ ↑↑) Water heat losses ↓
Reducing NO$_X$ emission		
Evaporation + heat capacity	Miller effect	Total heat capacity
Reducing exhaust gas temperature		
MFB$_{50\%}$ ↓ Evaporation + heat capacity	MFB$_{50\%}$ ↓ Reducing effective compression	MFB$_{50\%}$ ↓ Total heat capacity

combustion chamber leading to higher total heat capacity, even if the cooling effect in case of water injection occurs mainly by water evaporation. Thus, cooled EGR and water injection have a greater impact on the isentropic coefficient than just by lower temperature.

Cooled EGR is often used in reciprocating engines. In principle, different EGR system configurations are possible which are frequently known by their exhaust sampling point and the point where the gas is fed back to the intake air. Figure 9.46 schematically shows the two widely used EGR systems, namely, low-pressure EGR and high-pressure EGR. Among others, one key criterion for an EGR system is the pressure difference between sampling and feeding point as a measure for the maximum external EGR rate which is possible to use in real engine [95]. Considering this condition, the low-pressure EGR has an advantage in low engine speed region (Fig. 9.46), while high-pressure EGR can achieve higher EGR rates at high engine speeds, e.g., rated power.

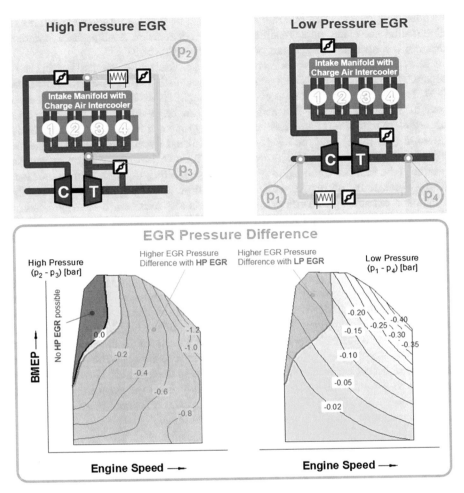

Fig. 9.46 Comparison of low- and high-pressure exhaust gas recirculation (EGR) in reciprocating engine [95]

9.7   Combustion Noise Determination

The estimation of combustion noise directly from the measured cylinder pressure data can be beneficial in the development of reciprocating engines especially in diesel engines and related fuel injection systems. Optimization of engine noise is becoming increasingly important in the development process and is often a calibration strategy target for diesel engines in addition to performance, economy, and emissions. The combustion noise measurement determined using cylinder pressure data helps to improve the understanding of the noise from combustion without expensive acoustic measuring equipment or the environment. Combustion noise plays a considerable role in the acoustic tuning of engines and vehicles.

Internal combustion engine radiated noise can be divided into mechanical noise, combustion noise, and noise resulting from the accessories. Among them, combustion noise contributes a large part of the overall engine noise.

The combustion process-related noise itself, i.e., the load-dependent noise, can be separated furthermore into direct and indirect combustion noise as well as flow noise [96]. The structure of an internal combustion engine is excited by the gas force and several additional forces. The acting forces can be divided into two categories according to their time characteristics. The first group of forces includes the cylinder pressure (gas force), and the resulting noise radiated by the structure is defined as direct combustion noise. The second group follows the rotary force curve that produces the indirect combustion noise, mainly resulting from the momentum load of the crankshaft and piston side force. The flow noise is typically produced by intake and exhaust system components, which is also ascribed to the (combustion) process noise, as it is a load-related noise, too [96]. The combustion noise is caused by load-dependent forces in the engine which cause the engine components to vibrate in their natural frequencies.

Typically, the combustion noise is known as the noise radiated from the surfaces of the engine structure as they vibrate and resonate in response to the fast pressure development in the combustion chamber during combustion. The high pressure developed in the cylinder deflects the engine structure, causing it to resonate in a number of vibration modes. The external surfaces of the engine that are in direct contact with the surrounding environment, and vibrations in the structure radiate energy as sound pressure [5]. Generally, the combustion noise can be directly related to the combustion event and the pressure rise rate in the cylinder. However, it is not possible to estimate the combustion noise directly from the pressure rise; experimental results have shown that the correlation is poor and cannot be relied upon for all combustion modes and engine operating conditions.

9.7.1 Combustion Noise Calculation and Metrics

In general, combustion noise is a result of the interaction between combustion and turbulence [97]. The contribution of both phenomena to the overall noise emissions may be completely different depending on the application [98]. In conventional diesel combustion (CDC) engines, the pressure instabilities generated during the premixed combustion phase mainly dominate the acoustic source, rendering the pressure oscillations induced by turbulence-combustion interaction [98, 99]. Therefore, a better fundamental understanding of noise is needed for evaluating the relationship between combustion and its corresponding acoustics. In addition to the pressure instability induced by combustion itself, the generated pressure waves resonate inside the cylinder, interacting with the cylinder walls, thus acting as an extra acoustic source. This complex phenomenon is generally known as combustion

chamber resonance. The combustion chamber resonance significantly affects the
radiated engine noise due to the characteristic excitation frequency span that is in the
highly sensitive human perception range [100, 101].

After the acoustic excitation, acoustic perturbations are transferred through the
engine block into the vehicle and the environment. The NVH (noise, vibration, and
harshness) analysis has shown the complexity of the propagation patterns of acoustic
energy [102]. During combustion, a sudden pressure rise is produced which leads to
the vibration of the engine block that in turn radiates aerial noise. The engine block
vibration is produced both by pressure forces exerted directly by the gas and
mechanical forces associated with piston slap, bearing clearances, elements deformation, and friction, which are powered by the pressure forces during combustion
[103]. Additionally, pressure forces strongly depend on the combustion process,
which is dependent on the combustion mode, fuel, as well as fuel injection
parameters.

Based on the physical mechanism of noise generation, two possible methods can
be used to estimate the combustion noise, i.e., (1) determining the transmission path
and the emitter characteristics and (2) a direct correlation between the source and the
radiated noise. Figure 9.47 schematically illustrates the physical mechanisms of
engine noise radiation analysis approaches. The first method is complex and needs
a high computational effort because of the time-variant and nonlinear features of the
block response [103].

The sound pressure level (SPL) of noise is assumed to be the sum of the
combustion noise level (CNL) and mechanical noise level (MNL). Combustion
noise level is assumed to be the cylinder pressure level (CPL) attenuated through

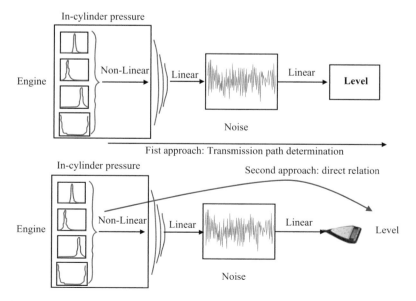

Fig. 9.47 Two approaches for determination of combustion noise level in diesel engines [103]

the engine structure in a linear relation, which is defined as structure attenuation (SA) [104]:

$$SPL = CNL + MNL = CPL/SA + MNL \qquad (9.40)$$

The SA indicates a subtraction of CNL from CPL with a unit of dB. Thus, SA is considered as an index for evaluating the noise and vibration performance of the engine structure for the combustion excitation. Separating MNL from SPL is very difficult on running an engine with combustion [104]. The attenuation properties of most engine structures are very similar, and a standard, average structural response curve was proposed (often known as SA1) [5]. Engine structure attenuation was found to increase with improved design, and because the increased attenuation was almost consistent at all frequencies, a new attenuation curve was proposed called SA1-7 (SA1being the original curve and −7 representing an extra 7 dB attenuation) [105].

Typically, combustion noise calculation from measured cylinder pressure data includes the prefiltering of pressure data, application of a filter to represent the attenuation of a standard engine structure, and application of a filter that represents the human ear response [5]. A study proposed the combustion noise calculation method by considering the structural attenuation and human hearing system response [106]. The algorithm of combustion noise calculation is schematically illustrated in Fig. 9.48. First, the measured pressure data is converted into the frequency domain using Fourier transform, and then the filter is applied to account

Fig. 9.48 Combustion noise level determination algorithm (adapted from [106])

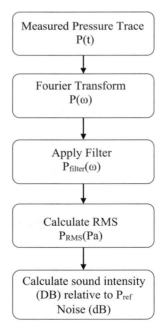

for structural attenuation and human hearing system response. The root mean square (RMS) value of the filtered pressure (P_{RMS}) is calculated, and the final noise level is obtained by comparing P_{RMS} to a reference sound level (20 µPa) [106].

The structure attenuation (SA) function implemented in Fig. 9.48 is given in Eq. (9.41):

$$SA(dB) = \begin{cases} \displaystyle\sum_{i=0}^{6} a_i f^i & 100 \leq f \leq 2300 \\ \displaystyle\sum_{i=0}^{6} b_i f^i & 2300 < f < 10000 \end{cases} \tag{9.41}$$

where f is the frequency in [Hz], SA is in units of dB, and the coefficients a_i, and b_i are constants. Outside of the frequency range given is fully attenuated. The equation for the A-weighting filter is given by Eqs. (9.42) and (9.43):

$$A(f) = 1 + 20*\log_{10}(R_A(f)) \tag{9.41}$$

$$R_A(f) = \frac{12200^2 * f^4}{(f^2 + 20.6^2) * \sqrt{(f^2 + 107.7^2) * (f^2 + 737.9^2) * (f^2 + 12200^2)}} \tag{9.42}$$

The total transmission filter, $T(f)$, is given by Eq. (9.43), and combustion noise is calculated by Eq. (9.44):

$$T(f) = 10^{(SA(f)+A(f))/20} \tag{9.43}$$

$$\text{Noise (dB)} = 20*\log_{10}\left(\frac{P_{RMS}}{20\,\mu pa}\right) \tag{9.44}$$

Standard attenuation curves from another study are presented in Fig. 9.49. The algorithm presented in Fig. 9.48 is implemented using MATLAB code, and results showed excellent agreement with AVL noise meter. A more complete detail can be found in the original study [106].

Combustion noise can also be estimated by in-cylinder pressure decomposition proposed by the study [107]. Based on this method, three frequency bands in the pressure spectrum are identified where each linked to one of the three parts of the engine cycle: compression-expansion phase, combustion event, and resonance phenomenon. This process of cylinder pressure decomposition allows determining the most influential parameters in each frequency band. Using this information, studies [98, 100, 101] have found cause-effect relationships between the noise source and both the objective and subjective effects of noise.

Figure 9.50 illustrates the pressure decomposition approach (top plot). The excess pressure can be obtained by subtracting the compression (motored) pressure from the

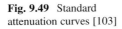

Fig. 9.49 Standard attenuation curves [103]

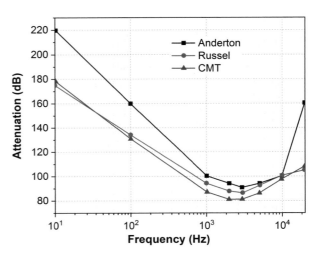

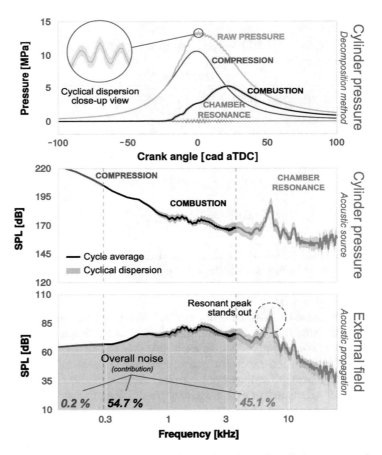

Fig. 9.50 Pressure decomposition technique illustration shown in-cylinder pressure along with each contribution in the time domain (top plot), frequency content (middle plot), and estimated radiated sound pressure levels (bottom plot) [108]

measured raw pressure data. The motored pressure can be obtained by switching off the fuel line and measuring the cylinder pressure. The subtraction of motored pressure from measured pressure during combustion provides the excess pressure due to combustion, which contains the pressure evolution due to fuel burning (combustion signal) as well as the resonant pressure fluctuations in the cylinder excited by pressure gradient during combustion [107]. Pressure fluctuation related to resonance can be determined by high-pass filtering of excess pressure and recomposing it in the time domain. The combustion pressure is obtained by subtracting the resonance signal from the excess pressure data (Fig. 9.50). The frequency content of each component is also shown in Fig. 9.50.

The relationship between the engine radiated noise or overall noise (ON) and three indicators are demonstrated [100]. Three indicators include the operation indicator that quantifies the effect of engine speed (I_n) and two combustion indicators (I_1 and I_2) indicating the in-cylinder pressure rise and the high-frequency gas oscillations inside the cylinder. The overall noise can be obtained by Eq. (9.45):

$$ON = C_0 + C_n I_n + C_1 I_1 + C_2 I_2 \qquad (9.45)$$

where C_i are coefficients which are dependent on the engine concept and size. The indicators can be calculated by Eqs. (9.46) to (9.49). The indicators I_i are considered as fundamental noise parameters and are linked to a specific bandwidth of frequency in response to the source [98, 100]:

$$I_n = \log\left[\frac{n}{n_{\text{idle}}}\right] \qquad (9.46)$$

The operation indicator (I_n) is a function of both the engine speed (n) and the idle engine speed (n_{idle}), and this parameter is associated with low frequencies. The combustion indicator (I_1) characterizes the sudden pressure rise due to combustion, and it is related to the medium bandwidth of frequencies [98]:

$$I_1 = \frac{n}{n_{\text{idle}}}\left[\frac{(dp/dt)^{\text{max}_1}_{\text{comb}} + (dp/dt)^{\text{max}_2}_{\text{comb}}}{(dp/dt)^{\text{max}}_{\text{comp}}}\right] \qquad (9.47)$$

where the two pressure derivative terms of Eq. (9.47) in the numerator are the two maximum peak values of the pressure rise rate (PRR) during combustion and the pressure derivative term in the denominator indicates the maximum peak value of the PRR of the pseudo-motored signal. The resonance indicator (I_2) indicates the contribution of the resonance phenomena inside the chamber and calculated by Eq. (9.48):

$$I_2 = \log\left[E_0\frac{E_{\text{res}}}{E_{\text{comp}}}\right] \qquad (9.48)$$

$$E_{res} = \int_{IVC}^{EVO} p(t)_{res}^2 dt \tag{9.49}$$

where E_0 is a convenient scaling factor and E_{res} is the signal energy of the resonance pressure oscillations.

Another study [101] used the same two combustion indicators to correlate with the perceived sound quality of the combustion noise. The sound quality is quantified by a mark ranging from 0 to 10 which indicates the customer satisfaction degree and represented by Eq. (9.50):

$$MARK = 10 - C_3 I_1 - C_4 I_2 \tag{9.50}$$

where C_i are other coefficients dependent on the engine size and concept.

Combustion noise variables are also investigated by statistical analysis, and empirical correlation for combustion noise determination is proposed in terms of maximum heat release rate (HRR$_{max}$), combustion duration (CD), and combustion phasing, which affects the combustion process [109]:

$$CNL[dB] = -0.544\left(CD\left[^\circ CA\right]\right) + 0.0275\left(HRR_{max}\left[J/^\circ CA\right]\right) + 85.526 \tag{9.51}$$

where combustion duration (CA) is calculated as the duration between 10% and 90% mass burn positions.

9.7.2 Combustion Noise Characteristics

Combustion noise characteristics depend on the different engine operating parameters and combustion modes. Figure 9.51 illustrates the effect of fuel injection timing, oxygen concentration, and fuel (gasoline and diesel blend) on overall noise (calculated using Eq. 9.45) and noise Mark (calculated using Eq. 9.50) in a PCCI engine. The figure illustrates that overall noise (ON) is inversely proportional to the mark which indicates sound quality. Additionally, the noise is increased as the injection timing is retarded, and there is a great impact of the oxygen concentration in the intake on the noise issues (Fig. 9.51). Combustion noise is improved with reduced oxygen concentration for all the injection timings and fuels.

Reducing the maximum rate of pressure rise is known as an empirical method to reduce the combustion noise of premixed diesel engines. Reductions in combustion noise are necessary for high-load diesel engine operation, and multiple fuel injections can achieve this with the resulting reductions in the maximum rate of pressure rise [111]. Combustion noise produced in the second combustion could assist in reducing the combustion noise of the first fuel injection and termed this "noise cancelling spike combustion (NCS combustion)" [112]. The start timing of the second fuel injection changes the noise reduction frequency range: the noise

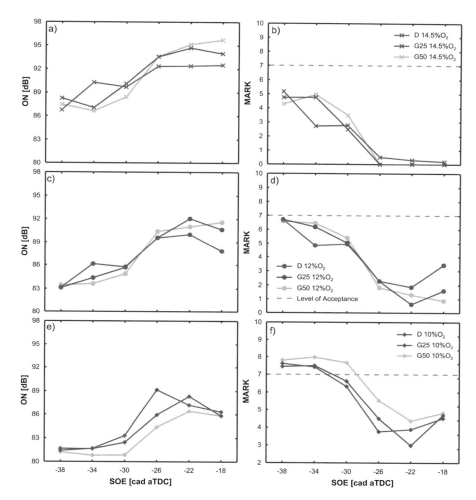

Fig. 9.51 Effect of injection timing, oxygen concentration, and fuel on engine noise: ON (left column) and Mark (right column) [110]

reduction frequency range moves less with the retarded timing of the second fuel injection. Thus, there is an optimum timing of the start of the second fuel injection achieving maximum reduction of the combustion noise generated by the first fuel injection [111]. The heating values in the first and second stage heat release do not change the noise reduction frequency range, but the intensity of the noise reduction effect changes. Another study showed that the multiple fuel injections near TDC are effective for combustion noise reduction maintaining a high degree of constant volume, because of the reduction in the maximum rates of heat release in each heat release and the noise reduction effect by the noise canceling spike (NCS) combustion [113]. The NCS processes become increasingly complex with more

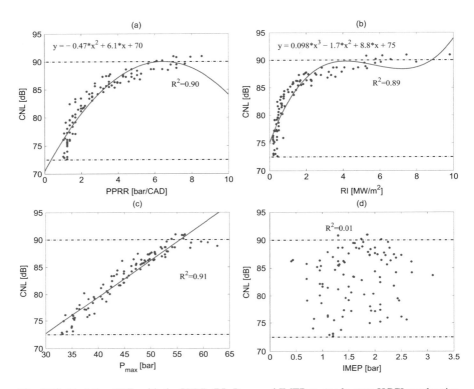

Fig. 9.52 Variation CNL with the PPRR, RI, P_{max}, and IMEP at steady-state HCCI combustion [114]

fuel injections, and the amplifying and canceling frequencies can be controlled by the length of time in crank angle between the peaks of the rates of pressure rise.

The combustion noise level (CNL) calculated using Eq. (9.44) is also used as a reliable indicator because it has been developed based on actual engine noise measurements. Figure 9.52 illustrates the variation of CNL with peak pressure rise rate (PPRR), ringing (RI), maximum cylinder pressure (P_{max}), and indicated mean effective pressure (IMEP) at steady-state conditions in HCCI engines. The solid red lines (Fig. 9.52) indicate the regression fit, and the two horizontal dashed lines represent the misfire and ringing limits. Figure 9.52a, breveals the quadratic and cubic polynomial regression curve to indicate the strong correlation ($R^2 = 0.9$) between CNL with PPRR and RI, respectively [114]. Additionally, the CNL becomes more than 90 dB for PPRR>8 bar/CAD (Fig. 9.52a), and HCCI operation at these high values should be avoided. The combustion community has long relied on the PPRR, as an indicator of engine noise. Citing discrepancies between observed sound levels and PPRR, the ringing intensity was introduced as an indicator of engine noise [64]. The higher PPRR engine operation leads to engine ringing and excessive heat transfer. Figure 9.52b shows almost constant RI with a large variation of CNL from 72 to 84 dB in steady-state data points. Figure 9.52c depicts a strong

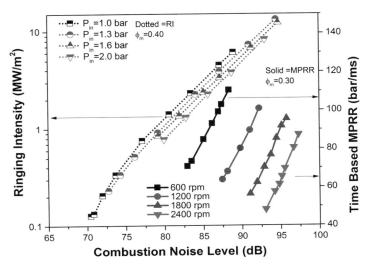

Fig. 9.53 CNL variation with RI and MPRR at different engine operating conditions in HCCI combustion (adapted from [115])

linear correlation between the CNL and P_{max} ($R^2 = 0.91$). No correlation is found between CNL and IMEP (Fig. 9.52d).

Figure 9.53 illustrates the relationship between RI and maximum pressure rise rate (MPRR) in time scale with CNL in the HCCI engine at different operating conditions. It can be noticed that the CNL increases linearly with log(RI) for CNL above 80 dB at particular intake pressure (P_{in}) condition. Additionally, the CNL increases by ~2.2 dB with an increase in P_{in} from 1.0 to 2.0 bar for any RI (~0.8 to 7 MW/m^2) as indicated by Fig. 9.53. The CNL significantly increases with an increase in engine speed (Fig. 9.53). It can also be noticed from the figure that the spacing between the data curves (CNL vs. MPRR) for various engine speeds is unequal, which suggests that slope of the increase in the CNL with engine speed decreases with higher engine speeds [115].

Ringing intensity is a metric that correlates with the sound produced by the resonating wave leading to knock. Thus, the RI appears to be a better criterion for the sake of avoiding unwanted knocking combustion regimes and the associated characteristic noise [115]. The RI is a better metric for controlling the detrimental effects of knocking such as irritating noise, loss of thermal efficiency, and engine damage. However, the CNL is valuable for estimating the overall loudness of an engine, particularly the noise induced by the normal combustion process. The RI and the CNL provide two distinctly different but complementary measurements. The CNL is not significantly affected by the high-frequency components related to the ringing/knock phenomenon [115]. Therefore, an appropriate metric needs to be selected for a particular application.

In the conventional diesel combustion (CDC), mid-frequency, broadband combustion noise dictated by MPRR and resonant noise characterized by higher frequency peaks can contribute equally to overall noise (ON) levels. The resonant peaks are dominating the spectral signature of the noise in CDC mode [108]. While in advanced partially premixed combustion (PPC), the contribution of resonance levels to ON levels is much lower. Thus, in optimizing the noise radiated from CDC engines, the reduction of wave resonance is a crucial factor, but this factor is of little importance in PPC noise [108].

Discussion/Investigation Questions

1. Discuss the difference between normal and abnormal combustion in spark ignition engine. Describe the various factors responsible for abnormal combustion in SI engine.
2. Describe the term "knock" in a reciprocating engine. Discuss the possible reasons for engine knocking in SI engine, and explain the adverse effects of engine knocking over a long period.
3. Write the engine or operating variables affecting the temperature and density of the unburned charge toward the end of combustion in SI engine.
4. Discuss the factors responsible for cycle-to-cycle variations in the engine knocking. Describe the typical distribution of knock intensity on a cyclic basis.
5. A PFI gasoline SI engine is designed to operate at compression ratio 9 using gasoline with octane number (ON) 90 (typical gasoline). Discuss the expected problems that arise due to the increase in compression ratio to 12 (with modified engine) while running at the same fuel. Write the possible solution to the expected problems. Can ethanol or methanol be used on the modified engine at a compression ratio of 12?
6. Discuss the methods that can be used for the detection of knock onset in SI engine. Explain the merit and limitations of the methods.
7. Discuss the different modes of knock combustion in the engines. Write the different parameters affecting the transition of knock mode from deflagration to developing detonation.
8. Why measurement and characterization of knocking are important in internal combustion engines? Discuss the reason for knocking in SI, CI, and LTC (HCCI, RCCI, etc.) modes of engine operation. Draw a typical knocking cylinder pressure curve for all three modes of engine combustion.
9. What is engine super-knock? Discuss reasons for super-knock and method of characterization and mitigation of super-knock.
10. Three possible positions of the spark plug (black circle) in a spark ignition (SI) engine are shown Fig. P9.1. Identify the intake and exhaust valves in the configuration shown. Arrange three configurations (A, B, and C) in ascending order for octane number (ON) of fuel required to run the SI engine in each configuration, and justify your answer with suitable reasons.

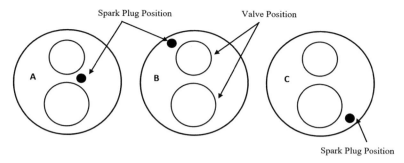

Fig. P9.1 Different spark plug installation configuration

11. Based on the two different designs of combustion chambers shown in Fig. P9.2, answer the following questions: **(a)** when combustion chamber design is changed from A to B, explain whether combustion rate will be faster, slower, or about the same rate. Justify your answer. **(b)** Which combustion chamber requires higher octane number fuel? If both engines are running on the same fuel, which engine can be operated at a higher compression ratio? Justify your answers.

Fig. P9.2 Effect of piston bowl on combustion characteristics in SI engine

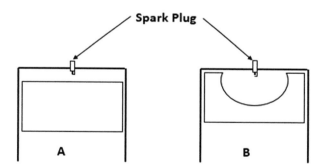

12. You are given the engine map (Fig. P9.3) for a conventional homogeneous charge SI engine. **(a)** Looking at the engine map, identify whether the engine is naturally aspirated or it is turbocharged/supercharged. Mark the regions in the map where engine is most (highest) and least susceptible to knocking combustion. Justify your answers for the highest and lowest susceptibility toward knocking. **(b)** Assume engine is operating in a region susceptible to knocking, suggest two ways (actions to be performed) to run the engine in the non-knocking combustion.

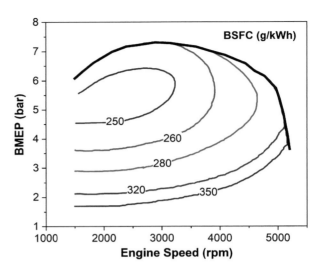

Fig. P9.3 Conventional SI engine map (adapted from [2])

13. Consider an SI engine in which the spark timing is maintained at the same crank angle and the air-fuel mixture is changed from stoichiometric to lean. What happens to the peak flame temperature as the air-fuel ratio is made lean from stoichiometric? Explain your answer. What would you expect to happen to the exhaust temperature for this situation? What about the tendency to knocking in such conditions?
14. Write the effect of engine operating variables on ignition, flame propagation, and knocking tendency by filling in the blanks in the Table P9.1 using the symbols provided. Discuss and justify your answers in terms of the governing phenomena or the factors responsible to it.

Table P9.1   Effect of ignition- and engine-related variables on flame propagation and knocking

	Effect		
Ignition system variables	Ignition	Flame propagation	Knocking tendency
Increased spark energy			
Hotter spark			
Increased electrode gap			
Sharp/thin electrodes of spark plug			
Engine variables			
Increased compression ratio			
Increased mixture swirl/turbulence			
Increased residual gas/EGR			
Close to stoichiometric mixture operation			

Helps ↑; hinders ↓; no effect –

15. Spark ignition engines' efficiency is mainly limited by three major factors, namely, knocking, fuel enrichment, and throttling. Figure P9.4 depicts the full-load curve of SI engine. Schematically identify zones corresponding to knocking, fuel enrichment, and throttling on engine operating map. You can draw few contour lines in the identified zones.

Fig. P9.4 Full-load torque curve of a SI engine

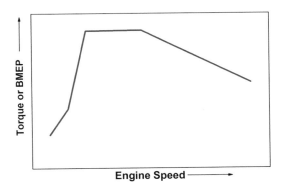

16. Knocking in SI engines is affected by several engine and operating parameters. Fill the following Table P9.2 showing the effect of increasing engine and operating variables on knock. Discuss these whether the engine parameters can be controlled by the engine operator or ECU to mitigate the knocking conditions.

Table P9.2 Effect of engine parameters on knock in conventional SI engine

Increase in engine parameter	Change in the state of end gases	Effect on knock ($\downarrow\uparrow\leftrightarrow$)
1. Compression ratio		
2. Spark timing advance		
3. Cylinder diameter		
4. Inlet air temperature		
5. Coolant temperature		
6. Engine speed		
7. Engine load		
8. Fuel octane number		
9. Humidity		
10. Altitude		

17. Discuss the differences in knocking in CI and SI engines. Write the typical characteristics of the parameters given in table to reduce the knocking tendency in SI and CI engines. Fill the Table P9.3 with qualitative values such as low/high and long/short.

Table P9.3 Characteristics tending to reduce engine knocking conditions

Characteristics	SI engines	CI engines
Ignition delay		
Intake temperature and pressure		
Compression ratio		
Autoignition temperature of fuel		
Cylinder wall temperature		
Engine speed		

18. Describe the difference between a conventional knock and super-knock. Discuss the sources of super-knock and mode of combustion of super-knock in SI engine.

19. Discuss the sources of preignition in SI engine. Write the possible methods to avoid the preignition in reciprocating SI engine. Explain whether preignition always leads to knocking, and discuss the severity of preignition.

20. Write the effect of knocking (over a long period of time) on engine performance and its state. Discuss the different methods of knock detection in combustion engines. How you can differentiate the pressure oscillations during a weak knock and pressure oscillation due to the non-flush mounting of the sensor (pipe oscillations).

21. Discuss the phenomena of acoustic wave generation during knocking conditions, and also write the equation to determine the frequency of acoustic mode oscillation. Explain why the axial mode of oscillations is typically not observed in reciprocating engines.

22. Calculate the characteristic frequency of oscillation modes: First circumferential (1,0), second circumferential (2,0), third circumferential (3,0), and first radial (0,1) with corresponding Bessel constants 1.841, 3.054, 4.201, and 3.832, respectively. Compute the frequency corresponding to temperatures 2000 K, 2500 K, and 3000 K. Discuss the effect of the equivalence ratio of the charge on the frequency of oscillations.

23. Discuss the different knock indices based on in-cylinder pressure and heat release used for characterization of knock in internal combustion engines.

24. Discuss how wavelets can be used for characterization of engine knocking under different engine operating conditions.

25. Assume a gasoline SI engine having bore 60 mm and displacement volume of 400 cc working on the stoichiometric mixture with spark timing 15° before TDC. The spark plug is located at the center of the engine head. Assuming constant turbulent flame speed of 8.56 m/s, determine whether or not engine knocking will occur in the combustion chamber. Unburned gas temperature and pressure can be assumed to be constant throughout combustion with value 1650 K and 5 bar, respectively. An empirical relation for autoignition delay of a stoichiometric gasoline-air mixture is given as

$$\tau_{\text{ignition delay}}[\text{ms}] = 0.08 \cdot \frac{1}{P^{1.5}[\text{MPa}]} \exp\left(\frac{3800}{T[K]}\right)$$

Clearly state your assumptions if any.

26. Discuss how the setting of knock threshold limits the efficiency of the engine. Describe the methods used for setting the knock threshold and how it can be optimized.
27. Fill the Table P9.4 by discussing the effect of particle and operating condition on preignition in SI engine. Write the effect on increasing the value of variable on preignition (increase or decrease) along with the mechanism responsible for preignition.

Table P9.4 Effect on preignition in SI engine

Variable	Effect on preignition	Responsible mechanism
Particle size		
Particle temperature		
Boost pressure		
Air-fuel ratio		
EGR/residuals		

28. Discuss the different sensing methodologies of engine knock in modern SI engine. Describe the merits and demerits of the method. Write the typical factors you will consider for selecting the knock detection methods.
29. Discuss the method for combustion noise level assessment in engines by means of in-cylinder pressure components? Explain various indices used for combustion noise determination base on cylinder pressure data.

References

1. Wang, Z., Liu, H., & Reitz, R. D. (2017). Knocking combustion in spark-ignition engines. *Progress in Energy and Combustion Science, 61*, 78–112.
2. Heywood, J. B. (1988). *Internal combustion engine fundamentals.* New York: McGraw-Hill Education.
3. Zhen, X., Wang, Y., Xu, S., Zhu, Y., Tao, C., Xu, T., & Song, M. (2012). The engine knock analysis—An overview. *Applied Energy, 92*, 628–636.
4. Wang, Z., Liu, H., Song, T., Qi, Y., He, X., Shuai, S., & Wang, J. (2015). Relationship between super-knock and pre-ignition. *International Journal of Engine Research, 16*(2), 166–180.
5. Rogers, D. R. (2010). *Engine combustion: Pressure measurement and analysis.* Warrendale: Society of Automotive Engineers.
6. Mahendar, S. K., Erlandsson, A., & Adlercreutz, L. (2018). *Challenges for spark ignition engines in heavy duty application: A review* (No. 2018-01-0907). SAE Technical Paper.
7. Mittal, V., Revier, B. M., & Heywood, J. B. (2007). *Phenomena that determine knock onset in spark-ignition engines* (No. 2007-01-0007). SAE Technical Paper.
8. Kawahara, N., Tomita, E., & Sakata, Y. (2007). Auto-ignited kernels during knocking combustion in a spark-ignition engine. *Proceedings of the Combustion Institute, 31*(2), 2999–3006.
9. Merola, S. S., & Vaglieco, B. M. (2007). Knock investigation by flame and radical species detection in spark ignition engine for different fuels. *Energy Conversion and Management, 48* (11), 2897–2910.

10. McKenzie, J., & Cheng, W. K. (2016). *The anatomy of knock* (No. 2016-01-0704). SAE Technical Paper.
11. Westbrook, C. K. (2000). Chemical kinetics of hydrocarbon ignition in practical combustion systems. *Proceedings of the Combustion Institute, 28*(2), 1563–1577.
12. Kalghatgi, G. (2018). Knock onset, knock intensity, superknock and preignition in spark ignition engines. *International Journal of Engine Research, 19*(1), 7–20.
13. Chen, L., Li, T., Yin, T., & Zheng, B. (2014). A predictive model for knock onset in spark-ignition engines with cooled EGR. *Energy Conversion and Management, 87*, 946–955.
14. Douaud, A. M., & Eyzat, P. (1978). *Four-octane-number method for predicting the anti-knock behavior of fuels and engines* (No. 780080). SAE Technical Paper.
15. Syed, I. Z., Mukherjee, A., & Naber, J. (2011). *Numerical simulation of autoignition of gasoline-ethanol/air mixtures under different conditions of pressure, temperature, dilution, and equivalence ratio* (No. 2011-01-0341). SAE Technical Paper.
16. Livengood, J. C., & Wu, P. C. (1955). Correlation of autoignition phenomena in internal combustion engines and rapid compression machines. In *Symposium (International) on Combustion* (Vol. 5, No. 1, pp. 347–356). Elsevier.
17. Kim, K. S., & Ghandhi, J. (2012). *Preliminary results from a simplified approach to modeling the distribution of engine knock* (No. 2012-32-0004). SAE Technical Paper.
18. Mittal, V., Heywood, J. B., & Green, W. H. (2010). The underlying physics and chemistry behind fuel sensitivity. *SAE International Journal of Fuels and Lubricants, 3*(1), 256–265.
19. Kalghatgi, G., Head, R., Chang, J., Viollet, Y., Babiker, H., & Amer, A. (2014). An alternative method based on toluene/n-heptane surrogate fuels for rating the anti-knock quality of practical gasolines. *SAE International Journal of Fuels and Lubricants, 7*(3), 663–672.
20. Boot, M. D., Tian, M., Hensen, E. J., & Sarathy, S. M. (2017). Impact of fuel molecular structure on auto-ignition behavior–Design rules for future high performance gasolines. *Progress in Energy and Combustion Science, 60*, 1–25.
21. Kalghatgi, G. T. (2001). *Fuel anti-knock quality-Part I. Engine studies* (No. 2001-01-3584). SAE Technical Paper.
22. Amer, A., Babiker, H., Chang, J., Kalghatgi, G., Adomeit, P., Brassat, A., & Günther, M. (2012). Fuel effects on knock in a highly boosted direct injection spark ignition engine. *SAE International Journal of Fuels and Lubricants, 5*(3), 1048–1065.
23. Kalghatgi, G., Risberg, P., & Ångstrom, H. E. (2003). *A method of defining ignition quality of fuels in HCCI engines* (No. 2003-01-1816). SAE Technical Paper.
24. Kalghatgi, G. T., & Head, R. A. (2004). *The available and required autoignition quality of gasoline-like fuels in HCCI engines at high temperatures* (No. 2004-01-1969). SAE Technical Paper.
25. Kalghatgi, G. T. (2005). *Auto-ignition quality of practical fuels and implications for fuel requirements of future SI and HCCI engines* (No. 2005-01-0239). SAE Technical Paper.
26. Kalghatgi, G., Babiker, H., & Badra, J. (2015). A simple method to predict knock using toluene, n-heptane and iso-octane blends (TPRF) as gasoline surrogates. *SAE International Journal of Engines, 8*(2), 505–519.
27. Shahlari, A. J., & Ghandhi, J. (2017). *Pressure-based knock measurement issues* (No. 2017-01-0668). SAE Technical Paper.
28. Xiaofeng, G., Stone, R., Hudson, C., & Bradbury, I. (1993). *The detection and quantification of knock in spark ignition engines* (No. 932759). SAE Technical Paper.
29. Burgdorf, K., & Denbratt, I. (1997). *Comparison of cylinder pressure based knock detection methods* (No. 972932). SAE Technical Paper.
30. Shahlari, A. J., & Ghandhi, J. B. (2012). *A comparison of engine knock metrics* (No. 2012-32-0007). SAE Technical Paper.
31. Bradley, D., Morley, C., Gu, X. J., & Emerson, D. R. (2002). *Amplified pressure waves during autoignition: Relevance to CAI engines* (No. 2002-01-2868). SAE Technical Paper.
32. Gu, X. J., Emerson, D. R., & Bradley, D. (2003). Modes of reaction front propagation from hot spots. *Combustion and Flame, 133*(1-2), 63–74.

33. König, G., Maly, R. R., Bradley, D., Lau, A. K. C., & Sheppard, C. G. W. (1990). Role of exothermic centres on knock initiation and knock damage. *SAE Transactions, 99*, 840–861.
34. Wang, Z., Qi, Y., He, X., Wang, J., Shuai, S., & Law, C. K. (2015). Analysis of pre-ignition to super-knock: Hotspot-induced deflagration to detonation. *Fuel, 144*, 222–227.
35. Zeldovich, Y. B. (1980). Regime classification of an exothermic reaction with nonuniform initial conditions. *Combustion and Flame, 39*(2), 211–214.
36. Qi, Y., Wang, Z., Wang, J., & He, X. (2015). Effects of thermodynamic conditions on the end gas combustion mode associated with engine knock. *Combustion and Flame, 162*(11), 4119–4128.
37. Bradley, D., & Kalghatgi, G. T. (2009). Influence of autoignition delay time characteristics of different fuels on pressure waves and knock in reciprocating engines. *Combustion and Flame, 156*(12), 2307–2318.
38. Kalghatgi, G. T., & Bradley, D. (2012). Pre-ignition and 'super-knock' in turbo-charged spark-ignition engines. *International Journal of Engine Research, 13*(4), 399–414.
39. Hurle, I. R., Price, R. B., Sugden, T. M., & Thomas, A. (1968). Sound emission from open turbulent premixed flames. *Proceedings of the Royal Society of London A, 303*(1475), 409–427.
40. Kim, K. S. (2015). *Study of engine knock using a Monte Carlo method* (Doctoral dissertation). The University of Wisconsin-Madison.
41. Bates, L., Bradley, D., Paczko, G., & Peters, N. (2016). Engine hot spots: Modes of auto-ignition and reaction propagation. *Combustion and Flame, 166*, 80–85.
42. Pasternak, M., Netzer, C., Mauss, F., Fischer, M., Sens, M., & Riess, M. (2017, December). Simulation of the effects of spark timing and external EGR on gasoline combustion under knock-limited operation at high speed and load. In *International Conference on Knocking in Gasoline Engines* (pp. 121–142). Cham: Springer.
43. Zhou, L., Kang, R., Wei, H., Feng, D., Hua, J., Pan, J., & Chen, R. (2018). Experimental analysis of super-knock occurrence based on a spark ignition engine with high compression ratio. *Energy, 165B*, 68–75.
44. Dahnz, C., Han, K. M., Spicher, U., Magar, M., Schießl, R., & Maas, U. (2010). Investigations on pre-ignition in highly supercharged SI engines. *SAE International Journal of Engines, 3*(1), 214–224.
45. Döhler, A., & Schaffner, P. (2017, December). Optical diagnostic tools for detection and evaluation of glow ignitions. In *International Conference on Knocking in Gasoline Engines* (pp. 55–70). Cham: Springer.
46. Kalghatgi, G., Algunaibet, I., & Morganti, K. (2017). On knock intensity and superknock in SI engines. *SAE International Journal of Engines, 10*(2017-01-0689), 1051–1063.
47. Kassai, M., Shiraishi, T., & Noda, T. (2017, December). Fundamental mechanism analysis on the underlying processes of LSPI using experimental and modeling approaches. In *International Conference on Knocking in Gasoline Engines* (pp. 89–111). Cham: Springer.
48. Morikawa, K., Moriyoshi, Y., Kuboyama, T., Imai, Y., Yamada, T., & Hatamura, K. (2015). *Investigation and improvement of LSPI phenomena and study of combustion strategy in highly boosted SI combustion in low speed range* (No. 2015-01-0756). SAE Technical Paper.
49. Jatana, G. S., Splitter, D. A., Kaul, B., & Szybist, J. P. (2018). Fuel property effects on low-speed pre-ignition. *Fuel, 230*, 474–482.
50. Spicher, U. (2017, December). Detection and analysis methods for irregular combustion in SI engines. In *International Conference on Knocking in Gasoline Engines* (pp. 225–242). Cham: Springer.
51. Vressner, A., Lundin, A., Christensen, M., Tunestål, P., & Johansson, B. (2003). Pressure oscillations during rapid HCCI combustion. *SAE Transactions*, 2469–2478.
52. Draper, C. S. (1935). *The physical effects of detonation in a closed cylindrical chamber.* Technical report, National Advisory Committee for Aeronautics.
53. Guardiola, C., Pla, B., Bares, P., & Barbier, A. (2018). An analysis of the in-cylinder pressure resonance excitation in internal combustion engines. *Applied Energy, 228*, 1272–1279.

54. Dahl, D., Andersson, M., & Denbratt, I. (2011). The origin of pressure waves in high load HCCI combustion: A high-speed video analysis. *Combustion Science and Technology, 183* (11), 1266–1281.
55. Wissink, M., Wang, Z., Splitter, D., Shahlari, A., & Reitz, R. D. (2013). *Investigation of pressure oscillation modes and audible noise in RCCI, HCCI, and CDC* (No. 2013-01-1652). SAE Technical Paper.
56. Zhang, Q., Hao, Z., Zheng, X., & Yang, W. (2017). Characteristics and effect factors of pressure oscillation in multi-injection DI diesel engine at high-load conditions. *Applied Energy, 195*, 52–66.
57. Millo, F., & Ferraro, C. V. (1998). *Knock in SI engines: a comparison between different techniques for detection and control* (No. 982477). SAE Technical Paper.
58. Brunt, M. F., Pond, C. R., & Biundo, J. (1998). *Gasoline engine knock analysis using cylinder pressure data* (No. 980896). SAE Technical paper.
59. Puzinauskas, P. V. (1992). *Examination of methods used to characterize engine knock* (No. 920808). SAE Technical Paper.
60. Lounici, M. S., Benbellil, M. A., Loubar, K., Niculescu, D. C., & Tazerout, M. (2017). Knock characterization and development of a new knock indicator for dual-fuel engines. *Energy, 141*, 2351–2361.
61. Brecq, G., Bellettre, J., & Tazerout, M. (2003). A new indicator for knock detection in gas SI engines. *International Journal of Thermal Sciences, 42*(5), 523–532.
62. Maurya, R. K. (2018). *Characteristics and control of low temperature combustion engines: Employing gasoline, ethanol and methanol*. Cham: Springer.
63. Checkel, M. D., & Dale, J. D. (1989). *Pressure trace knock measurement in a current si production engine* (No. 890243). SAE Technical Paper.
64. Eng, J. A. (2002). *Characterization of pressure waves in HCCI combustion* (No. 2002-01-2859). SAE Technical Paper.
65. Bares, P., Selmanaj, D., Guardiola, C., & Onder, C. (2018). A new knock event definition for knock detection and control optimization. *Applied Thermal Engineering, 131*, 80–88.
66. Hettinger, A., & Kulzer, A. (2009). A new method to detect knocking zones. *SAE International Journal of Engines, 2*(1), 645–665.
67. Galloni, E. (2012). Dynamic knock detection and quantification in a spark ignition engine by means of a pressure based method. *Energy Conversion and Management, 64*, 256–262.
68. Nilsson, Y., Frisk, E., & Nielsen, L. (2009). Weak knock characterization and detection for knock control. *Proceedings of the Institution of Mechanical Engineers, Part D: Journal of Automobile Engineering, 223*(1), 107–129.
69. Peyton Jones, J. C., Spelina, J. M., & Frey, J. (2014). Optimizing knock thresholds for improved knock control. *International Journal of Engine Research, 15*(1), 123–132.
70. Galloni, E. (2016). Knock-limited spark angle setting by means of statistical or dynamic pressure based methods. *Energy Conversion and Management, 116*, 11–17.
71. Shayestehmanesh, S., Jones, J. C. P., & Frey, J. (2018). *Stochastic characteristics of knock and IMEP* (No. 2018-01-1155). SAE Technical Paper.
72. Cavina, N., Brusa, A., Rojo, N., & Corti, E. (2018). *Statistical analysis of knock intensity probability distribution and development of 0-D predictive knock model for a SI TC engine* (No. 2018-01-0858). SAE Technical Paper.
73. Naber, J., Blough, J. R., Frankowski, D., Goble, M., & Szpytman, J. E. (2006). *Analysis of combustion knock metrics in spark-ignition engines* (No. 2006-01-0400). SAE Technical Paper.
74. Wei, H., Hua, J., Pan, M., Feng, D., Zhou, L., & Pan, J. (2018). Experimental investigation on knocking combustion characteristics of gasoline compression ignition engine. *Energy, 143*, 624–633.
75. Zhou, L., Hua, J., Wei, H., Dong, K., Feng, D., & Shu, G. (2018). Knock characteristics and combustion regime diagrams of multiple combustion modes based on experimental investigations. *Applied Energy, 229*, 31–41.

76. Ghandhi, J., & Kim, K. S. (2017). *A statistical description of knock intensity and its prediction* (No. 2017-01-0659). SAE Technical Paper.
77. Siano, D., & Bozza, F. (2013). *Knock detection in a turbocharged SI engine based on ARMA technique and chemical kinetics* (No. 2013-01-2510). SAE Technical Paper.
78. Tong, S., Yang, Z., He, X., Deng, J., Wu, Z., & Li, L. (2017). *Knock and pre-ignition detection using ion current signal on a boosted gasoline engine* (No. 2017-01-0792). SAE Technical Paper.
79. Abu-Qudais, M. (1996). Exhaust gas temperature for knock detection and control in spark ignition engine. *Energy Conversion and Management, 37*(9), 1383–1392.
80. Spelina, J. M. (2016). *Knock characterization, simulation, and control* (PhD Thesis). Villanova University.
81. Kowada, M., Azumagakito, I., Nagai, T., Iwai, N., & Hiraoka, R. (2015). *Study of knocking damage indexing based on optical measurement* (No. 2015-01-0762). SAE Technical Paper.
82. Spicher, U., Kroger, H., & Ganser, J. (1991). *Detection of knocking of combustion using simultaneously high speed Schlieren cinematography and multi-optical fibre technique*. SAE Technical Paper 912312.
83. Nagai, T., Hiraoka, R., Iwai, N., Kowada, M., & Azumagakito, I. (2015). *Development of highly durable optical probe for combustion measurement* (No. 2015-01-0759). SAE Technical Paper.
84. Philipp, H., Hirsch, A., Baumgartner, M., Fernitz, G., Beidl, C., Piock, W., & Winklhofer, E. (2001). *Localization of knock events in direct injection gasoline engines* (No. 2001-01-1199). SAE Technical Paper.
85. Matsura, K., Sato, Y., Yoshida, K., & Sono, H. (2017). Proposal of knock mitigation method through enhancement of local heat transfer. In *International Conference on Knocking in Gasoline Engines* (pp. 3–16). Cham: Springer.
86. Giglio, V., Police, G., Rispoli, N., di Gaeta, A., Cecere, M., & Della Ragione, L. (2009). *Experimental investigation on the use of ion current on SI engines for knock detection* (No. 2009-01-2745). SAE Technical Paper.
87. Laganá, A. A., Lima, L. L., Justo, J. F., Arruda, B. A., & Santos, M. M. (2018). Identification of combustion and detonation in spark ignition engines using ion current signal. *Fuel, 227*, 469–477.
88. Daniels, C. F., Zhu, G. G., & Winkelman, J. (2003). *Inaudible knock and partial-burn detection using in-cylinder ionization signal* (No. 2003-01-3149). SAE Technical Paper.
89. Patro, T. N. (1997). *Combustion induced powertrain NVH-a time-frequency analysis* (No. 971874). SAE Technical Paper.
90. Kiencke, U., & Nielsen, L. (2005). *Automotive control systems: For engine, driveline, and vehicle*. Berlin: Springer Science & Business Media.
91. Liu, C., Gao, Q., Jin, Y. A., & Yang, W. (2010). Application of wavelet packet transform in the knock detection of gasoline engines. In *2010 International Conference on Image Analysis and Signal Processing (IASP)* (pp. 686–690). IEEE.
92. Bi, F., Ma, T., & Wang, X. (2019). Development of a novel knock characteristic detection method for gasoline engines based on wavelet-denoising and EMD decomposition. *Mechanical Systems and Signal Processing, 117*, 517–536.
93. Souder, J. S., Mack, J. H., Hedrick, J. K., & Dibble, R. W. (2004, January). Microphones and knock sensors for feedback control of HCCI engines. In *ASME 2004 Internal Combustion Engine Division Fall Technical Conference* (pp. 77–84). New York: American Society of Mechanical Engineers.
94. Hunger, M., Böcking, T., Walther, U., Günther, M., Freisinger, N., & Karl, G. (2017). Potential of direct water injection to reduce knocking and increase the efficiency of gasoline engines. In *International Conference on Knocking in Gasoline Engines* (pp. 338–359). Cham: Springer.
95. Fischer, M., Günther, M., Berger, C., Troeger, R., Pasternak, M., & Mauss, F. (2017). Suppressing knocking by using CleanEGR–better fuel economy and lower raw emissions

simultaneously. In *International Conference on Knocking in Gasoline Engines* (pp. 363–384). Cham: Springer.

96. Alt, N. W., Nehl, J., Heuer, S., & Schlitzer, M. W. (2003). *Prediction of combustion process induced vehicle interior noise* (No. 2003-01-1435). SAE Technical Paper.

97. Schwarz, A., & Janicka, J. (Eds.). (2009). *Combustion noise*. Berlin: Springer Science & Business Media.

98. Broatch, A., Novella, R., Gomez-Soriano, J., Pal, P., & Som, S. (2018). *Numerical methodology for optimization of compression-ignited engines considering combustion noise control*. SAE Technical Paper 2018-01-0193.

99. Flemming, F., Sadiki, A., & Janicka, J. (2007). Investigation of combustion noise using a LES/CAA hybrid approach. *Proceedings of the Combustion Institute, 31*(2), 3189–3196.

100. Torregrosa, A. J., Broatch, A., Martín, J., & Monelletta, L. (2007). Combustion noise level assessment in direct injection diesel engines by means of in-cylinder pressure components. *Measurement Science and Technology, 18*(7), 2131.

101. Payri, F., Broatch, A., Margot, X., & Monelletta, L. (2009). Sound quality assessment of diesel combustion noise using in-cylinder pressure components. *Measurement Science and Technology, 20*(1), 015107.

102. Stanković, L., & Böhme, J. F. (1999). Time–frequency analysis of multiple resonances in combustion engine signals. *Signal Processing, 79*(1), 15–28.

103. Payri, F., Torregrosa, A. J., Broatch, A., & Monelletta, L. (2009). Assessment of diesel combustion noise overall level in transient operation. *International Journal of Automotive Technology, 10*(6), 761.

104. Ozawa, H., & Nakada, T. (1999). Pseudo cylinder pressure excitation for analyzing the noise characteristics of the engine structure. *JSAE Review, 20*(1), 67–72.

105. Russell, M. F., Palmer, D. C., & Young, C. D. (1984). Measuring diesel noise at source with a view to its control. *IMechE, C142*(84), 97–105.

106. Shahlari, A. J., Hocking, C., Kurtz, E., & Ghandhi, J. (2013). Comparison of compression ignition engine noise metrics in low-temperature combustion regimes. *SAE International Journal of Engines, 6*(1), 541–552.

107. Payri, F., Broatch, A., Tormos, B., & Marant, V. (2005). New methodology for in-cylinder pressure analysis in direct injection diesel engines—Application to combustion noise. *Measurement Science and Technology, 16*(2), 540.

108. Broatch, A., Novella, R., García-Tíscar, J., & Gomez-Soriano, J. (2019). On the shift of acoustic characteristics of compression-ignited engines when operating with gasoline partially premixed combustion. *Applied Thermal Engineering, 146*, 223–231.

109. Shibata, G., Shibaike, Y., Ushijima, H., & Ogawa, H. (2013). *Identification of factors influencing premixed diesel engine noise and mechanism of noise reduction by EGR and supercharging* (No. 2013-01-0313). SAE Technical Paper.

110. Torregrosa, A. J., Broatch, A., Novella, R., Gomez-Soriano, J., & Mónico, L. F. (2017). Impact of gasoline and diesel blends on combustion noise and pollutant emissions in premixed charge compression ignition engines. *Energy, 137*, 58–68.

111. Shibata, G., Nakayama, D., Okamoto, Y., & Ogawa, H. (2016). Diesel engine combustion noise reduction by the control of timings and heating values in two stage high temperature heat releases. *SAE International Journal of Engines, 9*(2), 868–882.

112. Fuyuto, T., Taki, M., Ueda, R., Hattori, Y., Kuzuyama, H., & Umehara, T. (2014). Noise and emissions reduction by second injection in diesel PCCI combustion with split injection. *SAE International Journal of Engines, 7*(4), 1900–1910.

113. Shibata, G., Ogawa, H., Okamoto, Y., Amanuma, Y., & Kobashi, Y. (2017). Combustion noise reduction with high thermal efficiency by the control of multiple fuel injections in premixed diesel engines. *SAE International Journal of Engines, 10*(3), 1128–1142.

114. Bahri, B., Shahbakhti, M., & Aziz, A. A. (2017). Real-time modeling of ringing in HCCI engines using artificial neural networks. *Energy, 125*, 509–518.

115. Dernotte, J., Dec, J. E., & Ji, C. (2014). Investigation of the sources of combustion noise in HCCI engines. *SAE International Journal of Engines, 7*(2014-01-1272), 730–761.

Chapter 10
Estimation of Engine Parameters from Measured Cylinder Pressure

Abbreviations and Symbols

AFR	Air-fuel ratio
BDC	Bottom dead center
BMEP	Brake mean effective pressure
CAD	Crank angle degree
CR	Compression ratio
CRDI	Common rail direct injection
ECR	Effective compression ratio
ECU	Electronic control unit
EGR	Exhaust gas recirculation
EVC	Exhaust valve closing
EVO	Exhaust valve opening
FC	Fresh charge
GCR	Geometric compression ratio
HCCI	Homogeneous charge compression ignition
IMEP	Indicated mean effective pressure
$IMEP_{error}$	Error in IMEP calculation
IVC	Intake valve closing
IVO	Intake valve opening
LTC	Low-temperature combustion
MFB	Mass fraction burned
NASA	National Aeronautics and Space Administration
NLLS	Nonlinear least squares
NVO	Negative valve overlap
OF	Overlap factor
PCCI	Premixed charge compression ignition
RG	Residual gas
RGF	Residual gas fraction

© Springer Nature Switzerland AG 2019
R. K. Maurya, *Reciprocating Engine Combustion Diagnostics*, Mechanical Engineering Series, https://doi.org/10.1007/978-3-030-11954-6_10

SACI	Spark-assisted compression ignition
SI	Spark ignition
SOI	Start of injection
TDC	Top dead center
TDC_{error}	TDC position error
VCR	Variable compression ratio
VVA	Variable valve actuation
VVT	Variable valve timing
a	Speed of sound
A_u	Area of the pressure cycle
$B_{i,j}$	Bessel constant
c_m	Mean piston speed
C_P	Constant pressure specific heat of the gas
C_V	Constant volume specific heat of the gas
D	Cylinder diameter
$F_{i,j}$	Resonance frequency
L_e	Exhaust valve lift
L_i	Intake valve lift
m	In-cylinder gas mass
M_f	Fuel mass
N	Engine speed
n	Polytrophic exponent
P	Cylinder pressure
$p(\theta_{inv})$	Gas pressure at the inversion crank angle
P_e	Exhaust pressure
P_i	Intake pressure
P_{im}	Intake manifold pressure
P_{max}	Maximum pressure
P_u	Unfired pressure
Q	Energy
S	Entropy
s	In-cylinder gas specific entropy
T	Cylinder temperature
T_{cyl}	In-cylinder gas temperature
T_{exh}	Exhaust temperature
T_{in}	Intake temperature
T_{max}	Maximum temperature
T_{RG}	Temperature of residual gas
T_w	Wall temperature
V	Volume of cylinder
$V(\theta_{inv})$	Volume at the inversion crank angle
V_c	Clearance volume
V_d	Displaced volume
θ_C	Centroid of the pressure distribution

°CA	Crank angle degree
θ	Crank angle position
θ_{EVO}	Crank position at EVO
θ_{IVC}	Crank position at IVC
θ_{TDC}	Crank position at TDC
ϕ	Equivalence ratio
θ_{ign}	Ignition angle
δF	Loss function
θ_0	Phasing error
Δp	Pressure difference
χ_{res}	Residual gas fraction
γ	Ratio of specific heat
θ_{loss}	Thermodynamic loss angle

10.1 TDC Determination

In-cylinder pressure measurement is one of the most important tools for investigating the internal processes in reciprocating engines. Pressure-volume phase lag is one of the major sources of error in thermodynamic calculations using experimental pressure data. Finding the correct absolute angular position of crankshaft is essential for calculation of combustion chamber volume as function of crank angle, which is required for calculation of indicated work and different combustion parameters. Incorrect cylinder pressure-crank angle phasing due to an erroneous determination of the top dead center (TDC) can lead to a very large error in the calculation of indicated work and heat release depending on the level of error in TDC measurement. A study proposed a relation giving the ratio between IMEP error (in %) and TDC position error (in °CA), which is given by Eq. (10.1) [1]:

$$\frac{\text{IMEP}_{error}(\%)}{\text{TDC}_{error}(°\text{CA})} = 9 \qquad (10.1)$$

The crank angle position measurement system is always present in all modern reciprocating combustion engine. Typically, a toothed wheel with an inductive sensor is used in production engines for crank angle measurement. The engine controller senses the crankshaft position when one of the teeth of the wheel is in the vicinity of the inductive sensor. However, test engine in the laboratory has the optical incremental encoder with a very high resolution for crank angle position measurement. Differential (incremental) crank position is determined quite accurately by both the sensor arrangements, but the absolute crankshaft position (or piston position) needs to be meticulously calibrated [2]. The TDC position can be determined both statically (with a dial gauge) and dynamically (with a capacitive or microwave probe). The TDC determination using capacitive or microwave probe

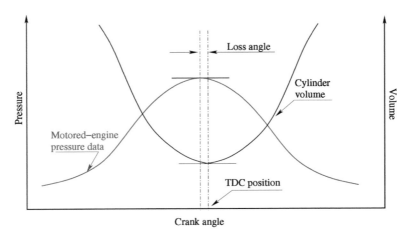

Crank angle

Fig. 10.1 Illustration of thermodynamic loss angle [4]

(see Chap. 5) is more accurate, but there are additional cost and mounting problems. The most convenient dynamical method is to use the signal from the pressure transducer which is mounted on the engine head in any case for combustion analysis. It would be possible to determine TDC very precisely in the absence of combustion in the cylinder and heat transfer from the walls of combustion chamber. The combustion in the cylinder can be avoided by motoring the engine. However, it is impossible to eliminate the heat transfer in a real engine during motoring also. Finally, the peak compression pressure occurs earlier than the minimum of combustion chamber volume during motoring of engine. This difference expressed in degrees of crank angle is called the "loss angle" [3]. Figure 10.1 illustrates the thermodynamic loss angle. The value of loss angle is unknown, and thus, it is determined based on calculations of compression pressure curve under the same conditions including the heat transfer and blowby losses. Therefore, the correct prediction of both heat and blowby losses in the vicinity of TDC is very important. The TDC should be determined with an accuracy of at least ±0.1 °CA, which is required for the calculation of indicated work and combustion parameters with sufficient accuracy [1]. There are several methods for dynamic evaluation of TDC based on cylinder pressure measurement, which are discussed in the following subsections.

10.1.1 Polytropic Exponent Method

The TDC position of the engine can be determined by solely the in-cylinder pressure signal by assuming the constant ratio of specific heats (γ), and there is neither heat transfer nor crevice flow. The maximum cylinder pressure will occur at minimum cylinder volume when these assumptions are true. Therefore, in this case, the TDC position is the angle corresponding to the maximum cylinder pressure. In recipro-cating engines, the temperature variations during compression cannot be avoided which leads to heat transfer and variation in γ. The variation in polytropic exponent near TDC can be used for determination of phase lag or TDC position. The polytropic exponent is calculated by Eq. (10.2) [5]:

$$n(\theta_i) = \frac{\ln P(\theta_i) - \ln P(\theta_{i+1})}{\ln V(\theta_{i+1}) - \ln V(\theta_i)} = \frac{\ln\left(\frac{P(\theta_i)}{P(\theta_{i+1})}\right)}{\ln\left(\frac{V(\theta_{i+1})}{V(\theta_i)}\right)} \qquad (10.2)$$

where n is a polytropic exponent, P is in-cylinder pressure, V is cylinder volume, and θ is crank angle position.

The polytropic exponent has the same value independent of crank angle when the crank angle is correctly calibrated, and assuming constant specific heat capacity and no heat transfer. It is found that for the correct correlation between pressure and volume (crank angle), the polytropic exponent should vary in the range between $1.35 < n < 1.4$ [1]. However, when the TDC is not positioned correctly (phase lag between volume and pressure), the polytropic exponent varies with crank angle as a pair of hyperbolic-like curves (Fig. 10.2). The curves lie in the quadrants II and IV when the phase lag error is positive and in the quadrants I and III when the error is negative. Therefore, a simple self-corrective procedure can be used to determine the correct position of TDC. The switch between quadrants gives the correct position of the crank angle.

To calculate the polytropic coefficient using Eq. (10.2), the measured pressure-crank angle data is transformed into adiabatic pressure-crank angle data. To com-pensate for blowby and heat losses, a pressure difference (Δp) is added to the

Fig. 10.2 Variation of polytropic exponent as a function of crank position for a motored engine cycle with no heat transfer and constant specific heat capacity (adapted from [1])

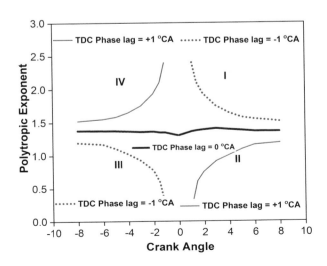

measured in-cylinder pressure data. The pressure difference between the measured and adiabatic data can be calculated using Eq. (10.3) [1]:

$$\Delta P = P_{\text{adi}} - P_{\text{mea}} = \frac{n-1}{V}(\Delta Q + \Delta I) \qquad (10.3)$$

where ΔQ is heat transfer loss and ΔI is the enthalpy loss due to blowby.

The heat transfer loss and blowby loss can be calculated using Eqs. (10.4) to (10.5) [1]:

$$\dot{Q} = hA(T - T_{\text{w}})$$
$$h = KD^{-0.2}P^{0.8}T^{-0.53}\left(2.588 \cdot 10^{-5}c_{\text{m}}\right)^{0.8} \qquad (10.4)$$

$$\dot{I} = F_{\text{v}}C_{\text{p}}P\sqrt{\frac{2T}{R}} \qquad (10.5)$$

where T is cylinder temperature, T_{w} is wall temperature, c_{m} is mean piston speed, D is cylinder diameter, and K and F_{v} are proportionality constants.

For correct determination of TDC position, it is crucial to determine the proportionality constants (K and F_{v}). The value of the proportional factors varies with engine size, design, and material. A statistical analysis is used for determination of the constants. If correctly calibrated, the predicted position of TDC should be independent of engine speed. Therefore, K and F_{v} should be selected to the value that gives the minimum standard deviation of crank angle corresponding to TDC, when it is calculated at various engine operating speeds [1].

10.1.2   Symmetry-Based Method

This TDC determination method compares the compression phase of the in-cylinder pressure with the expansion phase. The motoring cylinder pressure curve is symmetric with respect to TDC position when there is no heat transfer, and the volume curve is symmetric with respect to TDC [5]. The cylinder pressure on either side of TDC has the same value, which is illustrated by Eq. (10.6):

$$P(\theta_{\text{TDC}} - v) = P(\theta_{\text{TDC}} + v) \qquad (10.6)$$

Equation (10.6) is true for an arbitrary value of $v > 0$, where $\theta_{\text{TDC}}-v > \theta_{\text{IVC}}$ and $\theta_{\text{TDC}} + v < \theta_{\text{EVO}}$. Cylinder pressure data after intake valve closing (IVC) and before the exhaust valve opening (EVO) is considered for the TDC determination.

A crank angle varying offset $\Delta(\theta)$ is added to the expansion curve, to compensate for heat transfer (Fig. 10.3):

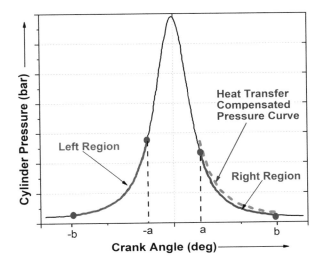

Fig. 10.3 Regions of investigation in motoring pressure for TDC determination using symmetry method (adapted from [5])

$$\Delta(\theta) = (P(\theta_{\mathrm{TDC}} - \theta_0) - P(\theta_{\mathrm{TDC}} + \theta_0)) \left(\frac{V(\theta_0)}{V(\theta_{\mathrm{TDC}} + \theta_0)} \right)^{\eta} \qquad (10.7)$$

where θ_{TDC} is the estimated TDC position, θ_0 is the median of the region $\{a, b\}$, and η is an exponential constant, which has to be determined through simulations. If all heat transfer would occur in the excluded region around TDC (Fig. 10.3), the value of η would be the same as the ratio between specific heats (γ). This method compares the pressure trace for the two regions instead of for two single points, which gives it a low sensitivity to noise.

The estimated position of TDC, θ_{TDC}, is the position that minimizes the following sum of squared errors, illustrated using Eq. (10.8):

$$\sum_{a < \theta_i < b} [P(\theta_{\mathrm{TDC}} - \theta_i) - (P(\theta_{\mathrm{TDC}} + \theta_i) + \Delta(\theta_{\mathrm{TDC}} + \theta_i))]^2 \qquad (10.8)$$

The limiting angles a and b must be selected in such a way that the investigated regions are enclosed in the compression and expansion phase, but exclude the highest cylinder temperatures to reduce the effects of heat transfer. To avoid an offset error, (between the beginning of the compression phase and end of expansion phase), only the region where the pressure changes rapidly should be included in the investigation (Fig. 10.3). The region where heat transfer is too dominating is excluded from the study. A more complete discussion and justification of relevant equations can be found in the original study [5]. This study also compared the other four methods of TDC determination with varying complexity and found that symmetry-based method is fairly accurate and robust. However, the demerit of the method is that it requires knowledge of the polytropic exponent or η, which needs to be estimated through simulations.

10.1.3   Loss Function-Based Method

The loss function method of TDC determination is proposed in the studies [6–8]. This method considers the influence of cylinder heat transfer as well as mass leakage and evaluates the TDC position by establishing the concept of loss function, which is shown in Eq. (10.9) [7]:

$$\delta F = C_p \frac{\delta V}{V} + C_v \frac{\delta P}{P} = \delta S + C_p \frac{\delta m}{m} \tag{10.9}$$

where δF is loss function, C_p is constant pressure specific heat of the gas, C_v constant volume specific heat of the gas, P is cylinder pressure, V is volume, S is in-cylinder gas specific entropy, and m is in-cylinder gas mass. Typical variations in the loss function and its consecutive terms as a function of crankshaft position with a phase lag of 1 CAD are presented in Fig. 10.4. The figure shows that the entropy variation

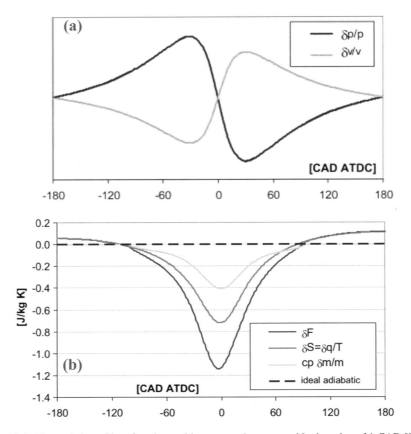

Fig. 10.4 The variation of loss function and its consecutive terms with phase lag of 1 CAD [8]

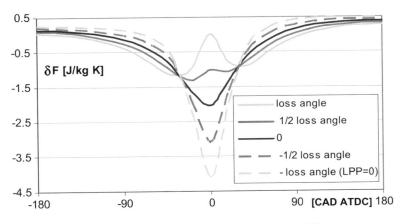

Fig. 10.5 The variation of loss function for different TDC phase errors [8]

begins with a decreasing positive value (as $T < T_{wall}$) and it crosses the zero line when the gas temperature (T) equals to wall temperature (T_{wall}). Entropy reaches to a minimum close to the TDC position and then starts increasing with a positive value near BDC position. The relative mass leakage (dm/m) follows the similar trend as it is related to the difference between in-cylinder pressure and outer pressure. The loss function is the addition of the two curves (Fig. 10.4).

It was found that the loss function is not significantly influenced by thermal loss angle at crank angle position near $\pm 30°$ CA, but it depends only on the geometry of the engine. The variation in the loss function with crankshaft position for different TDC phase errors (expressed as a fraction of the loss angle) is illustrated in Fig. 10.5. The figure shows that loss function undergoes great distortions even with small phase errors between cylinder pressure and cylinder volume curve. The loss function at peak pressure location (LPP) is directly related to the angular position θ_1 and θ_2 of the minimum and maximum $\delta V/V$, so the relationship between these two loss function can be derived. Then, the loss function at peak pressure position can be evaluated and heat loss angle computed using Eqs. (10.10) to (10.13):

$$\theta_{1,2} = \pm 76.307 \cdot \mu^{0.123} \cdot \rho^{-0.466} \tag{10.10}$$

$$\delta F_m = \frac{1}{2}(\delta F_1 + \delta F_2) \tag{10.11}$$

$$\delta F_{peak} = 1.95 \cdot \delta F_m \tag{10.12}$$

$$\theta_{loss} = \frac{2}{\rho - 1} \cdot \frac{\mu}{\mu + 1} \left[\frac{1}{C_p} \frac{\delta F}{\delta \theta} \right]_{peak} \tag{10.13}$$

where μ is the connecting rod-to-crank ratio and ρ is the engine compression ratio. The loss function increment at maximum pressure position can be determined from Eq. (10.12) by using the mean value calculated by Eq. (10.11). The loss angle θ_{loss} and thus the TDC location can be calculated using Eq. (10.13). A more complete discussion and justification of relevant equations can be found in the original study [8].

10.1.4   Temperature-Entropy Diagram Method

In this method, the TDC position can be determined by analyzing the temperature-entropy $(T - S)$ diagram during compression and expansion strokes of the engine cycle [9, 10]. When the TDC position is well calibrated, compression and expansion strokes under motoring conditions are symmetrical with respect to the peak temperature in the (T-S) diagram. In this method, the constant mass evolution (i.e., no gas leakage or blowby) is assumed during analysis. The progress of the $T - S$ curve in this diagram depends on the relative position between the cylinder pressure and the cylinder volume. The temperature and entropy can be calculated by Eqs. (10.14) to (10.15) under ideal gas assumptions [9]:

$$T = \frac{PV}{mR} \tag{10.14}$$

$$dS = C_p \frac{dT}{T} - R \frac{dP}{P} \tag{10.15}$$

$$C_p = 1403.06 - 360.72 \frac{1000}{T} + 108.24 \left(\frac{1000}{T}\right)^2 - 10.79 \left(\frac{1000}{T}\right)^3 \tag{10.16}$$

Figure 10.6 schematically illustrates the $T - S$ diagram shifting with respect to corrections on TDC position. The entropy must change symmetrically with respect

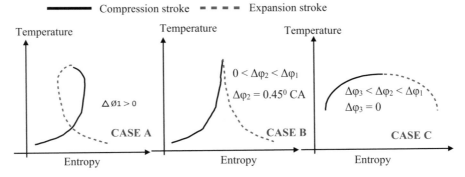

Fig. 10.6 Illustration of $T - S$ diagrams according to the correction on the TDC position (adapted from [9])

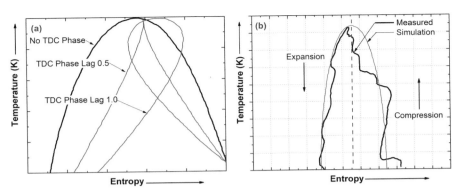

Fig. 10.7 (a) Simulated $(T - S)$ diagram with different TDC phase lags and (b) comparison of simulated and experimental $T - S$ diagram with a correct TDC calibration (adapted from [10])

to the peak temperature $(T_{\max})$ for a correctly phased cylinder pressure (in motoring condition) with the cylinder volume (also illustrated in Fig. 10.7b). A loop appears in the $T - S$ diagram under gross TDC position errors (Figs. 10.6 and 10.7a), which has no thermodynamic significance. Thus, the correct phasing between cylinder pressure and volume can be obtained by shifting the cylinder pressure curve until the loop disappears (Figs. 10.6 and 10.7a). An incorrect TDC introduces a nonphysical loop in the temperature-entropy diagram. A further shift of -0.45 CAD should be applied to find the correct pressure curve position (Case C in Fig.10.6), which is determined using simulations [9].

The main demerit of the $T - S$ method is the requirement of gas temperature calculation, which is often inaccurate. The calculated temperature is then used both directly and in entropy polynomials. Thus, inaccuracy of temperature calculation will propagate in TDC correction estimation.

10.1.5   Inflection Point Analysis-Based Method

TDC calibration can also be done using the inflection point analysis on the motoring cylinder pressure [3, 11]. In this method, first, the crank angle is phased in such a way that maximum cylinder pressure occurs at TDC position. With this calibration of TDC position, the polytropic exponent is calculated using Eq. (10.2). The polytropic exponent at the inflection points of the cylinder pressure curve is defined as m_1 and m_2, where m_1 is on the left hand of TDC. An inflection point is a point that is established by the condition shown in Eq. (10.17):

$$\frac{d^2 P_{\text{cyl}}(\theta)}{d\theta^2} = 0 \tag{10.17}$$

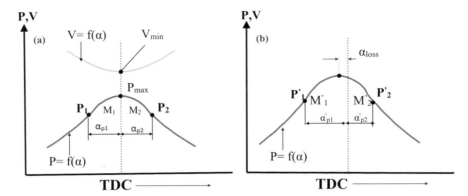

Fig. 10.8 Moving of the inflection points due to thermodynamic loss angle (adapted from [3])

After defining the inflection point and calculation of polytropic exponent, the phasing between the crank angle and cylinder pressure is adjusted. With the new adjusted setting, another pair of polytropic exponents at the inflection points, defined as m'_1 and m'_2 are calculated. Figure 10.8 schematically shows the movement of inflection points due to loss angle. It is suggested in the study [3] that when the crank angle is correctly adjusted, the value of the ratio (R) is in the range of $2.2 \leq R \leq 2.3$. The ratio (R) is calculated using Eq. (10.18). The crank angle phasing between cylinder pressure and volume should be adjusted until this equation is fulfilled.

$$R = \frac{m'_2 - m'_1}{m_2 - m_1} \tag{10.18}$$

The inflection point method is very sensitive to measurement noise because it relies on second derivatives of both pressure and volume [2]. However, this method does not requires cylinder temperature and wall heat transfer calculations. In another study [11], the author refined this method and concluded that the Vdp term must have an inflection point (second derivative equal to zero) at the maximum pressure position. This can be used to determine the correct TDC position, by adjusting the phasing to meet the inflection point condition. This improved method uses the third derivatives of experimental cylinder pressure, which makes it extremely sensitive to measurement noise. This is the main demerit of this method. To avoid this problem, a zero-phase lag smoothing is applied to the measured pressure signal. This method is applicable to both SI and CI engines [11].

10.1.6   TDC Calibration Using IMEP

The TDC calibration results from most of the dynamic methods depend on the heat transfer coefficient or heat transfer model. A study [12] proposed a calibration based on the IMEP calculation and the maximum pressure (P_{max}). The angular distance between the cylinder pressure maximum and the actual TDC, θ_{loss}, is defined as a function of $IMEP_{gross}$ and the maximum cylinder pressure, which is shown as Eq. (10.19) [5]:

$$\theta_{loss} = K\frac{IMEP_{gross}}{P_{max}} \qquad (10.19)$$

The error in TDC leads to a large error in IMEP calculations. Hence, the IMEP must be computed with the correct crank angle phasing which is unknown in this case. Thus, Eq. (10.20) is used for IMEP calculation:

$$IMEP_{gross} = \chi \cdot \theta_{loss} + IMEP_{gross,0} \qquad (10.20)$$

where $IMEP_{gross,0}$ is the calculated IMEP value when the crank angle is phased in such a way that maximum peak pressure occurs at TDC position. The correlation coefficients K and χ used in Eqs. (10.19) and (10.20) are estimated by simulation. An engine model for motored cycles using Woschni heat transfer correlation can be implemented for TDC determination [5] using this method.

10.1.7   Model-Based TDC Determination Method

Crank angle phasing can also be performed by model-based or heat release shaping-based method [2, 13], which is used to determine the TDC position. In this method, the net heat release model is used to decide specific heat ratio (γ), heat release rate (k), and TDC offset θ_0. This method of TDC determination utilizes the fact that pressure and temperature are fairly constant close to TDC position. The wall temperature and the in-cylinder flow are assumed to be constant, which means heat transfer (dQ/dt) is constant. Additionally, the crankshaft speed is also assumed constant, which results in $dQ/d\theta$ constant during a motoring cycle. This is a strong assumption; however, this method of TDC determination is resilient to measurement noise because it utilizes a range of points [14]. In this method, no separate simulation is required to select the model parameters, and it is easy to deploy on any engine. The study found that in a laboratory setting, this method can estimate TDC within 0.1 CAD of the true TDC [2].

Assuming the $dQ/d\theta = k$, the pressure rise rate can be written by Eq. (10.21):

$$\frac{dP}{d\theta} = k(\gamma - 1)\frac{1}{V(\theta)} - \gamma P(\theta)\frac{dV}{d\theta}\frac{1}{V(\theta)} \qquad (10.21)$$

By inserting the phasing error θ_0 in Eq. (10.21), the pressure rise rate can be written by Eq. (10.22):

$$\frac{dP}{d\theta} = k(\gamma - 1)\frac{1}{V(\theta + \theta_0)} - \gamma P(\theta)\frac{dV(\theta + \theta_0)}{d\theta}\frac{1}{V(\theta + \theta_0)} \qquad (10.22)$$

Given θ_0, the problem is linear in $C = k(\gamma - 1)$ and γ. The problem can be written on the form $y_i = \phi_i x$, and different terms can be written as Eqs. (10.23) to (10.25):

$$x = [C \ \gamma]^T \qquad (10.23)$$

$$y_i = \frac{dP(\theta_i)}{d\theta} \qquad (10.24)$$

$$\phi_i = \left[\frac{1}{V(\theta_i + \theta_0)} \quad p(\theta_i)\frac{dV(\theta_i + \theta_0)}{d\theta}\frac{1}{V(\theta_i + \theta_0)}\right] \qquad (10.25)$$

where i points the output, and regressor matrix is constructed as Eqs. (10.26) and (10.27) with the solution shown by Eq. (10.28):

$$Y = \begin{bmatrix} y_1(\theta_1/\theta_0) \\ y_2(\theta_2/\theta_0) \\ . \\ . \\ y_N(\theta_N/\theta_0) \end{bmatrix} \qquad (10.26)$$

$$\phi = \begin{bmatrix} \phi_1(\theta_1/\theta_0) \\ \phi_2(\theta_2/\theta_0) \\ . \\ . \\ \phi_N(\theta_N/\theta_0) \end{bmatrix} \qquad (10.27)$$

$$\hat{x} = \phi + Y \qquad (10.28)$$

With a new estimation $\hat{x}$, the phasing error θ_0 can be calculated by solving the nonlinear least-square (NLLS) problem. The detailed solution methodology can be found in [2, 14]. The calibrated phase angle can be validated further by using a $P - V$ diagram (compression and expansion lines should not cross each other) or by standard deviation based methods.

10.2 Compression Ratio Determination

Heat release analysis based on in-cylinder pressure measurements is a commonly used tool for combustion analysis in reciprocating engines. Measured in-cylinder pressure and calculated cylinder volume are the two main parameters used in heat release analysis. Crank angle phasing (determination of TDC) has a significant role in pressure analysis (discussed in Sect. 10.1). Cylinder volume calculation is directly affected by the engine compression ratio, which is dependent on displacement volume and clearance volume (volume at TDC position). The displacement volume can be calculated with tight limits (sufficient accuracy). Determination of clearance volume is difficult because it includes the crevices consisting of first piston ring and cylinder, the top-land region between piston, head-gaskets, spark plug thread, valve seats interstices, etc. In laboratory settings, depending on instrumentation such as pressure sensor installation (particularly in recess mounting) or optical engines, combustion chamber can jointly have significantly large volume at TDC position. Due to geometrical uncertainties, a spread in compression ratio among the different engine cylinders is inherent. To determine the compression ratio, the measurement of clearance volume (including crevices) is difficult and time-consuming. Thus, alternative determination procedure of compression ratio using cylinder pressure measurement (which is present in any case) can be helpful.

10.2.1 Temperature-Entropy-Based Method

The compression ratio (CR) is defined using the basic geometry of a reciprocating engine, which is described by Eq. (10.29):

$$CR = \frac{V_d + V_c}{V_c} \tag{10.29}$$

where V_d is the displaced volume and V_c is clearance volume. The V_d can be measured with good accuracy because it depends on the cylinder bore and the piston stroke. However, the measurement of the V_c can be imperfect, which leads to an error in the calculation of the compression ratio. The error in clearance volume can be determined using temperature-entropy diagram [10].

The error in clearance volume (ΔV) can be estimated as follows. The instantaneous cylinder volume can be calculated by Eq. (10.30):

$$V = V_c + V_d(\theta) \tag{10.30}$$

If an error exists on the clearance volume, the instantaneous cylinder volume can be written as

$$V' = V'_c + V_d(\theta) \tag{10.31}$$

where $V_c^{'}$ is the wrong clearance volume. The error on the clearance volume is defined by Eq. (10.32):

$$\Delta V = V - V^{'} = V_c - V_c^{'} \tag{10.32}$$

The error on the clearance volume can be expressed by Eq. (10.33), which is calculated using temperature-entropy diagram. The detailed derivation of the expression can be found in the original study [10]:

$$\Delta V = \frac{\left(V^{'C}\right)^2 \left(dS^{'C} - dS^{'e}\right)}{2C_{p'}dV^{'C} - V^{'C}\left(dS^{'C} - dS^{'e}\right)} \tag{10.33}$$

where S denotes entropy, $C_{p'}$ is constant pressure specific heat, and superscripts c and e denote compression and expansion stroke, respectively. Equation (10.33) can be used to correct an error on the clearance volume.

Figure 10.9 presents the variation of the temperature-entropy diagram for different clearance volume error, which is estimated by simulated motoring pressure curve. When ΔV is negative, the variation of entropy (ΔS) decreases during the compression stroke and increases during the expansion stroke. The reverse effect is obtained for positive ΔV [10]. Hence, for a negative clearance volume error ($\Delta V < 0$), the $T - S$ diagram leans toward the right, whereas it leans toward the left if the clearance volume error is positive ($\Delta V > 0$) (see Fig. 10.9). The asymmetry of the $T - S$ diagram is an easy criterion to detect if an error exists on the clearance volume.

Fig. 10.9 Illustration of clearance volume error on the simulated temperature-entropy diagram (adapted from [10])

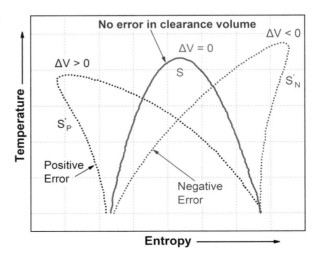

10.2.2   *Heat Loss Estimation Method*

The engine compression ratio can be estimated by a number of methods based on cylinder pressure measurement [15–18] or measured flow rates [19, 20]. Heat loss estimation-based method [15] of compression ratio determination utilizes the measured motored pressure trace. The motored data is preferred as it removes the uncertainties related to calculation of heat release during combustion. The motored pressure trace and gas temperature at IVC are the only input parameters in this method. Volume is considered as total volume including crevices. The heat loss rate is calculated using motored pressure by convective heat transfer coefficient model. Figure 10.10 shows the variations of heat loss rate, which is calculated at different compression ratios. When bulk gas temperature (T) is higher than the temperature of cylinder walls (T_w), heat transfer (loss) occurs from the gases to cylinder walls. Gas temperature is relatively higher near TDC position due to compression. Hence, heat addition is not possible in the region where $T > T_w$. In Fig. 10.10, the reasonable heat loss rate curve is only for CR 9 and 10, while other curves show consistent heat addition (heat loss rate negative) in zone $T > T_w$. Thus, roughly, it can be estimated that CR between 9 and 10 is possible as heat loss rate reaches the maximum point near TDC.

It can also be observed from Fig. 10.10 that the heat loss rate is independent of volume (or CR) at the peak pressure position, and all curves intersect at this point. At peak pressure position, the heat release equation can be represented as Eq. (10.34) as $dP/d\theta$ is zero at this position:

$$\frac{dQ}{d\theta} = \frac{\gamma}{\gamma - 1} P \frac{dV}{d\theta} \qquad (10.34)$$

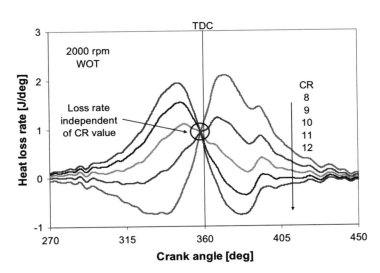

Fig. 10.10 Calculated heat loss rate at different compression ratio [15]

The ratio of specific heat (γ) is dependent on temperature, and it has a slight effect on heat release rate. However, it is not significant for one CR to another (Fig. 10.10). It can be noted that this condition (Eq. 10.34) is only valid when there is no blowby loss. Heat transfer model can be calibrated by selecting the heat loss rate at peak pressure location [15].

To reduce the effect of measurement noise, the pressure signal needs to be filtered, which leads to a more precise estimation of peak pressure location. Heat release rate curves are recalculated using the calibrated heat transfer model for different assumed compression ratios. The compression at which the root mean square error for dQ is minimum within $\pm$ 20 °CA interval near TDC position is considered as the correct compression ratio of the engine. It was found that engine speed had the greatest influence on compression ratio determinations, and estimation errors were lowest at lower engine speeds. A more complete discussion on compression ratio determination using this method can be found in the study [15].

10.2.3 Polytropic Model Method

The compression ratio can be determined using the polytropic model. The motored cycles with a high initial pressure as possible should be used for the accurate determination of the compression ratio [17]. The high initial pressure improves the signal-to-noise ratio of motored pressure data, while the effects of heat transfer and crevice flows remain the same. Additionally, all pressure information available is utilized for estimation of compression ratio when motored pressure data is used. The polytropic model used to determine the compression ratio is described in Eq. (10.35):

$$P(\theta)(V_{\mathrm{d}}(\theta) + V_{\mathrm{c}}(r_{\mathrm{c}}))^n = C \qquad (10.35)$$

where n is a polytropic exponent, r_{c} is compression ratio, and C is the constant. In this method, two problems are iteratively solved to find the "n" and clearance volume (V_{c}), which defines the compression ratio by Eq. (10.29).

This method is formulated as least-square problems in a set of unknown parameters. The residuals derived from Eq. (10.35) is presented by Eqs. (10.36) and (10.37) [17, 18]:

$$\varepsilon_{1a}(C_1, n) = \ln p(\theta) - (C_1 - n \ln (V_{\mathrm{d}}(\theta) + V_C)) \qquad (10.36)$$

$$\varepsilon_{1b}(C_2, V_C) = V_{\mathrm{d}}(\theta) - \left(C_2 p(\theta)^{-1/n} - V_C\right) \qquad (10.37)$$

Equation (10.36) is linear in the parameters $C_1 = \ln C$ and n, for fixed V_{c} (clearance volume). Equation (10.37) is linear in the parameters $C_2 = C^{1/n}$ and V_{c}, for a fixed value of n. The main idea is to use the two residuals, ε_{1a} and ε_{1b}, iteratively for determination of the parameters n, V_c, and C by solving two linear least-square problems. The study also derived the approximate relation between the residuals, which is shown in Eq. (10.38):

$$\varepsilon_{1a}(\theta, x) \approx \frac{n}{V_d(\theta) + V_C} \varepsilon_{1b}(\theta, x) \qquad (10.38)$$

To achieve the comparable norms in the least-square problems, the residual ε_{1a} is multiplied by the weight ($w = V_d(\theta) + V_c$). Detailed solution procedure along with another solution method (variable projection method) can be found in [17, 18].

10.2.4   Determination of Effective Compression Ratio

Engine designers are able to optimize the gas exchange process due to flexibility in the valve train similar to the optimization of the fuel injection process by common rail direct injection (CRDI) systems. Volumetric efficiency can be directly influenced by modulating valve timings because it directly controls the mass trapped in the engine cylinder. It is demonstrated that modulation of the valve timing can achieve volumetric efficiency three times larger than the range obtained by modulation of other engine actuators such as the exhaust gas recirculation (EGR) valve or the variable geometry turbocharger (VGT) [21]. The effective compression ratio (ECR) is one of the key characteristics of the gas exchange process, which is influenced by valve timing modulation. The ECR is a measure of the effective in-cylinder compression process above intake manifold conditions. Figure 10.11 illustrates the typical exhaust and intake valve lift profiles and intake profiles with intake valve closing (IVC) modulation. The compression ratio is usually fixed or has limited variability by using conventional engine hardware and actuators. However, there is a need for variable compression ratios (VCR) in automotive engines for improved performance [20–22]. Modulation of the ECR is a key enabler of advanced combustion strategies meant for emission reduction while maintaining fuel conversion efficiency. Effective compression ratio is directly controlled by the modulation of intake valve closing time (Fig. 10.11). Additionally, effective coordination of valve timings and the ECR is the key for closed-loop targeting of advanced combustion engines. The real-time employment of these strategies needs the knowledge of the effective compression ratio.

The trapped charge mass in the combustion chamber is compressed to a minimum volume (clearance volume) in the compression stroke. During compression, the pressure and temperature of the gas in the cylinder increase, which leads to rapid fuel evaporation, mixing, and combustion following direct fuel injection. The compression ratio is a commonly used parameter to indicate the extent of compression. The geometric compression ratio (GCR) is defined as the ratio of the combustion chamber volume when the piston is at bottom dead center (V_{BDC}) to the

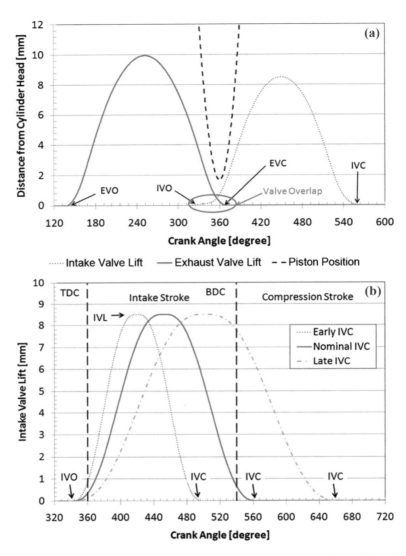

Fig. 10.11 (**a**) Typical intake and exhaust valve lift profiles and (**b**) intake valve profiles with IVC modulation [21]

combustion chamber volume when the piston is at the top dead center (V_{TDC}). The geometric compression ratio can be written as Eq. (10.39):

$$GCR = \frac{V_{BDC}}{V_{TDC}} \qquad (10.39)$$

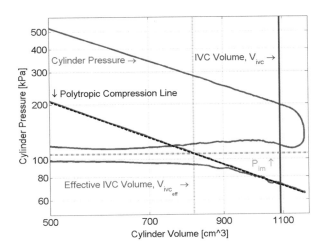

Fig. 10.12 Illustration of effective compression ratio for early IVC [21]

However, the actual compression process effectively starts after the closing of intake valves (IVC). Typically, the IVC does not occur exactly when the piston is at BDC. Therefore, a geometric ECR (GECR) is defined as the ratio of cylinder volume at IVC to cylinder volume at TDC, which is shown in Eq. (10.40) [20, 21]:

$$GECR = \frac{V_{IVC}}{V_{TDC}} \tag{10.40}$$

The overall compression process is also affected by both the piston-induced and gas momentum-induced compression, which is not accounted in the definition of GECR. A commonly used method for calculating the ECR is the pressure-based method, which captures the effective compression process more accurately [21–23].

Figure 10.12 illustrates the method of the effective compression ratio determination. First, a linear fit is applied to the compression process on the $\log(P)-\log(V)$ diagram. The slope of this line on a log-log scale is the polytropic coefficient. The linear fit line is extrapolated to the intake manifold pressure (P_{im}). The corresponding volume at the point of intersection is defined as the effective IVC volume (V_{IVCeff}). The effective compression ratio (ECR) is then defined as the ratio of the effective volume to the TDC volume, which is depicted by Eq. (10.41):

$$ECR = \frac{V_{IVCeff}}{V_{TDC}} \tag{10.41}$$

This ECR calculation method is relatively more accurate than using a GECR (Eq. (10.40)) because it considers the effect of flow momentum on cylinder charge compression, as well as compression by movement of piston. However, the accurate estimation of ECR needs a reliable method for in-cylinder pressure measurement [20].

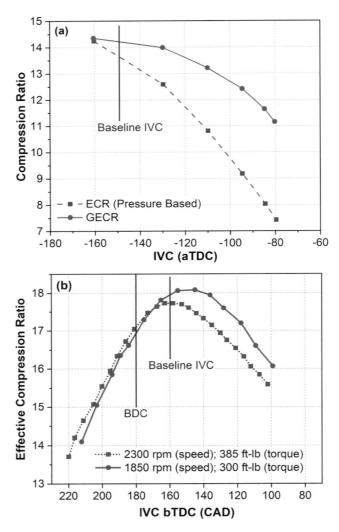

Fig. 10.13 (**a**) The variation of ECR and GECR with IVC (adapted from [23]) and (**b**) variation of ECR with IVC at different operating conditions (adapted from [22])

Figure 10.13 depicts the variations of ECR and GECR as a function of IVC timing. The GECR decreases almost linearly with IVC timings. However, very little change is observed in the ECR, when IVC changes from 160 to 130 bTDC (Fig. 10.13a). A significant change in ECR occurs when the IVC timing is retarded by more than 50 CADs. Figure 10.13b also shows that the ECR reaches to the peak value around the conventional IVC timing (~160 CAD bTDC). The ECR maximum occurs around 20 CAD after BDC because the more charge is continued to be pushed in the cylinder by momentum of the flow even as the piston starts moving upward. The ECR reduces with modulated timings, i.e., either early or late IVC (Fig. 10.13). Therefore, the ECR can be varied by IVC timing modulation, and a given ECR can be obtained by choosing suitable IVC timings.

10.3   Blowby Estimation

The engine performance and exhaust emissions are affected by blowby and gas flow through the cylinder-piston-ring crevices. Additionally, blowby flow rates affect the quantity of charge during engine cycle as well as cylinder pressure and temperature. Several researches have been performed to investigate the blowby phenomenon and gas flow through the cylinder-piston-ring crevices [24–27]. Blowby process plays an important role in the determination of hydrocarbon emissions and bore wear pattern analysis and for validation of combustion simulation against experimental data. One-dimensional orifice-volume method has been typically used for crevice flow analysis. Blowby models can be tuned/calibrated using heat release analysis based on cylinder pressure measurement. Measured cylinder pressure data for the closed part of the cycle (from IVC to EVO) are analyzed using the first law of thermodynamics for open systems, which is represented by Eq. (10.42) [15]:

$$dQ = dU + p \cdot dV - h \cdot dm \tag{10.42}$$

Equation (10.42) can be rewritten as Eq. (10.43) using the ideal gas equation for gas flowing out of the cylinder:

$$dQ = \frac{\gamma}{\gamma - 1} \cdot p \cdot dV + \frac{\gamma}{\gamma - 1} \cdot V \cdot dp - \frac{\gamma}{\gamma - 1} \cdot p \cdot V \cdot \frac{dm}{m} \tag{10.43}$$

A simplified compressible flow model can be used for calculating blowby flow velocity (w_{bb}) using Eqs. (10.44) and (10.45) for non-chocked flow and choked flow, respectively:

$$W_{bb} = \sqrt{2 \cdot \frac{\gamma}{\gamma - 1} \cdot \frac{R}{M} \cdot T \left[1 - \left(\frac{p_0}{p}\right)^{\frac{\gamma-1}{1}}\right]} \tag{10.44}$$

$$W_{bb} = \sqrt{2 \cdot \frac{\gamma}{\gamma + 1} \cdot \frac{R}{M} \cdot T} \tag{10.45}$$

The mass flow rates of blowby can be estimated using Eq. (10.46) with A_{bb} considered as an equivalent flow area, and N represents engine speed (rad/s):

$$dm = w_{bb} \cdot A_{bb} \cdot d\theta \cdot N^{-1} \qquad (10.46)$$

Assuming the correct prediction of heat loss rate by convective heat transfer mode (calibrated), dQ values can be calculated for different equivalent blowby flow areas [15]. The correct value for the effective blowby flow area was chosen as the one that minimizes dQ residuals around TDC, which is the similar procedure that is used for determination of compression ratio in Sect. 10.2.2. Crevice gas temperature can be assumed equal to the bulk gas temperature. The blowby rates are predicted with good accuracy using this method.

Similar blowby model is used in another study [24], and results confirmed a good agreement with experimental results. The blowby model showed that the maximum blowby mass loss position is not the same as the maximum pressure position, and the maximum mass loss position typically occurs after maximum pressure position. The model also predicted that the maximum mass loss increased with the increase of compression ratio and decreased with the increase of engine speed [24].

10.4   Wall Temperature Estimation

To control the large number of actuators in reciprocating engines, determination of the operating states and outputs is essential for the actual engine control systems in steady-state and transient operations. Several sensors are used for this purpose with an increase of costs and efforts for calibration and wiring complexity. Theoretically, most of the information given by the different sensors can be determined from the measured in-cylinder pressure signal. Some of the existing sensors may be replaced by cylinder pressure measurement. Additionally, the pressure signal reveals high dynamic features that are useful for more stringent control. However, appropriate methodologies should be used for in-cylinder pressure signal processing [28].

Among the many engine states and output variables of interest, the evaluation of the thermal state is particularly important for the engine control because thermal state of cylinder strongly affects the engine heat flow and engine performance. Different control strategies are applied based on the engine thermal state. The coolant water temperature is one of the options available to consider for control task. However, coolant temperature has poor dynamic features because of high thermal inertia involved with the heat transfer process between gas mixture, combustion chamber wall, and coolant fluid. Therefore, the estimation of cylinder wall thermal state can be beneficially utilized for the most engine combustion control problems [28, 29].

The cylinder wall temperature has great influence on the engine heat transfer and the exhaust emissions [30]. Typically, average cylinder wall temperature is higher than the cylinder charge temperature at the start of the compression stroke (except in the early stages of cold start of the engine). As the piston moves further upward, the cylinder charge temperature increases. The heat flux between the cylinder walls and the core regions of cylinder charge (across the thermal boundary layer) is reversed due to increase in charge temperature by piston movement. Due to very high gas temperature of burned gases during combustion process, the effect of cylinder wall temperature variations on the surface heat transfer can be neglected for the period where combustion is taking place. However, during intake and compression strokes, the effect of cylinder wall temperature is more significant and cannot be neglected because of the significantly lower gas temperature in this time zone. During intake and compression stroke, a reduction in the cylinder wall temperature leads to lower bulk temperature of the unburned gases before the initiation of combustion due to lower heat transfer. The lower bulk temperature prior to the combustion consequently leads to a lower burned gas temperature [29]. The exhaust emissions (particularly HC and NO_x) are also strongly affected by the cylinder wall temperature. The HC emission decreases from two sources (oil layer absorption and crevice volume) with an increase in wall temperature [31]. Lubricating oil layer thickness is directly related to the wall temperature which affects the absorption of fuel. The amount of fuel trapped in the crevice volume is also affected by temperature variations of the cylinder liner and piston. The increase in cylinder wall temperature substantially increases the NO_x emissions. The cylinder wall temperature affects the peak temperature in the core region of cylinder charge due to thermal flow effects at the start of the compression of the charge [29].

The direct measurement of the cylinder wall temperature using conventional experimental methods is not suitable for onboard application on production engines, due to costs and measurement complexity. The typical alternative of experimental measurement is simulation models for wall temperature estimation. Several simulation models are proposed to investigate the wall temperature including transient operation by wall periodic heat conduction model [32–34]. However, most of the modeling approaches seem to be not suitable for real-time application. Therefore, cylinder pressure-based techniques for cylinder wall temperature estimation can be beneficial. The next subsections will discuss the wall temperature estimations from measured cylinder pressure data.

10.4.1 Heat Transfer Inversion Method

In this method of wall temperature determination, the crank angle position (during the compression stroke) corresponding to the inversion of the heat transfer between gas and cylinder walls is estimated. The adiabatic condition occurs at the inversion point, which means the corresponding in-cylinder charge temperature is equal to the mean wall temperature (T_w) [28]. It can be noted that this assumption is true for steady-state heat transfer process. Detailed analysis has demonstrated that the time of zero temperature gradient is not necessarily the point at which a thermal flux can be neglected because of a time lag between temperature gradient (cause) and thermal flux (effect). However, the assumption of a quasi-steady heat transfer process is considered sufficiently accurate for most calculation purposes in reciprocating engines [35]. Additionally, a complex thermal field occurs in the combustion chamber with substantial spatial distribution as well as cyclic time fluctuations in the reciprocating engine. In particular, the "effective" wall temperature has a tendency to increase with the progress of compression stroke because of the increasing effect of the regions closer to the TDC position, where typically higher cylinder wall temperature is observed [28].

In this method, a lumped parameter model representing the interaction between piston, gas, and the cylinder is considered. The in-cylinder fluid is assumed as an ideal gas and its compression as a non-adiabatic internally reversible process (i.e., non-adiabatic polytropic process). The thermal energy exchanged between the wall and the gas δQ can be expressed as Eq. (10.47) by considering the energy conservation for a closed system with polytropic process:

$$\delta Q = -\frac{n - \gamma}{\gamma - 1} PdV \tag{10.47}$$

where "n" is the polytropic exponent of the compression process and "γ" is the ratio of specific heats, which is expressed as Eq. (10.48):

$$\gamma(\theta) = \frac{C_p(\theta)}{C_p(\theta) - R} \tag{10.48}$$

where R is universal gas constant and C_p is specific heat at constant pressure.

The polytropic exponent can be calculated using Eq. (10.2) using the measured pressure data. The ratio of specific heat can be calculated by using thermodynamic properties of gases present in the combustion chamber. The C_p can be calculated by NASA polynomials, which is represented as Eq. (10.49):

$$\frac{C_p}{R} = a_1 + a_2 T + a_3 T^2 + a_4 T^3 + a_5 T^4 \tag{10.49}$$

The value of "n" will be higher than "γ" when the gas temperature is lower than wall temperature and vice versa. Consequently, the crank angle at which the adiabatic condition $n = \gamma$ occurs corresponds to the heat flux inversion angle. After calculating the inversion angle (θ_{inv}), the wall temperature (T_w) can be calculated by Eq. (10.50):

$$T_w = \frac{P(\theta_{inv}) . V(\theta_{inv})}{m.R} \tag{10.50}$$

where $p(\theta_{inv})$ and $V(\theta_{inv})$ are the gas pressure and volume at the inversion crank angle and mass (m) incorporates the air mass, the residual gas fraction, and, eventually, the recirculated exhaust gas (i.e., EGR).

   Due to the pressure signal fluctuation, the polytropic index "n" does not exhibit a clear trend, especially in the first stages of the compression stroke. In fact, oscillating "n" values around the adiabatic index "γ" can be noted. The calculated "n" value can be zero or negative at some crank angles because of the pressure gradient measured at the corresponding $\Delta\theta$ range, evidently affected by sensor sensitivity or effective in-cylinder pressure fluctuation. In order to detect a well-defined intersection point between n and γ curves, suitable processing of the experimental pressure data is used [28, 29]. The following modeling equation (10.51) was used to approximate the measured pressure gradient $\Delta p(\theta)$:

$$\Delta\hat{p}(\theta) = ab^{\theta} + c \qquad (10.51)$$

   The parameters a, b, and c can be determined using a nonlinear least-square regression technique applied to the measured pressure data. The polytropic exponent can be calculated using Eq. (10.52):

$$n(\theta) = \frac{\log\left(1 + \frac{\Delta\hat{p}(\theta)}{p(\theta - \Delta\theta)}\right)}{\log\left(\frac{V(\theta - \Delta\theta)}{V(\theta + \Delta\theta)}\right)} \qquad (10.52)$$

   The plots of polytropic (n) and adiabatic (γ) indices calculated by using Eqs. (10.51) and (10.52) along the compression stroke are shown in Fig. 10.14. The intersection between the two curves (n and γ) corresponds to the thermal flux inversion angle. In some cases, more than one intersection between n and γ curves may occur, and it is reasonable to consider the first intersection as the one that identifies the crank angle at which the inversion of the heat flux takes place [28].

   Figure 10.15 depicts the inversion crank position and corresponding thermal state as a function of brake mean effective pressure (BMEP) for a naturally aspirated

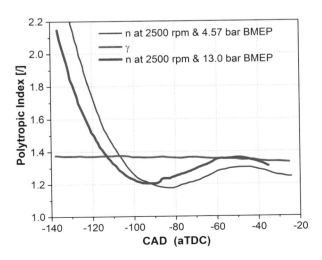

Fig. 10.14 Typical estimated polytropic index vs. crank angle (adapted from [28])

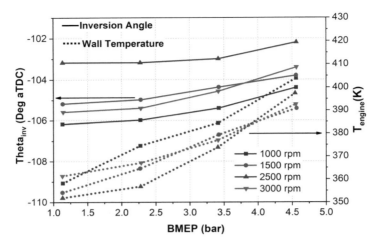

Fig. 10.15 Variation of estimated inversion angle and calculate wall temperature as a function of BMEP (adapted from [28])

engine. The figure illustrates that the estimated inversion angles show a slight variation toward the TDC as the load is increased. This results in a sensible increase of the engine temperature (Fig. 10.15). A more complete discussion on this method can be found in the original study [29, 30].

10.4.2 Observer Model-Based Method

The cylinder wall temperature can be estimated on cycle-to-cycle based on dynamic modeling using measured in-cylinder pressure [36]. In this method, the heat release model is used between IVC and start of injection (SOI). During the compression process, the heat release due to combustion is zero. Thus, net heat release can be written as a function of heat transfer, which is represented by Eq. (10.53):

$$\dot{Q}_n = \frac{\gamma}{\gamma - 1} P\frac{dV}{d\theta} + \frac{1}{\gamma - 1} V\frac{dP}{d\theta} = -\dot{Q}_{ht} = -hA(T - T_w) \tag{10.53}$$

The heat transfer coefficient can be calculated using Eq. (10.4). Equation (10.53) can be used to determine the wall temperature model as shown in Eq. (10.54):

$$\frac{dP}{d\theta} = -\frac{\gamma}{V} P\frac{dV}{d\theta} - \frac{hA(\gamma - 1)}{6NV} T + \frac{hA(\gamma - 1)}{6NV} T_w \tag{10.54}$$

The cylinder wall temperature estimation is transformed into a disturbance observer problem, which can be represented as Eq. (10.55):

$$\frac{dP}{d\theta} = \psi(\theta)P + d(\theta) \tag{10.55}$$

where

$$\psi(\theta) = -\frac{\gamma}{V}\frac{dV}{d\theta} - \frac{hA(\gamma - 1)}{6NmR} \tag{10.56}$$

$$d(\theta) = \frac{hA(\gamma - 1)}{6NV}T_{w} \tag{10.57}$$

The detailed procedure for observer-based solution methodology is provided in the study [36]. The disturbance $d(\theta)$ in Eq. (10.57) can be estimated using the proposed observer. Thus, the wall temperature can be determined. The wall temperature estimation affects the heat transfer calculation during heat release analysis. The heat transfer coefficient (including wall temperature), TDC phasing, and compression ratio can also be simultaneously tuned using motoring heat release curve [37].

10.5 Trapped Mass Estimation

Increasing stringent emission legislation governs the future engine control strategies to not only calculate the air mass inducted in the cylinder but also determines the in-cylinder charge composition at IVC position on a cycle-to-cycle and cylinder-by-cylinder basis. Additionally, the charge contained in the cylinder must be controlled to be able to guarantee a given engine load. This can be achieved using traditional methods (closed loop using the throttle and air flowmeter on the intake circuit), but the response delay of the flowmeter and intake piping dynamics doesn't allow for accurate feedback during transients [38]. To overcome these demerits, in-cylinder pressure can be used to determine the fresh charge. The cylinder pressure before the start of combustion is strongly correlated to the in-cylinder mass on a cycle-to-cycle basis. The fresh air charge is determined by subtracting the residual gas mass from the total mass in the cylinder. The residual gas fraction can be estimated using different methods (Sect. 10.6). A study summarizes the air charge estimation method in a spark ignition engine with a focus on experimental methods using mass flow estimation and speed density methods [39]. The trapped mass can be estimated by in-cylinder pressure measurement by several methods, discussed in the next subsections.

10.5.1 Fitting Cylinder Pressure Curve During Compression

In this method, the fresh charge inducted into the cylinder during the intake stroke is estimated using measured cylinder pressure data. The air charge is calculated by determining the total in-cylinder mass and subtracting the residual gas. The total in-cylinder mass is calculated by fitting cylinder pressure (to a polytropic process) considering a polytropic transformation during the compression stroke. The residual gas fraction is estimated using a model fitted on experiments. Mass estimation algorithm consists of fitting the cylinder pressure when the valves are closed (after IVC) and before the start of combustion [38]. The gases are considered ideal, and the in-cylinder mass is considered constant as long as the valves are closed (blowby is neglected). Cylinder pressure and manifold pressure are measured. Cylinder volume is computed from the engine geometry. The residual gas fraction is calculated using the correlation equation (10.58), which is derived using in-cylinder sampling experiments on a single-cylinder engine:

$$\chi_{\text{res}} = \frac{2.095}{N} \frac{\Gamma - 1}{\Gamma} \left(\frac{p_{\text{man}}}{p_{\text{exh}}}\right)^{0.285} \sqrt{|p_{\text{exh}} - p_{\text{man}}|} + 0.436 \frac{\phi^{0.295}}{\Gamma} \left(\frac{p_{\text{man}}}{p_{\text{exh}}}\right)^{0.719} \tag{10.58}$$

$$\Gamma = \frac{V_{\text{disp}} + V_{\text{clearance}}}{V_{\text{IVO}}(\text{OF})} \tag{10.59}$$

where OF is valve overlap factor and V_{IVO} is the volume at intake valve opening. The valve overlap factor is defined by Eq. (10.60) [38, 40]:

$$\text{OF} = \frac{(D_i A_i + D_e A_e)}{V_d} \tag{10.60}$$

where D_i and D_e are the inner seat diameters of the intake and exhaust valves and V_d is the displacement volume of the engine. The quantities, A_i and A_e, which have dimensions of length times crank angle degree, are areas under the valve-lift/crank angle curves defined by Eq. (10.61):

$$A_i = \int_{\text{IVO}}^{\text{IV=EV}} L_i d\theta \quad \text{and} \quad A_e = \int_{\text{IV=EV}}^{\text{EVC}} L_e d\theta \tag{10.61}$$

where L_i and L_e are the intake and exhaust valve lifts; IVO and EVC denote intake valve opening and exhaust valve closing crank angles, respectively; and IV = EV denotes the crank angle when the exhaust lift equals the inlet lift. The value OF has the dimension of crank angle per unit length.

The estimated cylinder pressure is fitted assuming a polytropic transformation as Eq. (10.62):

$$\hat{p}_{\text{cyl}} V_{\text{cyl}}^k = C \tag{10.62}$$

The polytropic exponent k and the variable C are supposed to be constant. While computing this equation in the logarithmic form, both values can be computed using a least-square method.

The estimated cylinder pressure is extended from the beginning of the computation zone to BDC position. The in-cylinder mass is computed using the ideal gas law by Eq. (10.63):

$$\hat{m}_{\text{cyl}} = \frac{CV^{1-k}(\text{IVC})}{r_{\text{cyl}}T(\text{IVC})} \tag{10.63}$$

The ideal gas constant of the fresh/residual gas mixture is computed from residual gas fraction and the ideal gas constant of fresh and residual gases (Eq. 10.64) ($r_{\text{man}} = 287$ for fresh charge and $r_{\text{exh}} = 291$ for burned gases). This parameter can be considered constant because the points considered for calculation are before the start of combustion.

$$r_{\text{cyl}} = (1 - \chi_{\text{res}})r_{\text{man}} + \chi_{\text{res}}r_{\text{exh}} \tag{10.64}$$

The in-cylinder temperature is computed from the first law of thermodynamics considering identical specific heat capacities.

$$T(\text{IVC}) = (1 - \chi_{\text{res}})T_{\text{man}} + \chi_{\text{res}}T_{\text{exh}} + \Delta T_{\text{ht}}(\text{OF}) \tag{10.65}$$

The term ΔT_{ht} corresponds to the fresh air warming up from the runners, which can be calibrated from experiments. This method presents the advantage that it doesn't need the heat transfer to the walls to be explicitly computed [38].

The estimated admitted gas mass (fresh charge) at cycle "n" can be calculated using Eq. (10.66) assuming the residual gas fraction is known (Eq. 10.58):

$$\hat{m}_{\text{air}}(n) = \hat{m}_{\text{cly}}(n)(1 - \chi_{\text{res}}(n)) \tag{10.66}$$

This method of trapped mass calculation is simple, fast to compute, and deterministic, does not need a lot of calibration (it only needs the cylinder pressure measurements), and is robust to measurement noise [38].

Another study proposed two observers, based on a physical approach, which determine the in-cylinder mass and composition using the measured cylinder pressure [41]. In this method m_{cyl} is calculated by iteratively reducing the error ε in Eq. (10.67):

$$\varepsilon(i) = \sum_{\alpha=1}^{n\alpha} \frac{\hat{p}_{\text{cyl}}(\alpha) - p_{\text{cyl}}(\alpha)}{n_\alpha} \tag{10.67}$$

$$\hat{p}_{\text{cyl}}(i, \alpha) = p_{\text{cyl}}(\alpha)\left(\frac{\hat{T}_{\text{cyl}}(i, \alpha)}{\hat{p}_{\text{ref}}(i)}\right)^{\frac{k(\alpha)}{k(\alpha)-1}} \tag{10.68}$$

$$\hat{T}_{cyl} = (i, \alpha) = \frac{p_{cyl}(\alpha) V_{cyl}(\alpha)}{\hat{m}_{cyl}(i) R} \tag{10.69}$$

where α is the crank angles between two points during the compression stroke and "i" is an iteration index. This method needs the reference temperature T_{ref} that is computed using the temperatures in the intake and exhaust manifolds, cylinder pressure, heat transfer, and a residual gas fraction estimate (determined using another model). Using a proportional feedback of the estimation error, the in-cylinder mass trapped (10.69) is calculated.

$$\hat{m}_{cyl}(i + 1) = \hat{m}_{cyl}(i) + k_p \cdot \varepsilon(i) \tag{10.69}$$

The computation is performed until the error is smaller than a predetermined threshold. A more complete detail can be found in the original study [41].

10.5.2 The ΔP Method

The ΔP (Delta P) method was first presented in the study [42] and has since been used with various modification [43–48]. This method is a fast and easy method for computing the trapped air mass on a cycle-to-cycle basis. This method starts by assuming the contents of the cylinder as an ideal gas, and mass is conserved during an adiabatic compression process. Two ideal gas equations can be formed (Eqs. 10.70 and 10.71) at two positions on pressure curve (P_1 and P_2) where the point $P1$ should be after IVC position and $P2$ should be before the start of combustion. Considering a polytropic compression process, it is possible to substitute the ratio of volumes for the ratio of temperatures. The mass of the total contents of the cylinder (i.e., fresh air, fuel, residual gas, water vapor, and EGR if any) can be written in a single equation (10.72) after some algebraic simplification:

$$P_1 = \frac{M \cdot R \cdot T_1}{V_1} \tag{10.70}$$

$$P_2 = \frac{M \cdot R \cdot T_2}{V_2} \tag{10.71}$$

$$M_{tot} = \frac{V_1 \cdot V_2^{\gamma} \cdot \Delta P}{R \cdot T_1 \cdot (V_1^{\gamma} - V_2^{\gamma})} \tag{10.72}$$

where γ is the ratio of specific heats (polytropic exponent). Polytropic coefficient γ is strongly dependent on the operating conditions, but it can be treated as constant in this approach. The polytropic exponent "γ" and the temperature at the starting point T_1 have to be known or estimated.

A study developed a black box model by expressing the in-cylinder temperature T_1 (reference temperature) as a function of the available engine variables (engine speed (N) and the injected fuel mass (m_f)) [45]. The model equation obtained using the most effective variables (N and m_f) is shown in Eq. (10.73):

$$T_1 = c_0 + c_1 N^2 + c_1 m_f^2 + c_1 \frac{1}{N^3} + c_4 m_f^3 N \qquad (10.73)$$

The fresh air mass can be calculated by subtracting the mass of fuel and the mass of residual gas (Eq. 10.74):

$$M_{fresh} = \frac{V_1 \cdot V_2^\gamma \cdot \Delta P}{R \cdot T_1 \cdot \left(V_1^\gamma - V_2^\gamma\right)} - M_{fuel} - M_{res} \qquad (10.74)$$

Equation (10.74) can be simplified by defining two variables: α represents everything that is multiplied by the ΔP term, while β represents the combined fuel and residual gas mass as shown in Eq. (10.75). The terms α and β can be determined experimentally during steady-state operation:

$$M_{fresh} = \alpha \cdot \Delta P - \beta \qquad (10.75)$$

$$\alpha = \frac{V_1 \cdot V_2^\gamma}{R \cdot T_1 \cdot \left(V_1^\gamma - V_2^\gamma\right)} \qquad (10.76)$$

$$\beta = M_{res} + M_{fuel} \qquad (10.77)$$

The relationship between mass and ΔP can be defined experimentally by plotting steady-state airflow-derived trapped air mass against the change in cylinder pressure between two points on the compression stroke. Once α and β values have been estimated, they can be used to calculate the fresh mass from any ΔP, even if the measured ΔP is during a transient maneuver [43].

The ΔP method is very sensitive to changes in engine speed and residual gas fraction; therefore the anticipated range of both of these parameters must be comprehended when the correlation between steady-state airflow and cylinder pressure is established. This method, real-time, provides good results but is sensitive to signal noise. Moreover, it needs a lot of calibration because a matrix of coefficients at steady states is required.

10.5.3 Mass Fraction Burned Method

In this method, the mass fraction burned (MFB) is used to compute the energy generated during the combustion and compare it to a cycle without combustion [49, 50]. An iterative method for determination of total mass, air charge, and residual gas fraction based on conditions at IVC and MFB50 is proposed in the study [50], which is described as follows. This method starts with an estimate of the thermodynamic states in the cylinder at intake valve closes (IVC). These states and the cylinder pressure trace are used for calculation of the burn rate and determine the crank angle for 50% MFB. The cylinder charge is a mixture of fuel, air, and burned gas. Thus, the total mass in the cylinder can be expressed by Eq. (10.78):

$$m_{Tot} = m_{fuel} + m_{air} + m_{RG} \qquad (10.78)$$

It is well established that the SI engines within normal operating conditions are running most efficiently if 50% of the fresh charge is burned at around 8° after TDC. Thus, for estimating common properties of SI engines, the 50% mass burned position is a good starting point for further investigations [50].

The total mass (m_{tot}) at 50% MFB position can be determined using ideal gas law (Eq. 10.79) where temperature (T_{50}) at 50% MFB position. When T_{50} is assumed to be 1850 K as there as only ±6% variation in mean gas temperature at 50% MFB:

$$m_{tot} = \frac{P_{50} \cdot V_{50}}{R \cdot T_{50}} \qquad (10.79)$$

The residual gas fraction (RGF) can be estimated by Eq. (10.80) by using the temperature at IVC and temperature of fresh charge (FC) and residual gas (RG):

$$X_{RG} = \frac{c_{VFC} \cdot (T_{IVC} - T_{FC})}{c_{VRG} \cdot (T_{RG} - T_{IVC}) + c_{VFC} \cdot (T_{IVC} - T_{FC})} \qquad (10.80)$$

The temperature of the mixture at IVC is determined by Eq. (10.81):

$$T_{IVC} = \frac{P_{IVC} \cdot V_{IVC}}{R \cdot m_{tot}} \qquad (10.81)$$

The temperature (T_{RG}) of residual gases is estimated by the correlation presented by Eq. (10.82):

$$T_{RG} = -(579.5 \cdot (m_{tot} \cdot N))^{-0.157} + 1963 \qquad (10.82)$$

The temperature (T_{FC}) of fresh charge from temperature (T_{im}) and pressure (P_{im}) at intake manifold by assuming polytropic compression of the air-fuel mixture is given by Eq. (10.83):

$$T_{FC} = T_{im} \cdot \left(\frac{P_{im}}{P_{IVC}}\right)^{\left(\frac{1-n}{n}\right)} \qquad (10.83)$$

The polytropic index (n) has been taken as 1.32. The specific heat capacity for residual gas is determined by Eq. (10.84):

$$c_{VRG} = 0.128 \cdot T_{RG} + 698 \qquad (10.84)$$

The specific heat capacity of fresh charge is considered a constant value ($C_{vFC} = 736$ J/KgK).

By using the above estimates, the temperature of the mixture at 50% MFB can again be calculated by Eq. (10.85):

$$T_{50} = \frac{\int_{IVC}^{V_{50}} PdV}{m_{tot} \cdot c_{V50}} + \frac{c_{VIVC} \cdot T_{IVC}}{c_{V50}}$$
$$+ \frac{\left(1 - q_{cooling}\right) \cdot 0.5 \cdot X_c \cdot (1 - X_{RG}) \cdot \frac{1}{1+L_{st}} \cdot H_u}{c_{V50}} \qquad (10.85)$$

where L_{st} is stoichiometric air requirement, H_u is lower heating value of fuel, X_c is the completeness of combustion (combustion efficiency), X_{RG} is the residual gas fraction, and $q_{cooling}$ is a factor which accounts to the heat transfer relative to total combustion energy.

The specific heat capacities at 50% mass burned and at IVC can be expressed by the following equations:

$$c_{V50} = (1 - X_{RG}) \cdot 0.0591 \cdot T_{50} + 719.4 + X_{RG} \cdot 0.2361 \cdot T_{50} + 627.5 \quad (10.86)$$

$$c_{VIVC} = (1 - X_{RG}) \cdot c_{VFC} + X_{RG} \cdot c_{VRG} \qquad (10.87)$$

$$q_{cooling} = 1.1137 \times 10^{-4} \cdot T_{50} - 0.229 \qquad (10.88)$$

The total mass at 50% MFB is again calculated using the updated value of T_{50}. These steps from Eq. (10.79) to Eq. (10.88) are iteratively repeated until there is no significant change in RGF or total mass. Thus, the total mass, as well as residual gas fraction, can be estimated by this method using cylinder pressure measurement.

This residual gas determination model is modified to account for external EGR and VVT system, and results are validated on a six-cylinder engine [51]. Figure 10.16 shows the variation of residual gas fraction (determined using modified model) as a function of the IVO advance and EVO delay at 2000 rpm and 600 mbar inlet pressure operation. The figure shows that the IVO-EVO combination that guarantees the minimum RGF is not the one characterized by the minimum overlap, but if the overlap is very high (as in the case of IVO advanced 40° and EVO delayed 40°), the amount of residual gas fraction increases [51].

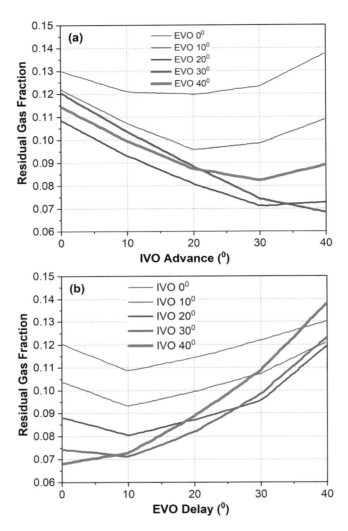

Fig. 10.16 RGF variation as a function of the (**a**) IVO advance and (**b**) EVO delay at 2000 rpm and 600 mbar inlet manifold pressure [51]

10.5.4   *Physical Model-Based Method*

In this method, the air mass is calculated with a physically based model using two points of cylinder pressure in the compression stroke [47]. The in-cylinder temperature is adapted with an adaptive Kalman filter based on airflow rate measurement. The algorithm used in the method determines the air mass in the cylinder of an SI engine in order to replace the conventional load sensor, a hot film air mass meter. The advantages of estimating the air mass by using a combustion pressure sensor are obvious: (1) determination of the air mass of each individual cylinder, (2) possible faults due to manifold pulsations and air leaks are eliminated, and (3) high redundancy for security concepts (depending on the number of sensors) [47]. Computation of the air mass from a cylinder pressure data can be done in different phases of the pressure curve. The combustion phase is typically not preferred due to the cyclic variations. The compression stroke is preferred because, in order to keep the prediction interval for the injection signal small, it is very important to do the calculation as soon as possible. The signal-to-noise ratio is not very high in the compression stroke. However the short-term drift is almost constant in this phase. The compression phase can be described by a polytropic state change, and the air mass can be calculated with Eq. (10.89):

$$m_{\text{air}} = \frac{p_2 V_2}{R \cdot T_{\text{AF}}} \cdot \left(\frac{V_2}{V_1}\right)^{n-1} - \frac{T_{\text{RG}}}{T_{\text{AF}}} \cdot m_{\text{RG}} - m_{\text{fuel}} \qquad (10.89)$$

where T_{AF} is the temperature of air mixed with fuel and T_{RG} is the residual gas temperature. Equation 10.89 consists of three parts: (1) the pressure proportional part, (2) the influence of the residual gas m_{RG}, and (3) the influence of the vaporized fuel m_{fuel}. The fuel mass in the cylinder can be determined by using a wall-wetting model [52]. The prediction of the residual gas fraction in the cylinder is done by the method presented in another study [40]. The temperature T_{AF} is estimated using adaptive Kalman filtering. The T_{AF} is adapted because it changes with operating conditions, influencing the first two parts of Eq. (10.89), and affects the sensitivity of the air mass calculation. The more details of adaptive Kalman filter algorithm based on the maximum likelihood method for estimating the air mass in the cylinder by using a combustion pressure signal can be found in the original study [47].

Another study developed an air charge estimation method using the full engine cycle in-cylinder pressure data for a SI engine equipped with intake variable valve timing [53]. Cylinder pressure data at specific cycle events is used in thermodynamics and heat transfer relationships, in an algorithm to estimate fresh air and residual gas mass in each cylinder and each cycle.

10.5.5 Resonance Frequency Analysis-Based Method

In this method of trapped mass determination, the resonance frequency in the measured cylinder pressure signal, excited by the combustion event, has been used [54–57]. Mostly the low-frequency pressure trace is used by mass determination algorithms. However, the cylinder pressure data can show significant contents over a broad band of the spectrum. The fast combustion event excites the resonance frequencies of the combustion chamber, which generates the cylinder pressure oscillations [54]. In-cylinder pressure resonance is a well-known phenomenon, extensively studied along the last decades in order to detect knock and to reduce combustion noise [55]. Pressure resonance consists of high-frequency (few kHz) oscillations produced by pressure waves in the combustion chamber. To estimate the resonance frequencies, Draper approach [58] is used by a cylindrical combustion chamber assumption. Draper solved the wave equation with Bessel functions for a cylindrical geometry, and the resonance frequency is determined using Eq. (10.90). This equation relates the resonance frequency ($f_{i,j}$) with the cylinder diameter (D), the speed of sound (a), and a Bessel constant ($B_{i,j}$). The axial modes can be neglected because the frequencies associated with h (height of the combustion chamber) are too high. The first circumferential and radial modes ($B_{i,j}$) are presented in Fig. 10.17.

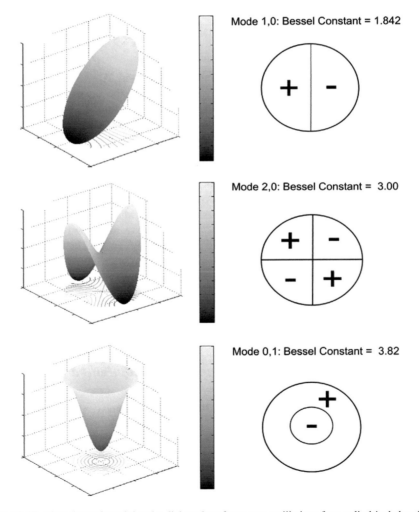

Fig. 10.17   First circumferential and radial modes of pressure oscillations for a cylindrical chamber [56]

Fig. 10.18 Theoretical
Bessel constants (black line)
over a spectral distribution
of the pressure signal [54]

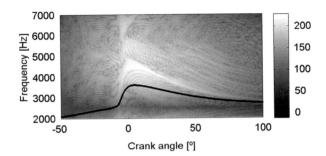

$$f_{i,j} = a\sqrt{\frac{B_{i,j}^{2}}{(\pi D)^{2}} + \frac{g^{2}}{(2h)^{2}}} = \frac{aB_{i,j}}{\pi D} \qquad (10.90)$$

Draper's equation is typically used to determine a reference frequency to record
(knock detection) or to eliminate the resonance (combustion diagnosis). The resonance frequency variation is governed by the variations of the cylinder geometry
with the angular rotation of crankshaft. Figure 10.18 depicts the theoretical resonance frequency of the first mode over a spectral distribution of experimental
pressure trace. The figure shows that the experimental resonance frequencies appear
to converge to the cylindrical approach far from the TDC position. However, when
the piston approaches the TDC, the cylindrical assumption in Draper's equation does
not valid due to the shape of combustion chamber at TDC position.

The issue of Draper's equation near TDC was addressed in the study [55], which
suggested a method for experimentally characterizing the cylinder resonance. A
possibility for correcting the piston bowl is by applying pre-calibrated crank angle
varying constants for the bowl (Eq. 10.91). In this method, although a cylindrical
solution is being used, the geometry shifting errors are taken into account by specific
constants ($B_{i,j}$). The main demerit of this method is the calibration of constants,
which need to perform with previous data when the mass must be precisely known
[55]:

$$B'_{i,j} = f(\alpha) = \frac{\pi D f_{i,j}}{a} \qquad (10.91)$$

where "a" is the speed of sound, which can be estimated by Eq. (10.92):

$$a = \sqrt{\gamma R T} = \sqrt{\gamma p V / m} \qquad (10.92)$$

Then, the total trapped mass can be estimated using Eq. (10.93):

$$m_{\text{total}} = \left(\frac{B_{i,j}(\alpha)\sqrt{\gamma P(\alpha) V(\alpha)}}{\pi D f_{i,j}(\alpha)} \right)^{2} \qquad (10.93)$$

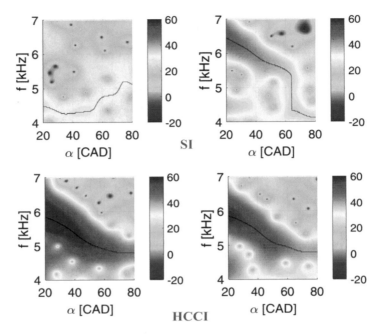

Fig. 10.19 Spectrogram of the in-cylinder pressure for two different cycles, for SI (top) and HCCI (bottom) operation (adapted from [57])

In this method, trapped mass can be estimated by measuring the cylinder pressure and evaluating the resonance frequencies. The level of the frequency excitation is highly dependent on the particular cycle and combustion mode considered [57]. Figure 10.19 shows the spectrogram from two individual cycles in SI (top) and HCCI (bottom) combustion. Power spectral density (PSD) is presented in logarithmic scale (dB/rad/sample). The highest excitation frequencies at a particular angular position have been marked with a black line in the spectrograms. The figure shows that the resonant frequency is sufficiently excited in most of the analyzed cycles, depicting a smooth trajectory from slightly above 6 kHz at 20 CAD to near 5 kHz at 80 CAD.

10.6   Residual Gas Fraction Estimation

The internal residual gases from the previous cycle significantly affect the engine combustion quality. It has been recognized that the requirement of external EGR systems can be eliminated by using variable valve actuation (VVA) to control and maximize internal residuals. The utilization of VVA also leads to significant improvement in fuel economy and NOx control. Internal dilution using residuals plays an important role in low temperature combustion strategies such as HCCI.

Considering these factors, estimation of the residual mass fraction is required for engine development.

Residual generation in the combustion chamber is a complex physical process. Pressure and velocity pulsations are produced in the intake and exhaust manifolds during the gas exchange process because of gas inertia and wave action. The pulsations generated in the manifolds strongly affect the gas flows through the valves, which determine the residual content of the trapped mass in the engine cylinder [59]. Due to the complexity of the process, several experimental methods including optical and gas sampling methods are used for measurement of the residual gas fraction. Several approaches are applied for gas sampling in the published literature. The residual gas fraction can be determined by sampling of hydrocarbon or nitrous oxide either from the exhaust port or directly from the combustion chamber in combination with skip firing [60–62]. Another most widely used method is a direct in-cylinder sampling of CO_2 for residual gas fraction determination [63–65]. All of the gas sampling techniques require elaborate and expensive instrumentation, and may not be feasible for routine engine calibration work, particularly for multicylinder engines by measuring in all cylinders [59].

The published literature contains several models for the determination of residual gas fraction. Direct measurement of the residual gas fraction (RGF) is not practical on the vehicle (onboard), which makes residual gas estimation models more interesting and required. Additionally, the proposed measurement techniques for laboratory tests are still quite complex and particularly expensive (the simplest ones are based on gas composition measurement, to be performed both (1) inside the cylinder, during compression phase, and (2) in the exhaust manifold). If the goal is onboard application of RGF determination, such models can be subdivided into two main categories: (1) the models which can work with inputs taken only from the ECU signals and (2) the models that use also experimental data normally not available onboard the vehicle, such as cylinder pressure of the various cylinders [66]. An iterative method for residual gas fraction determination using cylinder pressure data is already discussed in Sect. 10.5.3.

One of the earliest attempts to measure RGF in a spark ignition engine used a fast-response flame ionization hydrocarbon detector in the vicinity of the spark plug [67]. The results obtained were used to determine mean RGF as well as cycle-to-cycle variations in RGF. Later the model was updated by considering the backflow of exhaust gases during the valve overlap period [40], and also a zero-dimensional model was simulated for calculating RGF [68] and compared with the experimental results. To calculate the mean RGF in spark ignition engine, a model was developed considering trapped mass at IVO, and backflow overlap is presented by Eq. (10.94) [62]. This model was validated against the experimentally measured RGF using fast-response flame ionization detector:

$$x_r = (x_r)_{\text{IVO}} + (x_r)_{\text{backflow}} \tag{10.94}$$

$$(x_r)_{IVO} = \left(1 - \frac{P_{a,\phi}/P_i}{P_{a,\phi=1}/P_i}\left(1 - (x_r)^*_{IVO}\right)\right) \cdot \left(\frac{1/T_{e,\phi}}{1/T_{e,\phi=1}}\right) \qquad (10.95)$$

$$(x_r)^*_{IVO} = \frac{1}{r_c}\left(\frac{P_e}{P_i}\right)^{\frac{1}{\gamma}}\left(1 + \frac{Q^*}{c_v T_i r_c^{(\gamma-1)}}\right)^{-\frac{1}{\gamma}} \qquad (10.96)$$

$$Q^* = \frac{m_f Q_{LHV}}{m} = \frac{Q_{LHV}}{1 + \frac{F}{A}} \qquad (10.97)$$

$$(x_r)_{backflow} = \int_{IVO}^{EVC} \frac{m}{\rho_a \times V_d\left(\frac{r_c+1}{r_c}\right)}\frac{1}{6N}dt \qquad (10.98)$$

where P_e, P_i, and T_i are exhaust pressure and intake pressure and temperature, respectively, $P_{a,\phi=1}$ and $T_{e,\phi=1}$ are inlet pressure and exhaust temperature at equivalence ratio equal to 1, r_c is compression ratio, c_v is specific heat at constant volume, m_f is the mass of fuel injected, m is in-cylinder mixture mass, and ρ_a is air density at intake temperature.

Another empirical model typically used is presented in Eq. (10.99), which is illustrated in study [40]:

$$\chi_{res} = a_1 \frac{OF}{N}\left(\frac{P_i}{P_e}\right)^{a_2}\sqrt{|P_e - P_i|} + a_3 \frac{\phi^{a_4}}{r_c}\left(\frac{P_i}{P_e}\right)^{a_5} \qquad (10.99)$$

where χ_{res} is RGF; N is engine speed; P_i and P_e are inlet and exhaust pressure, respectively; r_c is compression/ratio; OF is valve overlap factor (Eq. 10.60); and ϕ is equivalence ratio. The study [40] estimated the empirical constants as $a_1 = 1.266$, $a_2 = -0.87$, $a_3 = 0.632$, $a_4 = 1$, and $a_5 = -0.74$. Another study [60] showed that these constants underperformed at low intake pressure and high fuel-air ratio in comparison to experimental data. Thus, improved empirical constants are proposed as $a_1 = 15.1849$, $a_2 = -0.7106$, $a_3 = 0.0381$, $a_4 = -8.4357$, and $a_5 = 0.155$.

Determination of the residual gas fraction on cycle-to-cycle basis is required for online engine management. The cylinder pressure-based RGF estimation methods are proposed in several studies [47, 50–53, 69–72]. The RGF estimation method based on cylinder pressure measurement is discussed in the next subsections.

10.6.1　Pressure Resonance Analysis Method

The Sect. 10.5.5 describes the estimation of trapped mass in the cylinder using the resonance frequency analysis based on measured in-cylinder pressure data. This method is well established in the previous studies [54–57]. After the estimation of trapped mass (m_{total}) in the cylinder using Eq. (10.93), the residual mass (m_{res}) can be determined using Eq. (10.100) [57]:

$$m_{\text{res}} = m_{\text{total}} \frac{V_{\text{EVC}}}{V_{\text{EVO}}} \left(\frac{P_{\text{EVC}}}{P_{\text{EVO}}} \right)^{\frac{1}{\gamma}} \qquad (10.100)$$

The pressure (P) and volume (V) at exhaust valve closing (EVC) and exhaust valve opening (EVO) can be determined from the measured pressure signal and calculated volume from engine geometry. The fresh air can be estimated by subtracting the residual mass (Eq. 10.100) from the total mass estimate (Eq. 10.93).

10.6.2   Iterative and Adaptive Methods

Section 10.5.3 discussed an iterative method for residual gas fraction estimation using a cylinder pressure signal in a spark ignition engine. In advanced low temperature combustion modes such as HCCI, the negative valve overlap (NVO) strategy is used for extending the lower engine load limit [37]. The NVO retains large residual fractions through early closing of the exhaust valve (s) and late opening of the intake valve (s). Additionally, recompression provides an opportunity to control charge temperature and reactivity by reacting and reforming fuel injected during NVO. The NVO fueling can both raise temperatures and alter the chemical composition of the charge before main compression. Accurately determining trapped mass can be particularly challenging for NVO operation since the amount of retained residuals is often on the same order as the inducted mass. Computing charge temperatures at IVC position for NVO operation is a complex process due to an unknown fraction of residual gases at an unknown temperature. To address this issue, a model is proposed for determination of cycle temperature and RGF by modeling the blowdown and recompression during exhaust valve opening and closing events, which makes it possible to estimate the in-cylinder charge temperatures based on exhaust port measurements [70].

In this method [70], in-cylinder temperatures at EVO and EVC are estimated based on exhaust gas temperatures measured in the exhaust port. In-cylinder mass is related to cylinder volume, pressure, and temperature using the ideal gas equation of state ($PV = mRT$). For the whole engine cycle, using measured pressures, calculated volumes, and estimated gas constants, the mass can be determined, and only cycle temperature is unknown. In this method, it is assumed that the same mass of air and fuel that enters the cylinder during intake exits during exhaust. Thus, for NVO operation during which the exhaust and intake valve events do not overlap, the difference in trapped mass between closed portions of the cycle can be written as Eq. (10.101):

$$m_1 - m_2 = m_{\text{EVO}} - m_{\text{EVC}} = \frac{P_{\text{EVO}} V_{\text{EVO}}}{R_{\text{EVO}} T_{\text{EVO}}} - \frac{P_{\text{EVC}} V_{\text{EVC}}}{R_{\text{EVC}} T_{\text{EVC}}} = m_{\text{air+fuel, in}} \qquad (10.101)$$

where m_1 is the trapped mass during main compression and expansion, m_2 is that during the NVO period, and m_{ex} is the mass exhausted from the cylinder each cycle. The T_{EVO} represents cylinder contents just before the EVO, and T_{EVC} represents contents just after EVC. The intake air and fuel masses ($m_{air\,+\,fuel,\,in}$) are measured. The residual gas fraction can be computed as the ratio of m_2 and m_1 ($RGF = m_2/m_1$). Knowing pressures, volumes, gas constants, inducted mass, and a single closed-system temperature is sufficient to compute RGF and temperatures throughout the closed portions of the cycle.

The relation between T_{EVO} and T_{EVC} with measured exhaust temperature (T_{exh}) is shown in Fig. 10.20. Considering ideal gas undergoing a reversible process during blowdown and recompression process, the heat loss is estimated using Eqs. (10.102) and (10.103):

$$q_{blowdown} = \int_{EVO}^{cyl,1\,atm} c_p dT - R \int_{EVO}^{cyl,1\,atm} \frac{T}{P} dP \qquad (10.102)$$

$$q_{recompression} = \int_{cyl,1\,atm}^{EVC} c_p dT - R \int_{cyl,1\,atm}^{EVC} \frac{T}{P} dP \qquad (10.103)$$

The ratio of heat loss per unit mass for blowdown and recompression is defined in Eq. (10.104):

$$C_{BD/RC} \equiv \frac{q_{blowdown}}{q_{recompression}} \qquad (10.104)$$

By assuming constant specific heats, the first integral in Eqs. (10.102) and (10.103) can be expressed in terms of temperature differences at EVO and EVC.

Fig. 10.20 Illustration of cycle temperature during blowdown and recompression process (adapted from [70])

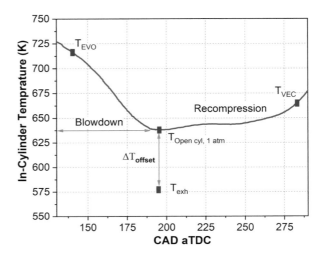

Average values of temperature are used to evaluate the remaining integrals. Thus, temperatures can be expressed as Eq. (10.105):

$$c_p \left(T_{cyl,1\,atm} - T_{EVO} \right) - R \frac{\left(T_{cyl,1\,atm} + T_{EVO} \right)}{2} \ln \left(\frac{1}{P_{EVO}} \right)$$

$$= C_{BD/RC} \left[c_p \left(T_{EVC} - T_{cyl,1\,atm} \right) - R \frac{\left(T_{cyl,1\,atm} + T_{EVO} \right)}{2} \ln \left(P_{EVC} \right) \right] \qquad (10.105)$$

Equations (10.105) and (10.101) are used for calculating the in-cylinder temperatures during the closed part of the cycle using the reference in-cylinder temperature from the exhaust period.

The $T_{cyl,1atm}$ temperature in Eq. (10.105) needs to be estimated from the measured exhaust port temperature (T_{exh}). A simple temperature offset ($\Delta T_{cyl/ex}$) is used to relate the temperatures as shown in Eq. (10.106). The study found that a value of $\Delta T_{cyl/ex} = 65$ K performs well for predicting cycle temperatures over this range-predicted temperatures [70].

$$T_{cyl,1\,atm} = T_{ex} + \Delta T_{cyl/ex} \qquad (10.106)$$

The $C_{BD/RC}$ is estimated by an iterative process in which an initial guess is made. The in-cylinder gas temperatures (T_{cyl}) throughout blowdown and recompression are computed using the polytropic coefficients. The heat transfer ratio $C_{BD/RC}$ is calculated from the convective heat transfer equation (10.107):

$$C_{BD/RC} \equiv \frac{\int_{blowdown} h_c A \left(T_{cyl} - T_w \right) dt}{\int_{recompression} h_c A \left(T_{cyl} - T_w \right) dt} \qquad (10.107)$$

where h_c is the convective heat transfer coefficient, A is the instantaneous cylinder surface area, and T_w is the temperature of the surrounding walls. Once $C_{BD/RC}$ is recomputed, the procedure is repeated to obtain convergence of $C_{BD/RC}$ for each test. The study found that the value of $C_{BD/RC}$ is between 1 and 1.2 for all the test conditions. A more complete detail regarding this method can be found in the original study [70].

An adaptive method of residual mass determination based on combustion pressure is developed [71, 72]. Figure 10.21 depicts the block diagram structure of the adaptive model for residual mass estimation. Using Eqs. (10.102) to (10.105), the residual mass equation can be written more compactly by grouping terms and lumping constant coefficients by Eq. (10.108) [71, 72]:

$$m_{res}(k+1) = \frac{\alpha(k) + \beta(k) m_{res}(k)}{A(k) + m_{res}(k)} \qquad (10.108)$$

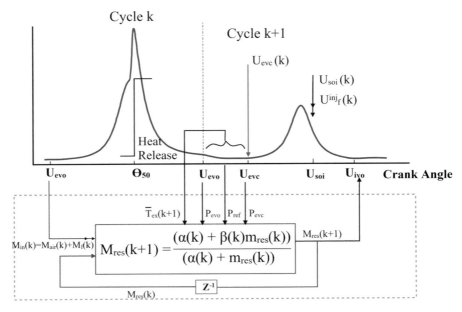

Fig. 10.21 A block diagram representation of an adaptive residual mass estimation model (adapted from [71], Courtesy of J.W. Larimore)

where

$$\alpha(k) = \left(\frac{-2r_{ex}c_p + r_{ex}R\ln\frac{P_{evc}}{P_{ref}}}{-2c_p(1 + r_{ex}) + R\left(\ln\frac{P_{ref}}{P_{evo}} - r_{ex}\ln\frac{P_{evo}}{P_{ref}}\right)}\right)\left(\frac{P_{evc}V_{evc}}{RT_{ex}}\right)m_{in}(k) \quad (10.109)$$

$$\beta(k) = \left(\frac{-2r_{ex}c_p + r_{ex}R\ln\frac{P_{evc}}{P_{ref}}}{-2c_p(1 + r_{ex}) + R\left(\ln\frac{P_{ref}}{P_{evo}} - r_{ex}\ln\frac{P_{evo}}{P_{ref}}\right)}\right)\left(\frac{P_{evc}V_{evc}}{RT_{ex}}\right) \quad (10.110)$$

$$A(k) = \left(\frac{2c_p + R\ln\frac{P_{ref}}{P_{evo}}}{-2c_p(1 + r_{ex}) + R\left(\ln\frac{P_{ref}}{P_{evo}} - r_{ex}\ln\frac{P_{evo}}{P_{ref}}\right)}\right)\left(\frac{P_{evc}V_{evc}}{RT_{ex}}\right) + m_{in}(k) \quad (10.111)$$

The values of P_x, T_x, and V_x are all from the $k + 1$ cycle. Eq. (10.108) predicts the amount of residual mass in cycle $k + 1$ based on previous measured data and the value of the residual mass on the previous cycle. Therefore, the only unknown is the initial guess of $m_{res}(0)$. It was shown that regardless of the initial guess, the difference equation will converge relatively quickly to a stable equilibrium. A more complete detail of the model can be found in [71].

A study estimated the residual gas mass and trapped air mass by thermodynamic analysis of full engine cycle [53]. To calculate the residual gas mass, equations similar to (10.94) and (10.99) are used. However, the temperature at different points is calculated using thermodynamic analysis and model of exhaust blowdown process.

10.7   Air-Fuel Ratio Estimation

Automotive engines and control systems are becoming more and more sophisticated due to increasingly restrictive environmental regulations. The emission legislation is the principal issue for vehicle manufacturers, together with the highly competitive market. Thus, complex engine architectures and configurations have been developed to face with these requirements, involving hard manufacturing and expensive hardware as well. The fast determination of fuel-air ratio over a wide range of engine operating conditions is desirable for better transient fuel control [73]. Air-fuel ratio (AFR) estimation by analyzing the constituents of the exhaust gas using oxygen sensor is used from the last few decades. However, one major demerit in calculating AFR by oxygen sensor is that it only measures mean AFR due to its time response and gas transport delay. Additionally, the oxygen sensor needs fully warm up engine conditions for its effective operation. This condition leads to an open-loop control of the engine during cold start or cold operation. Furthermore, the bandwidth limitation oxygen sensor will result in AFR excursions during fast transients. To overcome the limitations of the oxygen sensor, the estimation methods which can calculate cyclic AFR for all the states of the engine need to be developed and used. The in-cylinder pressure signal offers the opportunity to estimate with a high dynamic response almost all the variables of interest for an effective engine combustion control even in case of nonconventional combustion processes (e.g., PCCI, HCCI, LTC) [74]. The cylinder pressure signal is used for calculation of cylinder air-fuel ratio in several studies [73–79], which are discussed in the next subsections.

10.7.1   Statistical Moment Method

In-cylinder pressure can be used to determine most of the engine parameters with the adequate mathematical model. The AFR can be computed based on the shape of a cylinder pressure curve using the statistical moments of the curve. For a random variable distribution X or in-cylinder pressure data, an infinite number of moments will be able to characterize the entire shape of the distribution. The first four moments have been found enough to determine the shape of the in-cylinder pressure trace distribution due to the complexity and sensitivity of the model. The shape of the in-cylinder pressure trace varies with the variation of AFR at which engine is operated. Therefore, an empirical relation based on the quantified shape of the pressure trace can be used to estimate the cylinder air-fuel ratio. Figure 10.22 illustrates the second (M_2) and third (M_3) moment descriptors to capture the shape

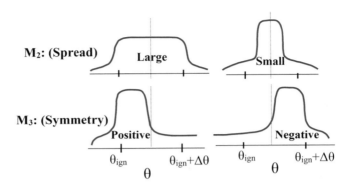

Fig. 10.22 Illustration of cylinder pressure trace shape descriptors M_2 and M_3 (adapted from [77], Courtesy of R.S. Patrick)

of the pressure curve. The pressure traces are exaggerated to emphasize descriptors in Fig. 10.22.

In-cylinder pressure time histories are used to estimate the mean air-fuel ratio [73]. The shape of the pressure curve is defined using pressure moments and other inputs (speed, throttle, EGR). An empirical relationship was developed to relate the air-fuel ratio to pressure moments and other inputs with linear or a nonlinear function. The study [73] used two empirical correlations (Eqs. 10.112 and 10.113) consisting of engine rpm, manifold pressure, and second and third pressure moments in linear and nonlinear form. Equation (10.113) performs better for estimation of equivalence ratio for the cycles used to determine the estimator coefficients. On the other hand, the simpler relation (Eq. 10.112) gives better results for other cycles, which are not used to determine the estimator coefficients. This observation is maybe because of the utilization of too many terms in the equation that leads to the model to become highly sensitive which results in higher RMS error:

$$\phi, \phi^* = a_1 rpm + a_2 P_{\text{man}} + a_3 M_2 + a_4 M_3 + a_5 \tag{10.112}$$

$$\phi, \phi^* = a_1 rpm + a_2 P_{\text{man}} + a_3 M_2 + a_4 (M_2)^2 + a_5 M_3 + a_6 (M_3)^2 + a_7 \tag{10.113}$$

$$M_2 = I_2 - 2\bar{\theta} I_1 + \bar{\theta}^2 I_0 \tag{10.114}$$

$$M_3 = I_3 - 3\bar{\theta} I_2 + 3\bar{\theta}^3 I_0 \tag{10.115}$$

$$\bar{\theta} = \frac{I_1}{I_0} \tag{10.116}$$

where a_n are constants to be determined by the least-square method, ϕ is the equivalence ratio, ϕ^* is the equivalence ratio with EGR, and I_n is the nth pressure moment.

$$\phi^* = \phi(1 - \text{EGR}) \qquad (10.117)$$

The nth central moment can be calculated by Eq. (10.118):

$$I_n = \int_{\theta_{\text{initial}}}^{\theta_{\text{final}}} (P(\theta) - P_{\text{bias}}(\theta))(\theta - \theta_{\text{ref}})^n d\theta \qquad (10.118)$$

The AFR can be calculated using a cylinder pressure signal using Eqs. (10.112) to (10.118). Another study [80] developed the similar correlation for estimation AFR in spark ignition engine in both stationary and transient regimes. The empirical correlation developed is shown in Eq. (10.119). The central moments of pressure curves are calculated up to the third order using a crank angle interval $\Delta\theta$ (120°CAD) beginning from the spark advance angle with the centroid of pressure distribution ($<\theta>$) estimated on the same angular window through Eq. (10.122). The moments of pressure curves are computed using Eq. (10.120). The central moments are divided by the area of the pressure cycle obtained under motoring condition (Eq. 10.121):

$$\text{AFR} = K_1(\beta) + K_2(\beta)M_2 + K_3(\beta)M_3 \qquad (10.119)$$

$$M_n = \frac{\int_{\theta_s}^{\theta_s + \Delta\theta} (\theta - \langle\theta\rangle)^n P(\theta)d\theta}{A_u} \qquad (10.120)$$

$$A_u = \int_{\theta_s}^{\theta_s + \Delta\theta} P_u(\theta)d\theta \qquad (10.121)$$

$$\langle\theta\rangle = \frac{\int_{\theta_s}^{\theta_s + \Delta\theta} P_u(\theta)\theta d\theta}{\int_{\theta_s}^{\theta_s + \Delta\theta} P_u(\theta)d\theta} \qquad (10.122)$$

where K_1, K_2, and K_3 are constants which can be calculated by least-square fit, M_n is a normalized central moment, A_u is an area of the pressure cycle under motoring condition, and $\langle\theta\rangle$ is the centroid of pressure distribution. The unfired pressure (P_u) cycle is reconstructed from the measured pressure value at spark advance crank angle with a classical polytropic relationship. It was found that this correlation overestimates cyclic AFR in the rich cycles during transient operating conditions [80].

A study used the pressure moment method to determine the AFR in a diesel engine using second and third moments [45]. Moment calculations start from ignition angle (θ_{ign}) and extend up to combustion duration ($\Delta\theta_{\text{comb}}$) in the pressure curve. The empirical correlation derived by the study is presented in Eq. (10.123). This correlation is able to predict the equivalence ratio with sufficient accuracy as confirmed by the correlation index R^2 equal to 0.996 and mean absolute and relative errors equal to 0.04 and 9.78%, respectively:

$$\phi = a_0 + a_1 M_2 + a_2 \frac{1}{M_2} + a_3 \frac{M_3}{M_2} + a_4 N^2 \tag{10.123}$$

$$M_2 = \frac{\displaystyle\int_{\theta_{\text{ign}}}^{\theta_{\text{ign}}+\Delta\theta_{\text{comb}}} (\theta - \theta_c)^2 P(\theta) d\theta}{A_u} \tag{10.124}$$

$$M_3 = \frac{\displaystyle\int_{\theta_{\text{ign}}}^{\theta_{\text{ign}}+\Delta\theta_{\text{comb}}} (\theta - \theta_c)^3 P(\theta) d\theta}{A_u} \tag{10.125}$$

$$\theta_c = \frac{M_1}{M_0} = \frac{\displaystyle\int_{\theta_{\text{ign}}}^{\theta_{\text{ign}}+\Delta\theta_{\text{comb}}} \theta \cdot P(\theta) d\theta}{\displaystyle\int_{\theta_{\text{ign}}}^{\theta_{\text{ign}}+\Delta\theta_{\text{comb}}} P(\theta) d\theta} \tag{10.126}$$

where M_0, M_1, M_2, and M_3 are zero, first, second, and third moment, respectively, θ_C is the centroid of the pressure distribution, A_u is the area of the pressure cycle under motoring condition, and ϕ is the equivalence ratio.

A recent study further introduced IMEP in the empirical correlation to account for the engine load in AFR prediction in transient operating conditions [75]. The empirical correlation used by this study is presented in Eqs. (10.127) and (10.128):

$$X_{\text{afr}} = (M_{n2})^{a_1} \cdot (M_{n3})^{a_2} \cdot (M_{n4})^{a_3} \cdot (\text{imep})^{a_4} \tag{10.127}$$

$$\overline{\text{AFR}} = b_0 + b_1 . X_{\text{afr}} \tag{10.128}$$

where M_{nn} is a normalized moment with respect to motoring curve. Constants a_1, a_2, a_3, and a_4 are calculated by optimization algorithm for maximum correlation between X_{afr} and the measured AFR along the transient. After estimating the coefficients a_n, a black box model expressing AFR as a function of X_{afr} was developed as Eq. (10.128). Figure 10.23 compared the measured and predicated AFR during transient operation. The figure shows that the model is able to predict AFR with sufficient accuracy as confirmed by a correlation coefficient of 0.96.

Another study further improved the prediction capability of empirical correlation during transient condition by introducing accelerator pedal travel instead of IMEP [74]. The improved correlation predicted the ARF with increased accuracy as confirmed by increased correlation index to 0.983 from 0.96 and mean relative error to 0.12%.

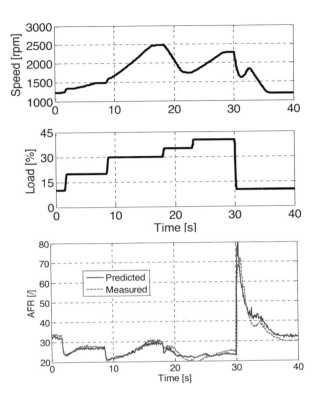

Fig. 10.23 Comparison between measured and predicted AFR for the transient operation [75]

10.7.2   G-Ratio Method

The *G*-ratio approach for determination of AFR using cylinder pressure is proposed in the reference [77, 78]. The *G*-ratio is defined as the ratio of average molecular weights before and after combustion. During the combustion of hydrocarbon fuels, the reaction occurring in the combustion chamber reduces the majority of fuel molecules to less molecular weight CO_2, CO, and H_2O as well as other combustion products in relatively smaller concentration. The net effect of the combustion process is the reduction in average molecular weights and increase in a number of molecules. The *G*-ratio estimator is an indirect type estimator which determines the cylinder AFR by using the relationship existing between the ratio of the average molecular weights (*G*-ratio) of the gases before and after the combustion process [77]. This method uses the properties of gases before and after the combustion and does not use the information from the period of the combustion. This acts as the merit of this method because the combustion process is subjected to disturbances and also dependent on the engine, which makes it difficult to model. A simple way to determine the AFR from *G*-ratio uses the linear relationship as shown in Eq. (10.129):

$$\hat{\phi} = \begin{cases} b_1(G\text{-ratio}) + b_2 & \text{for} \quad \phi \leq 1.0 \\ b_3(G\text{-ratio}) + b_4 & \text{for} \quad \phi > 1.0 \end{cases} \qquad (10.129)$$

where constants $b1$ to $b4$ are determined by fuel $H{:}C$ ratio and air $N{:}O$ ratios.

The calculation of G-ratio requires at least one pressure and one mean temperature to be measured before and after the combustion process. The G-ratio is calculated using ideal gas law and presented as Eq. (10.130):

$$G\text{-ratio} = \frac{G_4}{G_1} = \frac{P_1}{P_4} \cdot \frac{T_4}{T_1} \cdot \frac{V_1}{V_4} \cdot \frac{M_4}{M_1} \qquad (10.130)$$

where the subscripts "1" and "4" refers to before and after combustion. For port-injected engines $M1 \approx M4$, and if the points "1" and "4" are measured at the same volume, the equation can be simplified to Eq. (10.131):

$$G\text{-ratio} = \frac{P_1}{P_4} \cdot \frac{T_4}{T_1} \qquad (10.131)$$

The G-ratio is uniquely related to the AFR of the cylinder charge. The study showed the accuracy of AFR estimation was 4.85% RMS based on ten-cycle average during steady-state operation. The accuracy can improve substantially with accurate in-cylinder temperature measurements.

10.7.3   Net Heat Release-Based Method

A study developed an estimation model that utilizes the net heat release curve for determination of the cylinder air-fuel ratio of a spark ignition engine [76]. The net heat release curve is calculated from the measured cylinder pressure trace and quantifies the conversion of chemical energy of the reactants in the charge into thermal energy.

The relation developed to estimate AFR is given by Eq. (10.132):

$$\text{AFR} = c\frac{\Delta \alpha_b}{Q_{\text{tot}}} p_0^{1+\mu} T_0^{\beta-1} N^{-\eta} \qquad (10.132)$$

where c, μ, β, and η are unknown constants to be determined from experiments; $\Delta\alpha_b$ is rapid burn angle, i.e., 10–50% heat release angle; p_0 and T_0 are inlet pressure and temperature; and N is engine speed. Cycle-averaged AFR estimates over a range of engine speeds and loads show an RMS error of 4.1% compared to measurements in the exhaust [76].

10.7.4   Other Estimation Methods

Various studies have proposed different methods for AFR estimation by use of in-cylinder pressure frequency analysis [79], spark plug ionization current [81, 82], and a combination of ionization current and artificial neural network (ANN) [83] and by using a fiber optic probe [84]. The air-fuel ratio (AFR) of the mixture which burned in the combustion chamber can be estimated using the information hidden in the corresponding in-cylinder pressure signal. A novel approach based on in-cylinder pressure frequency analysis is presented [79]. In-cylinder pressure harmonic components are affected by AFR variations. The relationship derived for estimation of AFR is presented in Eq. (10.133):

$$AFR_e = C_0 + C_1 F_1 + C_2 F_2 + C_3 F_3 \qquad (10.133)$$

where F_1 and F_2 are the real part of the first two in-cylinder pressure frequency components and F_3 is the linear combination of the imaginary parts of the first two in-cylinder pressure frequency components. The results show that this method has the ability to predict AFR values, and average absolute AFR estimation error of around 1% is found. This error may still be too high to completely substitute the use of an oxygen sensor, even if its application together with a HEGO sensor should allow improving closed-loop engine control performance, which leads to removal of a more expensive UEGO sensor [79].

Ionization current signal is typically used as an alternative to the piezoelectric sensor for combustion diagnostics and analysis in reciprocating internal combustion engines. Thus, the AFR estimation methods are also developed using the ion current signal. A study [82] developed a correlation to estimate the AFR in a spark ignition engine which is given by Eq. (10.134):

$$\phi - \phi_s = a_0 + a_1 \theta_P^{1/2} + a_2 \theta_P + a_3 A + a_4 A^2 + a_5 \theta_P^{1/2} A + a_6 \theta_P A^2 \qquad (10.134)$$

where θ_P is the crank angle position of peak current, ϕ_s is the stoichiometric equivalence ratio, A is the area under the ionization current curve, and the constants a_n are estimated by the least-square regression of the experimental data. The study demonstrated that the AFR could be successfully determined using ion current signal at the spark plug for both average and individual engine cycle [82]. This method has the potential for designing real-time closed-loop control system while eliminating the need for conventional sensors such as oxygen sensor and a piezoelectric pressure sensor.

Discussion/Investigation Questions

1. A researcher uses the crank angle corresponding to peak motoring pressure as TDC position of the piston. Write the assumptions, when TDC position determination using this method is accurate. Justify your answer, whether it is possible to have TDC at peak motoring pressure in real reciprocating engines.
2. Describe the different dynamic methods of TDC determination using additional hardware as well as a thermodynamic method based on measured cylinder pressure data. Write the merits and demerits of both types of methods for TDC determination.
3. Write the possible sources of error in TDC determination using polytropic coefficient method, temperature-entropy method, and inflection point method.
4. Discuss how sampling rate of motoring pressure data can affect the TDC determination accuracy. Justify your answers. What is the typical encoder resolution you will recommend for TDC determination experiments?
5. Explain the reasons why thermodynamic loss angle varies with engine speed. What is the typical value of thermodynamic loss angle at rated diesel engine speed? Write the typical acceptable error (in terms of crank angle) in TDC determination.
6. Write the advantages of compression ratio determination using cylinder pressure measurement. Discuss methods for determination of compression ratio from motoring and firing in-cylinder pressure data. Explain the effectiveness of method with respect to accuracy, convergence speed, and overall convergence.
7. Engineers continuously strive to reduce fuel consumption and emissions from combustion engines through new designs. Some interesting technologies such as VVT and variable compression (VC) in combination with other technologies show to give a reduction in fuel consumption without compromising on driving performance. Utilization of these technologies leads to requirement of compression ratio or effective compression ratio determination. Explain how effective compression ratio can be determined from in-cylinder pressure data for the engine operating with VVT technology. Discuss the parameters which affect the effective compression ratio such as intake valve closing timings, engine speed, etc.
8. Calculate the effective compression ratio (ECR) and geometric ECR (GECR) for a $P - V$ diagram shown in Fig. P10.1 for a spark ignition engine operated at 1200 rpm. Assume engine bore/stroke in 86/86 mm and connecting rod dimension 146 mm. Intake valve closing can be assumed at 60 CAD aBDC. The baseline geometrical compression ratio can be assumed to be 9.3. Estimate the change in ECR for early exhaust valve closing (EVC) condition in comparison to normal valve timings.

Fig. P10.1 Pressure
volume curves at two valve
timings (adapted from [85])

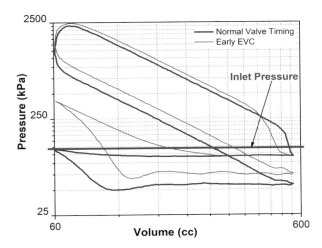

9. Write the effect of wall temperature on heat transfer and emissions (particularly on HC and CO). Discuss the mechanisms how wall temperature affects the HC and CO emissions in SI engines.
10. Write the advantages of wall temperature estimation by cylinder pressure measurement. Discuss the assumptions and sources of error in the heat transfer inversion method of wall temperature determination.
11. Write the classical method of air mass determination in spark ignition engines. Discuss the limitations of this method of trapped mass estimation. Explain how the cylinder pressure-based method can overcome the shortcomings, and also write the advantages of cylinder pressure-based estimation of trapped mass.
12. Discuss two methods for estimation of the trapped mass and residual gas fraction determination in a reciprocating engine using measured cylinder pressure?
13. Calculate the residual gas fraction using Eq. (10.58) for a spark ignition engine having bore/stroke dimensions 88/52 mm and connecting rod dimension 138 mm. A geometrical compression ratio of the engine is 10.1 and intake valve opening and closing at 20 CAD before TDC and 72 after BDC, respectively. Given OF $= 0.58°/m$, $P_{man} = 600$ mbar, $N = 2000$ rpm, and intake exhaust manifold pressure ratio is 0.6. Discuss the effect of manifold pressure ratio on the residual gas fraction.
14. Calculate the trapped mass and fresh inducted air for the engine conditions in question 13. Clearly state your assumptions. Given the pressure equations $PV^{1.28} = 3827$ and the pressure values in bar and cylinder volume in cc. Runner heating can be assumed zero.
15. Increasingly high demands on pollutant control on automotive engines have been pushing the level of sophistication of engine control modules higher and higher. In this context, explain the need for closed-loop control in automotive engines. Typically, for closed-loop control of spark ignition (SI) engines, the air-fuel ratio is used as feedback parameter, which is measured by the zirconia-

based oxygen sensor. Oxygen sensors used in SI engines have limited performance during engine transients. It is proposed to use the cylinder pressure-based
parameter as a feedback parameter. Explain and discuss the air-fuel ratio
estimation method using suitable equations and schematics from the following
inputs.

(a) Measured crank angle-based in-cylinder pressure data over an engine cycle.
(b) Calculated crank angle-based heat release data over an engine cycle.
(c) Measured exhaust gas species.

16. The values of constants in Eq. (10.112) is given as $a_1 = 0.0438$,
$a_2 = -2.6 \times 10^{-6}$, $a_3 = 0.0112$, $a_4 = -1.5 \times 10^{-5}$, and $a_5 = -1.04$. Calculate
the ϕ for engine operated at 1500 rpm at two different engine load conditions.
The cylinder pressure at two load conditions is presented in Fig. P10.2, and
intake manifold pressure can be assumed to be 0.92 bar. Clearly state your
assumptions, if any.

Fig. P10.2 Cylinder
pressure trace at two engine
operating conditions

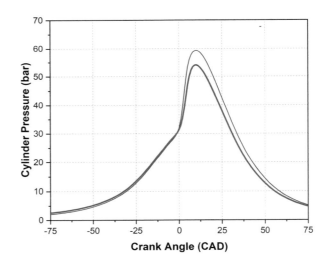

References

1. Hribernik, A. (1998). *Statistical determination of correlation between pressure and crankshaft angle during indication of combustion engines* (No. 982541). SAE Technical Paper.
2. Tunestål, P. (2011). TDC offset estimation from motored cylinder pressure data based on heat release shaping. *Oil & Gas Science and Technology–Revue d'IFP Energies Nouvelles, 66*(4), 705–716.
3. Stas, M. J. (1996). *Thermodynamic determination of TDC in piston combustion engines*. SAE paper, 960610.
4. Bueno, A. V., Velásquez, J. A., & Milanez, L. F. (2012). Internal combustion engine indicating measurements. In *Applied measurement systems*. Croatia: InTech.

5. Nilsson, Y., & Eriksson, L. (2004). *Determining TDC position using symmetry and other methods* (No. 2004-01-1458). SAE Technical Paper.
6. Pipitone, E., Beccari, A., & Beccari, S. (2007). *The experimental validation of a new thermodynamic method for TDC determination* (No. 2007-24-0052). SAE Technical Paper.
7. Pipitone, E., Beccari, A., & Beccari, S. (2008). *Reliable TDC position determination: A comparison of different thermodynamic methods through experimental data and simulations* (No. 2008-36-0059). SAE Technical Paper.
8. Pipitone, E., & Beccari, A. (2010). Determination of TDC in internal combustion engines by a newly developed thermodynamic approach. *Applied Thermal Engineering, 30*(14–15), 1914–1926.
9. Tazerout, M., Le Corre, O., & Rousseau, S. T. D. C. (1999). *TDC determination in IC engines based on the thermodynamic analysis of the temperature-entropy diagram* (No. 1999-01-1489). SAE Technical Paper.
10. Tazerout, M., Le Corre, O., & Stouffs, P. (1999). *Compression ratio and TDC calibrations using temperature-entropy diagram* (No. 1999-01-3509). SAE Technical Paper.
11. Staś, M. J. (2000). *An universally applicable thermodynamic method for TDC determination* (No. 2000-01-0561). SAE Technical Paper.
12. Pinchon, P. (1984). Calage thermodynamique du point mort haut des moteurs a piston. *Revue de l'institut francais du petrole, 39*(1), 93–111.
13. Tunestål, P. (2009). Model based TDC offset estimation from motored cylinder pressure data. *IFAC Proceedings Volumes, 42*(26), 241–247.
14. Johansson, T. (2015). *Virtual sensors for combustion parameters based on in-cylinder pressure* (Master's thesis). Linköping University, LiTH-ISY-EX--15/4913—SE.
15. Irimescu, A., Tornatore, C., Marchitto, L., & Merola, S. S. (2013). Compression ratio and blow-by rates estimation based on motored pressure trace analysis for an optical spark ignition engine. *Applied Thermal Engineering, 61*(2), 101–109.
16. Lamaris, V. T., & Hountalas, D. T. (2010). Validation of a diagnostic method for estimating the compression condition of direct injection diesel engines. *Proceedings of the Institution of Mechanical Engineers, Part A: Journal of Power and Energy, 224*(4), 517–532.
17. Klein, M., Eriksson, L., & Åslund, J. (2006). Compression ratio estimation based on cylinder pressure data. *Control Engineering Practice, 14*(3), 197–211.
18. Klein, M., & Eriksson, L. (2006). *Methods for cylinder pressure based compression ratio estimation* (No. 2006-01-0185). SAE Technical Paper.
19. Stricker, K., Kocher, L., Van Alstine, D., & Shaver, G. M. (2013). Input observer convergence and robustness: Application to compression ratio estimation. *Control Engineering Practice, 21*(4), 565–582.
20. Stricker, K., Kocher, L., Koeberlein, E., Van Alstine, D., & Shaver, G. M. (2012). Estimation of effective compression ratio for engines utilizing flexible intake valve actuation. *Proceedings of the Institution of Mechanical Engineers, Part D: Journal of Automobile Engineering, 226*(8), 1001–1015.
21. Kocher, L., Koeberlein, E., Van Alstine, D. G., Stricker, K., & Shaver, G. (2012). Physically based volumetric efficiency model for diesel engines utilizing variable intake valve actuation. *International Journal of Engine Research, 13*(2), 169–184.
22. Modiyani, R., Kocher, L., Van Alstine, D. G., Koeberlein, E., Stricker, K., Meckl, P., & Shaver, G. (2011). Effect of intake valve closure modulation on effective compression ratio and gas exchange in turbocharged multi-cylinder engines utilizing EGR. *International Journal of Engine Research, 12*(6), 617–631.
23. He, X., Durrett, R. P., & Sun, Z. (2009). Late intake valve closing as an emissions control strategy at Tier 2 Bin 5 engine-out NOx level. *SAE International Journal of Engines, 1*(1), 427–443.
24. Aghdam, E. A., & Kabir, M. M. (2010). Validation of a blowby model using experimental results in motoring condition with the change of compression ratio and engine speed. *Experimental Thermal and Fluid Science, 34*(2), 197–209.

25. Malagi, R. R. (2012). *Estimation of blowby in multi-cylinder diesel engine using finite element approach* (No. 2012-01-0559). SAE Technical Paper.
26. Armas, O., Rodríguez, J., Payri, F., Martín, J., & Agudelo, J. R. (2005). Effect of the trapped mass and its composition on the heat transfer in the compression cycle of a reciprocating engine. *Applied Thermal Engineering, 25*(17–18), 2842–2853.
27. Chang, H., Zhang, Y., & Chen, L. (2003). Gray forecast of diesel engine performance based on wear. *Applied Thermal Engineering, 23*(17), 2285–2292.
28. Arsie, I., Di Leo, R., Falco, S., Pianese, C., & De Cesare, M. (2015). *Estimation of the engine thermal state by in-cylinder pressure measurement in automotive diesel engines* (No. 2015-01-1623). SAE Technical Paper.
29. Arsie, I., Flauti, G., Pianese, C., & Rizzo, G. (1999). Cylinder thermal state detection from pressure cycle in SI engine. In *Proceedings of the 4th International Conference on Internal Combustion Engines: Experiments and Modeling*, Capri (pp. 581–588).
30. Jennings, M. J., & Morel, T. (1991). A computational study of wall temperature effects on engine heat transfer. *SAE Transactions, 910459*, 641–656.
31. Kaplan, J. A., & Heywood, J. B. (1991). *Modeling the spark ignition engine warm-up process to predict component temperatures and hydrocarbon emissions*. SAE paper 910302.
32. Rakopoulos, C. D., Giakoumis, E. G., & Rakopoulos, D. C. (2008). Study of the short-term cylinder wall temperature oscillations during transient operation of a turbo-charged diesel engine with various insulation schemes. *International Journal of Engine Research, 9*(3), 177–193.
33. Baker, D. M., & Assanis, D. N. (1994). A methodology for coupled thermodynamic and heat transfer analysis of a diesel engine. *Applied Mathematical Modelling, 18*(11), 590–601.
34. Rakopoulos, C. D., Rakopoulos, D. C., Mavropoulos, G. C., & Giakoumis, E. G. (2004). Experimental and theoretical study of the short term response temperature transients in the cylinder walls of a diesel engine at various operating conditions. *Applied Thermal Engineering, 24*(5–6), 679–702.
35. Heywood, J. B. (1988). *Internal combustion engines fundamentals*. New York: McGraw-Hill.
36. Yan, F., & Wang, J. (2012). Engine cycle-by-cycle cylinder wall temperature observer-based estimation through cylinder pressure signals. *Journal of Dynamic Systems, Measurement, and Control, 134*(6), 061014.
37. Maurya, R. K. (2018). *Characteristics and control of low temperature combustion engines: Employing gasoline, ethanol and methanol*. Cham: Springer.
38. Colin, G., Giansetti, P., Chamaillard, Y., & Higelin, P. (2007). *In-cylinder mass estimation using cylinder pressure* (No. 2007-24-0049). SAE Technical Paper.
39. Wang, Z., Zhu, Q., & Prucka, R. (2016). *A review of spark-ignition engine air charge estimation methods* (No. 2016-01-0620). SAE Technical Paper.
40. Fox, J. W., Cheng, W. K., & Heywood, J. B. (1993). *A model for predicting residual gas fraction in spark-ignition engines*. SAE 931025.
41. Giansetti, P., Colin, G., Chamaillard, Y., & Higelin, P. (2007). Two observers for in-cylinder mass estimation using cylinder pressure measurements. *IFAC Proceedings Volumes, 40*(10), 219–226.
42. Akimoto, A., Itoh, H., & Suzuki, H. (1989). Development of delta p method to optimize transient a/f-behavior in mpi engine. *JSAE Review, 10*(4).
43. Worm, J. (2005). *An evaluation of several methods for calculating transient trapped air mass with emphasis on the "delta p" approach* (No. 2005-01-0990). SAE Technical Paper.
44. Worm, J. (2005). *The effect of exhaust variable cam Phaser transients on equivalence ratio control in an SI 4 stroke engine* (No. 2005-01-0763). SAE Technical Paper.
45. Arsie, I., Di Leo, R., Pianese, C., & De Cesare, M. (2014). Estimation of in-cylinder mass and AFR by cylinder pressure measurement in automotive diesel engines. *IFAC Proceedings Volumes, 47*(3), 11836–11841.

46. Desantes, J. M., Galindo, J., Guardiola, C., & Dolz, V. (2010). Air mass flow estimation in turbocharged diesel engines from in-cylinder pressure measurement. *Experimental Thermal and Fluid Science, 34*(1), 37–47.

47. Hart, M., Ziegler, M., & Loffeld, O. (1998). *Adaptive estimation of cylinder air mass using the combustion pressure* (No. 980791). SAE Technical Paper.

48. Eriksson, L., & Thomasson, A. (2017). Cylinder state estimation from measured cylinder pressure traces—A survey. *IFAC-PapersOnLine, 50*(1), 11029–11039.

49. Ivansson, N. (2003). *Estimation of the residual gas fraction in an HCCI-engine using cylinder pressure* (Master's thesis). Linkopings Universitet, Sweden.

50. Mladek, M., & Onder, C. H. (2000). *A model for the estimation of inducted air mass and the residual gas fraction using cylinder pressure measurements* (No. 2000-01-0958). SAE Technical Paper.

51. Ponti, F., Piani, J. C., & Suglia, R. (2004). Residual gas model for on-line estimation for inlet and exhaust continuous VVT engine configuration. *IFAC Proceedings Volumes, 37*(22), 257–265.

52. Aquino, C. F. (1981). *Transient A/F control characteristics of the 5 liter central fuel injection engine.* SAE Transactions, (Paper No-81094) 1819-1833.

53. Yazdani, A., Naber, J., Shahbakhti, M., Dice, P., Glugla, C., Cooper, S., . . . & Huberts, G. (2017). *Air charge and residual gas fraction estimation for a spark-ignition engine using in-cylinder pressure* (No. 2017-01-0527). SAE Technical Paper.

54. Broatch, A., Guardiola, C., Pla, B., & Bares, P. (2015). A direct transform for determining the trapped mass on an internal combustion engine based on the in-cylinder pressure resonance phenomenon. *Mechanical Systems and Signal Processing, 62*, 480–489.

55. Guardiola, C., Pla, B., Blanco-Rodriguez, D., & Bares, P. (2014). Cycle by cycle trapped mass estimation for diagnosis and control. *SAE International Journal of Engines, 7*(3), 1523–1531.

56. Luján, J. M., Guardiola, C., Pla, B., & Bares, P. (2016). Estimation of trapped mass by in-cylinder pressure resonance in HCCI engines. *Mechanical Systems and Signal Processing, 66*, 862–874.

57. Guardiola, C., Triantopoulos, V., Bares, P., Bohac, S., & Stefanopoulou, A. (2016). Simultaneous estimation of intake and residual mass using in-cylinder pressure in an engine with negative valve overlap. *IFAC-PapersOnLine, 49*(11), 461–468.

58. Draper, C. S. (1935). *The physical effects of detonation in a closed cylindrical chamber.* Technical report, National Advisory Committee for Aeronautics.

59. Sinnamon, J. F., & Sellnau, M. C. (2008). *A new technique for residual gas estimation and modeling in engines* (No. 2008-01-0093). SAE Technical Paper.

60. Giansetti, P., Perrier, C., Higelin, P., Chamaillard, Y., Charlet, A., & Couet, S. (2002). *A model for residual gas fraction prediction in spark ignition engines* (No. 2002-01-1735). SAE Technical Paper.

61. Ford, R., & Collings, N. (1999). *Measurement of residual gas fraction using a fast response NO sensor* (No. 1999-01-0208). SAE Technical Paper.

62. Cho, H., Lee, K., Lee, J., Yoo, J., & Min, K. (2001). *Measurements and modeling of residual gas fraction in SI engines* (No. 2001-01-1910). SAE Technical Paper.

63. Schwarz, F., & Spicher, U. (2003). *Determination of residual gas fraction in IC engines* (No. 2003-01-3148). SAE Technical Paper.

64. Giansetti, P., & Higelin, P. (2007). *Residual gas fraction measurement and estimation in spark ignition engine* (No. 2007-01-1900). SAE Technical Paper.

65. Jannoun, A., Tauzia, X., Chesse, P., & Maiboom, A. (2017). *Experimental investigation of an in-cylinder sampling technique for the evaluation of the residual gas fraction* (No. 2017-24-0042). SAE Technical Paper.

66. Cavina, N., Siviero, C., & Suglia, R. (2004). *Residual gas fraction estimation: Application to a GDI engine with variable valve timing and EGR* (No. 2004-01-2943). SAE Technical Paper.

67. Galliot, F., Cheng, W. K., Cheng, C. O., Sztenderowicz, M., Heywood, J. B., & Collings, N. (1990). In-cylinder measurements of residual gas concentration in a spark ignition engine. *SAE Transactions*, 1144–1150.
68. Senecal, P. K., Xin, J., & Reitz, R. D. (1996). *Predictions of residual gas fraction in IC engines* (No. 962052). SAE Technical Paper.
69. Wu, Y., Kumar, M., & Shen, T. (2016). A stochastic logical system approach to model and optimal control of cyclic variation of residual gas fraction in combustion engines. *Applied Thermal Engineering, 93*, 251–259.
70. Fitzgerald, R. P., Steeper, R., Snyder, J., Hanson, R., & Hessel, R. (2010). Determination of cycle temperatures and residual gas fraction for HCCI negative valve overlap operation. *SAE International Journal of Engines, 3*(1), 124–141.
71. Larimore, J. W. (2014). *Experimental analysis and control of recompression homogeneous charge compression ignition combustion at the high cyclic variability limit* (PhD Thesis). University of Michigan, USA.
72. Larimore, J., Jade, S., Hellström, E., Stefanopoulou, A. G., Vanier, J., & Jiang, L. (2013, October). Online adaptive residual mass estimation in a multicylinder recompression HCCI engine. In *ASME 2013 Dynamic Systems and Control Conference* (pp. V003T41A005–V003T41A005). American Society of Mechanical Engineers.
73. Gilkey, J. C., & Powell, J. D. (1985). Fuel-air ratio determination from cylinder pressure time histories. *Journal of Dynamic Systems, Measurement, and Control, 107*(4), 252–257.
74. Arsie, I., Di Leo, R., Pianese, C., & De Cesare, M. (2017). *Air-fuel ratio and trapped mass estimation in diesel engines using in-cylinder pressure* (No. 2017-01-0593). SAE Technical Paper.
75. Arsie, I., Di Leo, R., Pianese, C., & De Cesare, M. (2016). Air-fuel ratio estimation along diesel engine transient operation using in-cylinder pressure. *Energy Procedia, 101*, 670–676.
76. Tunestål, P., & Hedrick, J. K. (2003). Cylinder air/fuel ratio estimation using net heat release data. *Control Engineering Practice, 11*(3), 311–318.
77. Patrick, R. S. (1989). *Air-fuel ratio estimation in an Otto cycle engine: two methods and their performance* (PhD Thesis). Stanford University, CA, USA.
78. Patrick, R. S., & Powell, J. D. (1990). A technique for the real-time estimation of air-fuel ratio using molecular weight ratios. *SAE Transactions*, 686–698.
79. Cavina, N., & Ponti, F. (2003). Air fuel ratio estimation using in-cylinder pressure frequency analysis. *Journal of Engineering for Gas Turbines and Power, 125*(3), 812–819.
80. Arsie, I., Pianese, C., & Rizzo, G. (1998). Estimation of air-fuel ratio and cylinder wall temperature from pressure cycle in si automotive engines. *IFAC Proceedings Volumes, 31*(1), 161–167.
81. Balles, E. N., VanDyne, E. A., Wahl, A. M., Ratton, K., & Lai, M. C. (1998). *In-cylinder air/fuel ratio approximation using spark gap ionization sensing* (No. 980166). SAE Technical Paper.
82. Lee, B., Guezennec, Y. G., & Rizzoni, G. (2001). Estimation of cycle-resolved in-cylinder pressure and air-fuel ratio using spark plug ionization current sensing. *International Journal of Engine Research, 2*(4), 263–276.
83. Wickström, N., Taveniku, M., Linde, A., Larsson, M., & Svensson, B. (1997). Estimating pressure peak position and air-fuel ratio using the ionization current and artificial neural networks. In *IEEE Conference on Intelligent Transportation Systems*, ITSC, Boston, MA, USA, 9–12 November, 1997 (pp. 927–977).
84. Hall, M. J., & Koenig, M. (1996, January). A fiber-optic probe to measure precombustion in-cylinder fuel-air ratio fluctuations in production engines. In *Symposium (International) on Combustion* (Vol. 26, no. 2, pp. 2613–2618).
85. Rodriguez, J. F., & Cheng, W. (2018). *Potential of negative valve overlap for part load efficiency improvement in gasoline engines*. SAE Technical Paper 2018-01-0377.

Bibliography

1. Maurya, R. K. (2018). *Characteristics and control of low temperature combustion engines: Employing gasoline, ethanol and methanol*. Cham: Springer.
2. Atkins, R. D. (2009). *An introduction to engine testing and development*. Warrendale, PA: Society of Automotive Engineers.
3. Zhao, H., & Ladommatos, N. (2001). *Engine combustion instrumentation and diagnostics*. Warrendale, PA: Society of Automotive Engineers.
4. Rogers, D. R. (2010). *Engine combustion: Pressure measurement and analysis*. Warrendale, PA: Society of Automotive Engineers.
5. Turner, J. (2009). *Automotive sensors*. Highland Park, NJ: Momentum Press.
6. Gossweiler C., Sailer W., & Cater C. (2006). *Sensors and amplifiers for combustion analysis. A guide for the user 100-403e-10.06.* © Kistler Instrumente AG, CH-8408 Winterthur, Okt.
7. AVL List GmBH. (2002). *Engine indicating user handbook*. Graz: AVL List GmBH.
8. Gautschi, G. (2002). *Piezoelectric sensorics: Force, strain, pressure, acceleration and acoustic emission sensors, materials and amplifiers*. Berlin: Springer.
9. Martyr, A. J., & Plint, M. A. (2012). *Engine testing: Theory and practice* (4th ed.). Oxford: Elsevier.
10. Heywood, J. B. (1988). *Internal combustion engine fundamentals*. New York, NY: McGraw-Hill.

© Springer Nature Switzerland AG 2019
R. K. Maurya, *Reciprocating Engine Combustion Diagnostics*, Mechanical
Engineering Series, https://doi.org/10.1007/978-3-030-11954-6

Index

A

Acoustic techniques, 300
Acoustic theory, 480, 494
Advanced spark ignition (ASI), 406, 407
Air-fuel ratio (AFR)
    emission legislation, 589
    G-ratio, 593, 594
    HEGO and UEGO sensor, 595
    in-cylinder pressure frequency analysis, 595
    ionization current signal, 595
    net heat release, 594
    nonconventional combustion processes, 589
    oxygen sensor, 589
    piezoelectric pressure sensor, 595
    statistical moment method, 589–592
Analog to digital (A/D) system, 164
Analog to digital converter (ADC), 162, 249
Apparent heat release based method, 293
Arrhenius equation, 315, 470
Artificial neural networks (ANN), 106,
    388, 595
Autocorrelation function (ACF), 410, 411
Autoregressive moving average techniques
    (ARMA), 106
Average Energy of Heat Release Oscillations
    (AEHRO), 503
Average energy of pressure oscillations
    (AEPO), 501
Average exhaust absolute pressure
    (AEAP), 71
Average mutual information (AMI), 428
Average signal envelope (ASE), 71

B

Battery electric vehicles (BEVs), 3
Binary coded decimal (BCD), 164
Blow-by estimation, 565, 566
Bottom dead center (BDC), 232
Brake mean effective pressure (BMEP), 569
Brake specific fuel consumption (BSFC), 268
Burn rate analysis, 284
Burned mass fraction (BMF)
    apparent heat release, 293
    CA position, 290
    Marvin's graphical method, 285, 286
    PRM and PDR, 289, 291, 292
    RW method, 286–289
    Wibes function, 292

C

Cetane number (CN), 376
Chapman-Jouguet type, 476
Chebyshev filters, 205
Chemical delay, 339
Chemical kinetic models, 470
Closed-loop engine control, 284
Coefficient of variation (COV), 216, 385, 390,
    402–405, 408, 409, 413, 510
Coefficient of variation of IMEP
    (COV_{IMEP}), 366
Coherent anti-Stokes Raman scattering
    (CARS), 300
Combustion duration (CD), 284, 343, 346,
    347, 528

Combustion engines, 283, 284, 300, 301, 322,
        349, 545, 561, 595
Combustion gas temperature estimation
    A*, 299
    air-fuel mixture, 295
    auto-ignition kinetics, 299
    EVC, 295
    HCCI, 294, 296–298
    IVC, 294, 295
    NOx formation, 295
    optical temperature measurement
        techniques, 300
    properties
        density, 294
        pressure, 294
        temperature, 294
    spectroscopic techniques, 300
    thermocouples, 300
    thermodynamics and optical analysis, 299
    thermometry, 300
    two-zone model, 297
    unburned hydrocarbon emissions, 296
    unburnt and burnt zone temperature, 299
Combustion noise
    acoustic tuning, 521
    acting forces, 522
    calculation and metrics, 522–525, 527, 528
    characteristic, 528–532
    optimization, 521
    vibration modes, 522
Combustion noise level (CNL), 523, 530, 531
Combustion parameters
    EOC and combustion duration, 343–348
    ID, 338–340
    phasing, 340–343
    SOC, 329–336
Combustion phasing, 284, 340–343, 528
Combustion stability
    autocorrelation and cross-correlation,
        409–412
    automotive market demands, 363
    characterization, 390
    COV and STD, 402–408
    cycle-to-cycle variation, 365
    EEGR, 366
    EGR operation, 364
    emission legislation, 363
    frequency distribution and histograms, 395,
        397, 398, 400
    LNV, 408
    nonlinear and chaotic analysis
        autocorrelation and Fourier
            transform, 424

    multifractal analysis, 449, 450
    1 Test, 446–448
    phase space reconstruction, 425–428
    Poincare section, 428, 430
    return maps, 430–434
    RP and RQA, 441, 442, 444, 445
    symbol sequence statistics (see Symbol
        sequence statistics method)
    normal distribution analysis, 400, 402
    parameters, 365
    partial burn and misfire cycles, 366
    PCA, 412, 413
    quantitative measures, 451
    spark timings, 365
    stable and unstable engine operating region,
        365, 366
    statistical parameters, 390
    stochastic and deterministic processes, 365
    time series analysis, 392–395
    unintended excursions, 365
    variability (see Combustion variability)
    wavelet (see Wavelets)
Combustion variability
    automotive engines, 367
    burning rate, 369
    CI (see Compression ignition (CI))
    emissions and fuel economy, 368
    engine combustion cycles, 367
    idle instability, 368
    power and thermodynamic efficiency, 368
    prior-cycle and same-cycle effects, 369
    SI (see Spark ignition (SI) engines)
    speed and torque fluctuations, 368
Common rail direct injection (CRDI), 17, 132,
        134, 561
Compression ignition (CI), 494
    CRDI system, 19
    cycle-to-cycle analysis, 377
    cyclic variation, 376
    diesel cold starting, 376
    diesel engines, 17, 376
    diesel fuel, 15
    electronic fuel injection systems, 18
    factors, 377
    fuel distribution, 15
    fuel injection system, 376
    fuel-air mixing, 376
    HCCI, 377, 378
    heat release rate and flame luminescence
        images, 16
    IMEP, 378
    injection characteristics, 17
    LTC, 378, 380

mechanical fuel injection system, 17, 18
parameters, 380
prolonged ignition delay, 376
SOC, 15
temperature stratification, 377
Compression ratio (CR), 301
determination, 561, 563–565
heat loss estimation, 559, 560
polytropic model, 560, 561
T-S method, 557, 558
Computation methodology, 350
Cone of influence (COI), 418
Continuous variable valve lift (CCVL), 261
Continuous wavelet transform (CWT), 415
Contour lines, 406
Controlled electronic ignition (CEI), 409
Conventional diesel combustion (CDC), 257,
        494, 522, 531
Corona ignition, 406, 407
Crank angle degree (CAD), 288
Crank angle encoder
    cylinder pressure, 107
    resolution requirement, 112–114
    thermodynamic analysis, 108
    working principle and output signal,
        108–112
Crank angle phasing, 557
Crank angle position measurement system, 545
Crevice effect, 315
Cross-correlation function (CCF), 411
Cumulative distribution function (CDF),
        350, 510
Cycle-to-cycle variation, 216
Cyclic combustion variations
    indicators, 381–383
    partial burn and misfire cycles
        ANN, 388
        constraints, 384
        COV, 385
        cylinder pressure measurement, 387
        engine cycles, 383
        ignition timing metric, 388
        IMEP, 384–386
        leaner engine operation, 386
        skewness and kurtosis, 388
        spark energy and duration, 386
        speed and torque fluctuations, 383
        TDC, 388
Cylinder pressure
    acquisitions, 335
    analysis, 168
    combustion monitoring, 329
    components, 287

crank angle position, 340
electric noises, 331
engine combustion process, 285
EOC, 343
firing cycle, 289
first law of thermodynamics, 293
heat release analysis, 349
IVC, 324
motoring pressure, 335
oscillations, 310
polytropic process, 346
SOC, 330
Cylinder pressure based knock analysis
    abnormal combustion process, 495
    knock indices (see Knock indices)
    signal processing, 498, 499
Cylinder pressure level (CPL), 523
Cylinder pressure signal, 210, 216

D
Data acquisition
    crank angle encoder, 158
    cylinder pressure measurement, 157
    data display and storage, 167–168
    designing of experiments, 154
    engine combustion measurement, 157
    experimental setup fabrication, 154
    experimental signal, 162–167
    factors, 158
    internal combustion engines, 154
    online data analysis, 155
    PC-based/microprocessor-based, 155
    principle, 155–158
    sample and hold (S/H) device, 157
    sampling and digitization, 162, 164,
        166, 167
    sensors and transducers, 158–160
    signal conditionings, 160–162
    signal processing, 154
Data compression methodology, 435
Data transmission unit, 156
Delta P (ΔP) method, 574, 575
Derivative at maximum pressure
        (DMP), 506
Detonation, 477, 481
Diesel spray, 338, 339
Difference pressure (DP), 342
Difference pressure integral (DPI), 250, 251
Diffusive combustion, 338
Digital signal processing
    absolute pressure correction, 185
    averaging method, 194

Digital signal processing (*cont.*)
  capacitive probes (TDC sensor), 175,
    176, 178
  charge amplifier, 197
  crank angle encoder, 173
  current-to-voltage conversion, 197
  cyclic variations, 196
  cylinder pressure curve, 182
  cylinder pressure signal, 194
  effect, referencing error, 183
  filtering method, 194, 197
  in-cylinder pressure, 172, 194
  inlet and outlet manifold pressure
    referencing, 185–188
  inter-cycle and intra-cycle drift, 182
  iterative method, 185
  low pass IIR (butterworth) filters, 204–208
  low pass FIR filter, 202–204
  LSM, 191–193
  measured pressure data, 179, 180
  microwave probe, 179
  moving average filters, 199–201
  oscillations, 196
  parameters, 183
  phasing methods, 179, 180
  polytropic coefficient estimation, 193–194
  power spectrum, 197
  pressure waves, 196
  quartz piezoelectric pressure
    transducers, 181
  real-time signal processing methods, 173
  signal data processing, 172
  signal noise, 196
  standard deviation variations
    crank angle (CA) position, 211
    diesel engine, 215
    HCCI engine, 212–214
  static determination, 174, 175
  statistical Levene's test, 216–217
  TDC determination, 173
  thermal shock, 185
  thermodynamic method, 196, 208–209
  two and three point referencing, 188–190
  voltage signal, 181
  wavelet filtering, 210
Dimensionless knock indicator (DKI), 501
Direct injection (DI), 315, 375
Discrete Fourier transform (DFT), 202, 204
Draper approach, 492, 580
Dual coil offset (DCO), 409
Dual-fuel combustion mode, 322, 323

E
Effective compression ratio (ECR), 561, 563
Electronic control unit (ECU), 42
Empirical mode decomposition (EMD), 518
End of combustion (EOC), 284, 286, 338,
  343, 344
End of injection (EOI), 332, 338
End of LTR (EoLTR), 332
Engine control unit (ECU), 41
Engine knock analysis, 470
Engine knock vibration, 517, 518
Engine management systems, 329
Engine performance analysis
  air flow rate, 264
  brake efficiency, 257, 260
  CDC, 257
  chemical energy, 254
  crank angle based events, 226
  cylinder-pressure based combustion
    diagnostics, 225
  cylinder pressure data, 244
  cylinder pressure measurements, 226
  cylinder pressure signal, 226
  diesel combustion, 267
  effect of, 229, 231
  EGR, 251
  electrical noise, 226
  energy flow, 254
  energy transfer, 254
  engine efficiency, 254, 255, 257
  engine flexibility, 266
  engine geometry and kinematics
    angular velocity (ω_c) and
      acceleration, 236
    diesel engine, 240
    equation, 233, 234
    mean piston speed, 239
    mechanical efficiency, 240
    peak piston speed, 238
    power stroke, 239
    reciprocating engine, 233
    Taylor series expansion, 235
    TDC and BDC position, 232
    velocity and acceleration vectors, 235
    and terminology, 232
  engine map, 268
  exergy destruction, 258
  fuel consumption map, 268
  fuel injection timings, 230
  gas exchange analysis, 260–262
  gasoline SI engine, 267

gross indicated efficiency, 256
HCCI, 257, 259
heat released (HR), 256
IMEP, 245, 246, 249, 253
measurable parameters, 225
mechanical efficiency, 265
MEP, 245
modern automotive engines, 265
net indicated efficiency, 257
performance parameters, 232
power, 263
P-V diagram, 227, 230, 245
RCCI, 257
reciprocating combustion engines, 244
tangential gas pressure, 250
TDC, 247
thermal shock, 226
thermodynamic air and actual cycles, 228
thermodynamic efficiency, 256
torque, 262, 265
torque calculation and analysis, 240–244
valve-closing noise, 226
volumetric efficiency, 265
Engine testing
combustion diagnostic, 27
cylinder pressure, 28
ion-current based diagnostics, 27
optical methods, 27
power and fuel consumptions, 25
product development and calibration, 26
test beds, 25, 26
Entropy determination method, 345
Entropy vs. gas temperature, 346
Exhaust and intake pressure sensors
acceleration sensitivity, 130
EGPS, 126
engine combustion process, 124
flow effects, 128
four-cylinder engine, 129
functions, 130
gas exchange process, 124
inlet pressure sensors, 130
instrumentation, 126
low-pressure measurement, 124
multicylinder engine, 129
packaging process, 126
piezoresistive sensors, 124
requirements, 127
switching adapter, 127
volume micromechanics, 125
Exhaust gas pressure sensor (EGPS), 126
Exhaust gas recirculation (EGR), 127, 251, 319,
    364, 374, 519, 561

Exhaust valve closing (EVC), 294, 313, 585
Exhaust valve opening (EVO), 368, 585
External combustion engines (ECEs), 5
External exhaust gas recirculation
    (EEGR), 379

F
False nearest neighbors (FNN), 428
Fast Fourier transform (FFT), 494, 505
Filtering, see Digital signal processing
Finite impulse response (FIR), 202
Flame surface visualization, 394
Fourier transform, 413–415
Frequency modulation (FM) principle, 136
Frequency response functions (FRF), 106
Fuel line pressure sensor
    CRDI technology, 132, 134
    factors, 131
    fuel injection system, 131
    line pressure measurement, 132
    measurements, 132
    mechanical fuel injection systems, 131
    rail pressure, 135

G
Gasoline compression ignition (GCI), 25,
    510–512
Gasoline direct injection (GDI), 13, 89,
    267, 363
Gasoline direct injection compression ignition
    (GDCI), 25
Gaussian distribution, 395
Generalized extreme value (GEV), 397
Geometric compression ratio (GCR), 561
Global wavelet spectrum (GWS), 418
G-ratio approach, 593, 594
Gross indicated efficiency (GIE), 378
Gross indicated mean effective pressure
    (GIMEP), 513

H
Heat loss estimation method, 559, 560
Heat release analysis, 283, 284
    characteristics, 319, 321–323
    cylinder pressure measurement, 324
    motoring pressure, 324–327
    real time estimation, 318, 319
    R-W model, 311, 312
    self-tuning, 327–329
    signal zone models, 313–318, 324

Heat release rate (HRR), 330, 332, 338, 466,
        475, 505, 510, 528
Heat release shaping based method, 555
Heat transfer inversion method, 567–570
High-speed Schlieren photography, 515
High-temperature combustion (HTC), 21, 24
Hohenberg model, 306
Homogeneous charge compression ignition
        (HCCI), 21–23, 98, 257, 364, 377,
        492–494, 502, 511, 512, 518,
        530, 531
Hybrid electric vehicle (HEV), 3

I

Ignition current sensor, 147
Ignition delay (ID), 332, 338–340
Ignition timings (IT), 289, 301
IHR after peak (IHRap), 342
IHR before peak (IHRbp), 342
In-cylinder pressure measurement
        calibration engineers, 28
        data acquisition system records, 41
        diesel engine, 39
        electrical pressure transducers, 39
        mechanical indicators, 39
        parameters and signal processing
                methods, 29
        performance and combustion parameters,
                30, 31
        piezoelectric sensor, 41
        piezoelectric transducer, 40
        piezoresistive and piezoelectric pressure
                sensor system, 40
        real-time analyses, 41
        supplementary characteristic variables, 40
        thermodynamic combustion analysis, 39
        top dead center (TDC) position, 40
Independent component analysis (ICA), 413
Indicated mean effective pressure (IMEP), 79,
        216, 241, 245, 530
Infinite impulse response (IIR), 202, 204
Inflection point analysis based method,
        553, 554
Initial heat release (IHR), 333, 341
Initial heat release rate (IHRR), 333
Injected gas, 301
Inlet valve closing (IVC), 287, 294, 313
Intake valve closing (IVC), 471, 548, 561
Integral of the modulus of pressure gradient
        (IMPG), 503
Integral of the modulus of pressure oscillations
        (IMPO), 501, 503
Integral-based methods, 501

Integrated electronics piezoelectric (IEPE),
        57, 160
Internal combustion engines (ICEs), 2, 5
Inverse Fourier transform, 413
Ion current sensor, 516, 517
        characteristic phases, 97
        current leakage measurement, 99
        gas temperature, 102
        HCCI and CI engines, 103
        hydrocarbons, 97
        methods, 101
        multicylinder engines, 99
        multielectrode spark plug, 100
        nitric oxide (NO), 96
        reciprocating engine, 96
        spark ignition (SI) engines, 95, 100

K

Knock detection methods, 495, 504, 505, 513
        direct methods, 514
        drawbacks, 514
        engine knock vibration, 517, 518
        indirect methods, 514
        ion current sensor, 516, 517
        microphone, 518
        optical, 515, 516
        resonant characteristics, 514
Knock frequencies
        acoustic modes, 489, 492
        CI diesel engine, 494
        HCCI engine, 492, 493
        heat release rate, 495
        MIAA, 495
        pressure oscillation, 489, 492
        pressure resonance, 492
        pressure waves, 489
        resonance modes, 494
        time-frequency distribution, 495–497
Knock indices
        classification, 499
        heat release analysis, 503, 504
        parameters, 499
        pressure based methods, 500–502
        PRR, 502
        RI, 502
Knock intensity (KI), 485, 499, 504, 506,
        510, 513
Knock intensity metrics, 499
Knock limited spark advance (KLSA), 510
Knock onset (KO), 504
        autoignition region, 468
        chemical kinetics, 468, 470
        detection methods, 473

Fourier transform, 473
fuel-air mixture, 470
non-paraffinic components, 471
OI, 471
ON, 470
operating conditions, 468
phenomenological models, 470
RON and MON, 471, 473
sensitivity, 471, 472
SEPO method, 474
spectroscopic measurements, 469
thermodynamic conditions, 470
TRF, 473
TVE method, 473
Knock sensor, 514, 515
Knocking
acoustic resonance, 465
adverse effects, 464
attributes, 504
autoignition process, 468
blast wave theory, 513
combustion cycles, 506, 510
cyclic variability, 468
deterministic control theory, 509
engine development and calibration
process, 464
EOC, 505
factors, 467
flame geometry, 468
flame travel distance and speed, 467
fuel chemical composition and air-fuel
ratio, 467
GCI, 510
HCCI, 512
intensifies, 506
ion sensors/accelerometers, 504
KI2 calculation, 509
limitation, 466
MAPO, 507, 510, 511
mitigation methods, 518–520
modes
autoignition region, 475
autoignitive deflagration, 480
burned and unburned gas density, 480
detonation, 478, 481
EGR, 481
end of compression and combustion
mode, 479
energy density, 478, 479
first law of thermodynamics, 474
flame velocity, 476
non-autoignition and sequential
autoignition combustion modes, 476

paraffinic hydrocarbons, 481
pressure trace, 476
pressure wave, 476, 477
RoHR-u, 481
stages, 475
temperature gradient, 479
temperature stratification, 474
thermal explosion, 474
octane number, 468
onset (see Knock onset (KO))
parameters, 467
preignition, 465
SACI, 513
SI engines, 466, 504
spark timing, 465
surface ignition, 465
threshold, 504, 508
unobservable effects, 504
Knocking damage index (KDI), 516
Kolmogorov-Smirnov test, 397

L
Laminarity (LAM), 444
Large eddy simulation (LES), 393
Laser Doppler vibrometer (LDV), 147
Laser ignition (LI), 319
Laser-induced fluorescence (LIF), 300
Laser Rayleigh scattering (LRS), 300
Late mixing controlled phase, 332
Least square methods (LSM), 191, 192
Light-emitting diodes (LED), 109
Least-square polynomial approximation
(LPA), 201
Line of identity (LOI), 442
Linear phase characteristics, 202
Linear vs. nonlinear signal processing, 424
Long-term drift (LTD), 77
Loss angle, 546
Loss function method, 550–552
Lower signal envelope (LSE), 71
Lowest normalized value (LNV), 366, 390,
391, 408, 410
Low-speed preignition (LSPI), 487, 488, 490
Low-temperature combustion (LTC), 364,
378, 432
compression ignition (CI), 20, 21
fuel injections strategies, 25
HC and CO emissions, 22
HCCI engines, 21, 22, 24
NO_x formation, 21, 22
spark ignition (SI), 20, 21
Low-temperature reaction (LTR), 332

M

Magnetoresistive (MR)-sensors, 147
Main injection advanced angle (MIAA), 495
Marvin's graphical method, 285, 286
Mass Flow Sensors
    air-box method, 140, 142
    airflow measurement, 140, 142
    Coriolis effect, 144
    fuel flow measurement methods, 143
    fuel mass flow measurement, 143
    gravimetric fuel flow measurement
        method, 144
    gravimetric measurement methods,
        142, 145
    performance and combustion
        parameters, 140
    rotameter, 144
Mass fraction burned (MFB), 293, 405, 406,
    575–577
Mass fraction burning rate (MFBR), 405
Matekunas diagrams, 393
Maximum Amplitude of Heat Release
    Oscillations (MAHRO), 503
Maximum amplitude of pressure oscillations
    (MAPO), 500, 501, 504, 506, 507,
    510–512
Maximum brake torque (MBT), 372
Maximum heat release rate (MHRR), 388
Maximum pressure rise rate (MPRR), 502, 531
McCuiston, Lavoie, and Kauffmann (MLK)
    method, 288, 289
Mean effective pressure (MEP), 245, 255
Megaknock, 482
Microphone, 518
Mixing controlled, 338
Mixing period (MP), 332
Motor octane number (MON), 471
Motoring pressure based method, 324–327
Multifractal analysis, 449, 450
Multi-optical fiber techniques, 515
Multiplexing system, 157

N

Needle lift sensor, 135, 136, 138, 140
Negative Temperature Coefficient (NTC), 145
Negative valve overlap (NVO), 585
Net mean effective pressure (NMEP), 245
Noise cancelling spike (NCS) combustion, 528
Noise, vibration and harshness (NVH), 523
Noise-to-signal ratio, 202
Nonlinear least squares (NLLS), 556
Normal distribution analysis, 400, 402

Normalized difference pressure (NDP),
    342, 343
Normalized zone temperature (NZT), 349
Nusselt number, 307, 310
Nyquist criterion, 498
Nyquist frequency, 163

O

Observer model based method, 570, 571
Octane index (OI), 471
Octane number (ON), 430, 468, 470, 519
Onboard diagnostics (OBD), 383
Optical methods, 515, 516
Otto cycle, 286
Overall noise (ON), 527, 528, 532
Oxygen detection and air/fuel ratio control, 148

P

Partial burns/misfires, 465
Partially premixed combustion (PPC), 25,
    332, 532
Peak pressure rise rate (PPRR), 378
Peak-to-peak (PTP), 500
Pearson chi-squared test statistics, 509
Phase space, 425–428
Phenomenological models, 470
Physical delay, 339
Physical model based method, 578, 579
Piezoelectric pressure transducer
    combustion sensors, 41
    electronics and electronic packaging, 42
    functional principle, 43–46
    intrusive mounting
        burned and unburned gas
            temperatures, 85
        compression ratio (CR), 86
        gas dynamics, 81
        installation passage types, 83
        pipe oscillation frequency, 83
        pressure sensor, 80, 81, 87
        pressure transducer, 79, 80
        transducer cavity, 84
        transducer radial locations, 84
        wide open throttle (WOT) conditions, 84
    materials and construction, 46–50
    modern sensors and computer-based high-
        speed data acquisition systems, 42
    mounting position, 79
    non intrusive mounting
        cylinder pressure measurement, 87, 89
        cylinder pressure sensor, 90

GDI engines, 89
 glow plug adapters, 90
 measuring glow plug, 89
 measuring spark plug, 88
 miniature pressure sensor, 91
 stages of development, 88
optical sensors, 104, 105
properties and specifications
 AEAP, 71
 chemical influences and deposits, 78
 cyclic heat flux, 59
 diaphragm, 65
 dynamic characteristics, sensor, 60
 gaskets and flame arrestors, 66
 heat shield design, 75
 IMEP thermal shock errors, 76
 in-cylinder pressure transducer, 59
 intra-cycle variability method, 74
 load change drifts, 77
 LTD, 77
 measurement range and operating
  life, 63
 pressure measurements, 73
 pressure sensors, 60, 64
 pressure transducer, 64
 reference and test sensors, 69
 sensor characteristics, 60
 structure-borne vibration, 78
 temperature distribution and
  deformation, 64, 65
 thermal characteristics, 64
 thermal load variations, 64
 thermal shock, 64, 66, 68, 76
 thermal shock reduction methods, 66
 transducer deformation effect, 68
 transmission performance, 61–63
 water-cooled transducers, 78
sensor installations, 42
strain gauges and indirect methods, 105, 106
transducer selection, 91–94
transduction elements, 43
PiezoStar®, 49
Pilot injection combustion (PIC), 495
Pohlhausen equation, 306
Poincare section, 428, 430
Polytropic coefficient estimation, 193, 194
Polytropic compression process, 188
Polytropic exponent method, 328, 546–548
Polytropic index vs. crank angle, 569
Polytropic model, 560, 561
Port fuel injection (PFI), 267, 369, 375, 519
Power spectral density (PSD), 206, 310, 582
Prandtl number, 307

Preignition
 combustion cycle, 489, 491
 DISI engines, 486
 fuel/lubricant droplets, 487
 intensity/combustion phasing, 487
 knock control system, 486
 LSPI, 487
 normal combustion cycle, 486
 origins, 489
 parameters, 489
 RON/MON, 486
 timing cycle, 486
Premixed combustion, 332, 338
Pressure decomposition technique, 525, 526
Pressure departure ratio (PDR), 291, 292
Pressure ratio management (PRM), 289, 292
Pressure resonance analysis method, 584
Pressure rise rate (PRR), 502, 527
Primary reference fuel (PRF), 430, 471, 473
Principal component analysis (PCA), 412, 413
Probability density function (PDF), 350, 395,
  509, 510
Probability distribution functions (PDF), 397
Pumping mean effective pressure (PMEP), 245
Pyroelectricity, 47

R
Rassweiler and Withrow (R-W) model,
  286–289, 311, 312
Rate of heat release (ROHR), 324, 325, 327
Rate of heat release in unburned charge
  (RoHR-u), 481, 483
Reactivity controlled compression ignition
  (RCCI), 25, 257, 378
Rebreathed exhaust gas recirculation
  (REGR), 379
Reciprocating engines
 BEVs, 3
 bore-to-stroke ratio, 8
 characteristics, 7
 classification, 7
 displacement volume, 7, 8
 downsizing, 4
 driving mechanism, 5
 electric vehicle and fuel cell, 3
 four-stroke cycle, 6, 7
 fuel-engine challenges, 4
 high pressure and temperature gases, 7
 high-performance applications, 9
 hybrid vehicles, 5
 ICEs, 2
 internal combustion engines, 5

Reciprocating engines (*cont.*)
  liquid fuels, 2
  spark ignition engine, 6
  transport energy system, 3
  transport policy, 3
  transportation energy, 3
  types of, 5
Recurrence plot (RP), 441, 442
Recurrence quantification analysis (RQA), 442,
    444, 445
Relative standard deviation (RSD), 510, 511
Research octane number (RON), 471
Residual gas fraction (RGF), 576
  categories, 583
  direct measurement, 583
  internal dilution, 582
  iterative and adaptive method, 585–589
  optical and gas sampling methods, 583
  pressure and velocity pulsations, 583
  pressure resonance, 584
  spark ignition engine, 583
  VVA, 582
Resonance frequency analysis based method,
    579, 581, 582
Response surface model (RSM), 380
Reynolds number, 307, 310
Ringing intensity (RI), 502, 531
Root mean square (RMS), 390, 391, 525

S
Sampling/acquisition frequency, 498
Savitzky-Golay filter, 199, 200
Self-corrective procedure, 547
Self tuning method, 327–329
Shannon entropy, 437–439
Short time Fourier transform (STFT), 413, 492,
    493, 505, 506
Signal energy (SE), 310
Signal Energy of Heat Release Oscillations
    (SEHRO), 503
Signal energy of pressure oscillations (SEPO),
    474, 501
Signal energy ratio (SER), 473
Signal processing, 498, 499
Signal-to-noise ratio (SNR), 160, 514, 518
Single zone thermodynamic model, 313–318
Smoothing, *see* Digital signal processing
Sound pressure level (SPL), 523
Spark ignition (SI), 319
Spark ignition (SI) engines, 365, 484, 485, 487,
    504, 509
  ambient and operating conditions, 374

categories, 375
charge motion, 9
cyclic variation, 370
diluent, 371
engine compression ratio, 464
factors, 370
flame kernel development phases, 10
flame kernel formation, 11
flame propagation, 10
flow pattern, 373
four-stroke cycles, 9
fuel injection system, 375
fuel-lean operation, 13
fuel transportation mechanism, 375
fuel type, 371
GDI, 13
ignition systems, 12
initial flame kernel development stage, 370,
    372, 373
mass flow analysis, 374
mixed-mode combustion, 14
mixture motion, 373
normal/abnormal combustion process, 464
PFI, 375
phases, 12
physical factors, 374
prior-cycle and same-cycle effects, 374
single spot ignition, 12
spark duration and energy, 372
spark ignition engine, 370
spark plug and discharge, 372
spark plug gaps and charge composition, 11
stages, 370
stoichiometric combustion, 13
stoichiometric ratio, 371
thin/sharply pointed electrodes, 372
tumble and swirling flow, 373
turbulence, 373
types of, 12
Spark timing (ST), 386, 481
Spark-assisted compression ignition (SACI),
    14, 24, 511
Spline function, 201
Spontaneous Raman scattering (SRS), 300
Standard deviation (STD), 406
Standard deviation of indicated mean effective
    pressure (SD_{IMEP}), 366
Start of combustion (SOC), 15, 284, 286, 297,
    338, 343
  closed-loop control, 330
  crank-angle, 330
  detection methods, 331
  electrical noise, 331, 335

EOI, 332
HRR, 330, 332, 333
IHR, 333
isentropic index, 335–337
phases, 332
pollutant formation and combustion
    noise, 329
polytropic process, 335
requirements, engine control, 330
self-adaptive strategy, 336
SOI, 332
threshold, 335, 336
timings, 329
Start of energizing (SOE), 251
Start of injection (SOI), 136, 332, 338, 339, 419
Start of vaporization (SOV), 339
Statistical moment method, 589–592
Stratified charge compression ignition
    (SCCI), 24
Structure attenuation (SA), 524, 525
Super-knock
    autoignition phenomenon, 482
    combustion cycle, 484, 485
    detonation, 484
    high-pressure fluctuation, 484
    mitigation techniques, 483
    preignition, 483
    pressure oscillation, 484
    pressure traces, 484, 485
    SI engines, 484
Symbol sequence histogram, 438–441
Symbol sequence statistics method
    deterministic and stochastic
        behaviors, 435
    modified Shannon entropy, 437–439
    symbol sequence histogram, 438–441
    symbolization, 435–437
    time irreversibility, 440, 441
Symbolization, 435–437
Symmetry based method, 548–549

T
TDC determination method
    capacitive/microwave probe, 545
    combustion parameters, 545
    crank angle position measurement
        system, 545
    dial gauge, 545
    IMEP, 554, 555
    in-cylinder pressure measurement, 545
    infection point analysis, 553–554
    loss angle, 546
    loss function based, 550–552

model-based, 555–556
polytropic exponent, 546–548
pressure-volume phase, 545
symmetry-based, 548–549
T-S diagram, 552, 553
Temperature sensor, 145
Temperature-entropy (T-S) method, 552, 553,
    557, 558
Thermal explosion, 476
Thermal stratification analysis (TSA)
    autoignition progression, 349
    burned and unburned gases, 349
    CDF, 350, 351
    chemical kinetics, 350
    methodology, 349
    multi-zone modeling, 349
    NZT, 349, 350
    PDF, 350, 351
    unburned temperature distributions,
        351, 352
Thermally stratified compression ignition
    (TSCI), 24
Thermocouples, 300
Thermodynamic analysis, 284
Thermodynamic loss angle, 546
Thermo-fluid dynamic processes, 211
Thermometry, 300
Threshold value exceeded (TVE), 473
Throttle positions (TP), 301, 382
Time irreversibility, 440, 441
Time series analysis, 392–395
Toluene Number (TN), 473
Toluene reference fuels (TRF), 473
Top dead center (TDC), 5, 232
    determination (see TDC determination
        method)
Transducer design
    cooled sensors, 53
    Double Shell™ design, 54, 55
    ground-isolated pressure, 55
    Kistler company, 56
    local supply water, 53
    passive acceleration sensitivity
        compensation technique, 56
    PE sensors, 58
    pressure sensor system, 58
    pressure sensors, 51
    ThermoComp® sensor, 54
    uncooled pressure sensor, 54
    water-cooled piezoelectric sensors, 53
Transistor coil ignition (TCI), 406, 407, 409
Trapped mass estimation
    fitting cylinder pressure curve, 571–574
    MFB, 575–577

Trapped mass estimation (*cont.*)
   physical model, 578, 579
   ΔP method, 574, 575
   resonance frequency, 579, 581, 582
Trapping time (TT), 444
Triggering autoignition, 338

U
Upper signal envelope (USE), 71

V
Valve lift sensor, 145, 147
Variable compression ratios (VCR), 377, 561
Variable frequency oscillator (VFO), 179
Variable geometry turbocharger (VGT),
     17, 561
Variable valve actuation (VVA), 582
Variable valve lift (VVL), 260
Variable valve timing (VVT), 260
Volumetric efficiency, 561

W
Wall heat transfer estimation
   boundary layer theory, 306
   combustion gases, 301
   compression ratio, 302
   engine control strategies, 300
   excess air ratio, 301, 303
   factors, heat flux, 301–303
   fuels, 304
   HCCI, 307, 308
   heat flux measurement, 301
   in-cylinder pressure oscillations, 310
   knocking and non-knocking cycle, 308
   motoring operating conditions, 301
   parameters, 301
   power law, 308
   radiation mode, 301
   regression, 307
   scaling factors, 306
   thermal management scheme, 300
   thermal predictions, 300
   transient process, 304
   variation, 304, 305

Wall temperature estimation
   coolant, 566
   HC emission, 567
   heat flux, 567
   heat transfer inversion, 567–570
   intake and compression stroke, 567
   observer model, 570, 571
   sensors, 566
   simulation models, 567
   thermal state, 566
Wavelet
   angular frequency, 416
   COI, 418
   CRDI system, 422
   CWT, 415
   Fourier transform, 413–415
   gasoline/diesel RCCI operation, 422, 423
   GWS, 418–422
   integral transform, 414
   oscillation/wave, 414
   power spectrum, 419
   scaling parameter, 415
   single and double injection strategy,
     423, 424
   statistical and chaotic methods, 413
   STFT/WFT, 413
   translating the function, 415
   WPS, 416, 418–422
Wavelet denoising, 518
Wavelet filtering, 210
Wavelet power spectrum (WPS), 416
Wavelet transform, 505
   daughter wavelet, 414
   mother wavelet, 414
Weak combustion events, 408
Wibes function based method, 292
Wide open throttle (WOT), 368
Wiener-Khinchin theorem, 410
Wigner distribution (WD), 492
Wild ping, 465
Windowed Fourier transform (WFT), 413
Woschini model, 306, 307
Woschni heat transfer correlation, 555

Z
Zero-one test, 446–448

Printed in the United States
By Bookmasters